A MANUAL OF ZOOLOGY

A

MANUAL OF ZOOLOGY

FOR THE USE OF STUDENTS

WITH A GENERAL INTRODUCTION ON THE PRINCIPLES OF ZOOLOGY

BY

HENRY ALLEYNE NICHOLSON

M.D., D.Sc., M.A., Ph.D. (Gött.), F.R.S.E., F.G.S.

PROFESSOR OF NATURAL HISTORY AND BOTANY IN UNIVERSITY COLLEGE, TORONTO;
FORMERLY LECTURER ON NATURAL HISTORY IN THE MEDICAL SCHOOL OF EDINBURGH;
VICE-PRESIDENT OF THE GEOLOGICAL SOCIETY OF EDINBURGH, ETC.
AUTHOR OF 'ADVANCED TEXT-BOOK OF ZOOLOGY FOR THE USE OF SCHOOLS;'
'TEXT-BOOK OF GEOLOGY;'
'ESSAY ON THE GEOLOGY OF CUMBERLAND AND WESTMORELAND;'
'GRAPTOLITES OF THE SKIDDAW SERIES,' ETC. ETC.

SECOND EDITION

REVISED AND CONSIDERABLY ENLARGED

NEW YORK:
D. APPLETON AND COMPANY,
549 & 551 BROADWAY.
1876.

PREFACE TO SECOND EDITION.

THE call for a new issue of this work within little more than six months after the appearance of the last, is a very gratifying proof that, in spite of its defects, the work supplies a recognised want, and that the Author has to some extent succeeded in the objects aimed at.

With regard to the present edition, the Author need only say that the entire work has been carefully revised, and all the more striking discoveries of recent date have been noticed, whilst some errors have been corrected. Considerable additions have also been made, especially in the department of Vertebrate Zoology. The Author, however, would ask his readers to remember that the compass of the work will not admit of the introduction of many details, which must be sought for in other more extensive treatises, and that only the more important facts of Natural History should be looked for in a work of such limited size.

Lastly, no change has been made in the plan of classification adopted in the former editions of this work, and based essentially upon the views put forth by Professor Huxley. It is true that this classification is a modern one, and that it departs widely from older arrangements. The Author, however, must seek his excuse for its adoption—if such be needed—in his firm belief that of all classifications this approximates most nearly to a natural one, and that though subsequent researches may compel its partial modification, its broad outlines will long endure unaltered.

TORONTO, *October* 2, 1871.

PREFACE TO FIRST EDITION OF PART I.

IN bringing out the present work, the Author has been mainly guided by the recollection of his own difficulties as a student, and by the belief that he is supplying a distinct want. Many excellent and original works on Natural History are extant, but they mostly labour under disadvantages which more or less disqualify them as text-books for students. So vast, for instance, have been the additions to our Zoological knowledge within the last few years, that no work on Natural History, except the most recent ones, represents adequately the present state of the Science. Under this inevitable disqualification all the older Manuals labour. Other works, again, of the most profound research, are unsuitable for ordinary students from their bulk, cost, and, more than all, from their very profundity.

The Author's aim, therefore, has simply been to present to the ordinary student those leading facts in Natural History, the knowledge of which is essential, but which lie scattered through the pages of other larger and more costly works, inaccessible to those who merely desire to learn the outlines of the Science. In carrying out this object, it is unnecessary for the Author to remark that he does not lay any claim to originality. He trusts, however, that he has succeeded in laying before his readers, not a mere mass of undigested facts, but something like an orderly and systematic review of the main points required to be known by the student. The Author is conscious of many imperfections in his plan, and also in the execution of his plan. The subject, however, is so extensive, and so constantly changing, that he can reasonably claim some indulgence, if the brief leisure-time of a busy life has not enabled him in every respect to keep abreast of the latest discoveries. Such defects as there may be, are, it is hoped, of such a nature as not to diminish the value of the work for ordinary students.

Amongst the sources upon which the Author has mainly drawn, it is, perhaps, invidious to mention one more than another. He feels, however, bound to acknowledge with gratitude the very great assistance which he has derived from the various works of Professor Huxley.

EDINBURGH, *November* 2, 1869.

PREFACE TO FIRST COMPLETE EDITION.

IN issuing the first part of the present work in a second edition, and in bringing out the second part, the Author has little to add to what he has already said.

The chief point upon which it may be desirable to say a few words is, as to the object aimed at in the Introductory portion of the work. The Introduction is intended to exhibit to the student, in as brief a form as possible, the leading principles of Zoological Science. These principles are of the highest importance, and no adequate knowledge of Zoology can be attained without their full comprehension. At the same time, the principles in question depend, in many cases, upon data which are only evolved during the systematic study of the subject. For this reason, it is not to be expected that the student should find himself fully able to comprehend the Introductory portion of the work, whilst still standing at the threshold of the subject. Whilst the student, therefore, will do well to glance over the Introduction before commencing the study of the systematic portion of the work, he must be prepared to find many points which he can only fully grasp after he has attained a knowledge of the leading modifications of structure exhibited in the Animal Kingdom. The Author has only to add that the first part of the work (on the Invertebrate Animals) has been carefully revised, and, as far as possible, brought up to the present level of the Science; whilst the illustrations, with very few exceptions, have been drawn upon the wood by himself.

EDINBURGH, *December* 1, 1870.

CONTENTS.

PART I.—INVERTEBRATE ANIMALS.

GENERAL INTRODUCTION.

CHAPTER I.

CHAPTER II.

CHAPTER III.

CHAPTER IV.

CHAPTER V.

CHAPTER VI.

CHAPTER VII.

CHAPTER VIII.

CHAPTER IX.

CHAPTER X.

CHAPTER XI.

CHAPTER XII.

CHAPTER XIII.

CHAPTER XIV.

CHAPTER XV.

CHAPTER XVI.

CHAPTER XVII.

CHAPTER XVIII.

CHAPTER XIX.

CHAPTER XX.

CHAPTER XXI.

CHAPTER XXII.

CHAPTER XXIII.

CHAPTER XXIV.

CHAPTER XXV.

CHAPTER XXVI.

CHAPTER XXVII.

CHAPTER XXVIII.

CHAPTER XXIX.

CHAPTER XXX.

CHAPTER XXXI.

CHAPTER XXXII.

CHAPTER XXXIII.

CHAPTER XXXIV.

CHAPTER XXXV.

CHAPTER XXXVI.

CHAPTER XXXVII.

CHAPTER XXXVIII.

CHAPTER XXXIX.

CHAPTER XL.

CHAPTER XLI.

CHAPTER XLII.

CHAPTER XLIII.

CHAPTER XLIV.

CHAPTER XLV.

CHAPTER XLVI.

CHAPTER XLVII.

CHAPTER XLVIII.

CHAPTER XLIX.

CHAPTER L.

CHAPTER LI.

CHAPTER LII.

PART II.—VERTEBRATE ANIMALS.

CHAPTER LIII.

CHAPTER LIV.

CHAPTER LV.

CHAPTER LVI.

CHAPTER LVII.

CHAPTER LVIII.

CHAPTER LIX.

CHAPTER LX.

CHAPTER LXI.

CHAPTER LXII.

CHAPTER LXIII.

CHAPTER LXIV.

CHAPTER LXV.

CHAPTER LXVI.

CHAPTER LXVII.

CHAPTER LXXX.

CHAPTER LXXXI.

CHAPTER LXXXII.

CHAPTER LXXXIII.

CHAPTER LXXIV.

CHAPTER LXXXV.

LIST OF ILLUSTRATIONS.

PART I.

INVERTEBRATE ANIMALS

MANUAL OF ZOOLOGY.

GENERAL INTRODUCTION.

1. Definition of Biology and Zoology.

Natural History, strictly speaking, and as the term itself implies, should be employed to designate the study of all natural objects indiscriminately, whether these are organic or inorganic, endowed with life, or exhibiting none of those incessant vicissitudes which collectively constitute vitality. So enormous, however, have been the conquests of science within the last century, that Natural History, using the term in its old sense, has of necessity been divided into several more or less nearly related branches.

In the first place, the study of natural objects admits of an obvious separation into two primary sections, of which the first deals with the phenomena presented by the inorganic world, whilst the second is occupied with the investigation of the nature and relations of all bodies which exhibit life. The former department concerns the geologist and mineralogist, and secondarily the naturalist proper as well; the latter department, treating as it does of living beings, is properly designated by the term *Biology* (from βίος, *life*, and λόγος, *a discourse*). Biology, in turn, may be split up into the sciences of Botany and Zoology, the former dealing with plants, the latter with animals; and it is really *Zoology* alone which is nowadays understood by the term Natural History.

In determining, therefore, the limits and scope of Biology, we are brought at the very threshold of our inquiry to the question, What are the differences between dead and living bodies? or rather, in the first place, what are the characteristics of an organised as compared with an unorganised body? *

* The differences between dead or inorganic bodies on the one hand, and living or organic bodies on the other, may be taken for all practical purposes as the same as those between *unorganised* and *organised* bodies. It is quite true that certain living beings (*Foraminifera*) cannot be said to

2. DIFFERENCES BETWEEN DEAD AND LIVING BODIES.

In determining this somewhat difficult point, it will be best to examine the differences between organised and unorganised bodies *seriatim*, and to compare them together systematically under the following heads :—

a. Chemical Composition.—Unorganised bodies are composed of many elements, which may be either simple or combined; but the combinations are mostly limited to a small number of elements (forming binary and ternary compounds), and these are united in low combining proportions. Thus, carbonate of lime, or common limestone, is an excellent example of an inorganic body,* being a ternary compound composed of one atom of the metal calcium, three of oxygen, and one of carbon.

Organised bodies, on the other hand, are composed of few chemical elements, and these are almost always combined. Furthermore, the combinations are always complex (ternary and quaternary compounds), and the elements enter into union in high combining proportions. Finally, the combinations are invariably characterised by the presence of water, and are prone to spontaneous decomposition. Thus, the great organic compound, albumen, is composed of 144 atoms of carbon, 110 of hydrogen, 18 of nitrogen, 2 atoms of sulphur, and 42 of oxygen. Iron, however, exists in the blood, very probably in its elemental condition; and copper has been detected in the liver of certain Mammalia, and largely in the red colouring-matter of the feathers of certain birds.

b. Arrangement of Parts.—Unorganised bodies are composed of an aggregation of homogeneous parts (when unmixed) which bear no definite and fixed relations to one another.

Organised bodies are composed of heterogeneous parts, the relations of which amongst themselves are more or less definite.

c. Form.—Unorganised bodies are either of no definite shape —when they are said to be "amorphous"—or they are crystalline, in which case they are almost invariably bounded by plane surfaces and straight lines. Organised bodies are always more or less definite in shape, presenting convex and concave surfaces, and being bounded by curved lines.

be "organised" in the proper sense of the term; still organisation is in such a vast proportion of cases the concomitant of vitality, that the purpose here in view will be fully served by assuming that all *living* bodies are *organised*, and all *dead* bodies are *unorganised.*

* In another sense limestone may be said to be organic—namely, when it has been produced by the operations of living beings; but this does not affect the above definition.

d. Mode of Increase.—When unorganised bodies increase in size, as crystals do, the increase is produced simply by what is called "accretion;" that is to say, by the addition of fresh particles from the outside.

Organised bodies increase by what is often called the "intussusception" of matter; in other words, by the reception of matter into their interior and its assimilation there. To this process alone can the term "growth" be properly applied.

e. Cyclical Change.—Unorganised bodies exhibit no actions that are not purely physical or chemical, and they show no tendency to periodical vicissitudes. Organised bodies are pre-eminently distinguished by the tendency which they show to pass through spontaneous and cyclical changes.

To sum up, all bodies which are composed of an aggregation of diverse but definitely related parts, which have a definite shape, bounded by curved lines and presenting concave and convex surfaces, which increase in size by the intussusception of foreign particles, and which pass through certain cyclical changes, are *organised;* and it is with the study of bodies such as these that Biology is concerned.

In the foregoing it has been assumed, for the sake of simplicity, that all living bodies exhibit organisation. It is to be remembered, however, that there are living bodies (*e.g.*, *Foraminifera*) to which the term of "organised," as above defined, cannot be applied. Such bodies are *living*, but they are not *organised.* In these cases the distinction from dead matter depends wholly upon the mode of growth, and upon the presence of vital activity as shown by the occurrence of various periodic changes.

The grand and fundamental characters by which living bodies are distinguished from dead bodies are these:—1. Every living body possesses the power of taking into its interior certain foreign materials, and converting these into the substances required to build up fresh tissue or repair waste. By this power of "assimilation," as it is called, a living body *grows.* 2. Living bodies, as they are constantly assimilating fresh matter, are incessantly losing portions of their substance—or, in other words, partial death is a constant accompaniment of life. 3. If our observation be continued for a sufficient length of time, we find that every living body has the power of reproducing its like. That is to say, every living body has, directly or indirectly, the power of giving origin to minute germs which are developed into the likeness of the parent. 4. The matter of a living body is subject to the same physical and chemical forces as those which affect dead matter; but it is further the seat of something in virtue of which the living body can override the physical laws which control all dead matter. The living body is the seat of *energy*, and can overcome the primary law of the *inertia* of matter. It has certain relations with the outer world other than those of mere passivity. However humble it may be, and even if it be permanently rooted to one place, some part or other of every living body possesses the power of spontaneous and independent movement—a power possessed by nothing that is dead.

3. Nature of Life.

We have next to determine—and the question is one of great difficulty—what connection exists between organisation and life. Is organisation, as we have defined it, essential to the manifestation of life, or can vital phenomena be exhibited by any body which is devoid of an organised structure? In other words, is life the *cause* of organisation, or the *result* of it? And first, what do we mean by life?

Life has been variously defined by different writers. Bichat defines it as "the sum total of the functions which resist death;" Treviranus, as "the constant uniformity of phenomena with diversity of external influences;" Duges, as "the special activity of organised bodies;" and Beclard, as "organisation in action." All these definitions, however, are more or less objectionable, since the assumption underlies them all that life is inseparably connected with organisation. In point of fact, no rigid definition of life appears to be at present possible, and it is best to regard it as being simply a tendency exhibited by certain forms of matter, under certain conditions, to pass through a series of changes in a more or less definite and determinate sequence.

As regards the connection between life and organisation, it appears that whilst all organised bodies exhibit this tendency to change, and are therefore alive, all living beings are not necessarily organised. Many of the lowest forms of life (such as the Foraminifera amongst the Protozoa) fail to fulfil one of the most essential conditions of organisation, being devoid of definite parts or organs of any kind. Nevertheless, they are capable of manifesting all the essential phenomena of life; they are produced from bodies like themselves; they eat, digest, and move, and exhibit distinct sensibility to many external impressions. Furthermore, many of these little masses of structureless jelly possess the power of manufacturing for themselves, of lime, or of the still more intractable flint, external shells of surpassing beauty and mathematical regularity. In the face of these facts we are therefore compelled to come to the conclusion that life is truly the *cause* and not the consequence of organisation; or, in other words, that organisation is not an intrinsic and indispensable condition of vital phenomena.

Such an intrinsic and indispensable condition is, however, to be found in the presence of a uniform "physical basis," to which has been applied the name of "protoplasm" (the "bioplasm" of Dr Beale). Without some such a material substratum, or medium upon which to work, no one vital phenomenon can be exhibited. The necessary forces may be there, but in the absence of this necessary vehicle there can be no outward and visible manifestation of their existence. Life, therefore, *as we know it, and as far as we know it*, may be said to be inseparably connected with protoplasm. In other words, protoplasm bears to life the same relation that a conductor does to the electric current. It is the sole medium through which life can be brought into relation with the external world. There is, however, as yet, no reason to believe that protoplasmic matter holds any other or higher relation to life, or that vital phenomena are in any way an inherent property of the matter by which alone they are capable of being manifested.

As regards its nature, protoplasm, though capable of forming the most complex structures, does not necessarily exhibit anything which can be looked upon as organisation, or differentiation into distinct parts; and its chemical composition is the only constant which can be approximately stated. It consists, namely, in all its forms, of the four elements, carbon, hydrogen, oxygen, and nitrogen, united into a proximate compound to which Mulder applied the name of "proteine," and which is very nearly identical with albumen or white-of-egg. It further appears probable that all forms of protoplasm can be made to contract by means of electricity, and "are liable to undergo that peculiar coagulation at a temperature of 40°—50° centigrade, which has been called 'heat-stiffening'" (Huxley).

If we admit, then, with Huxley—and the admission requires some qualifications—that "protoplasm, simple or nucleated, is the formal basis of all life,* there, nevertheless,

* It has not yet been shown that the living matter which we designate by the convenient term of "protoplasm" has universally and in all cases a constant and undeviating chemical composition; and there is, indeed, reason to believe that this is not the case. It is also certain that there are other materials, the exact use of which we do not at present know, which

remain certain conditions equally indispensable to the external manifestation of vital phenomena; though life itself, or the *power* of exhibiting vital phenomena, may be preserved for a longer or shorter period, even though these conditions be absent. These *extrinsic* conditions of vitality are, *firstly*, a certain temperature varying from near the freezing-point to 120° or 130°; *secondly*, the presence of water, which enters largely into the composition of all living tissues; *thirdly*, the presence of oxygen in a free state,—this, like water, appearing to be a *sine qua non* of life, though certain fungi are stated to offer an exception to this statement.*

The non-fulfilment of any of these conditions for any length of time, as a rule, causes *death*, or the cessation of vitality; but, as before remarked, life may sometimes remain in a dormant or "potential" condition for an apparently indefinite length of time. An excellent illustration of this is afforded by the great tenacity of life, even under unfavourable conditions, exhibited by the ova of some animals and the seeds of many plants; but a more striking example is to be found in the Rotifera, or Wheel-animalcules. These are minute, mostly microscopic creatures, which inhabit almost all our ponds and streams. Diminutive as they are, they are nevertheless, comparatively speaking, of a very high grade of organisation. They possess a mouth, masticatory organs, a stomach, and alimentary canal, a distinct and well-developed nervous system, a differentiated reproductive apparatus, and even organs of vision. Repeated experiments, however, have shown the remarkable fact, that, with their aquatic habits and complex organisation, the Rotifers are capable of submitting to an apparently indefinite deprivation of the necessary conditions of their existence, without thereby losing their vitality. They may be dried and reduced to dust, and may be kept in this state for a period of many years; nevertheless, the addition of a little water will at any time restore them to their pristine vigour and activity. It follows, therefore, that an organism may be deprived of all power of manifesting any of the phenomena which constitute what we call life, without losing its hold upon the vital forces which belong to it.

If, in conclusion, it be asked whether the term "vital force" is any longer permissible in the mouth of a scientific man, the question must, I think, be answered in the affirmative.

are absolutely essential to the maintenance of life, probably even in its humblest manifestations.

* Recent experiments, as ye unconfirmed, would go to prove that these conditions of vitality are not of such essential importance.

Formerly, no doubt, the progress of science was retarded and its growth checked by a too exclusive reference of natural phenomena to a so-called vital force. Equally unquestionable is the fact that the development of Biological science has progressed contemporaneously with the successive victories gained by the physicists over the vitalists. Still, no physicist has hitherto succeeded in explaining any fundamental vital phenomenon upon purely physical and chemical principles. The simplest vital phenomenon has in it something over and above the merely chemical and physical forces which we can demonstrate in the laboratory. It is easy, for example, to say that the action of the gastric juice is a chemical one, and doubtless the discovery of this fact was a great step in physiological science. Nevertheless, in spite of the most searching investigations, it is certain that digestion presents phenomena which are as yet inexplicable upon any chemical theory. This is exemplified in its most striking form, when we look at a simple organism like the Amœba. This animalcule, which is structurally little more than a mobile lump of jelly, digests as perfectly—as far as the result to itself is concerned—as does the most highly organised animal with the most complex digestive apparatus. It takes food into its interior, it digests it without the presence of a single organ for the purpose; and still more, it possesses that inexplicable selective power by which it assimilates out of its food such constituents as it needs, whilst it rejects the remainder. In the present state of our knowledge, therefore, we must conclude that even in the process of digestion as exhibited in the Amœba there is something that is not merely physical or chemical. Similarly, any organism when just dead consists of the same protoplasm as before, in the same forms, and with the same arrangement; but it has most unquestionably lost a something by which all its properties and actions were modified, and some of them were produced. What that something is, we do not know, and perhaps never shall know; and it is possible, though highly improbable, that future discoveries may demonstrate that it is merely a subtle modification of some physical force. In the meanwhile, as all vital actions exhibit this mysterious something, it would appear unphilosophical to ignore its existence altogether, and the term "vital force" may therefore be retained with advantage. In using this term, however, it must not be forgotten that we are simply employing a convenient expression for an unknown quantity, for that residual portion of every vital action which cannot at present be referred to the operation of any known physical force.

4. Differences between Animals and Plants.

We have now arrived at some definite notion of the essential characters of living beings in general, and we have next to consider what are the characteristics of the two great divisions of the organic world. What are the characters which induce us to place any given organism in either the vegetable or the animal kingdom? What, in fact, are the differences between animals and plants?

It is generally admitted that all bodies which exhibit vital phenomena are capable of being referred to one of the two great kingdoms of organic nature. At the same time it is often extremely difficult in individual cases to come to any decision as to the kingdom to which a given organism should be referred, and in many cases the determination is purely arbitrary. So strongly, in fact, has this difficulty been felt, that some observers have established an intermediate kingdom, a sort of no-man's-land, for the reception of those debatable organisms which cannot be definitely and positively classed either amongst vegetables or amongst animals. Thus, Dr Ernst Hæckel has proposed to form an intermediate kingdom, which he calls the *Regnum Protisticum*, for the reception of all doubtful organisms. Even such a cautious observer as Dr Rolleston, whilst questioning the propriety of this step, is forced to conclude that "there are organisms which at one period of their life exhibit an aggregate of phenomena such as to justify us in speaking of them as animals, whilst at another they appear to be as distinctly vegetable."

In the case of the higher animals and plants there is no difficulty; the former being at once distinguished by the possession of a nervous system, of motor power which can be voluntarily exercised, and of an internal cavity fitted for the reception and digestion of solid food. The higher plants, on the other hand, possess no nervous system or organs of sense, are incapable of independent locomotion, and are not provided with an internal digestive cavity, their food being wholly fluid or gaseous. These distinctions, however, do not hold good as regards the lower and less highly organised members of the two kingdoms, many animals having no nervous system or internal digestive cavity, whilst many plants possess the power of locomotion; so that we are compelled to institute a closer comparison in the case of these lower forms of life.

a. Form.—As regards external configuration, of all characters the most obvious, it must be admitted that no absolute distinction can be laid down between plants and animals.

Many of our ordinary zoophytes, such as the Hydroid Polypes, the sea-shrubs and corals—as, indeed, the name zoophyte implies—are so similar in external appearance to plants that they were long described as such. Amongst the Molluscoida, the common sea-mat (Flustra) is invariably regarded by sea-side visitors as a sea-weed. Many of the Protozoa are equally like some of the lower plants (Protophyta); and even at the present day there are not wanting those who look upon the sponges as belonging to the vegetable kingdom. On the other hand, the embryonic forms, or "zoospores," of certain undoubted plants (such as the Protococcus nivalis, Vaucheria, &c.) are provided with ciliated processes with which they swim about, thus coming so closely to resemble some of the Infusorian animalcules as to have been referred to that division of the Protozoa.

b. Internal Structure.—Here, again, no line of demarcation can be drawn between the animal and vegetable kingdoms. In this respect all plants and animals are fundamentally similar, being alike composed of molecular, cellular, and fibrous tissues.

c. Chemical Composition.—Plants, speaking generally, exhibit a preponderance of ternary compounds of carbon, hydrogen, and oxygen—such as starch, cellulose, and sugar—whilst nitrogenised compounds enter more largely into the composition of animals. Still both kingdoms contain identical or representative compounds, though there may be a difference in the proportion of these to one another. Moreover, the most characteristic of all vegetable compounds, viz., cellulose, has been detected in the outer covering of the sea-squirts, or Ascidian Molluscs; and the so-called "glycogen," which is secreted by the liver of the Mammalia, is closely allied to, if not absolutely identical with, the hydrated starch of plants. As a general rule, however, it may be stated that the presence in any organism of an external envelope of cellulose raises a strong presumption of its vegetable nature. In the face, however, of the facts above stated, the presence of cellulose cannot be looked upon as absolutely conclusive. Another highly characteristic vegetable compound is *chlorophyll*, the green colouring-matter of plants. Any organism which exhibits chlorophyll in any quantity, as a proper element of its tissues, is most probably vegetable. As in the case of cellulose, however, the presence of chlorophyll cannot be looked upon as a certain test, since it occurs normally in certain undoubted animals (*e.g.*, *Stentor*, amongst the *Infusoria*, and the *Hydra viridis*, or the green Fresh-water Polype, amongst the *Cœlenterata*).

d. Motor Power.—This, though broadly distinctive of animals, can by no means be said to be characteristic of them. Thus, many animals in their mature condition are permanently fixed, or attached to some foreign object; and the embryos of many plants, together with not a few adult forms, are endowed with locomotive power by means of those vibratile, hair-like processes which are called "cilia," and are so characteristic of many of the lower forms of animal life. Not only is this the case, but large numbers of the lower plants, such as the Diatoms and Desmids, exhibit throughout life an amount and kind of locomotive power which does not admit of being rigidly separated from the movements executed by animals, though the closest researches have hitherto failed to show the mechanism whereby these movements are brought about.

e. Nature of the Food.—Whilst all the preceding points have failed to yield a means of invariably separating animals from plants, a distinction which holds good almost without exception is to be found in the nature of the food taken respectively by each, and in the results of the conversion of the same. The unsatisfactory feature, however, in this distinction is this, that even if it could be shown to be, theoretically, invariably true, it would nevertheless be practically impossible to apply it to the greater number of those minute organisms concerning which alone there can be any dispute.

As a broad rule, all plants are endowed with the power of converting inorganic into organic matter. The *food* of plants consists of the inorganic compounds, carbonic acid, ammonia, and water, along with small quantities of certain mineral salts. From these, and from these only, plants are capable of elaborating the proteinaceous matter or protoplasm which constitutes the physical basis of life. Plants, therefore, take as food very simple bodies, and manufacture them into much more complex substances. In other words, by a process of deoxidation or unburning, rendered possible by the influence of sunlight only, plants convert the inorganic or stable elements—ammonia, carbonic acid, water, and certain mineral salts—into the organic or unstable elements of food. The whole problem of nutrition may be narrowed to the question as to the modes and laws by which these stable elements are raised by the vital chemistry of the plant to the height of unstable compounds. To this general statement, however, an exception must seemingly be made in favour of certain fungi, which require organised compounds for their nourishment.

On the other hand, no known animal possesses the power of converting inorganic compounds into organic matter, but all, mediately or immediately, are dependent in this respect upon plants. All animals, as far as is certainly known, require ready-made proteinaceous matter for the maintenance of existence, and this they can only obtain in the first instance from plants. Animals, in fact, differ from plants in requiring as food complex organic bodies which they ultimately reduce to very much simpler inorganic bodies. The nutrition of animals is a process of oxidation or burning, and consists essentially in the conversion of the energy of the food into vital work; this conversion being effected by the passage of the food into living tissue. Plants, therefore, are the great manufacturers in nature, —animals are the great consumers.

Just, however, as this law does not invariably hold good for plants, certain fungi being in this respect animals, so it is not impossible that a limited exception to the universality ot the law will be found in the case of animals also. Thus, in some recent investigations into the fauna of the sea at great depths, a singular organism, of an extremely low type, but occupying large areas of the sea-bottom, has been discovered, to which Professor Huxley has given the name of Bathybius. As vegetable life is extremely scanty, or is altogether wanting, in these abysses of the ocean, it has been conjectured that this organism is possibly endowed with the power—otherwise exclusively found in plants—of elaborating organic compounds out of inorganic materials, and in this way supplying food for the higher animals which surround it. The water of the ocean, however, at these enormous depths, is richly charged with organic matter in solution, and this conjecture is thereby rendered doubtful.

Be this as it may, there remain to be noticed two distinctions, broadly though not universally applicable, which are due to the nature of the food required respectively by animals and plants. In the first place, the food of all plants consists partly of gaseous matter and partly of matter held in solution. They require, therefore, no special aperture for its admission, and no internal cavity for its reception. The food of almost all animals consists of solid particles, and they are therefore usually provided with a mouth and a distinct digestive cavity. Some animals, however, such as the tape-worm and the Gregarinæ, live entirely by the imbibition of organic fluids through the general surface of the body, and many have neither a distinct mouth nor stomach.

Secondly, plants decompose carbonic acid, retaining the carbon and setting free the oxygen, certain fungi forming an exception to this law. The reaction of plants upon the atmosphere is therefore characterised by the production of free oxygen. Animals, on the other hand, absorb oxygen and emit carbonic acid, so that their reaction upon the atmosphere is the reverse of that of plants, and is characterised by the production of carbonic acid.

Finally, it is worthy of notice that it is in their lower and not in their higher developments that the two kingdoms of organic nature approach one another. No difficulty is experienced in separating the higher animals from the higher plants, and for these universal laws can be laid down to which there is no exception. It might, not unnaturally, have been thought that the lowest classes of animals would exhibit most affinity to the highest plants, and that thus a gradual passage between the two kingdoms would be established. This is not the case, however. The lower animals are not allied to the higher plants, but to the lower; and it is in the very lowest members of the vegetable kingdom, or in the embryonic and immature forms of plants little higher in the scale, that we find such a decided animal gift as the power of independent locomotion. It is also in the less highly organised and less specialised forms of plants that we find the only departures from the great laws of vegetable life, the deviation being in the direction of the laws of animal life.

5. Morphology and Physiology.

The next point which demands notice relates to the *nature* of the differences between one animal and another, and the question is one of the highest importance. Every animal—as every plant—may be regarded from two totally distinct, and, indeed, often apparently opposite, points of view. From the first point of view we have to look simply to the laws, form, and arrangement of the *structures* of the organism; in short, to its external shape and internal structure. This constitutes the science of morphology (μορφή, *form*, and λόγος, *discourse*). From the second, we have to study the vital actions performed by living beings and the *functions* discharged by the different parts of the organism. This constitutes the science of physiology.

A third department of zoology is concerned with the relations of the organism to the external conditions under which it is placed, constituting a division of the science to which the term "distribution" is applied.

Morphology, again, not only treats of the structure of living beings in their fully-developed condition (anatomy), but is also concerned with the changes through which every living being has to pass before it assumes its mature or adult characters (embryology or development). The term "histology" is further employed to designate that branch of morphology which is specially occupied with the investigation of minute or microscopical tissues.

Physiology treats of all the functions exercised by living bodies, or by the various definite parts or organs, of which most animals are composed. All these functions come under three heads:—1. *Functions of Nutrition,* divisible into functions of absorption and metamorphosis, comprising those functions which are necessary for the growth and maintenance of the organism. 2. *Functions of Reproduction,* whereby the perpetuation of the species is secured. 3. *Functions of Correlation,* comprising all those functions (such as sensation and voluntary motion) by which the external world is brought into relation with the organism, and the organism in turn reacts upon the external world.

Of these three, the functions of nutrition and reproduction are often collectively called the functions of organic or vegetative life, as being common to animals and plants; while the functions of correlation are called the animal functions, as being more especially characteristic of, though not peculiar to, animals.

6. Differences between different Animals.

All the innumerable differences which subsist between different animals may be classed under two heads, corresponding to the two aspects of every living being, morphological and physiological. One animal differs from another either *morphologically,* in the fundamental points of its structure; or *physiologically,* in the manner in which the vital functions of the organism are discharged. These constitute the only modes in which any one animal can differ from any other; and they may be considered respectively under the heads of Specialisation of Function and Morphological type.

a. Specialisation of Function.—All animals alike, whatever their structure may be, perform the three great physiological functions; that is to say, they all nourish themselves, reproduce their like, and have certain relations with the external world. They differ from one another physiologically in the *manner* in which these functions are performed. Indeed, it is only in the functions of correlation that it is possible that there should be any difference in the amount or perfection of the function performed by the organism, since nutrition and reproduction, as far as their results are concerned, are essentially the same in all animals. In the manner, however, in which the same results are brought about, great differences are observable in different animals. The nutrition of such a simple organism as the Amœba is, indeed, performed perfectly, as far as the result to the animal itself is concerned—as perfectly as in the case of the highest animal—but it is performed with the simplest possible apparatus. It may, in fact, be said to be performed without any *special* apparatus, since any part of the surface of the body may be extemporised into a mouth, and there is no differentiated alimentary cavity. And not only is the nutritive apparatus of the simplest character, but the function itself is equally simple, and is entirely divested of those complexities and separations into secondary functions which characterise the process in the higher animals. It is the same, too, with the functions of reproduction and correlation; but this point will be more clearly brought out if we examine the method in which one of the three primary functions is performed in two or three examples. Nutrition, as the simplest of the functions, will best answer the purpose.

In the simpler Protozoa, such as the Amœba, the process of nutrition consists essentially in the reception of food, its digestion within the body, the excretion of effete or indigestible matter, and the distribution of the nutritive fluid through the body. The first three portions of this process are effected without any special organs for the purpose, and for the last there is simply a rudimentary contractile cavity. Respiration, if it can be said to exist at all as a distinct function, is simply effected by the general surface of the body.

In a Cœlenterate animal, such as a sea-anemone, the function of nutrition has not advanced much in complexity, but the means for its performance are somewhat more specialised. Permanent organs of prehension (tentacles) are present, there is a distinct mouth, and there is a persistent internal cavity for the reception of the food; but this is not shut off from the general cavity of the body, and there are no distinct circulatory or respiratory organs.

In a Mollusc, such as the oyster, nutrition is a much more complicated process. There is a distinct mouth, and an alimentary canal which is shut off from the general cavity of the body, and is provided with a separate aperture for the excretion of effete and indigestible matters. Digestion is performed by a distinct stomach with accessory glands; a special contractile cavity, or heart, is provided for the propulsion of the nutritive products of digestion through all parts of the organism, and the function of respiration is performed by complex organs specially adapted for the purpose.

It is not necessary here to follow out this comparison further. In still higher animals the function of nutrition becomes still further broken up into secondary functions, for the due performance of which special organs are provided, the complexity of the organism thus necessarily increasing *pari passu* with the complexity of the function. This gradual subdivision and elaboration is carried out equally with the other two physiological functions—viz., reproduction and correlation—and it constitutes what is technically called the "specialisation of functions," though it has been more happily termed by Milne-Edwards "the principle of the physiological division of labour." It is needless, however, to remark that in the higher animals it is the functions of correlation which become most highly specialised—disproportionately so, indeed, when compared with the development of the nutritive and reproductive functions.

b. Morphological Type.—The first point in which one animal may differ from another is the degree to which the principle of the physiological division of labour is carried. The second point in which one animal may differ from another is in its "morphological type;" that is to say, in the fundamental plan upon which it is constructed. By one not specially acquainted with the subject it might be readily imagined that each species or kind of animal was constructed upon a plan peculiar to itself and not shared by any other. This, however, is far from being the case; and it is now universally recognised that all the varied species of animals—however great the apparent amount of diversity amongst them—may be arranged under no more than half-a-dozen primary morphological types or plans of structure. Upon one or other of these five or six plans every known animal, whether living or extinct, is constructed. It follows from the limited number of primitive types or patterns, that great numbers of animals must agree with one another in their morphological type. It follows also that all so agreeing can differ from one another only in the sole remaining element of the question—namely, by the amount of specialisation of function which they exhibit. Every animal, therefore, as Professor Huxley has well expressed it, is the resultant of two tendencies, the one morphological, the other physiological.

The six types or plans of structure, upon one or other of which all known animals have been constructed, are technically called "sub-kingdoms," and are known by the names

Protozoa, Cœlenterata, Annuloida, Annulosa, Mollusca, and Vertebrata. We have, then, to remember that every member of each of these primary divisions of the animal kingdom agrees with every other member of the same division in being formed upon a certain definite plan or type of structure, and differs from every other simply in the grade of its organisation, or in other words, in the degree to which it exhibits specialisation of function.

Von Baer's Law of Development.—As the study of living beings in their adult condition shows us that the differences between those which are constructed upon the same morphological type depend upon the degree to which specialisation of function is carried, so the study of development teaches us that the changes undergone by any animal in passing from the embryonic to the mature condition are due to the same cause. All the members of any given sub-kingdom, when examined in their earliest embryonic condition, are found to present the same fundamental characters. As development proceeds, however, they diverge from one another with greater or less rapidity, until the adults ultimately become more or less different, the range of possible modification being apparently almost illimitable. The differences are due to the different degrees of specialisation of function necessary to perfect the adult; and therefore, as Von Baer put it, *the progress of development is from the general to the special.*

It is upon a misconception of the true import of this law that the theory arose, that every animal in its development passed through a series of stages in which it resembles, in turn, the different inferior members of the animal scale. With regard to man, standing at the top of the whole animal kingdom, this theory has been expressed as follows:—"Human organogenesis is a transitory comparative anatomy, as, in its turn, comparative anatomy is a fixed and permanent state of the organogenesis of man" (Serres). In other words, the embryo of a Vertebrate animal was believed to pass through a series of changes corresponding respectively to the permanent types of the lower sub-kingdoms—namely, the Protozoa, Cœlenterata, Annuloida, Annulosa, and Mollusca—before finally assuming the true vertebrate characters. Such, however, is not truly the case. The ovum of every animal is from the first impressed with the power of developing in one direction only, and very early exhibits the fundamental characters proper to its sub-kingdom, never presenting the structural peculiarities belonging to any other morphological type. Nevertheless, the differences which subsist between the members of each sub-kingdom in their adult condition are truly referable to the degree to which development proceeds, the place of each individual in his own sub-kingdom being regulated by the stage at which development is arrested. Thus, many cases are known in which the younger stages of a given animal *represent* the permanent adult condition of an animal somewhat lower in the scale. To give a single example, the young Gasteropod (amongst the Mollusca) transiently presents all the essential characters which permanently distinguish the adult Pteropod. The development of the Gasteropod, however, proceeds beyond this point, and the adult is much more highly specialised than is the adult Pteropod.

7. Homology, Analogy, and Homomorphism.

When organs in different animals agree with one another in fundamental *structure*, they are said to be "homologous;" when they perform the same functions they are said to be "analogous." Thus the wing of a bird and the arm of a man are constructed upon the same fundamental plan, and they are therefore homologous organs. They are not analogous, however, since they do not perform the same function, the one being adapted for aerial locomotion, the other being an organ of prehension. On the other hand, the wings of a bird and the wings of an insect both serve for flight, and they are therefore analogous, since they perform the same function. They are not homologous, however, as they are constructed upon wholly dissimilar plans. There are numerous cases, however, in which organs correspond with one another both structurally and functionally, in which case they are both homologous and analogous.

A form of homology is often seen in a single animal in which there exists a succession of parts which are fundamentally identical in structure, but are variously modified to fulfil different functions. Thus a Crustacean—such as the lobster—may be looked upon as being composed of a succession of rings, each of which bears a pair of appendages, these appendages being constructed upon the same type, and being therefore homologous. They are, however, variously modified in different regions of the body to enable them to fulfil special functions, some being adapted for swimming, others for walking, others for prehension, others for mastication, and so on. This succession of fundamentally similar parts in the same animal constitutes what is known as *serial homology*. When, however, the successive parts are similar to one another, both in structure and in function, the case becomes rather one of what is called "vegetative" or "irrelative repetition." An excellent instance of this is seen in the common Millipede (Iulus).

Homomorphism.—Many examples occur, both among animals and among plants, in which families widely removed from one another as to their fundamental structure, nevertheless present a singular, and sometimes extremely close, resemblance in their external characters. Thus the composite Hydroid Polypes and the Polyzoa are singularly like one another—so much so, that they have often been classed together; whereas, in reality, they belong to different sub-kingdoms. Many other cases of this resemblance of different animals might be adduced, and in many cases these "representative forms" appear to be able to fill each other's places in the general economy of nature. This is so far true, at any rate, that "homomorphous" forms are generally found in different parts of the earth's surface. Thus, the place of the Cacti of South America is taken by the Euphorbiæ of Africa; or, to take a zoological illustration, many of the different orders of Mammalia are *represented* in the single order Marsupialia in Australia, in which country this order has almost alone to discharge the functions elsewhere performed by several orders. Many homomorphous forms, however, live peacefully side by side, and it is difficult to say whether in this case the resemblance between them is for the advantage or for the disadvantage of either. In other cases we find certain animals putting on the external characters of certain other animals, to which they may be closely related, or from which they may be widely separated in zoological position. Such cases are said to be examples of "mimicry," and such animals are said to be "mimetic." Excellent examples of this may be found amongst certain Butterflies, or in the close resemblance of the clear-winged Moths to Bees and Hornets. In all these cases it appears that the mimetic species is protected from some enemy by its outward similarity to the form which it mimics. Finally, there are numerous cases in which animals mimic certain natural objects, and thus greatly diminish their chances of being detected by their natural foes. Excellent instances of this are afforded by the insects known as Walking-leaves (*Phyllium*) and Walking-sticks (*Phasmidæ*), which respectively present the most singular resemblance to leaves and dried twigs.

8. Correlation of Growth.

This term is employed by zoologists to express the empirical law, that certain structures, not necessarily or usually connected together by any visible link, invariably occur in association with one another, and never occur apart,—so far, at any rate, as human observation goes.

Thus, all animals which possess two condyles on the occipital bone, and possess non-nucleated red blood-corpuscles, suckle their young. Why an animal with only one condyle on its occipital bone should not suckle its young we do not know, and perhaps we shall at some future time find mammary glands associated with a single occipital condyle. Again, the feet are cleft in all animals which ruminate, but not in any other. In other cases the correlation is even more apparently lawless, and is even amusing. Thus all, or almost all, cats which are entirely white and have blue eyes, are at the same time deaf. With regard to these and similar generalisations we must, however, bear in mind the following three points:—

1. The various parts of the organisation of any animal are so closely interconnected, and so mutually dependent upon one another, both in their growth and development, that the characters of each must be in *some* relation to the characters of all the rest, whether this be obviously the case or not.

2. It is rarely possible to assign any reason for correlations of structure, though they are certainly in no case accidental.

3. The law is a purely empirical one, and expresses nothing more than the result of experience; so that structures which we now only know as occurring in association, may ultimately be found dissociated, and conjoined with other structures of a different character.

9. Classification.

Classification is the arrangement of a number of diverse objects into larger or smaller groups, according as they exhibit more or less likeness to one another. The excellence of any given classification will depend upon the nature of the points which are taken as determining the resemblance. Systems of classification, in which the groups are founded upon mere external and superficial points of similarity, though often useful in the earlier stages of science, are always found in the long-run to be inaccurate. It is needless, in fact, to point out that many living beings, the structure of which is fundamentally different, may nevertheless present such an amount of adaptive external resemblance to one another, that they would be grouped together in any "artificial" classification. Thus, to take a single example, the whale, by its external characters, would certainly be grouped amongst the fishes, though widely removed from them in all the essential points of its structure. "Natural" systems of classification, on the other hand, endeavour to arrange animals into divisions founded upon a due consideration of *all* the essential and fundamental points of structure, wholly irrespective of external similarity of form and habits. Philosophical classification depends upon a due appreciation of what constitute the true points of difference and likeness amongst animals; and we have already seen that these are morphological type and specialisation of function. Philosophical classification, therefore, is a formal expression of the facts and laws of Morphology and Physiology. It follows that the more fully the programme of a philosophical and strictly natural classification can be carried out, the more completely does it afford a condensed exposition of the fundamental construction of the objects classified. Thus, if the whale were placed by an artificial grouping amongst the fishes, this would simply express the facts that its habits are aquatic and its body fish-like. When, on the contrary, we obtain a natural classification, and we learn that the whale is placed amongst the Mammalia, we then know at once that the young whale is born in a comparatively helpless condition, and that its mother is provided with special mammary glands for its support; this expressing a fundamental distinction from all fishes, and being associated with other equally essential correlations of structure.

The entire animal kingdom is primarily divided into some half-a-dozen great plans of structure, the divisions thus formed being called "sub-kingdoms." The sub-kingdoms are, in turn, broken up into classes, classes into orders, orders into families, families into genera, and genera into species. We shall examine these successively, commencing with the consideration of a species, since this is the zoological unit of which the larger divisions are made up.

Species.—No term is more difficult to define than "species," and on no point are zoologists more divided than as to what should be understood by this word. Naturalists, in fact, are not yet agreed as to whether the term species expresses a real and permanent distinction, or whether it is to be regarded merely as a convenient, but not immutable, abstraction, the employment of which is necessitated by the requirements of classification.

By Buffon, "species" is defined as "a constant succession of individuals* similar to and capable of reproducing each other."

* In using the term "individual," it must be borne in mind that the

De Candolle defines species as an assemblage of all those individuals which resemble each other more than they do others, and are able to reproduce their like, doing so by the generative process, and in such a manner that they may be supposed by analogy to have all descended from a single being or a single pair.

M. de Quatrefages defines species as "an assemblage of individuals, more or less resembling one another, which are descended, or may be regarded as being descended, from a single primitive pair by an uninterrupted succession of families."

Müller defines species as "a living form, represented by individual beings, which reappears in the product of generation with certain invariable characters, and is constantly reproduced by the generative act of similar individuals."

According to Woodward, "all the specimens, or individuals, which are so much alike that we may reasonably believe them to have descended from a common stock, constitute a *species*."

From the above definitions it will be at once evident that there are two leading ideas in the minds of zoologists when they employ the term species; one of these being a certain amount of resemblance between individuals, and the other being the proof that the individuals so resembling each other have descended from a single pair, or from pairs exactly similar to one another. The characters in which individuals must resemble one another in order to entitle them to be grouped in a separate species, according to Agassiz, "are only those determining size, proportion, colour, habits, and relations to surrounding circumstances and external objects."

On a closer examination, however, it will be found that these two leading ideas in the definition of species—external resemblance and community of descent—are both defective, and liable to break down if rigidly applied. Thus, there are in nature no assemblages of plants or animals, usually grouped together into a single species, the individuals of which *exactly* resemble one another in every point. Every naturalist is compelled to admit that the individuals which compose any so-called species, whether of plants or animals, differ from one another to a greater or less extent, and in respects which may be regarded as more or less important. The existence of such individual differences is attested by the universal employment of the terms "varieties" and "races." Thus a "variety" comprises all those individuals which possess some

"zoological individual" is meant; that is to say, the total result of the development of a single ovum, as will be hereafter explained at greater ength.

distinctive peculiarity in common, but do not differ in other respects from another set of individuals sufficiently to entitle them to take rank as a separate species. A "race," again, is simply a permanent or "perpetuated" variety. The question, however, is this—How far may these differences amongst individuals obtain without necessitating their being placed in a separate species? In other words: How great is the amount of individual difference which is to be considered as merely "*varietal*," and at what exact point do these differences become of "*specific*" value? To this question no answer can be given, since it depends entirely upon the weight which different naturalists would attach to any given individual difference.* Distinctions which appear to one observer as sufficiently great to entitle the individuals possessing them to be grouped as a distinct species, by another are looked upon as simply of varietal value; and, in the nature of the case, it seems impossible to lay down any definite rules. To such an extent do individual differences sometimes exist in particular genera—termed "protean" or "polymorphic" genera—that the determination of the different species and varieties becomes an almost hopeless task.

Besides the individual differences which ordinarily occur in all species, other cases occur in which a species consists normally and regularly of two or even three distinct forms, which cannot be said to be mere varieties, since no intermediate forms can be discovered. When two such distinct forms exist, the species is said to be "dimorphic," and when three are present, it is called "trimorphic." Thus, in dimorphic plants a single species is composed of two distinct forms, similar to one another in all respects except in their reproductive organs, the one form having a long pistil and short stamens, the other a short pistil with long stamens. In trimorphic plants, the species is composed of three such distinct forms, which differ in like manner in the conformation of their reproductive organs, though they are otherwise undistinguishable.—(Darwin.) Similar cases are known in animals, but in them the differences, though apparently connected with reproduction, are not confined to the reproductive organs. Thus the females of certain butterflies normally appear under two or three entirely different forms, not connected by any intermediate links; and the same thing occurs in some of the Crustacea.

* As an example of this, it is sufficient to allude to the fact that hardly any two botanists agree as to the number of species of Willows and Brambles in the British Isles. What one observer classes as mere varieties, another regards as good and distinct species.

As regards, therefore, the first point in the definition of species—namely, the external resemblance of assemblages of individuals—we are forced to conclude that no two individuals are exactly alike; and that the amount and kind of external resemblance which constitutes a species is not a precise and invariable quantity, but depends upon the value attached to particular characters by any given observer.

The second point in the definition of species—namely, community of descent—is hardly in a more satisfactory condition, since the descent of any given series of individuals from a single pair, or from pairs exactly similar to one another, is at best but a probability, and is in no case capable of proof. In the case of the higher animals it can doubtless be shown that certain assemblages of individuals possess amongst themselves the power of fecundation and of producing fertile progeny, and that this power does not extend to the fecundation of individuals belonging to another different assemblage. Amongst the higher animals, "crosses" or "hybrids" can only be produced between closely-allied species, and when produced they are sterile, and are not capable of reproducing their like. In these cases, therefore, we may take this as a most satisfactory element in the definition of "species." The sterility, however, of hybrids is not universal, even amongst the higher animals; and amongst plants no doubt can be entertained but that the individuals of species universally admitted to be distinct are capable of mutual fertilisation; the hybrid progeny thus produced being likewise fertile, and capable of reproducing similar individuals. That this fertility is often irregular, and may be destroyed in a few generations, admits of explanation, and hardly alters the significance of these undoubted facts.

Upon the whole, then, it seems in the meanwhile safest to adopt a definition of species which implies no theory, and does not include the belief that the term necessarily expresses a fixed and permanent quantity. Species, therefore, may be defined as *an assemblage of individuals which resemble each other in their essential characters, are able, directly or indirectly, to produce fertile individuals, and which do not (as far as human observation goes) give rise to individuals which vary from the general type through more than certain definite limits.* The production of occasional monstrosities does not, of course, invalidate this definition.

Genus is a term applied to groups of species which possess a community of essential details of structure. A genus may include a single species only, in cases where the combination of characters which make up the species are so peculiar that no other species exhibits similar structural characters; or, on the other hand, it may contain many hundreds of species.

Families are groups of genera which agree in their general characters. According to Agassiz, they are divisions founded upon peculiarities of "form as determined by structure."

Orders are groups of families related to one another by structural characters common to all.

Classes are larger divisions, comprising animals which are formed upon the same fundamental plan of structure, but differ in the method in which the plan is executed (Agassiz).

Sub-kingdoms are the primary divisions of the animal kingdom, which include all those animals which are formed upon the same structural or morphological type, irrespective of the degree to which specialisation of function may be carried.

Impossibility of a Linear Classification.—It has sometimes been thought that the animal kingdom can be arranged in a linear series, every member of the series being higher in point of organisation than the one below it. As we have seen, however, the *status* of any given animal depends upon two conditions—one its morphological type, the other the degree to which specialisation of function is carried. Now, if we take two animals, one of which belongs to a lower morphological type than the other, no degree of specialisation of function, however great, will place the former above the latter, as far as its *type of structure* is concerned, though it may make the former a more highly organised animal. Every Vertebrate animal, for example, belongs to a higher morphological type than every Mollusc; but the higher Molluscs, such as cuttle-fishes, are much more highly organised, as far as their type is concerned, than are the lowest Vertebrata. In a linear classification, therefore, the cuttle-fishes should be placed above the lowest fishes—such as the lancelet—in spite of the fact that the type upon which the latter are constructed is by far the highest of the two.

It is obvious, therefore, that a linear classification is not possible, since the higher members of each sub-kingdom are more highly organised than the lower forms of the next sub-kingdom in the series, at the same time that they are constructed upon a lower morphological type.

10. Reproduction.

Reproduction is the process whereby new individuals are generated and the perpetuation of the species insured. The methods in which this end may be attained exhibit a good deal of diversity, but they may be all considered under two heads.

I. *Sexual Reproduction.* — This consists essentially in the production of two distinct elements, a germ-cell or ovum, and a sperm-cell or spermatozoid, by the contact of which the ovum—now said to be "fecundated"—is enabled to develop itself into a new individual. As a rule, the germ-cell is produced by one individual (female) and the spermatic element by another (male); in which case the sexes are said to be distinct, and the species is said to be "diœcious." In other cases the same individual has the power of producing both the essential elements of reproduction; in which case the sexes are said to be united, and the individual is said to be "hermaphrodite," "androgynous," or "monœcious." In the case of hermaphrodite animals, however, self-fecundation—contrary to what might have been expected—rarely constitutes the reproductive process; and, as a rule, the reciprocal union of two such individuals is necessary for the production of young. Even amongst hermaphrodite plants, where self-fecundation may, and certainly does, occur, provisions seem to exist by which perpetual self-fertilisation is prevented, and the influence of another individual secured at intervals. Amongst the higher animals sexual reproduction is the only process whereby new individuals can be generated.

II. *Non-sexual Reproduction.*—Amongst the lower animals fresh beings may be produced without the contact of an ovum and a spermatozoid; that is to say, without any true generative act. The processes by which this is effected vary in different animals, and are all spoken of as forms of "asexual" or "agamic" reproduction. As we shall see, however, the true "individual" is very rarely produced otherwise than sexually, and most forms of agamic reproduction are really modifications of growth.

a. Gemmation and Fission.—Gemmation, or budding, consists in the production of a bud, or buds, generally from the exterior, but sometimes from the interior, of the body of an animal, which buds are developed into independent beings, which may or may not remain permanently attached to the parent organism. Fission differs from gemmation solely in the fact that the new structures in the former case are produced by a division of the body of the original organism into separate parts, which may remain in connection, or may undergo detachment.

The simplest form of gemmation, perhaps, is seen in the power possessed by certain animals of reproducing parts of their bodies which they may have lost. Thus, the Crustacea possess the power of reproducing a lost limb, by means of a bud which is gradually developed till it assumes the form and takes the place of the missing member. In these cases, however, the process is not in any way generative, and the product of gemmation can in no sense be spoken of as a distinct being (or zoöid).

Another form of gemmation may be exemplified by what takes place in the Foraminifera, one of the classes of the Protozoa. The primitive form of a Foraminifer is simply a little sphere of sarcode, which has the power of secreting from its outer surface a calcareous envelope; and this condition may be permanently retained (as in Lagena). In other cases a process of budding or gemmation takes place, and the primitive mass of sarcode produces from itself, on one side, a second mass exactly similar to the first, which does not detach itself from its parent, but remains permanently connected with it. This second mass repeats the process of gemmation as before, and this goes on—all the segments remaining attached to one another — until a body is produced, which consists of a number of little spheres of sarcode in organic connection with one another, and surrounded by a shell, often of the most complicated description. In this case, however, the buds produced by the primitive spherule are not only not detached, but they can only remotely be regarded as independent beings. They are, in all respects, identical with the primordial segment, and it is rather a case of "vegetative" repetition of similar parts.

Another form of gemmation is exhibited in such an organism as the common sea-mat (Flustra), which is a composite organism composed of a multitude of similar beings, each of which inhabits a little chamber, or cell; the whole forming a structure not unlike a sea-weed in appearance. This colony is produced by gemmation from a single primitive being ("polypide"), which throws out buds, each of which repeats the process, apparently almost indefinitely. All the buds remain in contact and connected with one another, but each is, nevertheless, a distinct and independent being, capable of performing all the functions of life. In this case, therefore, each one of the innumerable buds becomes an independent being, similar to, though not detached from, the organism which gave it birth. This is an instance of what is called "continuous gemmation."

In other cases—as in the common fresh-water polype or Hydra—the buds which are thrown out by the primitive organism become developed into creatures exactly resembling the parent, but, instead of remaining permanently attached, and thus giving rise to a compound organism, they are detached to lead an entirely independent existence. This is a simple instance of what is termed "discontinuous gemmation."

The method and results of fission may be regarded as essentially the same as in the case of gemmation. The products of the division of the body of the primitive organism may either remain undetached, when they will give rise to a composite structure (as in many corals), or they may be thrown off and live an independent existence (as in some of the Hydrozoa).

We are now in a position to understand what is meant, strictly speaking, by the term "individual." In zoological language, an *individual* is defined as "*equal to the total result of the development of a single ovum.*" Amongst the higher animals there is no difficulty about this, for each ovum gives rise to no more than one single being, which is incapable of repeating itself in any other way than by the production of another ovum; so that an individual is a single animal. It is most important, however, to comprehend that this is not necessarily or always the case. In such an organism as the sea-mat, the ovum gives rise to a primitive polypide, which repeats itself by a process of continuous gemmation until an entire colony is produced, each member of which is independent of its fellows, and is capable of producing ova. In such a case, therefore, the term "individual" must be applied to the entire colony, since this is the result of the development of a single ovum. The separate beings which compose the colony are technically called "zoöids." In like manner the Hydra, which produces fresh and independent Hydræ by discontinuous gemmation, is not an "individual," but is a zoöid. Here the zoöids are not permanently united to one another, and the "individual" Hydra consists really of the primitive Hydra, *plus* all the detached Hydræ to which it gave rise. In this case, therefore, the "individual" is composed of a number of disconnected and wholly independent beings, all of which are the result of the development of a single ovum. It is to be remembered that both the parent zoöid and the "produced zoöids" are capable of giving rise to fresh Hydræ by a true generative process. It must also be borne in mind that this production of fresh zoöids by a process of gemmation is not so essentially different to the true sexual process of reproduction as might at first sight appear, since the ovum itself may be regarded merely as a highly specialised bud. In the Hydra, in fact, where the ovum is produced as an external process of the wall of the body, this likeness is extremely striking. The ovarian bud, however, differs from the true gemmæ or buds in its inability to develop itself into an independent organism, unless previously brought into contact with another special generative element. The only exceptions to this statement are in the rare cases of true "parthenogenesis," to be subsequently alluded to.

b. Reproduction by Internal Gemmation.—Before considering the phenomena of "alternate generations," it will be as well to glance for a moment at a peculiar form of gemmation exhibited by some of the Polyzoa, which is in some respects intermediate between ordinary discontinuous gemmation and alternation of generations. These organisms are nearly allied to the sea-mat, already spoken of, and, like it, can reproduce themselves by continuous gemmation (forming colonies), by a true sexual process, and rarely by fission. In addition to all these methods they can reproduce themselves by the formation of peculiar internal buds, which are called "statoblasts." These buds are developed upon a peculiar cord, which crosses the body-cavity, and is attached at one end to the fundus of the stomach. When mature they drop off from this cord, and lie loose in the cavity of the body, whence they are liberated on the death of the parent organism. When thus liberated, the statoblast, after a longer or shorter period, ruptures and gives exit to a young Polyzoön, which has essentially the same structure as the adult. It is, however, simple, and has to undergo a process of continuous gemmation before it can assume the compound form proper to the adult.

As regards the nature of these singular bodies, "the invariable absence of germinal vesicle and germinal spot, and their never exhibiting the phenomena of yelk-cleavage, independently of the conclusive fact that true ova and ovary occur elsewhere in the same individual, are quite decisive against their being eggs. We must then look upon them as *gemmæ* peculiarly encysted, and destined to remain for a period in a quiescent or pupa-like state."—(Allman.)

c. Alternation of Generations.—In the case of the Hydra and the sea-mat, which we have considered above, fresh zoöids are produced by a primordial organism by gemmation; the beings thus produced (as well as the parent) being capable not only of repeating the gemmiparous process, but also of producing new individuals by a true generative act. We have now to consider a much more complex series of phenomena, in which the organism which is developed from the primitive ovum produces by gemmation *two* sets of zoöids, one of which is destitute of sexual organs, and is capable of performing no other function than that of nutrition, whilst the other is provided with reproductive organs, and is destined for the perpetuation of the species. In the former case the produced zoöids all resembled each other, and the parent organism which gave rise to them; in the latter case, the produced zoöids are often utterly unlike each other and unlike the parent, since their functions are entirely different.

The simplest form of the process is seen in certain of the Hydroid Polypes, such as Sertularia. The ovum of Sertularia is a free-swimming ciliated body, which, after a short locomotive existence, attaches itself to some submarine object, develops a mouth and tentacles, and commences to produce zoöids like itself by a process of continuous gemmation. These remain permanently attached to one another, with the result that a compound organism is produced, consisting of a number of zoöids, or "polypites," organically connected together, but enjoying an independent existence. None of the zoöids, however, are provided with sexual organs; and though there is theoretically no limit to the size which the colony may reach by gemmation, its buds are not detached, and the species would therefore die out, unless some special provision were made for its preservation. Besides these nutritive zoöids, however, other buds are produced which differ considerably in appearance from the former, and which have the power of generating the essential elements of reproduction. These generative zoöids derive their nourishment from the materials collected by the nutritive zoöids, but only live until the ova are matured in their interior and liberated, when they disappear. The ova thus produced become free-swimming ciliated bodies, such as the one with which the cycle began.

In this case, therefore, the "individual" Sertularia consists of a series of nutritive zoöids, collectively called the "trophosome," and another series of reproductive zoöids, collectively called the "gonosome," the entire series often remaining in organic connection.

In other forms nearly allied to Sertularia (such as Coryne) the process advances a step further. In Coryne the generative buds, or zoöids, do not produce the reproductive elements as long as they remain attached to the parent colony; but they require a preliminary period of independent existence. For this purpose they are specially organised, and when sufficiently matured they are detached from the stationary colony. The generative zoöid now appears as an entirely independent being, described as a species of jelly-fish (or Medusa) under the name of Sarsia. It consists of a bell-shaped disc, by means of which it is enabled to swim freely; from the centre of this disc depends a nutritive process, with a mouth and digestive cavity, whereby the organism is able to increase considerably in size. The substance of the disc is penetrated by a complex system of canals, and from its margin hangs a series of tentacular processes. After a period of independent locomotive existence, the Medusa attains its full growth, when it develops ova and spermatozoa. By the contact of these embryos are produced; but these, instead of resembling the jelly-fish by which they were immediately generated, proceed to develop themselves into the fixed Hydroid colony by which the Medusa was originally produced.

Still more extraordinary phenomena have been discovered in other Hydrozoa, as in many of the Lucernarida. In these the ovum gives rise (as in Sertularia) to a locomotive ciliated body, which ultimately fixes itself, becomes trumpet-shaped, and develops a mouth and tentacles at its expanded extremity, when it is known as the "hydra-tuba," from its resemblance to the fresh-water polype, or Hydra. The hydra-tuba has the power of multiplying itself by gemmation, and it can produce large colonies in this way; but it does not obtain the power of generating the essential elements of reproduction. Under certain circumstances, however, the hydra-tuba enlarges, and, after a series of preliminary changes, divides by transverse fission into a number of segments, each of which becomes detached and swims away. These liberated segments of the little hydra-tuba (it is about half an inch in height) now live as entirely independent beings, which were described by naturalists as distinct animals, and were called Ephyræ. They are provided with a swimming-bell, or "umbrella," by means of which they propel themselves through the water, and with a mouth and digestive cavity. They now lead an active life, feeding eagerly, and attaining in some instances a perfectly astonishing size (the Medusoids of some species are several feet in circumference). After a while they develop the essential elements of reproduction, and after the fecundation and liberation of their ova they die. The ova, however, are not developed into the free-swimming and comparatively gigantic jelly-fish by which they were immediately produced, but into the minute, fixed, sexless hydra-tuba.

We thus see that a small, sexless zoöid, which is capable of multiplying itself by gemmation, produces by fission several independent locomotive beings, which are capable of nourishing themselves and of performing all the functions of life. In these are produced generative elements, which give rise by their development to the little fixed creature with which the series began.

To the group of phenomena of which the above are examples, the name "alternation of generations" was applied by Steenstrup; but the name is not an appropriate one, since the process is truly an alternation of generation with gemmation or fission. The only generative act takes place in the reproductive zoöid, and the production of this from the nutritive zoöid is a process of gemmation or fission, and not a process of generation. The "individual," in fact, in all these cases, must be looked upon as a double being composed of two factors, both of which lead more or less completely independent lives, the one being devoted to nutrition, the other to reproduction. The generative being, however, is in many cases not at first able to mature the sexual elements, and is therefore provided with the means necessary for its growth and nourishment as an independent organism. It must also be remembered that the nutritive half of the "individual" is usually, and the generative half sometimes, *compound*—that is to say, composed of a number of zoöids produced by continuous gemmation; so that the zoological individual in these cases becomes an extremely complex being.

These phenomena of so-called "alternation of generations," or "metagenesis," occur in their most striking form amongst the Hydrozoa; but they occur also amongst many of the intestinal worms (Entozoa), and amongst some of the Tunicata (Molluscoida).

d. Parthenogenesis.—"Parthenogenesis" is the term employed to designate certain singular phenomena, resulting in the production of new individuals by virgin females without the intervention of a male. By Professor Owen, who first employed the term, parthenogenesis is applied also to the processes of gemmation and fission, as exhibited in sexless beings or in virgin females; but it seems best to consider these phenomena separately. Strictly, the term parthenogenesis ought to be confined to the production of new individuals from virgin females by means of *ova*, which are enabled to develop themselves without the contact of the male element. The difficulty in this definition is found in framing an exact definition of an ovum, such as will distinguish it from an internal gemma or bud. No body, however, should be called an "ovum" which does not exhibit a germinal vesicle and germinal spot, and which does not exhibit the phenomenon known as segmentation of the yelk. Moreover, ova are almost invariably produced by a special organ, or ovary.

As examples of parthenogenesis we may take what occurs in plant-lice (Aphides) and in the honey-bee; but it will be seen that in neither of these cases are the phenomena so unequivocal, or so well ascertained, as to justify a positive assertion that they are truly referable to parthenogenesis in the above restricted sense of the term.

The Aphides, or plant-lice, which are so commonly found parasitic upon plants, are seen towards the close of autumn to consist of male and female individuals. By the sexual union of these true ova are produced, which remain dormant through the winter. At the approach of spring these ova are hatched; but instead of giving birth to a number of males and females, all the young are of one kind, variously regarded as neuters, virgin females, or hermaphrodites. Whatever their true nature may be, these individuals produce *viviparously* a brood of young which resemble themselves; and this second generation, in like manner, produces a third,—and so the process may be repeated, for as many as ten or more generations, throughout the summer. When the autumn comes on, however, the viviparous Aphides produce—in exactly the same manner—a final brood; but this, instead of being composed entirely of similar individuals, is made up of males and females. Sexual union now takes place, and ova are produced and fecundated in the ordinary manner.

The bodies from which the young of the viviparous Aphides are produced are variously regarded as internal buds, as "pseudova" (*i.e.*, as bodies intermediate between buds and ova), and as true ova.

Without entering into details, it is obvious that there is only one explanation of these phenomena which will justify us in regarding the case of the viviparous Aphides as one of true parthenogenesis, as above defined. If, namely, the spring broods are true females, and the bodies which they produce in their interior are true ova, then the case is one of genuine parthenogenesis, for there are certainly no males. The case might still be called one of parthenogenesis, even though the bodies from which these broods are produced be regarded as internal buds, or as "pseudova;" for a true ovum is essentially a bud. If, however, Balbiani be right, and the viviparous Aphides are really hermaphrodite, then, of course, the phenomena are of a much less abnormal character.

In the second case of alleged parthenogenesis which we are about to examine—namely, in the honey-bee—the phenomena which have been described cannot be said to be wholly free from doubt. A hive of bees consists of three classes of individuals—1. A "queen," or fertile female; 2. The "workers," which form the bulk of the community, and are really undeveloped or sterile females; and, 3. The "drones," or males, which are only produced at certain times of the year. We have here three distinct sets of beings, all of which proceed from a single fertile individual, and the question arises, In what manner are the differences between these produced? At a certain period of the year the queen leaves the hive, accompanied by the drones (or males), and takes what is known as her "nuptial flight" through the air. In this flight she is impregnated by the males, and it is immaterial whether this act occurs once in the life of the queen, or several times, as asserted by some. Be this as it may, the queen, in virtue of this single impregnation, is enabled to produce fresh individuals for a lengthened period, the semen of the males being stored up in a receptacle which communicates by a tube with the oviduct, from which it can be shut off at will. The ova which are to produce workers (undeveloped females) and queens (fertile females) are fertilised on their passage through the oviduct, the semen being allowed to escape into the oviduct for this purpose. The subsequent development of these fecundated ova into workers or queens depends entirely upon the form of the cell into which the ovum is placed, and upon the nature of the food which is supplied to the larva. So far there is no doubt as to the nature of the phenomena which are observed. It is asserted, however, by Dzierzon and Siebold, that the males or drones are produced by the queen from ova which she does not allow to come into contact with the semen as they pass through the oviduct. This assertion is supported by the fact that if the communication between the receptacle for the semen and the oviduct be cut off, the queen will produce nothing but males. Also, in crosses between the common honey-bee and the Ligurian bee, the queens and workers alone exhibit any intermediate characters between the two forms, the drones presenting the unmixed characters of the queen by whom they were produced.

If these observations are to be accepted as established—and, upon the whole, there can be no hesitation in accepting them as in the main correct—then the drones are produced by a true process of parthenogenesis; but some observers maintain that the development of any given ovum into a drone is really due —as in the case of the queens and workers—to the special circumstances under which the larva is brought up.*

* In the case of *Polistes Gallica*, Von Siebold appears to have proved

There are various other cases in which parthenogenesis is said to occur, but the above will suffice to indicate the general character of the phenomena in question. The *theories* of parthenogenesis appear to be too complex to be introduced here; and there is the less to regret in their omission, as naturalists have not yet definitely adopted any one explanation of the phenomena to the exclusion of the rest.

First Law of Quatrefages.—From the phenomena of asexual reproduction in all its forms, M. de Quatrefages has deduced the following generalisation:—

"The formation of new individuals may take place, in some instances, by gemmation from, or division of, the parent-being; but this process is an exhaustive one, and cannot be carried out indefinitely. When, therefore, it is necessary to insure the continuance of the species, the sexes must present themselves, and the germ and sperm must be allowed to come in contact with one another."

It should be added that the act of sexual reproduction, though it insures the perpetuation of the *species*, is very destructive to the life of the *individual.* The formation of the essential elements of reproduction appears to be one of the highest physiological acts of which the organism is capable, and it is attended with a corresponding strain upon the vital energies. In no case is this more strikingly exhibited than in the majority of insects, which pass the greater portion of their existence in a sexually immature condition, and die almost immediately after they have become sexually perfect, and have consummated the act whereby the perpetuation of the species is secured.

11. Development, Transformation, and Metamorphosis.

Development is the general term applied to all those changes which a germ undergoes before it assumes the characters of the perfect individual; and the chief differences which are observed in the process as it occurs in different animals consists simply in the extent to which these changes are external and visible, or are more or less completely concealed from view. For these differences the terms "transformation" and "metamorphosis" are employed; but they must be regarded as essentially nothing more than variations of development.

beyond reasonable doubt that the males are produced by a process of parthenogenesis. Landois, however, asserts that the eggs of insects are of no sex, that sex is only developed in the larva after its emergence from the egg, and that in each individual larva the sex is determined wholly by the nature of the food upon which it is brought up; abundant nourishment producing females, and scanty diet giving rise to males.

Transformation is the term employed by Quatrefages to designate "the series of changes which every germ undergoes in reaching the embryonic condition; those which we observe in every creature still within the egg; those, finally, which the species born in an imperfectly developed state present in the course of their external life."

Metamorphosis is defined by the same author as including the alterations which are "undergone after exclusion from the egg, and which alter extensively the general form and mode of life of the individual."

Though by no means faultless, these terms are sufficiently convenient, if it be remembered that they are merely modifications of development, and express differences of degree and not of kind. An insect, such as a butterfly, is the best illustration of what is meant by these terms. All the changes which are undergone by a butterfly in passing from the fecundated ovum to the condition of an imago, or perfect insect, constitute its *development.* The egg which is laid by a butterfly undergoes a series of changes which eventuate in its giving birth to a caterpillar, these preliminary changes constituting its *transformation.* The caterpillar grows rapidly, and after several changes of skin becomes quiescent, when it is known as a "chrysalis." It remains for a longer or shorter time in this quiescent and apparently dead condition, during which period developmental changes are going on rapidly in its interior. Finally, the chrysalis ruptures, and there escapes from it the perfect winged insect. To these changes the term *metamorphosis* is rightly applied. These changes, however, do not differ in kind from the changes undergone by a Mammal; the difference being that in the case of a Mammal the ovum is retained within the body of the parent, where it undergoes the necessary developmental changes, so that at birth it has little to do but grow, in order to be converted into the adult animal.

From these considerations we arrive at the second law laid down by Quatrefages:—"Those creatures whose ova—owing to an insufficient supply of nutritious contents, and an incapacity on the part of the mother to provide for their complete development within her own substance—are rapidly hatched, give birth to imperfect offspring, which, in proceeding to their definitive characters, undergo several alterations in structure and form, known as metamorphoses."

Retrograde Development.—Ordinarily speaking, the course of development is an ascending one, and the adult is more highly organised than the young; but there are cases in which there is an apparent reversal of this law, and the adult is to all appearance a degraded form when compared with the embryo. This phenomenon is known as "retrograde" or "recurrent" development; and well-marked instances are found amongst the Cirripedia and Lernææ, both of which belong to the Crustacea.

Thus, in the Cirripedes (acorn-shells, &c.) and in the parasitic Lernææ the embryo is free-swimming and provided with organs of vision and sensation, being in most respects similar to the permanent condition of certain other Crustacea, such as the Copepods. The adult, however, in both cases, is degraded into a more or less completely sedentary animal, more or less entirely deprived of organs of sense, and leading an almost vegetative life. As a compensation, reproductive organs are developed in the adult, and it is in this respect superior to the locomotive, but sexless, larva.

12. Spontaneous Generation.

Spontaneous or Equivocal generation is the term applied to the alleged production of living beings without the pre-existence of germs of any kind, and therefore without the pre-existence of parent organisms. The question is one which has been long and closely disputed, and is far from being settled; so that it will be sufficient to indicate the facts upon which the theory rests.

If an animal or vegetable substance be soaked in hot or cold water, so as to make an organic infusion, and if this infusion be exposed for a sufficient length of time to the air, the following series of changes is usually observed :—

1. At the end of a longer or shorter time, there forms upon the surface of the infusion a thin scum, or pellicle, which, when examined microscopically, is found to consist of an incalculable number of extremely minute molecules.

2. In the next stage these molecules appear, many of them, to have increased in size by endogenous division, till they form short staff-shaped filaments, called "bacteria." These increase in length by the same process until we get long filamentous bodies produced, which are termed "vibriones."* Both the bacteria and the vibrios now exhibit a vibratile or serpentine movement through the surrounding fluid.

3. After a varying period, the bacteria and vibrios become

* By some authorities it is believed that the bacteria are produced by the fusion together of the primitive molecules in twos and threes ; and that the vibrios are produced out of the bacteria by the addition of fresh molecules to the extremities of the latter, or by their uniting with one another.

motionless, and disintegrate so as to produce again a finely molecular pellicle.

4. Little spherical bodies now appear, each of which is provided with a vibratile cilium with which it moves actively through the infusion. (Monas lens.)

5. Varied forms of ciliated Infusoria—some which possess a mouth and are otherwise highly organised—make their appearance in the fluid.

The above is the general sequence of the phenomena which have been observed, and the following are the two theories which have been advanced to account for them :—

a. By the advocates of spontaneous generation, or "Heterogeny," it is affirmed that the Infusoria, which finally appear in the infusion, are produced spontaneously out of the molecular pellicle, the molecules of which are also of spontaneous origin, and are not derived from any pre-existing germs.

b. By the "panspermists," or the opponents of spontaneous generation, it is alleged, on the other hand, that the production of Bacteria, Vibrios, Monads, and Infusoria, in organic infusions, is due simply to the fact that the atmosphere, and probably the fluid itself, is charged with innumerable germs—too minute, perhaps, to be always detectable by the microscope—which, obtaining access to the fluid, and finding there favourable conditions, are developed into living beings.

A large number of elaborate experiments have been carried out to prove that atmospheric air is absolutely necessary for the production of these living beings, and that if the air be properly purified by passage through destructive chemical reagents, no such organisms will be produced, provided that the infusion have been previously boiled. As the results of all these experimental trials have hitherto proved more or less contradictory, it is unnecessary to enter into the question further, and it will be sufficient to indicate the following general considerations :—

a. The primary molecules which appear in the fluid are extremely minute, and if they are developed from germs, these may be so small as to elude any power of the microscope yet known to us. As they subsequently become converted into bacteria and vibrios, and as there can be little dispute as to these being truly living organisms, we are obliged to believe that they must have had *some* definite origin. It appears, however, to be hardly philosophical to assume that they form themselves out of the inorganic materials of the infusion; since this implies the sudden appearance, or creation, of new force, for which there seems to be no means of accounting.

b. The nature of the vibrios and bacteria must be looked upon as quite uncertain. To say the least of it, they are quite as likely to be plants as animals; and the most probable hypothesis would place the former near the filamentous Confervæ, or would regard them as the mycelium of various species of Moulds (*Penicillium*).

c. What has been said above with regard to the origin of the bacteria and vibrios applies equally to the origin of the Monads, which appear in the infusion subsequently to the death of the vibrios.

d. These Monads, as shown by recent researches, are probably to be looked upon as the embryonic, or larval, forms of the higher Infusoria which succeed them.

e. Many of the Infusoria which finally appear are of a comparatively high grade of organisation, being certainly the highest of the Protozoa, and being placed by some competent observers in the neighbourhood of the Trematode Worms (Annuloida). It is therefore very unlikely that these should be generated spontaneously; since, if this ever occurs, it is reasonable to suppose that the creatures thus produced will be of the lowest possible organisation (such as the Gregarinidæ or the Monera, for example), and will be far below the Infusoria in point of structure.

f. The reproductive process in many of these same Infusoria is perfectly well known, and it consists either in a true sexual process, for which proper organs are provided (as in Paramœcium), or in a process of gemmation or fission. It is therefore improbable that they should be generated in the manner maintained by the heterogenists, since this mode of reproduction would appear to be superfluous.

g. In the absence of any direct proof to the contrary, it is safer to adopt an explanation of the observed phenomena which does not have recourse to laws with which we are as yet unacquainted. Thus, it is not at variance with any known law to suppose that the primary molecules are the result of the development of germs which find in the inorganic infusion a suitable *nidus;* that these primary molecules and the vibrios which they produce are referable to the Protophyta, and should probably be placed near the filamentous Confervæ; that by the death of these vegetable organisms the fluid is prepared for the reception and development of the germs of the Protozoa, for which the former serve as *pabulum;* and that many of the forms which are observed are the larval stages of the higher Infusoria.*

* Recent researches, especially those of Dr Bastian, have established

13. ORIGIN OF SPECIES.

It is impossible here to do more than merely indicate in the briefest manner the two fundamental ideas which are at the bottom of all the various theories as to the origin of species. The opinions of scientific men are still divided upon this subject; and it will be sufficient to give an outline of the two leading theories, without adducing any of the reasoning upon which they are based.

I. *Doctrine of Special Creation.* — On this doctrine of the origin of species it is believed that species are immutable productions, each of which has been specially created at some point within the area in which we now find it, to meet the external conditions there prevailing, subsequently spreading from this spot as far as the conditions of life were suitable for it.

II. *Doctrine of Development.*—On the other hand, it is believed that species are not permanent and immutable, but that they "undergo modification, and that the existing forms of life are the descendants by true generation of pre-existing forms." —(Darwin.)

On Lamarck's theory of the development of species, the means of modification were ascribed to the action of external physical agencies, the inter-breeding of already existing forms, and the effects of habit.

The doctrine of the development of species by variation and

some new facts as to the possibility of Heterogeny, but they can by no means be said to have settled the question, if only upon the ground that they require confirmation by other experimentalists. The chief fact which appears to have been established upon a tolerably firm basis is, that living beings, vegetable or animal, may make their appearance in organic infusions which have been subjected to a temperature of considerably over the boiling-point, even though the said infusions have been hermetically sealed in a flask from which all atmospheric air has been previously withdrawn. The chief deduction which appears to flow from this—assuming its correctness—is, that there are low organisms which can exist, for a certain length of time at any rate, with an extremely small amount of air; for it is to be remembered that the production of a theoretically perfect vacuum is probably practically impossible. If it were conceded, in fact, that a perfect vacuum *had* been formed in the experiments in question, the sole result would be that we should have to alter all our beliefs as to the conditions under which life is a possibility. The only tangible result of these experiments, so far, is, that any supposed "pre-existent germs" must have been contained, if present at all, in the infinitesimal portion of air which could not be expelled from the flasks experimented on; or, they must have been able to withstand without injury a temperature of over 212°. Neither of these hypotheses is wholly incredible; but the question ought to be regarded as still *sub judice*, and there is grave doubt as to the reliability and accuracy of the experiments above alluded to.

natural selection — propounded by Darwin, and commonly known as the Darwinian theory—is based upon the following fundamental propositions :—

1. The progeny of all species of animals and plants exhibit variations amongst themselves in all parts of their organisation; no two individuals being exactly alike. In other words, in every species the individuals tend by variation to diverge from the parent-type, in some particular or other.

2. These variations can be transmitted to future generations under certain definite and discoverable laws of inheritance.

3. By artificial selection and breeding from individuals possessing any particular variation, man, in successive generations, can produce a breed in which the variation is permanent; the races thus produced being often as widely different as are distinct species of wild animals.

4. The world in which all living beings are placed is one not absolutely unchanging, but is liable to subject them to very varying conditions.

5. All animals and plants give rise to more numerous young than can by any possibility be preserved.

6. As these young are none of them exactly alike, a process of "Natural Selection" will ensue, whereby those individuals which possess any variation favourable to the peculiarities of the life of the species will be preserved. Those individuals which do not possess such a favourable variation will be placed at a disadvantage in the "struggle for existence," and will tend to be gradually exterminated.

7. Other conditions remaining the same, the individuals which survive in the struggle for existence will transmit the variations, to which their preservation is due, to future generations.

8. By a repetition of this process "varieties" are first established; these become permanent, and "races" are produced; finally, in the lapse of time, the differences become sufficiently great to constitute distinct species.

14. Distribution.

Under this head come all the facts which are concerned with the external or objective relations of animals—that is to say, their relations to the external conditions in which they are placed.

The *geographical* distribution of animals is concerned with the determination of the areas within which every species of animal is at the present day confined. Some species are found almost everywhere, when they are said to be "cosmopolitan;"

but, as a rule, each species is confined to a limited and definite area. Not only are species limited in their distribution, but it is possible to divide the globe into a certain number of geographical regions or "zoological provinces," each of which is characterised by the occurrence in it of certain associated forms of animal life. It is to be remembered, however, that the zoological provinces of the present day by no means correspond with those of former periods, and that they have only existed as such since comparatively recent times.

The *vertical* or *bathymetrical* distribution of animals relates to the limits of depth within which each marine species of animals is confined. As a rule it is found that each species has its own definite bathymetrical zone, and that its existence is difficult or impossible at depths greater or less than those comprised by that zone. Generalising on a large number of facts, naturalists have been able to lay down and name certain definite zones, each of which has its own special fauna.

The four following zones are those generally accepted:—

1. The Littoral zone, or the tract between tide-marks.
2. The Laminarian zone, from low water to 15 fathoms.
3. The Coralline zone, from 15 to 50 fathoms.
4. The deep-sea Coral zone, 50 to 100 fathoms or more.
5. To these must now be certainly added a fifth zone, extending from 100 fathoms to a depth of 2500 fathoms or more.

Recent researches, however, have rendered it certain that after a certain depth, say 100 fathoms, the bathymetrical distribution of animals is conditioned not by the *depth*, but by the *temperature* of the water at the bottom of the sea. Similar forms, namely, are always found inhabiting areas in which the bottom-temperature is the same, wholly irrespective of the depth of water in the particular locality in question. The supply of food, also, and the nature of the habitat, are important elements of the case. In the light, therefore, of these recent facts, it would perhaps be advisable to adopt the views of Mr Gwyn Jeffreys, and to consider that there are only two principal bathymetrical zones—namely, the *littoral* and the *submarine.*

In addition to the preceding forms of distribution, the zoologist has to investigate the condition and nature of animal life during past epochs in the history of the world

The laws of *distribution in time*, however, are, from the nature of the case, less perfectly known than are the laws of lateral or vertical distribution, since these latter concern beings which we are able to examine directly. The following are the chief facts which it is necessary for the student to bear in mind:

1. The rocks which compose the crust of the earth have been formed at successive periods, and may be roughly divided into aqueous or sedimentary rocks, and igneous rocks.

2. The igneous rocks are produced by the agency of heat, are mostly *unstratified* (*i.e.*, are not deposited in distinct layers or *strata*), and, with few exceptions, are destitute of any traces of past life.

3. The sedimentary or aqueous rocks owe their origin to the action of water, are *stratified* (*i.e.*, consist of separate layers or *strata*), and mostly exhibit "fossils"—that is to say, the remains or traces of animals or plants which were in existence at the time when the rocks were deposited.

4. The series of aqueous rocks is capable of being divided into a number of definite groups of strata, which are technically called "formations."

5. Each of these definite rock-groups, or "formations," is characterised by the occurrence of an assemblage of fossil remains more or less peculiar and confined to itself.

6. The majority of these fossil forms are "extinct"—that is to say, they do not admit of being referred to any species at present existing.

7. No fossil, however, is known, which cannot be referred to one or other of the primary subdivisions of the Animal Kingdom, which are represented at the present day.

8. When a species has once died out, it never reappears.

9. The older the formation, the greater is the divergence between its fossils and the animals and plants now existing on the globe.

10. All the known formations are divided into three great groups, termed respectively Palæozoic or Primary, Mesozoic or Secondary, and Kainozoic or Tertiary.

The Palæozoic or Ancient-life period is the oldest, and is characterised by the marked divergence of the life of the period from all existing forms.

In the Mesozoic or Middle-life period, the general *facies* of the fossils approaches more nearly to that of our existing fauna and flora; but—with very few exceptions—the characteristic fossils are all specifically distinct from all existing forms.

In the Kainozoic or New-life period, the approximation of the fossil remains to existing living beings is still closer, and some of the forms are now specifically identical with recent species; the number of these increasing rapidly as we ascend from the lowest Kainozoic deposit to the Recent period.

Subjoined is a table giving the more important subdivisions

of the three great geological periods, commencing with the oldest rocks and ascending to the present day.

I. PALÆOZOIC OR PRIMARY ROCKS.

1. Laurentian. (Lower and Upper.)
2. Cambrian. (Lower and Upper, with Huronian Rocks?)
3. Silurian. (Lower and Upper.)
4. Devonian, or Old Red Sandstone. (Lower, Middle, and Upper.)
5. Carboniferous. (Mountain-limestone, Millstone Grit, and Coal-measures.)
6. Permian. (= the lower portion of the New Red Sandstone.)

II. MESOZOIC OR SECONDARY ROCKS.

7. Triassic Rocks. (Bunter Sandstein, or Lower Trias; Muschelkalk, or Middle Trias; Keuper, or Upper Trias.)
8. Jurassic Rocks. (Lias, Inferior Oolite, Great Oolite, Oxford Clay, Coral Rag, Kimmeridge Clay, Portland Stone, Purbeck beds.)
9. Cretaceous Rocks. (Wealden, Lower Greensand, Gault, Upper Greensand, White Chalk, Maestricht beds.)

III. KAINOZOIC OR TERTIARY ROCKS.

10. Eocene. (Lower, Middle, and Upper.)
11. Miocene. (Lower and Upper.)
12. Pliocene. (Older Pliocene and Newer Pliocene.)
13. Post-tertiary. (Post-pliocene and Recent.)

INVERTEBRATE ANIMALS.

PROTOZOA.

CHAPTER I.

1 GENERAL CHARACTERS OF THE PROTOZOA.
2. CLASSIFICATION. 3. GREGARINIDÆ.

1. *General Characters.*—The sub-kingdom *Protozoa*, as the name implies, includes the most lowly organised members of the animal kingdom. From this circumstance it is difficult, if not impossible, to give an exhaustive definition, and the following is, perhaps, as exact as the present state of our knowledge will allow:—

The *Protozoa* may be defined as *animals, generally of minute size, composed of a nearly structureless jelly-like substance (termed "sarcode"), showing no composition out of definite parts or segments, having no definite body-cavity, presenting no traces of a nervous system, and having either no differentiated alimentary apparatus, or but a very rudimentary one.*

The *Protozoa* are almost exclusively aquatic in their habits, and are mostly very minute, though they sometimes form colonies of considerable size. They are composed of a more or less contractile, jelly-like substance, called "sarcode" or "animal protoplasm," which is semi-fluid in consistence, and is composed of an albuminous base with oil-globules scattered through it. Granules are generally developed in the sarcode, and in many cases there is a definite internal solid particle, termed the "nucleus."

In no *Protozoön* are any traces known of anything like the nervous and vascular arrangements which are found in animals of a higher grade. A nervous system is universally and entirely absent, and the sole circulatory apparatus consists in certain clear spaces called "contractile vesicles," which are found in some species, and which doubtfully perform the functions of a heart. A distinct alimentary aperture is present in the higher *Protozoa*, but in many there is none; and in all, the digestive apparatus is of the simplest character. Organs of generation, or at any rate differentiated portions of the body which act as these, are sometimes present; but in many cases true sexual reproduction has not hitherto been shown to exist.

The "sarcode," which forms such a distinctive feature in all the *Protozoa*, is a structureless albuminous substance, not possessing "permanent distinction or separation of parts," but nevertheless displaying all "the essential properties and characters of vitality," being capable of assimilation and excretion, of irritability, and of the power of contraction, so as to produce movements, strictly analogous, in many cases, to the muscular movements of the higher animals. In some, too, the sarcode possesses the power of producing an external case or envelope, usually of carbonate of lime or flint, and often of a very complicated and mathematically regular structure.

The power of active locomotion is enjoyed by a great many of the *Protozoa;* but in some cases this is very limited, and in other cases the animal is permanently fixed in its adult condition. The apparatus of locomotion in the *Protozoa* is of a very varied nature. In many cases, especially in the higher forms, movements are effected by means of the little hair-like processes which are known as "cilia," and which have the power of lashing to and fro or vibrating with great rapidity. In other cases the cilia are accompanied or replaced by one or more long whip-like bristles, which act in the same fashion, and are known as "flagella." The most characteristic organs of locomotion amongst the lower *Protozoa* are known as "pseudopodia," and consist simply of prolongations of the sarcodic substance of the body, which can usually be emitted from the greater portion of the general surface of the body, and are capable of being again retracted, and of fusing completely with the body-substance.

2. *Classification of the Protozoa.* The sub-kingdom *Protozoa* is divided into three classes—viz., the *Gregarinidæ*, the *Rhizopoda*, and the *Infusoria*. In the Infusoria only is a mouth present, and hence these are sometimes spoken of as the "*Stomatode*" *Protozoa*, whilst the two former classes collectively constitute the "*Astomata*."

The following is a tabular view of the divisions of the *Protozoa*:—

Class I. GREGARINIDÆ.

Class II. RHIZOPODA.

Order 1. *Monera.*
" 2. *Amœbea.*
" 3. *Foraminifera.*
" 4. *Radiolaria.*
" 5. *Spongida.*

Class III. INFUSORIA.

Order 1. *Suctoria.*
" 2. *Ciliata.*
" 3. *Flagellata.*

3. CLASS I. GREGARINIDÆ.—The *Gregarinidæ* may be defined as *parasitic Protozoa, which are destitute of a mouth, and do not possess the power of emitting "pseudopodia."* They constitute the lowest class of the *Protozoa*, and comprise certain microscopic animals which are parasitic in the alimentary canal of both Invertebrate and Vertebrate animals. They have, however, a special liking for the intestines of certain insects, being commonly found abundantly in the cockroach. As we shall see hereafter, in all probability a great deal of the degraded character of the *Gregarinidæ* is due to the fact that they are internal parasites, and are therefore not dependent upon their own exertions for food.

Nothing anatomically could be more simple than the structure of a *Gregarina*, since it is almost exactly that of the unimpregnated ovum (fig. 1, *a*). An adult Gregarina, in fact, may be said to be a single cell, consisting of an ill-defined membranous envelope filled with a more or less granular sarcode with fatty particles, which contains in its interior a vesicular nucleus, this in turn enclosing a solid particle, or nucleolus. In some the body exhibits an approach to a more complex structure by the presence of internal septa; but it is doubtful whether this appearance may not be due to the apposition and fusion of two separate individuals. A separate order, however, has been founded upon individuals of this kind, under the name of *Dicystidea;* the name *Monocystidea* being retained for the ordinary forms. As regards the size of the *Gregarinæ*, they vary from about the size of the head of a small pin up to as much as half an inch in length, when they assume the aspect of small worms. The integument or cuticle with which the protoplasmic body is enclosed may be quite smooth or striated, or it may be furnished with bristles or spines, or even in some cases with cilia. Sometimes one end of the body is furnished with uncinate processes, very similar in appearance to the hooked "head" of the common tape-worm (*Tænia solium*). Essentially, however, the structure of all appears to be the same. No differentiated organs of any kind beyond the nucleus and nucleolus exist, and both assimilation and excretion must be performed simply by the general surface of the body. The body is, nevertheless, contractile, and slow movements can be effected, not, however, by pseudopodia.

In spite of their exceedingly simply structure, the following very interesting reproductive phenomena have been observed,

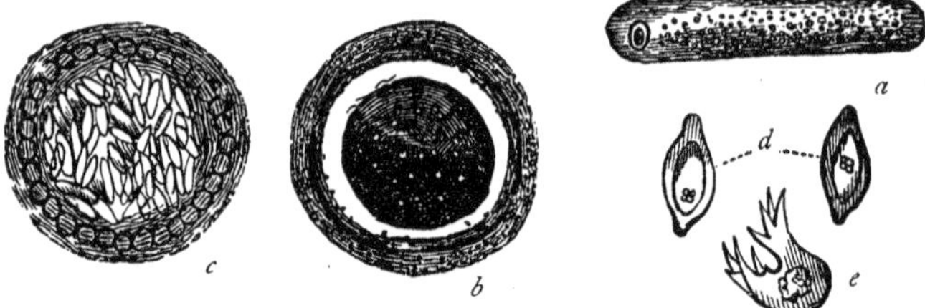

Fig. 1.—Gregarina of the earth-worm. *a* Adult *Gregarina*; *b* The same encysted; *c* With the contents divided into pseudonavicellæ; *d* Free pseudonavicellæ; *e* Free amœbiform contents of the pseudonavicellæ. (After Lieberkühn.)

sometimes in a single *Gregarina* without apparent cause, sometimes as the result of the apposition and coalescence of two individuals—the exact nature of the process being in either case obscure. The *Gregarina*—or it may be two individuals which have come into contact and adhered together—assumes a globular form, becomes motionless, and develops round itself a structureless envelope or cyst, when it is said to be "encysted" (fig. 1, *b*). The central nucleus then disappears, apparently by dissolution, whereupon the granular contents of the cyst break up into a number of little rounded masses, which gradually elongate and become lanceolate, when they are termed "pseudonavicellæ". (or "pseudonaviculæ") (fig. 1, *c*). The next step in the process consists in the liberation of the pseudonavicellæ, which escape by the rupture of the enclosing cyst (fig. 1, *d*). If they now find a congenial habitat, they give origin to little albuminous or sarcodic masses, which exhibit lively movements, and are endowed with the power of throwing out and retracting little processes of the body which closely resemble the "pseudopodia" of the

Rhizopoda; so that the pseudonavicella in this condition is very similar to an adult *Amœba* (fig. 1, *e*). Finally, these amœbiform bodies are developed into adult *Gregarinæ.* It will be seen from the above that the formation of the pseudo-navicellæ out of the granular contents of the body, subsequent to the disappearance of the nucleus, presents some analogy to the segmentation of the impregnated ovum which follows upon the dissolution of the germinal vesicle.

PSOROSPERMIÆ.—There occur as parasites on and within the bodies of fishes certain vesicular, usually caudate, bodies, termed *Psorospermiæ,* the exact nature of which is very problematical. According to Lieberkühn they occasionally give origin to amœbiform bodies, similar to those which are liberated from the pseudonavicellæ of *Gregarinidæ.* In this case they should probably be regarded as the embryonic forms of some *Gregarina.* By Balbiani, however, they are looked upon as properly belonging to the vegetable kingdom.

CHAPTER II.

RHIZOPODA.

GENERAL CHARACTERS OF THE RHIZOPODA.—The *Rhizopoda* may be defined as *Protozoa which are destitute of a mouth, are simple or compound, and possess the power of emitting "pseudopodia."* They are mostly small, but some of the composite forms, such as the sponges, may attain a very considerable size. Structurally, a typical Rhizopod—as an *Amœba*—is composed of almost structureless sarcode, without any organs appropriated to the function of digestion, and possessing the power of throwing out processes of its substance so as to constitute adventitious limbs. These are termed "pseudopodia," or false feet, and are usually protrusible at will from different parts of the body, into the substance of which they again melt when they are retracted. They are merely filaments of sarcode, sometimes very delicate and of considerable length, at other times more like finger-shaped processes; and they are somewhat analogous to the little processes which are occasionally thrown out by the white corpuscles of the blood and by pus-cells. Indeed it has been remarked by Huxley that an *Amœba* is structurally "a mere colourless blood-corpuscle, leading an independent life."

The class *Rhizopoda* is divided into five orders—viz., the *Monera*, the *Amœbea*, the *Foraminifera*, the *Radiolaria*, and the *Spongida*, of which the last is occasionally considered as a separate class.

ORDER I. MONERA.—This name has been proposed by Haeckel for certain singular organisms which may provisionally be regarded as the lowest group of the *Rhizopoda.* They are very minute in size, and are distinguished by the fact that the body is composed of structureless sarcode, capable of emitting thread-like prolongations or pseudopodia, but destitute of either nucleus or contractile vesicle. The pseudopodia are in the form of delicate filamentous processes of sarcode, which interlace and anastomose with one another in every direction, and which exhibit a circulation of minute molecules and granules in their interior, and along their edges. The body, when at rest, is more or less nearly circular in form, but it is capable of undergoing manifold changes of figure. No hard covering or "test" is ever developed. Reproduction is mostly by fission, with or without precedent encystation and quiescence. So far as is known, all the *Monera* are marine, and their systematic position is still doubtful. From the absence of a nucleus and contractile vesicle, and from the nature of the pseudopodia, they would appear upon the whole to be most closely allied to the *Foraminifera*, from which they differ, chiefly if not entirely, in the absence of a shell defending the soft sarcode of the body.

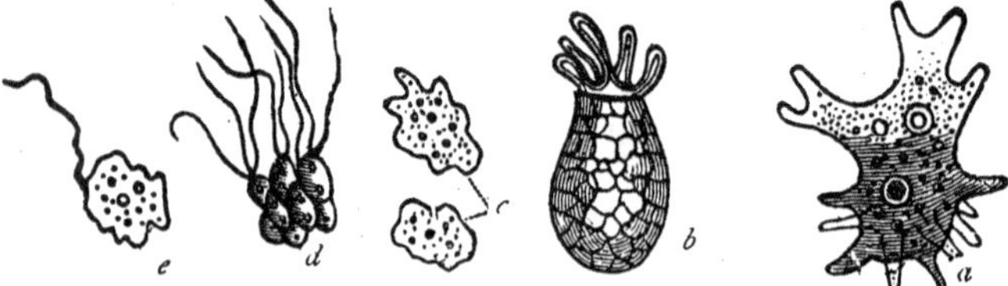

Fig. 2.—Morphology of Rhizopoda. *a Amœba radiosa*, showing the pseudopodia, contractile vesicle, and nucleus; *b Difflugia*, with the pseudopodia protruded from the anterior end of the carapace; *c* Individual sponge-particles, or "sarcoids;" *d* Ciliated sponge-particles of *Grantia*, showing the resemblance to flagellate Infusorians; *e* Mono-ciliated sarcoid of *Spongilla*. (After Carter.)

ORDER II. AMŒBEA.—This order comprises those *Rhizopoda which are, with one or two exceptions, naked, have usually short, blunt, lobose pseudopodia, which do not anastomose with one another, and contain a "nucleus," and one or more "contractile vesicles."*

The *Amœba*, or Proteus-animalcule, may be taken as the type, and a description of it will be sufficient to indicate the leading points of interest in the order. The *Amœba* (fig. 2, *a*) is a microscopic animalcule which inhabits fresh water, and is composed of gelatinous sarcode, which admits of a separation into two distinct layers: an outer transparent layer, termed the "ectosarc;" and an inner, more fluid and mobile, molecular layer, called the "endosarc." The "ectosarc" is highly extensile and contractile, and is the layer of which the pseudopodia are mainly composed; whilst the "endosarc" contains the only organs possessed by the animal—viz., the "nucleus" and "contractile vesicle" or vesicles, along with certain fortuitous cavities termed "food-vacuoles."

It is believed by some that the ectosarc is surrounded by a colourless and structureless investing membrane or cuticle; but this is denied by others. Be this as it may, there is no oral cavity, so far as has ever been certainly observed, and the food is merely taken into the interior of the body by a process of intussusception—any portion of the surface being chosen for this purpose, and acting as an extemporaneous mouth. When the particle of food has been received into the body, the aperture by which it was admitted again closes up, and the discharge of solid excreta is effected in an exactly similar manner. In this case, however, the area of the general surface within which an anus may be extemporised, appears to be more restricted, and to comprise a portion only of the body ("villous region").

The "nucleus" is a solid granular body, one or more of which is present within the endosarc of every *Amœba*, but its function is not known with any certainty. The "contractile vesicles" are cavities within the endosarc, of which ordinarily one only is present in the same individual, though sometimes there are more. In structure it is a little cavity or vesicle filled with a colourless fluid apparently derived from the digestion, and exhibiting rhythmical movements of contraction (*systole*) and dilatation (*diastole*). In some cases radiating tubes are said to have been seen proceeding from the vesicle at the moment of contraction. Regarded functionally, the contractile vesicle must be looked upon as a circulatory organ, and it offers therefore the most rudimentary form of a vascular system with which we are as yet acquainted.

Besides these proper organs, the endosarc usually contains clear spaces, which are called "vacuoles," or, more properly, "food-vacuoles." These spaces are of a merely temporary character, and are simply produced by the presence of particles of food, usually with a little water taken into the body along with the food.

There are no traces of any organs of sense, or of a nervous system, or, indeed, of any other organs in addition to those already described. Locomotion is effected, with moderate activity, but in an irregular manner, by means of the blunt, finger-shaped processes of sarcode, or pseudopodia, which can be protruded at will from any part of the body, and can be again retracted within it. The pseudopodia also serve as prehensile organs; but they do not interlace and form a network, nor do they exhibit any circulation of granules derived from the endosarc, as in many others of the *Rhizopoda.*

As regards the reproductive process in the *Amœba*, no differentiated sexual organs have hitherto been discovered, and the true sexual form of the process is therefore unknown. Fresh individuals, however, may be produced in three ways :—*Firstly*, by simple fission, the animal dividing into two parts, each of which becomes an independent organism. *Secondly*, by the detachment of a single pseudopodium, which becomes developed into a fresh *Amœba. Thirdly*, by the production of little spherical masses of sarcode which may be derived from the nucleus by fission, or may be produced by a segmentation of the endosarc, the animal having previously become torpid, and the nucleus and contractile vesicle having disappeared. These little masses, however produced, develop themselves when liberated into ordinary *Amœbæ.* This last method of reproduction is obviously very closely analogous to the production of "pseudonavicellæ" in an encysted *Gregarina.* It has been doubted, apparently with considerable reason, whether the so-called *Amœbæ* are distinct species of animals, or whether they are not rather transitory stages in the life-history of other organisms. It is quite certain that several of the *Protozoa* pass through an Amœboid stage, and it is also certain that vegetable matter not uncommonly assumes similar characters (*e.g.*, the mycelium of certain fungi). It is therefore not impossible that the forms known to the microscopist as *Amœbæ* may be ultimately discovered not to be permanent and distinct species; but the evidence on this head is still defective.

The remaining members of the *Amœbea* are constructed more or less closely after the type of the *Amœba* itself. In the nearly allied *Difflugia*, the sarcode forming the body of the animal is invested with a membranous envelope or "carapace," strengthened by grains of sand and other adventitious solid particles, and having a single aperture at one extremity, through which the pseudopodia are protruded (fig. 2, *b*). The animal generally creeps about head-downwards, so to speak;

that is to say, with the closed end of the carapace elevated above the surface on which it is moving. In *Arcella* there is a discoid or basin-shaped carapace, secreted by the animal itself, and likewise possessing but a single pseudopodial aperture, placed in this case on the flat surface of the body.

In *Pamphagus* there is no carapace, but the pseudopodia are nevertheless protrusible from one extremity only of the body, the remainder of the surface appearing to be of too resistant a consistence to allow of this. The common *sun-animalcule* (*Actinophrys sol*) is another well-known Rhizopod

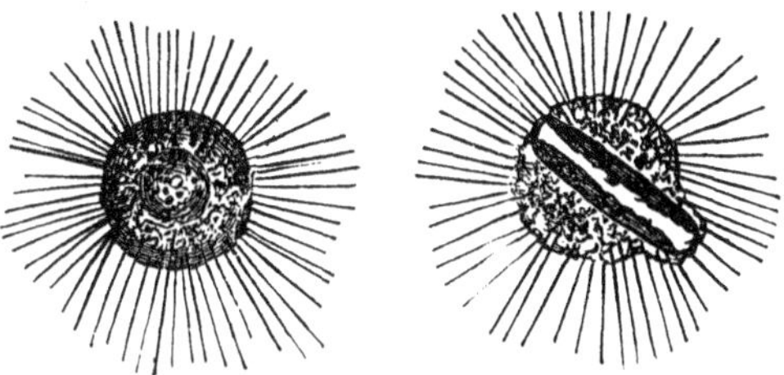

Fig. 3.—*Actinophrys sol:* showing the radiating pseudopodia. One specimen has swallowed a Diatom.

which is usually placed in this order (fig. 3). It consists of a spherical mass of sarcode, about 1-1300 of an inch in diameter, and usually covered with long, radiating, filamentous pseudopodia, which are much less mobile than in the case of the *Amœba.* The division of the substance of the body into ectosarc and endosarc is tolerably evident, and the latter contains numerous granules and vacuoles. The pseudopodia are derived from the ectosarc alone, the endosarc not passing into them, and they exhibit a circulation of granules along their edges, though this is not nearly so marked a feature as in the case of the *Foraminifera.* A nucleus and contractile vesicle are also present. The long filamentous pseudopodia of *Actinophrys* make a decided approach to the *Foraminifera*, and for this reason the sun-animalcule is sometimes placed with the latter in a single order.

The *Amœbea* may be divided into two sub-orders: 1. *Amœbina*, including those forms which have the body naked; and 2. *Arcellina*, comprising those in which the body is protected by a carapace.

CHAPTER III.

FORAMINIFERA.

ORDER III. FORAMINIFERA.—The *Foraminifera* may be defined as *Rhizopoda in which the body is protected by a shell or "test," usually composed of carbonate of lime; there is no distinct separation of the sarcode of the body into ectosarc and endosarc, and the nucleus and contractile vesicle are both absent. The pseudopodia are long and filamentous, and interlace with one another to form a network.*

The *Foraminifera* are specially characterised by the possession of a "test" or external shell, which is usually composed of carbonate of lime, but is often composed of grains of sand or other adventitious solid particles cemented together by animal matter, or which, as in *Gromia*, may be simply chitinous. (If *Lieberkühnia* is to be regarded as a *Foraminifer*, the possession of a test cannot be looked upon as essential, since this animalcule is naked. The *Monera*, also, differ from the present group mainly, if not altogether, by their naked and unprotected bodies.) The test is usually composed of an aggregation of chambers or "loculi" (fig. 4, *c*), and its walls are usually pierced by numerous pores or "foramina" through which the pseudopodia are protruded; the place of these being in some forms supplied by the large size of the terminal, or "*oral*," aperture of the shell (fig. 4, *b*). The presence or absence of foramina in the shell-walls is believed to constitute a genuine structural distinction, and the *Foraminifera* may be thereby divided into two great groups (*Perforata* and *Imperforata*).

As regards the soft parts of the *Foraminifera*, the body is composed of extensile and contractile sarcode—usually reddish or yellowish in colour—which not only fills the interior of the shell, but generally invests its outer surface also with a thin film, from which the pseudopodia are emitted. The test, therefore, in this case, is not a true cuticular secretion, like that of the *Mollusca*, but it is truly immersed within the sarcode of the body. The sarcode is not differentiated into a distinct ectosarc and endosarc, and is devoid of a nucleus and contractile vesicle, and, indeed, of any organs or specialised parts of any kind. From this uniformity in its composition there seems some reason to conclude that the *Foraminifera*—in spite of the complexity and mathematical regularity of many of their shells

—should be looked upon as the lowest forms of the *Rhizopoda*, or even of the *Protozoa*.

The pseudopodia in all the *Foraminifera* (fig. 4, *b*, *c*) are filamentous and protrusible to a great length, and they possess the singular property of uniting together in various directions so as to form a kind of network, like an "animated spider's web." (Hence the name *Reticulosa* applied to the order by Dr Carpenter.) This property, however, is not peculiar to members of this order, but is seen also in *Actinophrys* and in the *Thalassicollida*, though to a less extent. Further, throughout the entire network formed by the inosculating pseudo-

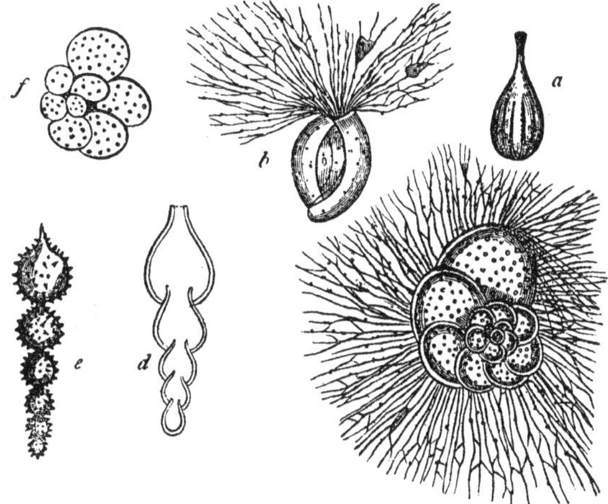

Fig. 4.—Morphology of Foraminifera. *a Lagena vulgaris*, a monothalamous Foraminifer; *b Miliola* (after Schultze), showing the pseudopodia protruded from the oral aperture of the shell; *c Discorbina* (after Schultze), showing the nautiloid shell with the foramina in the shell-wall giving exit to pseudopodia; *d* Section of *Nodosaria* (after Carpenter); *e Nodosaria hispida*; *f Globigerina bulloides*.

podia there is a constant circulation of granules in different directions. This singular phenomenon is in many respects analogous to the circulation of granules which is seen in many vegetable cells, and it is believed by Dr Carpenter that "the conditions of the two sets of phenomena are essentially the same."

The shells of *Foraminifera* may be classed in three divisions, termed respectively the "porcellanous," the "hyaline" or "vitreous," and the "arenaceous." The porcellanous shell is quite homogeneous in its composition, is opaque-white when seen by reflected light, and is not perforated by pseudopodial foramina. In these forms (*e.g.*, *Miliola*, fig. 4, *b*) the pseudopodia are emitted solely from the mouth of the last-formed segment of the shell. The vitreous shell is transparent and glassy in texture, and its walls are perforated by numerous pseudopodial apertures. The arenaceous shell is, properly speaking, not a true shell secreted by the animal, since it is simply composed of particles of sand united together by some unknown cement. Its walls may or may not be traversed by pseudopodial foramina.

As regards the form of the shell, the *Foraminifera* may be conveniently, though arbitrarily, divided into two sections: the *Monothalamia* and the *Polythalamia*. In the first of these sections (fig. 4, *a*), comprising the so-called "simple" or "unilocular" *Foraminifera*, the shell consists of a single chamber, and the animal is, in fact, nothing more than a little mass of sarcode enveloped in a calcareous covering. *Lagena*, with its beautiful flask-shaped shell, may be taken as the type of this division. Another well-known unilocular form is *Entosolenia*, which is like *Lagena* in shape, but has the tubular neck reversed, so as to be inserted into the interior of the test. In the *Polythalamia*, or "multilocular" *Foraminifera*, the shell is composed of many chambers separated from one another by divisional walls or "septa" (fig. 4, *c*, *d*, *e*), each of which is perforated by one or more openings, "septal apertures," by means of which the sarcode occupying the different chambers is united into a continuous and organic whole, the connecting bands being called "stolons." Complex as their structure often is, the compound *Foraminifera* are, nevertheless, formed by a process of continuous gemmation or budding from a single "primordial segment" in every respect identical with the permanent condition of a *simple* species. They commence their existence, therefore, as *Monothalamia*, and are converted into *Polythalamia* merely by a process of "vegetative" or "irrelative repetition." As their development proceeds, the primitive mass of sarcode, or "primordial segment," throws out fresh segments in the form of buds according to a determinate law; and it is upon the direction in which these segments are evolved that the ultimate form of the shell depends. The more important variations in this respect are as follows:—If the additional segments are added to the primordial chamber in a linear series, so as to form a straight or slightly curved line, we obtain respectively a *Nodosaria* (fig. 4, *d*, *e*) or a *Dentalina*. When the new chambers are added in a spiral direction, each being a little larger than the one which preceded it, and the coils of the spiral lying in one plane, then we get the "nautiloid" shell, so common amongst the *Foraminifera* (fig. 4, *c*). This type of shell is so closely similar to the shape of the Pearly Nautilus, that the older naturalists were long in the habit of classing these forms along with the *Cephalopoda*, or Cuttle-fish order. In the true nautiloid shell the convolutions of the spiral lie in a single plane, as in *Rotalina*, and the shell is said to be "equilateral." In other cases, however, the spiral passes obliquely round a central axis, and the shell becomes conical or turreted, when it is said to be "inequilateral" or "trochoid." In other forms, such as *Nummulites* (fig. 5) and *Orbitolites*, the structure of the shell, though regular, is much more complicated. Besides these symmetrical forms, there exist others in which the arrangement of the segments is very irregular, as is seen in *Globigerina*, *Acervulina*, &c. (fig. 4, *f*).

Besides the true pseudopodial foramina with which the walls of the test in most of the *Foraminifera* are pierced, there exists in some forms an additional system of complicated branching and anastomosing tubes, which are distributed between the laminæ of the shell, and establish a communication between its external and internal surfaces.

CLASSIFICATION OF FORAMINIFERA.—The classification of the *Foraminifera* has hitherto proved a matter of extreme difficulty, and probably none of the arrangements as yet proposed can be considered as more than provisional. The following is the classification adopted by Dr Carpenter, who is one of the greatest living authorities upon the group :—

ORDER RETICULOSA. (= FORAMINIFERA.)—*Rhizopods showing no differentiation, or a very imperfect one, into ectosarc and endosarc; no nucleus or contractile vesicle; pseudopodia filamentous, minutely subdivided, and inosculating freely to form a network.*

Section 1. *Imperforata.*—Envelope membranous or calcareous, the walls not perforated by apertures for the pseudopodia, which are emitted solely from the single or multiple aperture of the shell.

Families. 1. *Gromida.* Test membranous.
2. *Miliolida.* Test porcellanous.
3. *Lituolida.* Test arenaceous.

Section 2. *Perforata.*—Envelope calcareous (hyaline or vitreous) or rarely arenaceous, its walls traversed by numerous foramina for the emission of pseudopodia.

The following classifications by D'Orbigny and Schultze are founded merely upon the form of the shell, and, as such, are purely arbitrary. Of the two, Schultze's arrangement is probably the more satisfactory.

TABLE OF D'ORBIGNY'S ARRANGEMENT OF THE FORAMINIFERA.

Order 1. *Monostega.*—Body consisting of a single segment; the shell of a single chamber.

Order 2. *Stichostega.*—Segments arranged in a single row, in a straight or slightly curved line.

Order 3. *Helicostega.*—Segments arranged in a spiral, the shell forming a number of convolutions. (The "nautiloid" *Foraminifera.*)

Order 4. *Entomostega.*—Segments arranged on two alternating axes, forming a spiral.

Order 5. *Enallostega.*—Segments arranged on two or three alternating axes, not forming a spiral.

Order 6. *Agathistega.*—Chambers wound round an axis, each segment embracing half the entire circumference.

TABLE OF SCHULTZE'S ARRANGEMENT OF THE FORAMINIFERA.

Section 1. *Helicoidea.*—Segments arranged in a convolute series.

Section 2. *Rhabdoidea.*—Segments placed in a direct line.

Section 3. *Soroidea.*—Segments disposed in an irregular manner.

AFFINITIES OF FORAMINIFERA.—The *Foraminifera* are related on the one hand to the *Amœbea,* and on the other to the *Spongida.* From the former the "unilocular" *Foraminifera* differ, both in the possession of an external envelope, and in the much less highly differentiated characters of their sarcode; but the points of resemblance are obvious, and in such forms as *Actinophrys* and *Lieberkühnia* we are presented with an apparent transition between the two orders. From the shelled *Amœbea,* such as *Arcella,* the *Foraminifera* are broadly separated by the absence in the former of pseudopodial pores, and are fundamentally distinguished by the different nature of the sarcode-body.

To the Sponges the *Foraminifera* are related in various ways, one of the most striking links being found in *Carpenteria,* a singular attached form of *Foraminifer.* The shell, namely, of *Carpenteria* is conical and calcareous, composed of an aggregation of chambers arranged in a spiral, and having its walls perforated by numerous foramina of minute size. The interior of the chambers, however, is filled with "a fleshy, sponge-like body," strengthened by numerous spicula. Another curious link between the *Foraminifera* and the Sponges is the *Squamulina scopula* of Carter, which is truly a *Foraminifer,* though originally referred to the latter. It consists of an arenaceous test, forming a pedestal surmounted by an obversely conical column. Both pedestal and column are more or less perfectly chambered, and are filled with semi-transparent yellowish sarcode. At the summit of the column is a minute aperture surrounded by a brush of spicules, and the whole structure is fixed by the pedestal to some solid object.

To the *Polycystina*, the *Foraminifera* are obviously and closely allied. They agree in the nature of the sarcode-body, in the filamentous, inosculating pseudopodia, and in the phenomenon of a pseudopodial circulation of granules. They differ solely in the nature of the "test," which is calcareous or arenaceous in the *Foraminifera*, but is always siliceous in the *Polycystina*.

BATHYBIUS, COCCOLITHS, AND COCCOSPHERES.—It may be as well to notice here a singular organism which is certainly referable to the *Rhizopoda*, though its exact affinities are doubtful. Certain minute oval or rounded bodies have long been known as occurring attached to the surface of the shells of *Foraminifera*, and they were originally described by Professor Huxley under the name of *coccoliths*. Subsequently it was discovered by Dr Wallich that these singular bodies occur not only in the free condition, but also attached to the external surface of little spherical masses of sarcode to which he gave the name of *coccospheres*. The *coccospheres* are enclosed in a delicate envelope apparently of a calcareous nature, and are studded at nearly regular intervals by the *coccoliths*. More recently still, it has been discovered by Professor Huxley that both the *coccoliths* and the *coccospheres* are embedded in masses of protoplasmic or sarcodic substance, covering wide areas of the sea-bottom, to which they bear the same relation that the spicules of sponges or of *Radiolaria* do to the soft parts of these animals. To this undefined and diffused protoplasm with its contained *coccoliths* and *coccospheres* the name *Bathybius* has been applied by Professor Huxley. Its exact position, as already said, is doubtful; but it is believed by Dr Carpenter to be a rudimentary form of the *Foraminifera*, and to be somewhat allied to the ancient *Eozoön*. A curious point as regards the *coccoliths* has recently been brought to light by Dr Gümbel, the celebrated palæontologist, who believes that he has succeeded in demonstrating in them the existence of *cellulose*, or of some substance closely allied to cellulose. He has also shown that bodies similar to, if not identical with, *coccoliths*, occur in formations as old as the Potsdam Sandstone (Lower Silurian) of North America. More recently still, Mr Carter has shown that *coccoliths* occur in great numbers in the Laminarian zone, and he asserts them to be solitary, unicellular, calcareous *Algae*. He describes them under the name of *Melobesia unicellularis* and *M. discus*, according as they are oval or round; and he believes that the *coccospheres* are most probably their "sporangia." Upon this view the term "coccoliths" would be restricted to the fossil forms.

DISTRIBUTION OF FORAMINIFERA IN SPACE.—The *Foraminifera* are mostly marine, and are found in almost all seas, though more abundantly in those of the warmer parts of the globe. It is concluded by Dr Carpenter that "the foraminiferous *fauna* of our own seas probably presents a greater range of variety than existed at any preceding period; but there is no indication of any tendency to elevation towards a higher type." One of the most remarkable facts about their distribution at the present day, is the existence of a deposit at great depths in the Atlantic, but only in areas traversed by heated currents, formed almost entirely of the shells of *Foraminifera*, and very closely resembling chalk. It has further been quite recently established that there coexist with these *Foraminifera* various animals of a higher grade, some of which closely resemble, or are even specifically inseparable from, well-known Cretaceous species. There is therefore some reason to conclude that the bottom of the sea at great depths is peopled at the present day by a fauna which is very closely allied to that of the Chalk. Most living *Foraminifera* are very minute, but some of the extinct forms attained a size of as much as three inches in circumference (*e. g.*, the *Nummulite*, fig. 5), and spheres of *Parkeria* may attain a circumference of nearly four inches. Some forms may be obtained adhering to the roots of tangle at or near low-water mark, but they are mostly to be dredged from tolerably deep water. They have been found, in fact, in great abundance in the deepest parts of the ocean which have as yet been examined by the dredge—at a depth, namely, of nearly three miles.

DISTRIBUTION OF FORAMINIFERA IN TIME.—Remains of *Foraminifera* have been found in Palæozoic, Mesozoic, and Kainozoic formations. In the oldest stratified rocks with which we are acquainted—viz., the Laurentian rocks of Canada —there occurs a singular body which has been described as the remains of a gigantic *Foraminifer*, under the name of *Eozoön Canadense*. If truly organic, as is doubted by some, it is the oldest fossil as yet discovered. It appears to have grown in reef-like masses resembling the sessile patches of *Polytrema**

* *Polytrema* is a little branched coral-like Foraminifer, composed of a calcareous test forming a number of irregular chambers, which communicate with one another by wide orifices, and are filled with colourless sarcode. The walls of the chambers are also penetrated by an extensive system of capillary canals.

and *Carpenteria*, to both of which, as well as to the extinct *Nummulites*, it shows a decided affinity. In the Silurian rocks, remains of *Foraminifera*, some of which are apparently identical with existing forms, have been detected in various places, and it is not improbable that the large Silurian fossils known as *Receptaculites* and *Stromatopora* should really be referred to this order. In the Carboniferous rocks of Russia whole beds are composed of a species of *Fusulina*. In the Secondary rocks *Foraminifera* occur in great abundance, the widely-spread formation known as the Chalk being crowded with these organisms. Chalk itself, in fact, is almost entirely composed of the cases of *Foraminifera*, some of which are identical with species now existing.

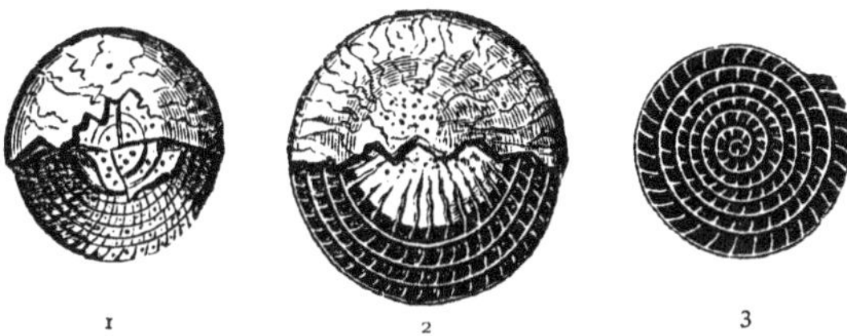

Fig. 5.—*Nummulites lævigatus.* Eocene.

In the Tertiary rocks the *Foraminifera* attain their maximum of development, both as regards the size and the number of the forms which characterise them. The period of the Middle Eocene is especially distinguished by a very widely spread and easily recognised rock known as the Nummulitic Limestone, so called from the abundance in it of a large coin-shaped *Foraminifer* termed the *Nummulite* (fig. 5). The Nummulitic Limestone stretches from the west of Europe to the frontiers of China; but in some cases, in place of *Nummulites* proper, it contains the remains of a mimetic form termed *Orbitoides*. Upon the whole, Dr Carpenter concludes that "there is no evidence of any fundamental modification or advance of the foraminiferous type from the Palæozoic period to the present time."

CHAPTER IV.

RADIOLARIA.

ORDER IV. RADIOLARIA.—The order *Radiolaria* was founded by Müller to include the *Polycystina*, the *Acanthometrina*, and the *Thalassicollida*, to which Dr Carpenter adds *Actinophrys* and its allies, chiefly on account of the form of the pseudopodia. Here, however, the term will be employed to designate the first three of these, and *Actinophrys* will be placed amongst the *Amœbea*, to which its alliance appears to be more decided. Most of the *Radiolaria* are marine, but some few forms of *Thalassicollida* have been described as occurring in fresh water.

The order *Radiolaria* may be defined as comprising those *Rhizopods which possess a siliceous test or siliceous spicules, and are provided with pseudopodia which stand out like radiating filaments, and occasionally run into one another.*

1. FAMILY ACANTHOMETRINA.—The *Acanthometræ* (fig. 6, *a*) are all minute, and are found floating near the surface in the open ocean, sometimes in great numbers. They consist of sarcode-bodies which are supported by a framework of radiating siliceous spines, the extremities of which usually project

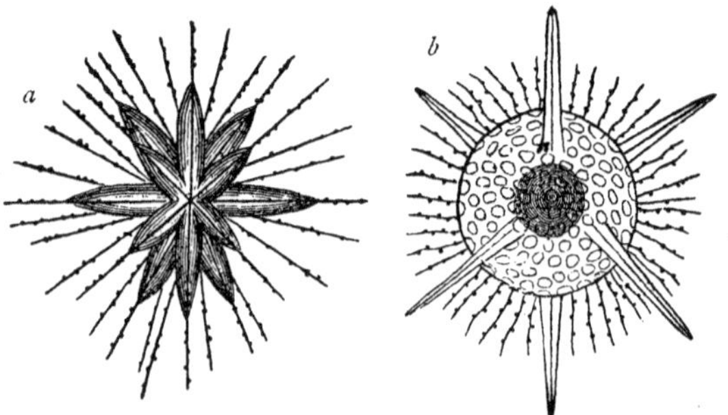

Fig. 6.—*a Acanthometra lanceolata*; *b Haliomma hexacanthum*, one of the *Polycystina*, showing the radiating pseudopodia. (After Müller.)

considerably beyond the body. The substance of the body admits of division into an outer membranous layer, or "ectosarc," and an internal granular layer, or "endosarc." The siliceous spines are hollow, being grooved at the base by a gutter, which is continued further up the spine by a canal terminating at the apex of the spine by a distinct aperture. The spines, in consequence of this structure, are able to serve for the transmission of the pseudopodia, which gain the exterior by running through the canals and escaping at their apices. Many of the pseudopodia, however, do not occupy the canals of the spines.

II. FAM. POLYCYSTINA.—The members of this family are closely related to the *Foraminifera*, differing from them chiefly in the fact that their shells are composed of flint instead of carbonate of lime, as in most of the latter. They possess a body of sarcode, which is enclosed in a foraminated siliceous shell, which is often furnished with spine-like processes, and is usually of great beauty (fig. 6, *b*). The sarcodic substance of the body is olive-brown in colour, with yellow globules, and often does not entirely fill the shell. The pseudopodia are emitted through the foramina in the test, and are long, ray-like filaments, which display a slow movement of granules along their borders.

The *Polycystina* are all microscopic, and are all inhabitants of the sea, having a very wide distribution. They are also found abundantly in certain Tertiary deposits, being often erroneously described as *Diatomaceæ*.

III. FAM. THALASSICOLLIDA.—The *Thalassicollida* have been defined as being *Rhizopoda which are* "*provided with structureless cysts containing cellular elements and sarcode, and surrounded by a layer of sarcode, giving off pseudopodia, which commonly stand out like rays, but may and do run into one another, and so form networks.*"—(Huxley.)

The *Thalassicollida* may be simple or composite, the latter consisting essentially of aggregations of the former; whilst these are fundamentally composed of a mass of granular protoplasm, containing a nucleus, but without a contractile vesicle, "enclosed in a membranous capsule, which is in turn protected by a more or less thick gelatinous exudation, whilst numerous sarcoblasts occur scattered through the endosarc, and occasionally a few may be seen suspended within the external gelatinous stratum."—(Wallich.) The whole organism is supported by more or less extensively developed skeletal structures. The skeleton may be simple, consisting of a delicate fenestrated shell; or may be compound, consisting of a number of spicular masses.

The three best-known genera of the family are *Sphærozoüm*, *Collosphæra*, and *Thalassicolla*. They are all marine, being found floating passively at the surface of most seas; and they vary in size from an inch in diameter downwards. *Sphærozoüm* consists essentially of a number of spherical sarcode-bodies (sometimes called "cellæform bodies") with distinct nuclei, surrounded by a zone of siliceous spicules, the whole being embedded in a common gelatinous matrix. The centre of the mass is vacuolated, sometimes to such an extent that it becomes a hollow sphere.

In *Collosphæra*, the spherical body—which is very like that of the preceding form—is enclosed in a transparent siliceous

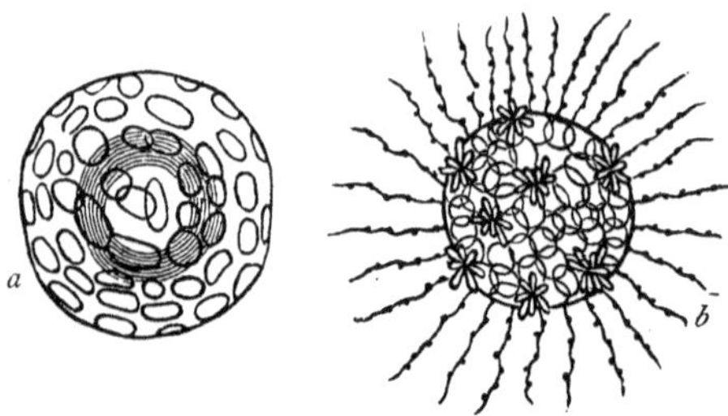

Fig. 7.—Morphology of Radiolaria. *a* Siliceous fenestrated test of *Collosphæra Huxleyi*; *b* *Thalassicolla morum*, showing cellæform bodies, compound groups of spicules, and radiating pseudopodia.

envelope, which is perforated by numerous rounded apertures or "fenestræ." This form, therefore, approaches very closely to the *Polycystina*, especially to those in which the foramina are so large that the test is reduced to a mere reticulate framework (fig. 7, *a*).

Thalassicolla differs little from either of the above in fundamental structure, but it contains a number of compound siliceous spicules embedded in its ectosarc (fig. 7, *b*).

CHAPTER V.

SPONGIDA.

Order V. Spongida.—The true nature of sponges has long been a matter of dispute, but they are now almost universally referred to the animal kingdon, and placed either in or near the *Rhizopoda*. Some observers still maintain the vegetable nature of sponges, but this opinion has no real grounds for its support, and is chiefly founded upon loose analogies, and upon a certain similarity in outward form.

The *Spongida* may be defined as "*sarcode-bodies, destitute of a mouth, and united into a composite mass, which is traversed by canals opening on the surface, and is almost always supported by a framework of horny fibres, or of siliceous or calcareous spicula.*"—(Allman.)

From the above definition it will be seen that a sponge is composed essentially of two elements—a soft, gelatinous, investing "flesh," and an internal supporting framework or "skeleton."

Taking an ordinary horny sponge as the type of the order, we find it to be composed of a skeleton (fig. 9, *d*) of horny reticulated fibres which interlace in every direction, and are pierced by numerous apertures, the whole surrounded externally and internally by a gelatinous glairy substance, like white-of-egg, the so-called "sponge-flesh." The horny skeleton is composed of a substance called "keratode," and is usually strengthened by spicula of lime or flint, which also occur less abundantly in the sponge-flesh. These must not, however, be confounded with the skeleton of the true calcareous or siliceous sponges in which the keratode is wanting. Of the apertures which penetrate the substance of the sponge in every direction, some are large crateriform openings, and are termed "oscules," or "exhalant apertures;" whilst others, which occur in much greater numbers, are greatly smaller in size, and are termed "pores," or "inhalant apertures." Both the oscula and pores can be closed at the will of the animal; but the oscula are permanent apertures, whereas the pores are not constant, but can be formed afresh whenever and wherever required. The "sponge-flesh," which invests the entire skeleton, is found upon a microscopical examination to be entirely composed of an aggregation of rounded amœbiform bodies—the so-called "sponge-particles" or "sarcoids" (fig. 2, *c*, *d*, *e*). Some of these are ciliated; whilst others are capable of emitting pseudopodia from all parts of their surface, and are provided with nuclei, thus coming closely to resemble so many *Amœbæ*. Regarding the skeleton as something superadded, we may, in fact, look upon a sponge as being essentially nothing more than an aggregation of *Amœbæ*, since each "sarcoid" is capable of procuring and assimilating food for itself in a manner strictly analogous to what we have seen in the *Amœba*. This view becomes still more easily comprehensible when we consider the simplest condition in which a sponge occurs in nature (as exemplified, for instance, in certain of the *Calcispongiæ*); the condition, namely, in which the entire sponge consists of a colony of amœbiform sarcoids, secreting a common skeleton, but provided with only a single "osculum," and a greater or less number of inhalant "pores." There are, in fact, many who hold that the more complex sponges are merely produced by the aggregation together of a number of these simpler colonies.

In a living sponge a constant circulation of water is maintained by means of an aquiferous system (fig. 8), which is

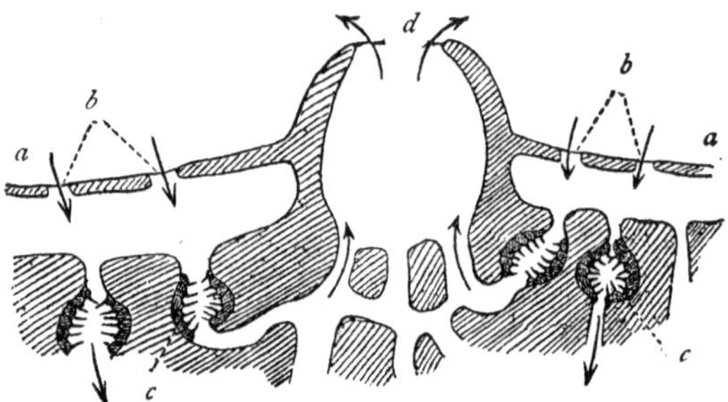

Fig. 8.—Diagrammatic section of *Spongilla* (after Huxley). *a a* Superficial layer or "dermal membrane;" *b b* Inhalant apertures or "pores;" *c c* Ciliated chambers; *d* An exhalant aperture or "osculum." The arrows indicate the direction of the currents.

constituted by the oscula and pores—already alluded to—and by a system of canals excavated in the substance of the sponge, and uniting the two sets of apertures. The water passes in by the "pores" or inhalant apertures, and is conveyed by a series of canals—the "incurrent" or "afferent" canals—to a second series of tubes—the "excurrent" or "efferent" canals—by which it reaches the "oscula" and is finally expelled from the body. These processes are regularly performed, and their mechanism was long a subject of speculation. It is now known, however, that beneath the superficial layer or "dermal membrane" of the sponge there exist chambers lined with sponge-particles which are provided with vibratile filaments or cilia (fig. 8, *c*, *c*). The pores open into these chambers, and from them proceed the incurrent canals, each being dilated at its commencement into a sac, which is also lined with ciliated sponge-particles. By the vibratile action of these cilia, currents of water are caused to set in by the pores; and as out-going currents proceed from the oscula, a constant circulation of fresh water is maintained through the entire sponge. In this way each individual sponge-particle is enabled to obtain nutriment: the process being at the same time not improbably a rudimentary form of respiration. The chambers or sacs lined with ciliated sarcoids have been shown by Mr Carter to be the essential element in the organisation of the fresh-water and marine sponges, and to be the fundamental expression of the alimentary system.

The reproduction of sponges may be effected either asexually or sexually, the following being a brief outline of the phenomena which have been observed in the common fresh-water sponge (*Spongilla*), in which the process has been most accurately noticed.

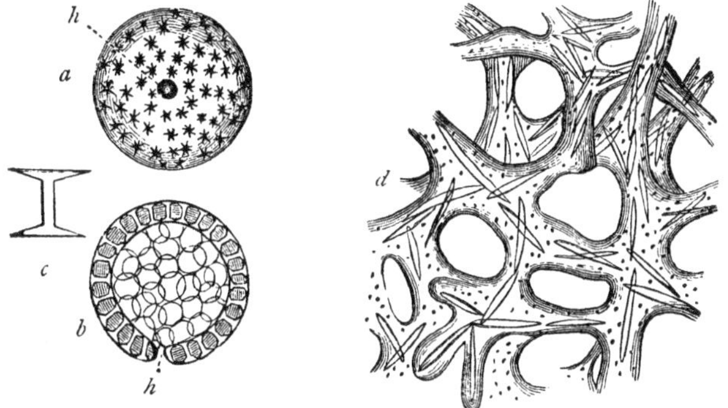

Fig. 9—*a* Gemmule of *Spongilla*; *h* Hilum; *b* Diagrammatic section of the gemmule, showing the outer layer of amphidiscs and the inner mass of cells; *c* One of the amphidiscs seen in profile; *d* Fragment of the skeleton of a horny sponge (after Bowerbank), showing the interlacing horny fibres with spicula. All much magnified.

In the first or asexual method of reproduction, which takes place in the winter, the deeper portions of the sponge are found to be filled with small seed-like rounded bodies, termed "gemmules" or "spores," each of which possesses a small aperture or "hilum" at one point (fig. 9, *h*). Each gemmule is composed of an outer coriaceous capsule surrounded by a layer of peculiar asteroid spicula, resembling two toothed wheels united by an axle, and termed "amphidiscs" (fig. 9, *b*, *c*). These amphidiscs are embedded in sarcode, whilst their inner surfaces rest upon the tesselated capsule already mentioned. In the interior of the capsule thus formed is a mass of cells, of which the central ones contain numerous germs. When the spring comes, these masses of "ovi-bearing cells" are discharged through the "hilum" of the gemmule into the water, and are developed into new *Spongillæ*.

Spongilla also appears to reproduce itself in a somewhat analogous manner by means of what are termed "swarm-spores." These are small bodies, containing reproductive germs, and provided with numerous cilia by which they move about actively, becoming finally attached to some solid body, and developing themselves into the adult sponge.

In the second or sexual method of reproduction, certain of the sponge-particles or "sarcoids" separate themselves and become nucleolo-nucleated, thus coming to resemble ova. At the same time other sarcoids become motionless, and their contents become molecular, and are finally converted into spermatozoa. By the rupture of these, and by the consequent contact of the different elements, embryos are produced, which are at first ciliated and move about freely, becoming eventually stationary, and developing into new individuals.

CLASSIFICATION OF THE SPONGES.—The *Spongida* have been variously classed, and a good natural arrangement is still a *desideratum*. By Dr Bowerbank they are somewhat arbitrarily arranged in three orders—viz., the *Keratosa*, the *Silicea*, and the *Calcarea*, of which the first is believed to hold the lowest place. In the *Keratosa* the skeleton is composed of interlacing horny fibres, usually strengthened by spicula either of flint or lime. In the *Calcarea* or *Calcispongiæ* the skeleton is composed of carbonate of lime; whilst in the *Siliceous* sponges it is composed either of spicules of silex, or "of solid, laminated, and continuous siliceous fibre." By Professor Wyville Thomson the siliceous sponges are arranged in a separate order under the name of the "vitreous sponges" (*Vitrea*). The nature of the skeleton thus varies considerably, whilst the spicules show almost indefinite modifications of shape, though they are constant for any given species, in any given part of its organisation. The sponge-flesh is much more uniform in its nature and composition. It may be noticed, however, that in *Spongilla* the sponge-particles are filled with green granules, which are apparently identical in chemical composition with the green colouring matter of plants (*chlorophyll*). In *Grantia*, too, the sarcoids are furnished with long filamentous appendages or cilia (fig. 2, *d*). The siliceous sponges are mostly inhabitants of the deep sea, and many of them are remarkable for their long and slender spicules of flint. In *Hyalonema*, or the glass-rope, long placed amongst the Zoophytes (*Zoantharia sclerobasica*), there is a cup-shaped sponge-body, supported by a rope of long twisted siliceous fibres, which are sunk in the mud of the sea-bottom. In other "anchoring sponges," such as *Pheronema* and *Holtenia*, the body is sessile or stemless, and is moored to the mud by a beard of long delicate spicules. These sponges, in their single, long, chimney-like osculum, show a curious resemblance to the fossil *Siphonias* of the greensand.

Similar root-fibres of flint, traversing the mud in every direction, occur in the beautiful Venus' Flower-basket (*Euplectella*), without any exception one of the most exquisite of all organic structures known to us.

Distribution of Sponges in Space.—Sponges are almost exclusively marine, the *Spongillæ* alone being inhabitants of fresh water; and they are of almost universal occurrence. The sponges of commerce are mostly obtained from the Grecian Archipelago and the Bahama Islands. Recently the existence of numerous siliceous sponges at great depths in the ocean has been demonstrated by Drs Carpenter and Wyville Thomson. They are associated with numerous *Foraminifera* and with *Crinoidea*, the whole assemblage bearing a singularly close resemblance to the fauna of the Cretaceous epoch. The common marine sponges are mostly found attached to some solid object between tide-marks or in deep water. The vitreous or siliceous sponges appear to be exclusively inhabitants of the deeper parts of the ocean. One genus (*Cliona*) inhabits branching cavities in shells, which the sponge excavates for itself apparently by means of its siliceous spicula. Fossil shells mined by a boring-sponge, allied to the recent *Clionæ*, are found from the Silurian rocks upwards.

Distribution of Sponges in Time.—Remains of sponges are known to occur in formations belonging to the Palæozoic, Mesozoic, and Kainozoic epochs. The keratose or horny sponges are obviously incapable of leaving any evidence of their existence, otherwise than by the preservation of the spicula with which the skeleton is furnished; and such are occasionally found, though they are of rare occurrence. The calcareous sponges are found from the Silurian rocks upwards, attaining their maximum in the seas of the Secondary epoch, the Chalk being especially characterised by their presence. The most important group of fossil sponges is that known as the *Petrospongiadæ*, characterised by the possession of a stony reticulate framework or skeleton, and by the absence of spicula. The most important genera of this group are *Sparsispongia* (Devonian) and *Ventriculites* (Chalk).

Of the Palæozoic sponges, *Archæocyathus* is found in the Potsdam sandstone of North America (Upper Cambrian?); *Palæospongia* and *Acanthospongia* are familiar Lower Silurian forms; and *Amphispongia* and *Favospongia* occur along with other forms in the Ludlow rocks. In the Devonian rocks sponges occur pretty frequently, *Sparsispongia* being the commonest genus. (The Devonian *Steganodictyum* is really the cephalic buckler of a *pteraspidean* fish.) The most important

Mesozoic genera of sponges are *Ventriculites* and *Siphonia;* and the order appears, upon the whole, to attain its maximum in the Cretaceous epoch. There seems no reason to doubt but that many of the chalk-flints owe their origin to sponges; and in some sections of flint are found minute "spherical bodies covered with radiating and multicuspid spines," which have been termed *Spiniferites* or *Xanthidia*, and are probably the "gemmules" of sponges. (By some, however, these bodies are regarded as being the "sporangia" of *Desmidiæ*, an order of the *Protophyta.*) Many Cretaceous and Tertiary shells are found to be mined by a species of boring-sponge, which is nearly allied to the recent *Cliona.*

AFFINITIES OF SPONGES.—As already pointed out, the sponges are allied both to the *Amœbea* and to the *Foraminifera.* Indeed the individual "sarcoids" or sponge-particles can scarcely be distinguished, when detached, from *Amœbæ.* The sponges show likewise a decided relationship to the *Radiolaria;* and by Professor James-Clark they are believed to be nearly allied to the "flagellate" *Infusoria.* This observer, in fact, states his "conviction that the true ciliated *Spongiæ* are not *Rhizopoda* in any sense whatever, nor even closely related to them, but are genuine compound *flagellate Protozoa.*" To prove this view, however, it should be shown that each sponge-particle possesses at any rate a distinct mouth, with or without a rudimentary alimentary canal. More recently Dr Ernst Hæckel and others have endeavoured to show that the sponges are most nearly allied to the Sea-anemones (*Actinozoa*); but this seems to have arisen from a misconception as to the compound nature of the former. Three views, namely, may be held as to the "individuality" of a sponge. *Firstly*, it may be held that the entire organism which we call a sponge is a single animal. The microscope has rendered this view wholly untenable. *Secondly*, it may be held that the entire sponge-mass is a single "zoological individual," of which each sarcoid is a single "zoöid." As each sponge-mass is certainly in most cases the product of a single ovum, this is the most probable and reasonable view. *Thirdly*, it may be held that each sponge-mass consists of a number of aggregated "individuals," each of which is constituted by a single exhalant "osculum," together with the greater or less number of inhalant "pores" thereto appertaining. Upon no other view than this does there appear to be any relation of affinity between the sponges and the *Cœlenterata;* and even on this view the general affinities between the two are not of a very striking nature. In some cases (such as *Euplectella* amongst the *Silicispongiæ*, and *Sycum* and *Ute* amongst the *Calcispongiæ*) the adult sponge never consists of more than a single osculum and the pores belonging to it, constituting what Hæckel terms a single "person." Most sponges, however, form a "stock," consisting of several such "persons" united together.

CHAPTER VI.

INFUSORIA.

THE *Infusoria* of many writers comprise many of the lowest forms of plants—such as the *Diatoms*—together with the *Rotifera*, a class of minute animals now known to belong to the *Annuloida*. By modern writers, however, the term *Infusoria* is used strictly to designate those *Protozoa* which possess a mouth and rudimentary digestive cavity. They are, for this reason, often called collectively the "stomatode" *Protozoa*, in contradistinction to the remaining members of the sub-kingdom, which are all "astomatous." The so-called "suctorial" *Infusoria* (*Acinetæ*), however, appear to have no definite oral aperture; and the same is the case with the parasitic *Opalina*, though there is great doubt as to the propriety of placing this in the *Infusoria* at all. The name *Infusoria* itself is derived from the fact that the members of the class are often developed in organic infusions.

The *Infusoria*, or *Stomatode Protozoa*, may be defined as *Protozoa which are mostly provided with a mouth and rudimentary digestive cavity, which do not possess the power of emitting pseudopodia, but which are furnished with vibratile cilia, or with contractile filaments. They are mostly microscopic in size, and their bodies usually consist of three distinct layers.*

The *Infusoria* may be divided into three orders—viz., *Suctoria*, *Ciliata*, and *Flagellata*, of which the second comprises the majority of the members of the class, and alone requires much consideration.

I. ORDER CILIATA.—This order comprises those *Infusoria in which the outer layer of the body is more or less abundantly furnished with vibratile cilia, which serve either for locomotion or for the procuring of food.* Besides *cilia*, properly so called, some of the ciliated *Infusoria* are provided with *styles* or jointed bristles, which are movable, and subserve locomotion; whilst others have little hooks or *uncini*, with which they can attach themselves to foreign bodies. As types of the order, *Paramæ-*

cium and *Vorticella* may be selected, the former being free, whilst the latter is permanently fixed in its adult condition.

Paramœcium (fig. 10, *c*) is a slipper-shaped animalcule, composed externally of a structureless transparent pellicle—the "cuticle"—which is lined by a layer of firm and consistent sarcode, which is termed the "cortical layer," or the "parenchyma

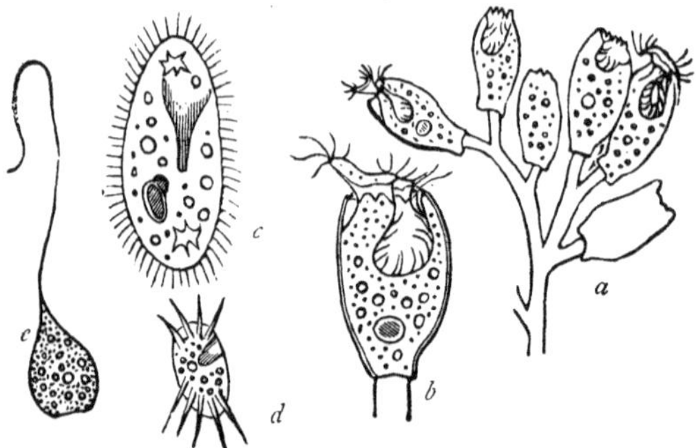

Fig. 10.—Morphology of Infusoria: *a Epistylis*, a stalked Infusorian; *b* A single calyx of the same greatly magnified, showing the ciliated disc which protrudes at will, and the ciliated internal cavity into which the particles of food are received. In the substance of the body are the contractile vesicle and smaller food-vacuoles. *c* Diagrammatic representation of *Paramœcium*, showing the funnel-shaped gullet, the nucleus and nucleolus, food-vacuoles, and two contractile vesicles. *d Aspidisca lynceus*; *e Peranema globulosa*, a flagellate Infusorian.

of the body," this in turn passing into a central mass of softer and more diffluent sarcode, known as the "chyme-mass," or "abdominal cavity." The "cuticle" is covered with vibratile cilia, and is perforated by the aperture of the mouth. The mouth leads into a funnel-shaped gullet, which is not continued into any distinct digestive sac, but is lost in the central "chyme-mass." Within the "cortical layer" are the "nucleus" and "nucleolus," and the "contractile vesicle" (or vesicles). The nucleus is usually a solid band or rod-shaped body, having a small spherical particle applied to its exterior, or immersed in its substance. This latter is the so-called "nucleolus," which must be carefully distinguished from the nucleolus of a cell, which occurs in the *interior* of the nucleus. The contractile vesicles are clear spaces, which contract and dilate at intervals, and occasionally exhibit radiating canals passing into the surrounding sarcode. Ordinarily one contractile vesicle is present, or at most two, but in some cases there may be several. It has also been maintained that the contractile vesicles communicate with the exterior of the body, but proofs are wanting on this point. Whether this should ultimately be established or not, there can be little doubt but that the vesicles are a rudimentary form of vascular apparatus. Others, however, hold, with some probability, that the contractile vesicles are to be regarded as excretory in function, and that they correspond more with the water-vascular system of the *Annuloida* than with the true blood-vascular system of higher animals. Certain other spaces termed "vacuoles" are generally visible in addition to the contractile vesicles. These, however, are probably merely collections of water surrounding the particles of ingested food, and performing with them a circulation in the abdominal cavity, something like the circulation of granules which is seen in certain vegetable cells. It was the appearance of these "vacuoles"—which are certainly not permanent organs of any kind—which induced Ehrenberg to term the *Infusoria* the "Polygastrica," upon the belief that they were so many stomachs.

Paramœcium obtains its food by means of the currents of water which are set up by the constantly-vibrating cilia. The nutritive particles thus brought to the mouth pass into the central abdominal cavity, along with the contents of which they undergo the circulation above spoken of. Indigestible and fæcal particles appear to be expelled by a distinct anal aperture, which is situated near the mouth.

Reproduction in *Paramœcium* is effected either non-sexually by fission (*i.e.*, by a simple division of its substance), or by a true sexual process. In this latter method two *Paramœcia* come together, and adhere closely to one another by their ventral surfaces. The "nucleus," which is truly an *ovary*, enlarges, and a number of ovules are formed in its interior. In like manner, the "nucleolus" of each, which is really a *testis* or *spermarium*, also enlarges, and develops in its interior a number of fusiform or rod-like bodies, which are believed to be spermatozoa. The nucleolus of each then passes into the body of the other, the act of transference being effected through the mouth. Contact of the two reproductive elements then takes place, and a number of germs are produced, which, after their liberation from the body of the parent, are developed into adult *Paramœcia*.

Vorticella (fig. 11, *c*) is a beautiful flower-like Infusorian which is commonly found in fresh water, adhering to the stems of aquatic plants. It consists of a bell-shaped body or "calyx," supported upon the extremity of a slender contractile stem or "pedicle." The other extremity of the pedicle is fixed to some foreign body, and its power of contraction is due to the presence in its interior of a spiral contractile fibre, which is sometimes called the "stem-muscle." The edge of the bell or calyx is surrounded by a projecting rim or border, called the "peristome," within which is a circular surface, the "disc," forming the upper extremity of the so-called "rotatory organ." The disc is surrounded by a fringe of vibratile cilia, forming a spiral line which is prolonged into the commencement of the digestive canal. Near the edge of the disc is situated the mouth, which conducts by its entrance or "vestibulum" into a fusiform canal or "pharynx," which terminates abruptly in the abdominal cavity. The particles of food are taken in at the mouth, descend through the short alimentary canal, and enter the abdominal cavity, where they are subjected to the general rotation of the "chyme-mass," being finally excreted by an anal aperture which is situated near the mouth. As in *Paramœcium*, the body in *Vorticella* is composed of an outer "cuticle," a central "chyme-mass," and an intermediate "cortical layer," which contains a contractile vesicle and a band-like nucleus.

Reproduction in *Vorticella* may take place by fission, or by gemmation, or by a process of encystation and endogenous division. In the first of these modes the calyx becomes indented in a longitudinal direction—viz., from the pedicle to the disc; and the groove thus formed becomes gradually deeper until the calyx is finally divided into two halves supported upon the same pedicle. On one of these cups a "posterior" circlet of cilia is then formed in addition to the "anterior" circlet already existing (*i. e.*, a fringe of cilia is developed round that end of the calyx which is nearest the attachment of the pedicle and furthest from the disc). The cup (fig. 11, *d*), thus furnished with a circlet of cilia at both extremities, is then detached, and swims about freely. Finally, the *anterior* circlet of cilia disappears, and this end of the calyx puts forth a pedicle and becomes attached to some foreign object. A new mouth is now formed within what was before the *posterior* circlet of cilia; so that the position and function of the two extremities of the calyx are thus reversed.

In the second mode of reproduction—namely, that by gemmation—exactly the same phenomena take place, with this single difference, that in this case the new individual is not produced by a splitting into two of the adult calyx, but by means of a bud thrown out from near its proximal extremity. This bud is composed of a prolongation of the cuticular and cortical layers of the adult with a cæcal diverticulum of the abdominal cavity or chyme-mass. It soon develops a posterior circlet of cilia, the connection with the parent is rapidly constricted until complete separation is effected, and then the process differs in no respect from that described as occurring in the fissiparous method of reproduction.

In the third mode of reproduction the *Vorticella* encysts itself in a capsule, the cilia and pedicle disappear, and the nucleus breaks up into a number of rounded germs which are ultimately liberated by the rupture of the cyst, and, after a short locomotive stage, develop themselves into fresh *Vorticellæ*. How far this process may be truly sexual is not known, and no form of unequivocal sexual reproduction has hitherto been shown to occur in the case of *Vorticella*.

Epistylis is a not uncommon form of fixed Infusorian which is nearly allied to *Vorticella*, and differs chiefly in the fact that the pedicle is much branched, and is rigid and not contractile. *Epistylis* (fig. 10, *a*) usually occurs in the form of a greyish-white nap on the stems of water-plants, or on the head of the common water-beetle, the *Dytiscus marginalis*. It consists of a plant-like branching and re-branching frond, the stems of which are quite transparent and faintly striated, but are not contractile, though capable of movement from side to side. Each branch of the entire colony terminates in an oval calyx, articulated to the stem by a distinct joint, upon which it can move from side to side. The calyces are oval or somewhat campanulate, but have the power of altering their dimensions, and especially of contracting so as to shorten their antero-posterior diameter. Each calyx terminates distally in a slightly-elevated annular aperture, the margins of which are regularly toothed. The calyx appears to be formed by a hardening of the cuticle, and to form a distinct case, with a double margin, enclosing the animal. The sarcode-body enclosed within this outer envelope is of a light-brown colour, and full of minute granules, with larger food-vacuoles and a well-marked contractile vesicle, which contracts and dilates two or three times a minute. The animal can retract itself entirely within its cup, and can at will exsert a ciliated disc. This disc (fig. 10, *b*) is inversely conical, and acts as a kind of plug, and it is provided with two tufts of long cilia, one on each side. On one side of the protrusible disc is the oral aperture, which is continued by a distinct and well-marked gullet into a central ill-defined cavity. Both the entrance of the gullet and the bottom of the central cavity are provided with very long, actively-vibrating cilia, some of which are almost setiform. The entire granular contents of the abdominal cavity undergo a constant though slow rotation.

Carchesium is another form which is like *Epistylis* in consisting of a number of calyces supported upon a branched pedicle, but differs from *Epistylis* and agrees with *Vorticella* in the fact that the pedicle is contractile.

Stentor, or the trumpet-animalcule (fig. 11, *b*), is another common Infusorian which is closely related to *Vorticella*. It

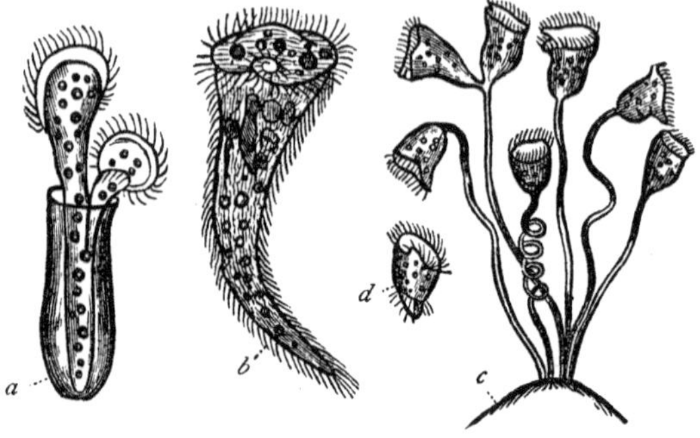

Fig. 11.—*a Vaginicola crystallina*; *b Stentor Mülleri*; *c* Group of *Vorticellæ*; *d* Detached bud of *Vorticella*, showing the posterior circlet of cilia.

consists of a trumpet-shaped calyx, devoid of a pedicle, but possessing the power of attaching and detaching itself at will. When detached it swims by means of the anterior circlet of cilia, just as the calyx of *Vorticella* will, if broken from its stalk. In *Vaginicola* (fig. 11, *a*) the essential structure is much the same as in *Vorticella*, but the body is protected by a membranous or horny case ("carapace" or "lorica"), within which the animal can retire. In this beautiful Infusorian the carapace is certainly a cuticular secretion, but it appears at the same time to be quite distinct from the true cuticle itself.

Amongst the structures of the *Infusoria* which require some notice are the "pigment spot" and the "trichocysts." The pigment spot is a brightly-coloured solid particle, generally red, of very common occurrence in many *Infusoria*, but of quite unknown function. The "trichocysts" are vesicular bodies, capable of emitting thread-like filaments, and greatly resembling the urticating cells of many of the *Cœlenterata*. They have been detected in *Bursaria*, as well as in various other members of this order; and they are very like certain cells which are found in the integument of many *Planarians*.

II. Order Suctoria.—This order includes a series of *Infusoria* of a very anomalous nature. In *Acineta*, which may be taken as the type, the body is covered with a number of radiating filamentous tubes, which are furnished at their extremities with suctorial discs, and are capable both of exsertion and retraction. These retractile tubes both seize the prey and serve as vehicles for the ingestion of food; hence the term "polystome," or many-mouthed, has been proposed for the order by Professor Greene.

III. ORDER FLAGELLATA.—This order comprises those *Infusoria* which, like *Peridinium*, find their means of locomotion in long, flexible, lash-like filaments, termed "flagella;" cilia occasionally being present as well. In some, as in *Peranema* (fig. 10, *e*), there is only a single one of these appendages; in others, as in *Anisonema*, there are two flagella; whilst in *Heteromastix* and *Pleuronema* we have forms apparently transitional between the *Ciliata* and the *Flagellata*, since both cilia and flagella are present in these genera. In all their other essential characters the flagellate *Infusoria* do not differ from the more typical members of the class. In one singular form (*Phalansterium intestinale*), the organism consists of numerous zoöids, each with a single flagellum and projecting membranous collar, enveloped basally in slimy sarcode, so as to form a cylindrical colony.

NOCTILUCA.—Amongst the numerous organisms which contribute to the phosphorescence of the sea,* one of the commonest is the animalcule known as *Noctiluca*, the true position of which has not yet been determined. It is nearly spherical in shape, having an indention, or "hilum," at one side, close to which is fixed a long filament, probably used in locomotion. The body consists of a "cuticle" and "cortical layer," enclosing a central mass of sarcode. Near the filament there is a minute oral aperture leading into a short digestive cavity. A nucleus and vacuoles are also present. From the presence of a mouth, and from its general structure, *Noctiluca* should probably be looked upon as a flagellate *Infusorian*, but it is placed by M. de Quatrefages amongst the *Rhizopoda*.

* The *diffused* luminosity of the sea is mainly due to the *Noctiluca miliaris*; but its partial luminosity is due to various phosphorescent animals, amongst which are the *Physalia utriculus* (the Portuguese man-of-war), *Medusæ*, *Tunicata*, *Annelides*, &c. The cause of phosphorescence is variously stated, being supposed very generally to be caused by a process of slow combustion analogous to that which takes place in phosphorus when exposed to the atmosphere. Upon the whole, however, it appears that the phenomenon is a vital process, consisting essentially in the conversion of nervous force (or vital energy) into light; just as the same force can be converted by certain fishes into electricity. This transformation often requires a special apparatus for its production, but it appears to be sometimes effected by the entire organism.

AFFINITIES OF THE INFUSORIA.—Though generally placed amongst the *Protozoa*, of which they form the highest division, the position of the *Infusoria* cannot be looked upon as definitely settled. There is a growing opinion amongst competent authorities that the *Infusoria* should be entirely removed from the *Protozoa*, and that they should be placed amongst the *Annuloida*, having their nearest allies in the *Turbellarian* Worms. If this change be carried out, the *Infusoria* and *Rotifera*, which older naturalists grouped together, and which modern observers have placed widely apart, will be again brought nearly together. If the sponges also should be removed from the *Protozoa*, as is maintained by some modern observers, then the entire sub-kingdom of the *Protozoa* would contain only the *Gregarinida*, *Amœbea*, *Foraminifera*, and *Radiolaria*. At present, however, there certainly do not appear to be sufficiently decisive grounds for such a step; and the affinities of the *Spongida* are certainly more intimate with the typical *Rhizopods* than with the *Cœlenterata*.

CŒLENTERATA.

CHAPTER VII.

THE SUB-KINGDOM CŒLENTERATA.

1. CHARACTERS OF THE SUB-KINGDOM. 2. DIVISIONS. 3. GENERAL CHARACTERS OF THE HYDROZOA. 4. EXPLANATION OF TECHNICAL TERMS.

THE sub-kingdom *Cœlenterata* (Frey and Leuckhart) may be considered as the modern representative of the *Radiata* of Cuvier. From the *Radiata*, however, the *Echinodermata* and *Scolecida* have been removed to form the *Annuloida*, the entire sub-kingdom of the *Protozoa* has been taken away, and the *Polyzoa* have been relegated to their proper place amongst the *Mollusca*. Deducting these groups from the old *Radiata*, the residue, comprising most of the animals commonly known as Polypes or Zoophytes, remains to constitute the modern *Cœlenterata*.

The *Cœlenterata* may be defined as *animals whose alimentary canal communicates freely with the general cavity of the body* ("*somatic cavity*"). *The substance of the body is made up of two fundamental membranes—an outer layer, called the "ectoderm," and an inner layer, or "endoderm." There are no distinct neural and hæmal regions, and in the great majority of the members of the sub-kingdom there are no traces of a nervous system. Peculiar urticating organs, or "thread-cells," are usually present; and, generally speaking, a radiate condition of the organs is perceptible, especially in the tentacles with which most are provided. In all the Cœlenterata distinct reproductive organs have been shown to exist.* By Professor Allman the *Cœlenterata* have been defined as follows:—"Animals composed of numerous merosomes (body-segments), which are disposed radially round a longitudinal (antero-posterior) axis; frequently with a determinable antero-posterior and dorso-ventral plane (bilateral); a distinct body-cavity, which always communicates with the outer world through the mouth."

The leading feature which distinguishes the *Cœlenterata*, and the one from which the name of the sub-kingdom is derived, is the peculiar structure of the digestive system. In the *Protozoa*, as we have seen, a mouth is only present in the higher forms, and in no case is there any definite internal cavity bounded by the walls of the body to which the name of "body-cavity" or "somatic cavity" could be properly applied. In animals higher than the *Cœlenterata*, on the other hand, there is not only generally a permanent mouth, but the walls of the body usually enclose a permanent chamber or "body-cavity." Further, in most cases, the mouth conducts into an alimentary canal, which is always distinct from the body-cavity, never opening into it, but usually passing through it to open on the surface by another distinct aperture (the anus). In most cases, therefore, the alimentary canal is a tube which communicates with the outer world by two apertures—a mouth and anus—but which simply passes through the body-cavity without in any way communicating with it. In the *Cœlenterata* there is an intermediate condition of parts. There is a distinct and permanent mouth, and a distinct and permanent body-cavity, but the mouth opens into, and communicates freely with, the body-cavity. In some cases (*Hydrozoa*) the mouth opens directly into the general body-cavity, which then serves as a digestive cavity as well (fig. 12). In other cases there intervenes between the mouth and the body-cavity a short alimentary tube, which communicates externally with the outer world through the mouth, and opens below by a wide aperture into the general cavity of the body (*Actinozoa*, fig. 28). In no case is there a distinct intestinal canal which runs through the body and opens on the surface by a mouth at one end and an excretory aperture or anus at the other.

With regard to the fundamental tissues of the *Cœlenterata*, there exist two primary membranes, of which one forms the outer surface of the body, and is called the "ectoderm;" whilst the other lines the alimentary canal, the general cavity of the body, and the tubular tentacles, and is termed the "endoderm." These membranes correspond with the primitive serous and mucous layers of the germinal area, and become differentiated in opposite directions, the ectoderm growing from within outwards, the endoderm from without inwards. Each consists of numerous nuclear bodies, or "endoplasts," embedded in a granular "intercellular substance" or "periplast;" and each may be rendered more or less complex by vacuolation or fibrillation.

In connection with the integument of the *Cœlenterata*, the organs termed "thread-cells" ("cnidæ," or "nematocysts") must be noticed. These are peculiar cellular bodies, of various shapes, which probably serve as weapons of offence and defence, and which communicate to many members of the sub-kingdom (*e.g.*, the Sea-blubbers) their well-known power of stinging. In the common *Hydra* the thread-cells consist of "oval elastic sacs, containing a long coiled filament, barbed at its base, and serrated along its edges. When fully developed the sacs are tensely filled with fluid, and the slightest touch is sufficient to cause the retroversion of the filament, which then projects beyond the sac for a distance, which is not uncommonly equal to many times the length of the latter." *—(Huxley.) (Fig. 12, *d.*) Many beautiful modifications of shape are known in the thread-cells of different Cœlenterates, but their essential structure in all cases is much the same as in the Hydra. It is only in few cases, comparatively speaking, that the thread-cells have the power of piercing and irritating the human skin; but even in the diminutive *Hydra* it is probable that they exercise some benumbing and deleterious influence on the living organisms which may be captured as prey. The *Cœlenterata* are divided into two classes, termed respectively the *Hydrozoa* and the *Actinozoa.*

CLASS I. HYDROZOA.

The *Hydrozoa* are defined as *Cœlenterata in which the walls of the digestive sac are not separated from that of the general body-cavity, the two coinciding with one another; the reproductive organs are in the form of external processes of the body-wall.* (Fig. 12, *a*, *b.*)

It follows from the above, that, since there is but a single internal cavity, the body of a *Hydrozoön* on transverse section appears as a single tube, the walls of which are formed by the limits of the combined digestive and somatic cavity.

The *Hydrozoa* are all aquatic, and the great majority are marine. The class includes both simple and composite organisms, the most familiar examples being the common Fresh-water Polype (*Hydra*), the Sea-firs (*Sertularida*), the Jelly-fishes (*Medusæ*), and the Portuguese man-of-war (*Phy-*

* Thread-cells, though very commonly, if not universally, present in the *Cœlenterata*, are nevertheless not peculiar to them. Similar organs have been shown to exist in several of the *Nudibranchiate Mollusca*, as well as in some Annelides (*Spio seticornis*). There likewise exist analogous organs (*trichocysts*) in several of the *Infusoria*, and in the *Planarida.*

salia). Owing to the great difficulty which is ordinarily experienced by the student in mastering the details of this class of animals, it has been thought advisable to introduce here a short explanation of some of the technical terms which are in more general use in describing these organisms.

General Terminology of the Hydrozoa.

Individual.—We have already seen (*see* Introduction) that the term "individual," in its zoological sense, must be restricted to "the entire result of the development of a single fertilised ovum," and that in this sense an individual may either be simple, like an *Amœba*, or may be composite, like a Sponge, which is produced by an aggregation of amœbiform particles. If all the parts composing an individual remain mutually connected, its development is said to be "continuous;" but if any of these parts become separated as independent beings, the case becomes one of "discontinuous" development. We have seen, also, that however long zoöidal multiplication may go on, there ultimately arrives in the history of every individual a period at which sexual reproduction must be called in to insure the perpetuation of the species throughout time.

Amongst the *Hydrozoa*, the individual may be either simple or compound, and the development may be either continuous or discontinuous, the following terms being employed to denote the phenomena which occur.

Hydrosoma.—This is the term which is employed to designate the entire body of a *Hydrozoön*, whether it be simple, as in the *Hydra*, or composite, as in a *Sertularian*.

Polypite.—The alimentary region of a *Hydrozoön* is called a "polypite;" the term "polype" being now restricted to the same region in the *Actinozoa*. In the simple *Hydrozoa* the entire organism may be called a "polypite;" but the term is more appropriately applied to the separate nutritive factors which together make up a compound *Hydrozoön*.

Distal and Proximal.—These are terms applied to different extremities of the hydrosoma. It is found that one extremity grows more quickly than the other, and to this free-growing end—at which the mouth is usually situated—the term "distal" is applied. To the more slowly growing end of the hydrosoma—which is at the same time usually the fixed end—the term "proximal" is applied. These terms may be used either in relation to a single polypite in the compound *Hydrozoa*, or to the entire hydrosoma, whether simple or compound.

Cœnosarc.—This is the term which is employed to designate the common trunk, which unites the separate polypites of any compound *Hydrozoön* into a single organic whole.

Polypary.—The term "polypary" or "polypidom" is applied to the horny or chitinous outer covering or envelope with which many of the *Hydrozoa* are furnished. These terms have also not uncommonly been applied to the very similar structures produced by the much more highly organised Sea-mats and their allies (*Polyzoa*), but it is better to restrict their use entirely to the *Hydrozoa.*

Zoöids.—In continuous development the partially independent beings which are produced by gemmation or fission by the primitive organism, to which they remain permanently attached, are termed "zoöids."

In discontinuous development, where certain portions of the "individual" are separated as completely independent beings, these detached portions are likewise termed "zoöids;" that which is first formed being distinguished as the "producing zoöid," whilst that which separates from it is known as the "produced zoöid." In a great number of *Hydrozoa* there exist two distinct sets of zoöids, one of which is destined for the nutrition of the colony, and has nothing to do with generation, whilst the functions of the other, as far as the colony is concerned, are wholly reproductive. For *the whole assemblage of the nutritive zoöids of a Hydrozoön* Professor Allman has proposed the term "trophosome," applying the term "gonosome" to *the entire assemblage of the reproductive zoöids.* In such *Hydrozoa*, therefore, as possess these two distinct sets of zoöids, the "individual," zoologically speaking, is composed of a trophosome and a gonosome. It follows from this that neither the trophosome nor the gonosome, however apparently independent, and though endowed with intrinsic powers of nutrition and locomotion, can be looked upon as an "individual," in the scientific sense of this term. As a rule, the zoöids of the trophosome are all like one another, or are "homomorphic;" but there are some cases (as in *Hydractinia*, and in the nematophores of the *Plumularidæ*) in which some of the zoöids of the trophosome are unlike the others. The zoöids of the gonosome, on the other hand, are normally unlike, or are "heteromorphic," consisting of two or three different sets of zoöids, each with its special duty in the generative functions of the Hydroid Colony.

CHAPTER VIII.

DIVISIONS OF THE HYDROZOA.

SUB-CLASS HYDROIDA.

THE *Hydrozoa* are divided into four sub-classes—viz., the *Hydroida*, the *Siphonophora*, the *Lucernarida*, and the *Discophora*.

SUB-CLASS I. HYDROIDA.—This sub-class comprises those *Hydrozoa which consist of an alimentary region or "polypite," which is provided with an adherent disc or "hydrorhiza," and prehensile tentacles.*

In some few cases the hydrosoma is composed of a single polypite only, as in the *Hydrida* and in some of the *Corynida;* but usually there are several polypites united together by means of a common trunk or "cœnosarc," as in most of the *Corynida* and in the orders *Sertularida* and *Campanularida*. Further, in the great majority of cases the "hydrorhiza" is permanently attached to some foreign object.

The *Hydroida* comprises four orders—viz., the *Hydrida*, the *Corynida*, the *Sertularida*, and the *Campanularida*.

ORDER I. HYDRIDA (*Gymnochroa*, Hincks).—This order comprises those *Hydrozoa whose "hydrosoma" consists of a single locomotive polypite, with tentacles and "hydrorhiza," and with reproductive organs which appear as simple external processes of the body-wall. The hydrorhiza is discoid, and no hard cuticular layer is at any time developed.*

The order *Hydrida* comprises a single genus only (*Hydra*), including the various species of "Fresh-water Polypes," as they are often called. The common *Hydra* (fig. 12, *c*) is found abundantly in this country, and consists of a tubular cylindrical body, the "proximal" extremity of which is expanded into an adherent disc or foot—the "hydrorhiza"—by means of which the animal can attach itself to some foreign body. It possesses, however, the power of detaching the hydrorhiza at will, and thus of changing its place. At the opposite or "distal" extremity of the body is placed the mouth, surrounded by a circlet of tentacles, which arise a little distance below the margin of the oral aperture. The tentacles vary in number from five to twelve or more, and they vary considerably in length in different species, being much shorter than the body in the *Hydra viridis*, but being extremely long and filamentous in *Hydra fusca*. They are highly extensile and contractile, and serve as organs of prehension, being capable of retraction till they appear as nothing more than so many warts or tubercles, and of being extended to a length which is in some species many times longer than the body itself. (In the *Hydra fusca* the tentacles can be protruded to a length of more than eight inches.) Each con-

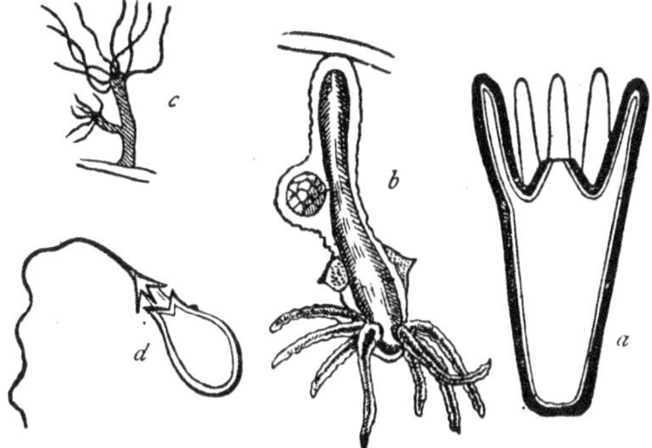

Fig. 12.—Morphology of Hydrozoa. *a* Diagrammatic section of *Hydra*. The dark line is the ectoderm, the fine line and clear space adjacent are the endoderm. *b Hydra viridis*, showing a single ovum contained in the body-wall near the proximal extremity, and two elevations containing spermatozoa near the bases of the tentacles; *c Hydra vulgaris*, with an undetached bud; *d* Thread-cell of the *Hydra*, greatly magnified.

sists of a prolongation of both ectoderm and endoderm, enclosing a diverticulum of the somatic cavity, and they are abundantly furnished with thread-cells. The cylindrical hydrosoma is excavated into a single large cavity, lined by the endoderm, and communicating with the exterior by the mouth. This—the "somatic cavity"—is the sole digestive cavity with which the *Hydra* is provided, the indigestible portions of the food being rejected by the mouth.

The *Hydra* possesses a most extraordinary power of resisting mutilation, and of multiplying artificially when mechanically divided. Into however many pieces a *Hydra* may be divided, each and all of these will be developed gradually into a new and perfect polypite. The remarkable experiments of Trembley upon this subject are well known, and have been often repeated, but space will not permit further notice of them here. Reproduction is effected in the *Hydra* both asexually by gemmation, and sexually,—the former process being followed in summer, and the latter towards the commencement of winter, few individuals surviving this season. In the first method the *Hydra* throws out one or more buds, generally from near its proximal extremity. These buds at first consist simply of a tubular prolongation of the ectoderm and endoderm, enclosing a cæcal diverticulum of the body-cavity; but a mouth and tentacles are soon developed, when the new being is usually detached as a perfect independent *Hydra*. The *Hydræ* thus produced throw out fresh buds, often before they are detached from the parent organism, and in this way reproduction is rapidly carried on.

In the second or sexual mode of reproduction, ova and spermatozoa are produced in outward processes of the body-wall (fig. 12, *b*). The spermatozoa are developed in little conical elevations, which are produced near the bases of the tentacles, and the ova are enclosed in sacs of much greater size, situated nearer the fixed or proximal extremity of the animal. Ordinarily there is but one of these sacs containing a single ovum, but sometimes there are two. When mature, the ovum is expelled through the body-wall, and is fecundated by the spermatozoa, which are simultaneously liberated. The embryo appears as a minute, free-swimming, ciliated body. The serous and mucous layers of the blastoderm (germinal area) correspond to the ectoderm and endoderm, and for the formation of the perfect *Hydra* nothing further seems wanting than the modification of one end of the body into a hydrorhiza, and the formation of a mouth and tentacles at the other.

ORDER II. CORYNIDA (= TUBULARIDA, the *Athecata* of Hincks).—The order *Corynida* comprises those *Hydrozoa whose hydrosoma is fixed by a hydrorhiza, and consists either of a single polypite, or of several united by a cœnosarc, which usually develops a firm outer layer or "polypary." No "hydrothecæ" are present. "The reproductive organs are in the form of gonophores, which vary much in structure, and arise from the sides of the polypites, from the cœnosarc, or from gonoblastidia."*—(Greene.)

The hydrosoma of the *Corynida* may consist of a single polypite, as in *Coryomorpha* and *Vorticlava*, or it may be composed of several united by a cœnosarc, as in *Cordylophora* (fig. 13, *a*). The order is entirely confined to the sea, with the single exception of *Cordylophora*, which inhabits fresh water. In *Tubularia* and its allies the organism is protected by a well-developed external chitinous envelope or "polypary;" but in the other genera belonging to the order the polypary is either rudimentary or is entirely absent. The polypary of the *Corynida*, when present, is readily distinguished from that of the *Sertularida*, by the fact that in the former it extends only to the base of the polypites; whereas in the latter it expands to form little cups for the reception of the polypites, these cups being called "hydrothecæ."

As regards the reproductive process in the *Corynida*, the reproductive elements are developed in distinct buds or sacs,

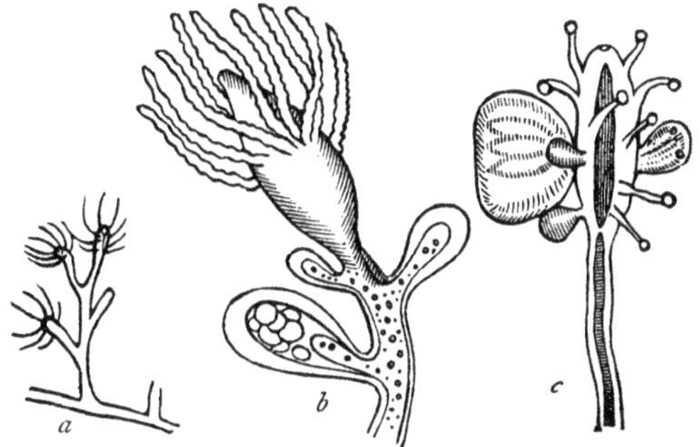

Fig. 13.—Morphology of Corynida. *a* Fragment of *Cordylophora lacustris*, slightly enlarged; *b* Fragment of the same considerably enlarged, showing a polypite and three gonophores in different stages of growth, the largest containing ova; *c* Portion of *Syncoryne Sarsii* with medusiform zoöids budding from between the tentacles.

which are external processes of the body-wall, and have been aptly termed "gonophores" by Professor Allman. Strictly speaking, Dr Allman understands by the term "gonophore" only the *ultimate* generative zoöid, that which *immediately* produces the generative elements. Great variations exist in the form and development of these generative buds, and an examination of these leads us to some of the most singular phenomena in the entire animal kingdom. In some species of *Hydractinia* and *Coryne*, the generative buds or "gonophores" exist in their simplest form—namely, as sacciform protuberances of the endoderm and ectoderm, enclosing a diverticulum of the somatic cavity. In this form they are attached to the "trophosome" by a short stalk, and they are termed "sporosacs" (fig. 14, *a*). They are exactly like the buds which we have already seen to exist in the *Hydra*, with this difference, that they are not themselves developed into fresh polypites, but are simply receptacles in which the essential elements of generation—the ova and spermatozoa—are prepared, by the union of which the young *Corynid* is produced.

In *Cordylophora* (fig. 13, *b*) a further advance in structure is perceptible. The gonophore now consists of a closed sac, from the roof of which depends a hollow process or peduncle —the "manubrium"—which gives off a system of tubes which run in the walls of the sac. For reasons which will be immediately evident, the gonophore in this case is said to have a "disguised" medusoid structure (fig. 14, *b*).

In certain *Corynida*, however, we meet with a still higher form of structure, the gonophores being now said to be "medusoid." In these cases the generative bud is primitively a simple sac—such as the "sporosac"—but ultimately develops itself into a much more complicated structure. The gonophore (fig. 13, *c*) is now found to be composed of a bell-shaped disc, termed the "gonocalyx," which is attached by its base to the parent organism (the trophosome), and has its cavity turned outwards. From the roof of the gonocalyx, like the clapper of a bell, there depends a peduncle or "manubrium," which contains a process of the somatic cavity. The manubrium gives out at its fixed or proximal end four prolongations of its cavity, in the form of radiating lateral tubes, which run to the margin of the bell, where they communicate with one another by means of a single circular canal which surrounds the mouth of the bell. This system of tubes constitutes what is known as the system of the "gonocalycine canals." The gonophore, thus constituted, may remain permanently attached to the parent organism, as in *Tubularia indivisa* (fig. 14, *c*); but in other cases still further changes ensue. In the higher forms of development (fig. 14, *d*) the

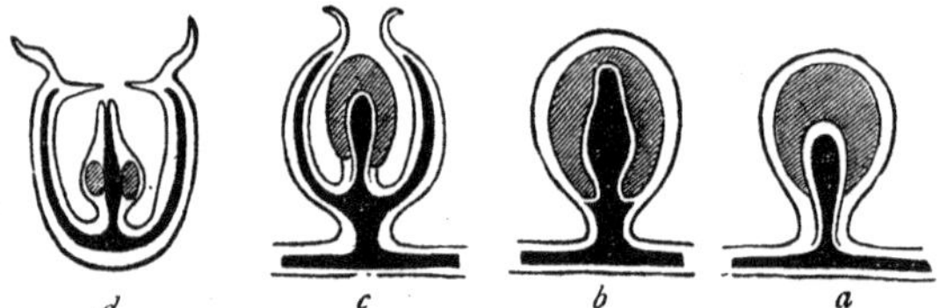

Fig. 14.—Reproductive processes of Hydrozoa. *a* Sporosac; *b* Disguised medusoid; *c* Attached medusiform gonophore; *d* Free medusiform gonophore. The cross shading indicates the reproductive organs, ovaria or spermaria. The part completely black indicates the cavity of the manubrium and the gonocalycine canals.

manubrium acquires a mouth at its free or distal extremity, and the gonocalyx becomes detached from the parent. The gonophore is now free, and behaves in every respect as an independent being. The gonocalyx is provided with marginal tentacles and with an inward prolongation from its margin, which partially closes the mouth of the bell, and is termed the "veil" or "velum." By the contractions of the gonocalyx, which now serves as a natatorial organ, the gonophore is propelled through the water. The manubrium, with the shape, assumes the functions of a polypite, and its cavity takes upon itself the office of a digestive sac. Growth is rapid, and the gonophore may attain a comparatively gigantic size, being now absolutely identical with one of those organisms which are commonly called "jelly-fishes," and are technically known as *Medusæ* (fig. 15). In fact, as we shall afterwards see, most if not all of the *gymnophthalmate Medusæ*, originally described as a distinct order of free-swimming Hydrozoa, are in truth merely the liberated generative buds, or 'medusiform gonophores," of the permanently-rooted *Hydroids*. Finally, the essential generative elements—the ova and spermatozoa—are developed in the walls of the manubrial sac, between its endoderm and ectoderm, and embryos are produced. These embryos, however, instead of resembling the organism which immediately gave them birth, develop themselves into the fixed *Corynid* from which the gonophore was produced, thus completing the cycle.

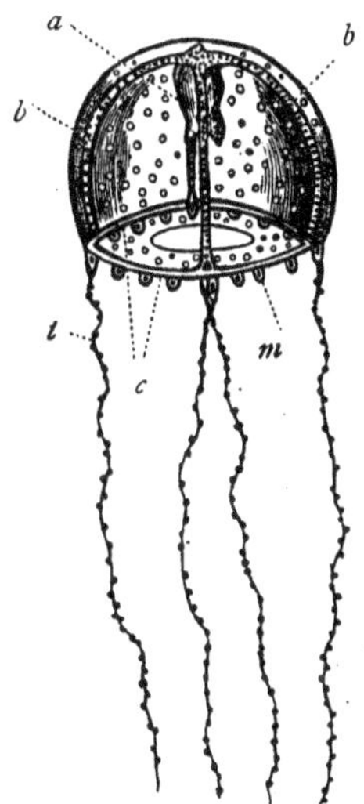

Fig. 15. — Free medusiform gonophore of *Clytia Johnstoni* (after Hincks). *a* Central polypite or manubrium; *bb* Radiating gastrovascular canals; *c* Circular canal; *m* Marginal bodies; *t* Tentacles.

As we have seen, the generative buds of the *Corynida* may exist in the following forms:—1. As "sporosacs," or simple closed sacs, consisting of ectoderm and endoderm, with a central cavity in which ova and spermatozoa are produced. 2. As "disguised medusoids," in which there is a central manubrial process and a rudimentary system of gonocalycine canals; but the gonocalyx remains closed. 3. As complete medusoids, which have a central manubrium, a complete system of gonocalycine canals, and an open gonocalyx; but which never become detached. 4. As perfect medusiform gonophores (fig. 15), which are detached, and lead an independent existence for a time, until the generative elements are matured. In whichever of these forms the gonophore may be present, the place of its origin from the trophosome may vary in different species of the order. 1. They may arise from the sides of the polypites, as in *Coryne* and *Stauridia*; 2. They may be produced from the cœnosarc, as in *Cordylophora*; 3. They may be produced upon certain special processes, which are termed "gonoblastidia," as in *Hydractinia* and *Dicoryne.* These gonoblastidia are processes from the body-wall or cœnosarc, which closely resemble true polypites in form, but differ from them in being usually devoid of a mouth, and in having shorter tentacles.

As regards the development of the *Corynida*, the embryo is very generally, though not always, ciliated at first, and becomes developed into a hydra-form polypite, which fixes itself to some foreign body, and then (if not belonging to one of the simple forms) proceeds to produce by gemmation the composite adult. The development of the *Corynida* (as well as that of the *Sertularida* and *Lucernarida*) obeys the general law, that the new polypites are developed at, or near, the distal end of the hydrosoma; so that the distal polypites are the youngest, the reverse of this obtaining amongst the oceanic *Hydrozoa.*

The subject of the reproduction of the *Corynida* having been treated at some length, so as to apply to the remaining *Hydroida*, we shall now give a brief description of the two leading types of structure exhibited by the order.

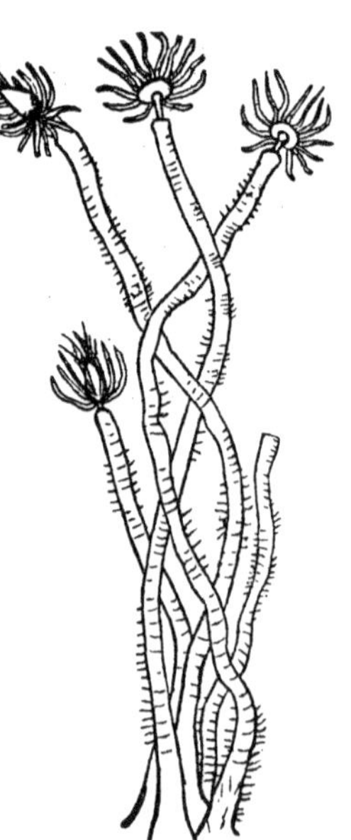

Fig. 16.—Corynida. Fragment of *Tubularia indivisa*, natural size.

Eudendrium, a genus of the *Corynida*, which is not uncommonly found attached to submarine objects, usually in tolerably deep water, may be taken as a good example of the fixed and composite division of the order. The hydrosoma consists of numerous polypites, united by a cœnosarc, which is more or less branched, and is defended by a horny tubular polypary. The polypites are borne at the ends of the branches and branchlets, and are not contained in "hydrothecæ," the polypary ending abruptly at their bases. The polypites are non-retractile, of a reddish colour, and provided with about twenty tentacles, arranged round the mouth in a single row. *Tubularia* (fig. 16) is very similar to *Eudendrium*, but the hydrosoma is either undivided or is very slightly branched. The hydrosoma consists of clustered horny tubes, of a straw colour, and not unlike straws to look at; hence the common name of pipe-coralline given to this zoophyte.

Each tube is filled with a soft, semi-fluid, reddish cœnosarc, and gives exit at its distal extremity to a single polypite. The polypites are bright red in colour, and are not retractile within their tubes, the horny polypary extending only to their bases. The polypites are somewhat conical in shape, the mouth being placed at the apex of the cone, and they are furnished with two sets of tentacles. One set consists of numerous short tentacles placed directly round the mouth; the other is composed of from thirty to forty tentacles of much greater length, arising from the polypite about its middle or near the base. Near the insertion of these tentacles the generative buds are produced at proper seasons.

Coryomorpha nutans may be taken to represent those *Corynida* in which there is no polypary and the hydrosoma is simple. It is about four inches in length, and is fixed by filamentous roots to the sand at the bottom of the sea. It consists of a single whitish polypite, striped with pink, and terminating upwards in a pear-shaped head, round the thickest part of which is a circlet of from forty to fifty long white tentacles. Above these comes a series of long branching gonoblastidia, bearing gonophores, and succeeded by a second shorter set of tentacles which surround the mouth. The gonophores become ultimately detached as free-swimming medusoids.

ORDER III. SERTULARIDA (*Thecaphora*, Hincks). — This order comprises those Hydrozoa "*whose hydrosoma is fixed by a hydrorhiza, and consists of several polypites, protected by hydrothecæ, and connected by a cœnosarc, which is usually branched and invested by a very firm outer layer. Reproductive organs in the form of gonophores arising from the cœnosarc or from gonoblastidia.*"—(Greene.)

The *Sertularida* resemble the *Corynida* in becoming permanently fixed after their embryonic condition by a hydrorhiza, which is developed from the proximal end of the cœnosarc; but they differ in the fact that the polypites are invariably protected by "hydrothecæ," or little cup-like expansions of the polypary (fig. 17, *a*, *b*); whilst the hydrosoma is in all cases composed of more than a single polypite. The cœnosarc generally consists of a main stem—or "hydrocaulus"—with many branches; and it is so plant-like in appearance that the common Sertularians are almost always mistaken for sea-weeds by visitors at the seaside. It is invested by a strong corneous or chitinous covering, often termed the "periderm."

The polypites are sessile or subsessile, hydra-form, and in all essential respects identical with those of the *Corynida*, though usually smaller. Each polypite consists of a soft, contractile, and extensile body, which is furnished at its distal extremity with a mouth and a circlet of prehensile tentacles, richly furnished with thread-cells. The tentacles have an indistinctly alternate arrangement. The mouth opens into a chamber which occupies the whole length of the polypite, and is to be regarded as the combined body-cavity and digestive sac. At its lower end this chamber opens by a constricted aperture into a tubular cavity which is everywhere excavated in the substance of the cœnosarc (fig. 17, *b*). The nutrient particles

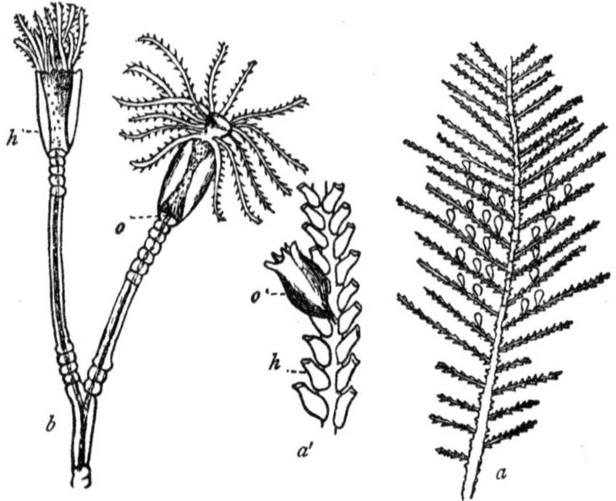

Fig. 17.—*a Sertularia* (*Diphasia*) *pinnata*, natural size; *a'* Fragment of the same enlarged, carrying a male capsule (*o*), and showing the hydrothecæ (*h*); *b* Fragment of *Campanularia neglecta* (after Hincks), showing the polypites contained in their hydrothecæ (*h*), and also the point at which the cœnosarc communicates with the stomach of the polypite (*o*).

obtained by each polypite thus serve for the support of the whole colony, and are distributed throughout the entire organism. The nutritive fluid prepared in the interior of each polypite gains access through the above-mentioned aperture to the cavity of the cœnosarc, which by the combined exertions of the whole assemblage of polypites thus becomes filled with a granular nutritive liquid. This cœnosarcal fluid is in constant movement, circulating through all parts of the colony, and thus maintaining its vitality, the cause of the movement being probably due, in part, at any rate, to the existence of vibrating cilia. The generative buds (gonophores or ovarian vesicles) are usually supported upon gonoblastidia, and seldom, if ever, become detached in the true Sertularids. They are often developed in chitinous receptacles known as "gonothecæ" (fig. 17, *o*). The young Sertularian on escaping from the ovum appears as a free-swimming ciliated body, which soon loses its cilia, fixes itself, and develops a young cœnosarc, by gemmation from which the branching hydrosoma of the perfect organism is produced.

In *Plumularia* and some of its allies there occur certain peculiar organs, probably offensive, to which the name of "nematophores" has been applied. Each of these consists of a process of the cœnosarc, which is invested by the horny polypary, with the exception of the distal extremity, which remains uncovered, and contains many large thread-cells embedded in it.

ORDER IV. CAMPANULARIDA.—The members of this order are closely allied to the *Sertularida;* so closely, indeed, that they are very often united together into a single group. The chief difference consists in the fact that the hydrothecæ of the *Campanularida* with their contained polypites are supported upon conspicuous stalks, thus being terminal in position (fig. 17, *b*); whilst in the *Sertularida* they are sessile or subsessile, and are placed laterally upon the branchlets. The gonophores also in the *Campanularida* are usually detached as free-swimming medusoids, whereas they remain permanently attached in the *Sertularians.* Each medusoid consists of a little transparent glassy bell, from the under surface of which there is suspended a modified polypite, in the form of a "manubrium" (fig. 15). The whole organism swims gaily through the water, propelled by the contractions of the bell or disc (*gonocalyx*); and no one would now suspect that it was in any way related to the fixed plant-like zoophyte from which it was originally budded off. The central polypite is furnished with a mouth at its distal end, and the mouth opens into a digestive sac. From the proximal end of this stomach proceed four radiating canals which extend to the circumference of the disc, where they all open into a single circular vessel surrounding the mouth of the bell. From the margins of the disc hang also a number of delicate extensile filaments or tentacles; and the circumference is still further adorned with a series of brightly-coloured spots, which are probably organs of sense. The mouth of the bell is partially closed by a delicate transparent membrane or shelf, the so-called "veil." Thus constituted, these beautiful little beings lead an independent and locomotive existence for a longer or shorter period. Ultimately, the essential elements of reproduction are developed in special organs, situated in the course of the radiating canals of the disc. The resulting embryos are ciliated and free-swimming, but ultimately fix themselves, and develop into the plant-like colony from which fresh medusoids may be budded off. The ova in the medusiform gonophores are usually developed in the course of the gonocalycine canals, and not between the ectoderm and endoderm of the manubrium, as is the case in the *Corynida*. Examples of the order are *Campanularia Laomedea*, &c The distinctions between the *Sertularida* and *Campanularida* are certainly insufficient to justify their being placed in separate orders. If united together, it would probably be best to adopt the name *Thecaphora* (Hincks) for the order, and to employ the names *Sertularida* and *Campanularida* for the sub-orders.

CHAPTER IX.

SIPHONOPHORA.

SUB-CLASS II. SIPHONOPHORA.—The members of this sub-class constitute the so-called "Oceanic *Hydrozoa*;" and are characterised by the possession of a "*free and oceanic hydrosoma, consisting of several polypites united by a flexible, contractile, unbranched or slightly-branched cœnosarc, the proximal end of which is usually furnished with 'nectocalyces,' and is dilated into a 'somatocyst' or into a 'pneumatophore.'*"—(Greene.)

All the *Siphonophora* are unattached, and permanently free, and all are composite. They are singularly delicate organisms, mostly found at the surface of tropical seas, the Portuguese man-of-war (*Physalia*) being the most familiar member of the group. The sub-class is divided into two orders—viz., the *Calycophoridæ* and the *Physophoridæ*.

ORDER I. CALYCOPHORIDÆ.—This order includes those *Siphonophora* whose *hydrosoma is free and oceanic, and is propelled by "nectocalyces" attached to its proximal end. The hydrosoma consists of several polypites, united by an unbranched cœnosarc, which is highly flexible and contractile, and never develops a hard cuticular layer. The proximal end of the hydrosoma is modified into a peculiar cavity called the "somatocyst." The reproductive organs are in the form of medusiform gonophores produced by budding from the peduncles of the polypites.*

In all the *Calycophoridæ* the cœnosarc is filiform, cylindrical, unbranched, and highly contractile, this last property being due to the presence of abundant muscular fibres. "The proximal end of the cœnosarc dilates a little, and becomes ciliated internally, forming a small chamber" which communicates with the nectocalycine canals. "At its upper end this chamber is a little constricted, and so passes, by a more or less narrowed channel, into a variously-shaped sac, whose walls are directly continuous with its own, and which will henceforward be termed the *somatocyst* (fig. 18, 3 *b*). The endoderm of this sac is ciliated, and it is generally so immensely vacuolated as almost to

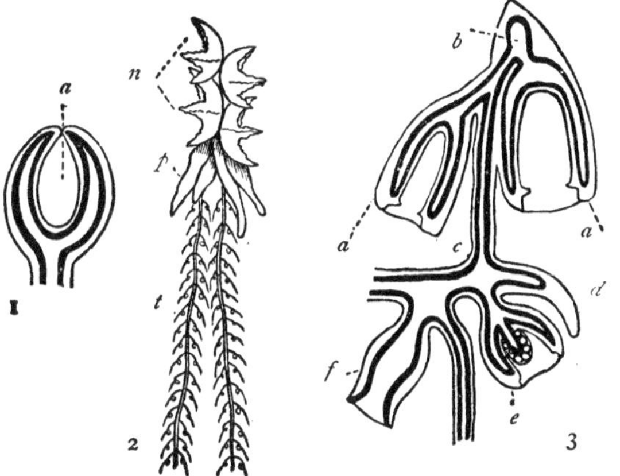

Fig. 18.—Morphology of the Oceanic Hydrozoa. 1. Diagram of the proximal extremity of a *Physophorid*. *a* Pneumatocyst. 2. *Vogtia pentacantha*, one of the *Calycophoridæ*. *n* Nectocalyces; *p* Polypites; *t* Tentacles. 3. Diagram of a *Calycophorid*. *a a'* Proximal and distal nectocalyces; *b* Somatocyst; *c* Cœnosarc; *d* Hydrophyllium or bract; *e* Medusiform gonophore; *f* Polypite. The dark lines in figs. 1 and 3 indicate the endoderm, the light line with the clear space indicates the ectoderm. (After Huxley.)

obliterate the internal cavity, and give the organ the appearance of a cellular mass."—(Huxley.) The polypites in the *Calycophoridæ* often show a well-marked division into three portions, termed respectively the proximal, median, and distal divisions. Of these, the "proximal" division is somewhat contracted, and forms a species of peduncle, which often carries appendages. The "median" portion is the widest, and may be termed the "gastric division," as in it the process of digestion is carried on. It is usually separated from the proximal division by a valvular inflection of the endoderm, which is known as the "pyloric valve." The polypites have only one tentacle "developed near their basal or proximal ends, and provided with lateral branches ending in saccular cavities," and furnished with numerous thread-cells. The proximal ends of the polypites usually bear certain overlapping plates, of a protective nature, which are termed "hydrophyllia," or "bracts." They are composed of processes of both ectoderm and endoderm (fig. 18, 3 *d*), and they always contain a diverticulum from the somatic cavity, which is called a "phyllocyst." The *Calycophoridæ* always possess swimming-bells, or "nectocalyces," by the contractions of which the hydrosoma is propelled through the water (fig. 18, 2). The nectocalyx in structure is very similar to the "gonocalyx" of a medusiform gonophore, as already described; but the former is devoid of the gastric or genital sac—the "manubrium"—possessed by the latter. Each nectocalyx consists of a bell-shaped cup, attached by its base to the hydrosoma, and provided with a muscular lining in the interior of its cavity, or "nectosac." There is also always a "velum" or "veil," in the form of a membrane attached to the mouth of the nectosac round its entire margin, and leaving a central aperture. The peduncle by which the nectocalyx is attached to the hydrosoma conveys a canal from the somatic cavity, which dilates into a ciliated chamber, and gives off at least four radiating canals, which proceed to the circumference of the bell, where they are united by a circular vessel; the entire system constituting what is known as the system of the "nectocalycine canals." In the typical *Calycophoridæ* two nectocalyces only are present, but in some genera there are more. In *Praya* the two nectocalyces are so apposed to one another that a sort of canal is formed by the union of two grooves, one of which exists on the side of each nectocalyx. This chamber, which is present in a more or less complete form in all the genera, is termed the "hydrœcium," and the cœnosarc can be retracted within it for protection.

The reproductive bodies in the *Calycophoridæ* are in the form of medusiform gonophores, which are budded from the peduncles of the polypites, becoming, in many instances, detached to lead an independent existence. In some *Calycophoridæ*, as in *Abyla*, "each segment of the cœnosarc, provided with a polypite, its tentacle, reproductive organ, and hydrophyllium, as it acquires a certain size, becomes detached, and leads an independent life—the calyx of its reproductive organ serving it as a propulsive apparatus. In this condition it may acquire two or three times the dimensions it had when attached, and some of its parts may become wonderfully altered in form."—(Huxley.) To these detached reproductive portions of adult *Calycophoridæ* the term "Diphyozoöids" has been applied.

As regards the development of the *Calycophoridæ*, "not

only the new polypites, but the new nectocalyces and reproductive organs, and even the branches of the tentacles, are developed on the proximal side of the old ones; so that the distal appendages are the oldest."—(Huxley.) The process of development is therefore the reverse of what obtains amongst the *Hydroida*.

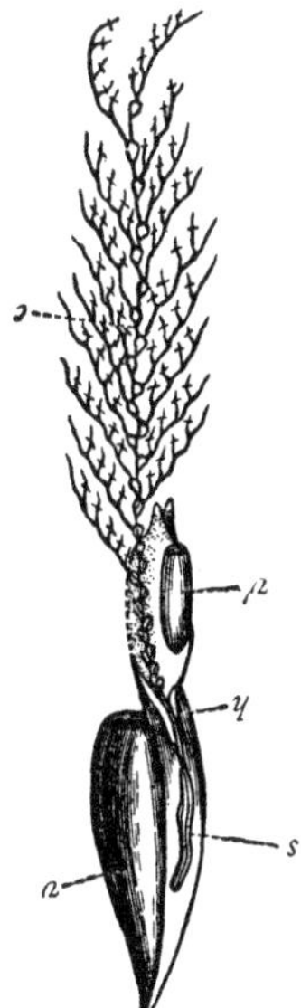

Fig. 19. — Calycophoridæ. *Diphyes appendiculata* (after Kölliker). *v* Proximal nectocalyx; *v'*, Distal nectocalyx; *h* Hydrœcium; *c* Cœnosarc, carrying polypites each with its bract and tentacle.

Diphyes (fig. 19), which may be taken as the type of the *Calycophoridæ*, consists of a delicate filiform cœnosarc, provided proximally with two large mitre-shaped nectocalyces (*v*, *v'*), of which one lies entirely on the distal side of the other. The pointed apex of the distal nectocalyx is received into a special cavity in the proximal nectocalyx. The "hydrœcium" (*h*) is formed partially by this chamber in the nectocalyx, and partially by an arched groove prolonged upon the inner surface of the distal nectocalyx, within which the cœnosarc moves freely up and down, and can be entirely retracted if necessary. The upper part of the cœnosarc dilates into a small ciliated cavity, from which are given off two tubes, which proceed respectively to the distal and proximal nectocalyces, where they open into the central chamber from which the nectocalycine canals take their rise. The upper portion of this small ciliated cavity is prolonged proximally into the larger chamber of the "somatocyst." The cœnosarc (*c*) bears polypites, each of which is protected by a delicate glassy "hydrophyllium."

DIVISIONS OF THE CALYCOPHORIDÆ.—(AFTER HUXLEY.)

Fam. I. *Diphydæ.*—Nectocalyces not more than two in number, and of a polygonal shape. Hydrœcium of the proximal nectocalyx complete, or closed posteriorly. Hydrophyllia well developed.

Fam. II. *Sphæronectidæ.*—Nectocalyces probably not more than two in number; the proximal nectocalyx spheroidal, with a complete hydrœcium. No hydrophyllia (?).

Fam. III. *Prayidæ.*—Nectocalyces two in number; hydrœcia incomplete and groove-like. Polypites protected by hydrophyllia.

Fam. IV. *Hippopodidæ.*—Nectocalyces numerous; hydrœcia incomplete. Polypites not protected by hydrophyllia.

ORDER II. PHYSOPHORIDÆ. — This second order of the

Oceanic Hydrozoa comprises those *Siphonophora in which the hydrosoma consists of several polypites united by a flexible, contractile, unbranched or very slightly branched cœnosarc, the proximal extremity of which is modified into a "pneumatophore," and is sometimes provided with "nectocalyces." The polypites have either a single basal tentacle, or the tentacles arise directly from the cœnosarc. "Hydrophyllia" are commonly present. The reproductive bodies are developed upon gonoblastidia.*

The cœnosarc in the *Physophoridæ*, like that of the *Calycophoridæ*, is perfectly flexible and contractile; but it is not necessarily elongated, being sometimes spheroidal or discoidal. The proximal end of the cœnosarc "expands into a variously-shaped enlargement, whose walls consist of both ectoderm and endoderm, and which encloses a wide cavity in free communication with that of the cœnosarc, and, like it, full of the nutritive fluid. From the distal end, or apex, of this cavity depends a sac, variously shaped, but always with tough, strong, and elastic walls, composed of a substance which is stated to be similar to chitine in composition, and more or less completely filled with air."—(Huxley.) The large proximal dilatation of the cœnosarc is termed the "pneumatophore," whilst the chitinous air-sac which it contains is termed the "pneumatocyst" (fig. 18, 1). The pneumatocyst is held in position by the reflection of the endoderm of the pneumatophore over it, and it doubtless acts as a buoy or "float." In the Portuguese man-of-war (*Physalia*) the pneumatocyst communicates with the exterior by means of an aperture in the ectoderm of the pneumatophore. In *Velella* and *Porpita* the pneumatocyst communicates with the exterior by means of several openings called "stigmata;" and from its distal surface depend numerous slender processes, containing air, and known as "pneumatic filaments."

The polypites of the *Physophoridæ* resemble those of the *Calycophoridæ* in shape, but the tentacles have a much more complicated structure, and are sometimes many inches in length, as in *Physalia*. The "hydrophyllia" have essentially the same structure as those of the former order. There occur also in the *Physophoridæ* certain peculiar bodies, termed "hydrocysts" or "feelers" ("fühler" and "taster" of the Germans). These resemble immature polypites in shape, consisting of a prolongation of both ectoderm and endoderm, usually with a tentacle, and containing a diverticulum of the somatic cavity, the distal extremity being closed, and furnished with numerous large thread-cells. They are looked upon as "organs of prehension and touch," and they are somewhat analogous to the "nematophores" of some of the *Sertularida*.

As regards the reproductive organs, they are developed upon special processes or "gonoblastidia," and they may remain permanently attached, or they may be thrown off as free-swimming medusoids. In many of the *Physophoridæ* the male and female gonophores differ from one another in form and size, and they are then termed respectively "androphores" and "gynophores." As regards their development, the *Physophoridæ* obey the same general law as the *Calycophoridæ*.

In *Physophora* the hydrosoma consists of a filiform cœnosarc, which bears the polypites and their appendages, and dilates proximally into a pneumatophore. Below this point the cœnosarc bears a double row of nectocalyces, which are channelled on their inner faces to allow of their attachment to the cœnosarc. There are no hydrophyllia, but there is a series of "hydrocysts" on the proximal side of the polypites.

Physalia, or the Portuguese man-of-war (fig. 20, *a*), is composed of a large, bladder-like, fusiform "float" or pneumatophore—sometimes from eight to nine inches in length—upon the under surface of which are arranged a number of polypites, together with highly contractile tentacles of great length, "hydrocysts," and reproductive organs. *Physalia* is of common occurrence, floating at the surface of tropical seas; and fleets of it are not uncommonly driven upon our own shores.

In *Velella* (fig. 20, *b*) the hydrosoma consists of a widely-expanded pneumatophore of a rhomboidal shape, carrying upon its upper surface a diagonal vertical crest. Both the horizontal disc and the vertical crest are composed of a soft marginal "limb," and a central more consistent "firm part." "To the distal surface of the firm part of the disc are attached the several appendages, including, 1. a single large polypite, nearly central in position; 2. numerous small gonoblastidia, which resemble polypites, and are termed "phyogemmaria;" and, 3. the reproductive bodies to which these last give rise. The tentacles are attached, quite independently of the polypites, in a single series along the line where the firm part and limb of the disc unite. There are no hydrocysts, nectocalyces, or hydrophyllia. . . . On all sides the limb is traversed by an anastomosing system of canals, which are ciliated, and communicate with the cavities of the phyogemmaria and large central polypite."—(Greene.) *Velella* is about two inches in length by one and a half in height. It is of a beautiful blue colour and semi-transparent, and it floats at the surface of the sea, with its vertical crest exposed to the wind as a sail.

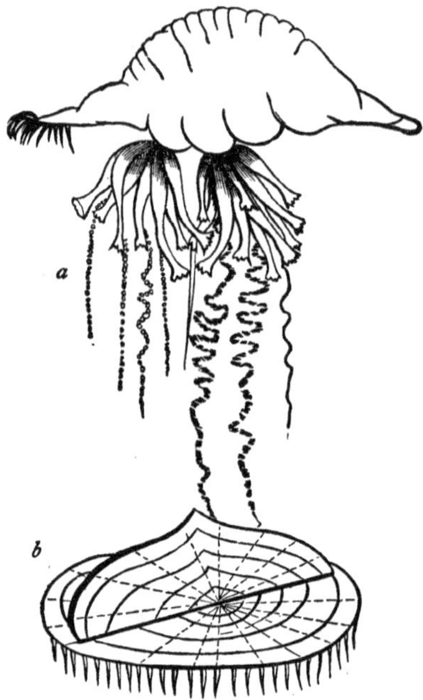

Fig. 20.—*Physophoridæ.* *a* Portuguese man-of-war (*Physalia utriculus*), showing the fusiform float and the polypites and tentacles (after Huxley); *b* *Velella vulgaris* (after Gosse).

Divisions of the Physophoridæ.—(After Huxley.)

Fam. I. *Apolemiadæ.*—Hydrosoma with nectocalyces and hydrophyllia, the latter united with the other organs into groups which are arranged at considerable intervals along the cœnosarc. Cœnosarc filiform. Pneumatocyst small.

Fam. II. *Stephanomiadæ.*—Hydrosoma with nectocalyces and hydrophyllia, the latter arranged with the other organs in a continuous series. Cœnosarc filiform. Pneumatocyst small.

Fam. III. *Physophoriadæ.*—Hydrosoma with nectocalyces, but without hydrophyllia. Distal end of the filiform cœnosarc dilated. Pneumatocyst small.

Fam. IV. *Athorybidæ.* — Hydrosoma without nectocalyces, but with hydrophyllia. Pneumatocyst occupying almost the whole of the globular cœnosarc.

Fam. V. *Rhizophysiadæ.* — Hydrosoma without either nectocalyces or hydrophyllia. Cœnosarc filiform. Pneumatocyst small.

Fam. VI. *Physaliadæ.*—Pneumatocyst occupying almost the whole of the thick and irregularly fusiform cœnosarc. No nectocalyces or hydrophyllia.

Fam. VII. *Velellidæ.*—Hydrosoma without nectocalyces or hydrophyllia; with short, simple, or branched submarginal tentacles. A single central principal polypite. Pneumatocyst flattened, divided into chambers by numerous concentric partitions, and occupying almost the whole of the discoidal cœnosarc.

CHAPTER X.

DISCOPHORA.

SUB-CLASS III. DISCOPHORA (*Acalephæ** in part).—Since this sub-class contains only a single order, that of the *Medusidæ*, a single definition necessarily suffices for both. The *Medusidæ* are defined as "*Hydrozoa whose hydrosoma is free and oceanic, consisting of a single nectocalyx, from the roof of which a single polypite is suspended. The nectocalyx is furnished with a system of canals. The reproductive organs are as processes either of the sides of the polypite or of the nectocalycine canals.*"—(Greene.)

The *Medusidæ* comprise most of the organisms commonly known as Jelly-fishes or Sea-nettles, the last name being derived from the property which some of them possess of severely stinging the hand, this power being due to the presence of numerous thread-cells. As employed by modern naturalists, the order is very much restricted, and it is by no means improbable that it will ultimately be entirely done away with, very many of its members having been shown to be really the free generative buds of other *Hydrozoa*. As used here, it corresponds to part of the *Gymnophthalmate Medusæ* of Professor E. Forbes, the *Steganophthalmate Medusæ* of the same author being now placed in the sub-class *Lucernarida.*

The hydrosoma of one of the *Discophora* (= a Gymnophthalmate *Medusa*) is composed of a single gelatinous bell-shaped

* The old sub-class of the *Acalephæ* contained the *Gymnophthalmate Medusæ* (= the *Discophora*,) and the *Steganophthalmate Medusæ* (= the *Lucernarida* in part), the two being placed in a single order under the name of *Pulmograda*. The *Acalephæ* also contained the *Ctenophora* and the *Calycophoridæ* and *Physophoridæ*, of which the former constituted the order *Ciliograda*, whilst the two latter made up the order *Physograda*. The *Ctenophora*, however, are now generally placed amongst the *Actinozoa*, whilst the *Calycophoridæ* and *Physophoridæ* constitute the Hydrozoal sub-class *Siphonophora*.

swimming organ, the "nectocalyx" or "disc," from the roof of which a single polypite is suspended (fig. 21). The interior of the nectocalyx is often called the "nectosac," and the term "codonostoma" has been proposed to designate the open

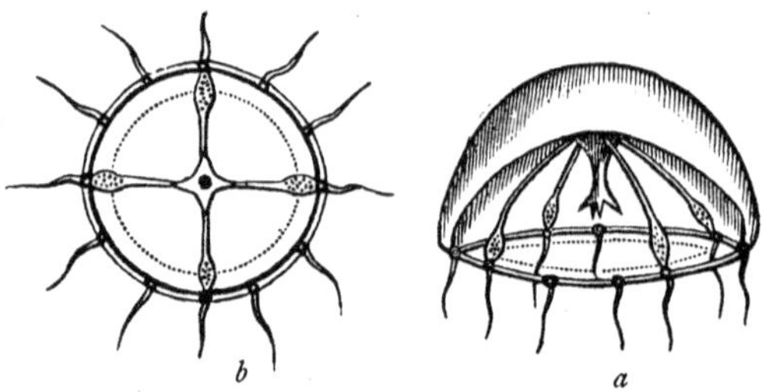

Fig. 21.—Morphology of Medusidæ. *a* A Medusid (*Thaumantias*) seen in profile, showing the central polypite, the radiating and circular gonocalycine canals, the marginal vesicles and tentacles, and the reproductive organs; *b* The same viewed from below. The dotted line indicates the margin of the velum.

mouth of the bell. The margin of the nectocalyx is produced inwards to form a species of shelf, running round the margin of the mouth of the bell, and termed the "veil" or velum," by the presence of which the nectocalyx is distinguished from the somewhat similar "umbrella" of the *Lucernarida*. The endodermal lining of the central polypite or "manubrium" (sometimes called the "proboscis") is prolonged into four radiating canals, which run to the periphery of the nectocalyx, where they are connected by a circular canal which runs round its circumference, the whole constituting the system of the "nectocalycine canals" (formerly called the "chylaqueous canals"). From the circumference of the nectocalyx depend marginal tentacles, which are usually hollow processes, composed of both ectoderm and endoderm, and in immediate connection with the canal system. Also round the circumference of the nectocalyx are disposed certain "marginal bodies," of which two kinds may be distinguished. Of these the first are termed "vesicles," and consist of rounded sacs lined by epithelium, and containing one or more solid, motionless concretions—apparently of carbonate of lime—immersed in a transparent fluid. The second class of marginal bodies, variously termed "pigment-spots," "eye-specks," or "ocelli," consists of little aggregations of pigment enclosed in distinct cavities. The "vesicles" are probably rudimentary organs of hearing, and possibly the eye-specks are a rudimentary form of visual apparatus. The oral margin of the polypite may be simple, or it may be produced into lobes, which are most frequently four in number. The essential elements of generation are produced in simple expansions either of the wall of the manubrium or of the radiating nectocalycine canals.

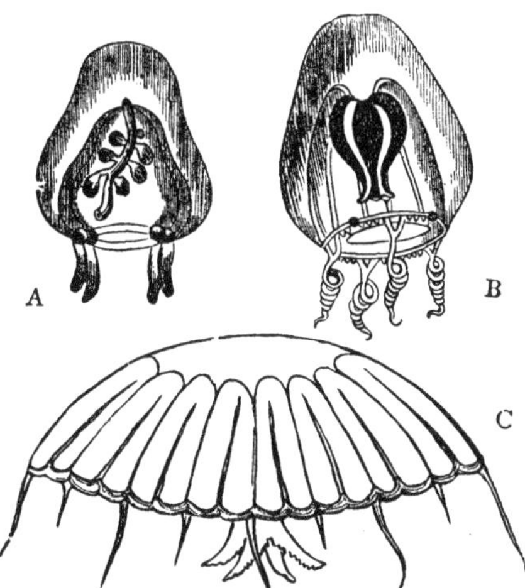

Fig. 22.—Group of naked-eyed *Medusæ*. A *Sarsia gemmifera*, with medusoids arising from the sides of the central polypite (after Greene); B *Modeeria formosa* (after Forbes); C *Polyxenia Alderi* (after Gosse).

From the above description it will be evident that the *Medusa* is in all essential respects identical in structure with the free-swimming generative bud or gonophore of many of the fixed and oceanic *Hydrozoa*. Indeed, a great many forms which were previously included in the *Medusidæ* have now been proved to be really of this nature, and it may fairly be doubted if this will not ultimately be found to apply to all. As to the value, however, of the order *Medusidæ*, the present state of our knowledge is well expressed by the following conclusions which have been drawn up by Professor Greene:—

"1. That several of the organisms formerly described as *Medusidæ* are the free gonophores of other orders of *Hydrozoa*.

"2. That the homology of these free gonophores with those simple expansions of the body-wall which in *Hydra* and some other genera are known to be reproductive organs by their contents alone, is proved alike by the existence of numerous transitional forms and by an appeal to the phenomena of their development.

"3. That many of the so-called *Medusidæ* may, from analogy, be regarded as, in like manner, medusiform gonophores.

"4. But that there may exist, nevertheless, a group of *Medusid* forms which may give rise by true reproduction to organisms directly resembling their parents, and therefore worthy of being placed in a separate order under the name *Medusidæ*."

The same authority concludes by remarking that to the order as above defined "may be referred provisionally that large assemblage of forms anatomically similar to true *Medusidæ*, but whose development is unknown." Besides the large group of forms thus temporarily admitted, all the *Trachynemidæ* and *Æginidæ* are stated by Gegenbauer to fulfil the conditions of the above definition, and should, therefore, be looked upon as true *Medusidæ*.

As to the development of these true *Medusidæ*, little is known for certain. It appears, however, that in *Trachynema*, *Æginopsis*, and other genera, the embryo is directly developed into a form resembling its parent, without passing through any intermediate changes of form. It is hardly necessary to remark that this is not the case with the embryos of a medusiform gonophore, these being developed into the sexless *Hydrozoön* by which the medusoid was produced.

In this connection, allusion may be made to the long-known fact that certain medusiform gonophores *are* capable of producing independent forms directly resembling themselves, but this is by a process of gemmation, and not by one of true reproduction. Technically these are called "tritozoöids," as being derived from organisms which are themselves but the generative zoöids of another being. This singular phenomenon has been observed in various medusiform gonophores (*e.g.*, *Sarsia gemmifera*, fig. 22, A), the buds springing in different species from the gonocalycine canals, from the tentacles, or from the sides of the polypite or manubrium.

The "naked-eyed" *Medusæ*, though mostly very diminutive in point of size, are exceedingly elegant and attractive when examined in a living condition, resembling little bells of transparent glass, adorned here and there with the most brilliant colours. They occur in their proper localities and at proper seasons in the most enormous numbers. They are mostly phosphorescent, or capable of giving out light at night, and they appear to be one of the principal sources of the luminosity of the sea. It does not seem, however, that they phosphoresce, unless irritated or excited in some manner.

CHAPTER XI.

LUCERNARIDA AND GRAPTOLITIDÆ.

SUB-CLASS IV. LUCERNARIDA (*Acalephæ*, in part).—The members of this sub-class may be defined as *Hydrozoa* "*whose hydrosoma has its base developed into an 'umbrella,' in the walls of which the reproductive organs are produced.*"—(Greene.)

A large number of forms included in the *Lucernarida* were described by Edward Forbes under the name of *Steganophthalmate Medusæ*, being in many external characters closely similar to the *Medusidæ*. These "hidden-eyed" *Medusæ* are familiar to every one as "sea-blubbers" or "sea-jellies," and they occur in great numbers round our coasts during the summer months. The resemblance to the little jelly-fishes is especially strong between the disc or "nectocalyx" of the true *Medusidæ* and the "umbrella" of the *Lucernarida*, the latter being often a bell-shaped swimming organ, with marginal tentacles, and containing one or more polypites. These analogous structures (figs. 22 and 25) are, however, distinguished as follows:—1. The "umbrella" of the *Lucernarida* is never furnished with a "velum," as is the nectocalyx of the *Medusidæ*. 2. The radiating canals in the former are never less than eight in number, and they send off numerous anastomosing branches, which join to form an intricate network; whereas in the latter they are rarely more than four in number, and though they may subdivide, they do not anastomose. 3. In the place of the separate and unprotected "vesicles" and "ocelli" of the *Medusidæ*, the marginal bodies of the *Lucernarida* consist of these bodies combined together into single organs, which are termed "lithocysts," and which are protected externally by a sort of hood.

The *Lucernarida* admit of being divided into three orders—viz., the *Lucernariadæ*, the *Pelagidæ*, and the *Rhizostomidæ*.

ORDER I. LUCERNARIADÆ.—This order includes those *Lucernarida which have only a single polypite, are fixed by a proximal hydrorhiza, and possess short tentacles on the margin of the umbrella. The reproductive elements* "*are developed in the primitive hydrosoma without the intervention of free zoöids.*"—(Greene.)

In *Lucernaria* (fig. 23), which may be taken as the type of the order, the body is campanulate or cup-shaped, and is attached proximally at its smaller extremity by a hydrorhiza, which, however, like that of the *Hydra*, is not permanently fixed. When detached, the animal is able to swim with tolerable rapidity by means of the alternate contraction and expansion of the umbrella. Around the margin of the umbrella are tufts of short tentacular processes, and in its centre is a polypite with a quadrangular, four-lobed mouth. "In transverse section the polypite may be described as somewhat quadrilateral, with a sinuous outline, which expands at its four angles to form as many deep longitudinal folds, within which the simple generative bands are lodged."—(Greene.) Wide longitudinal canals are formed by septa passing from the walls of the polypite to the inner surface of the cup, and a circular canal runs immediately beneath the insertion of the tentacles. The reproductive elements are produced within the body of *Lucernaria* itself, without the intervention of any generative zoöid.

Fig 23.—Lucernariadæ. *Lucernaria auricula* attached to a piece of sea-weed (after Johnston).

ORDER II. PELAGIDÆ.—This order is defined as including *Lucernarida which possess a single polypite only, and an umbrella with marginal tentacles. The reproductive elements "are developed in a free umbrella, which either constitutes the primitive hydrosoma, or is produced by fission from an attached Lucernaroid."*—(Greene.)

Two types, therefore, exist in the *Pelagidæ*. The one type is represented by a fixed "trophosome," resembling *Lucernaria*, but distinguished from it by the fact that the generative elements are not developed in the primitive hydrosoma, but in a free "gonosome," which is produced for the purpose. The second type, represented by *Pelagia* itself, is permanently free, thereby differing from *Lucernaria*, which it approaches, on the other hand, in the fact that its generative elements are produced in its own umbrella without the intervention of free generative zoöids. *Pelagia*, however, differs considerably in structure from *Lucernaria*, and in all essential characters is not anatomically separable from a *Steganophthalmate Medusid*. The process of reproduction as displayed in the first section of the *Pelagidæ* will be considered when treating of that of the *Rhizostomidæ*, there being no important difference between the two, except as concerns the structure of the generative zoöids.

ORDER III. RHIZOSTOMIDÆ.—The members of this order are defined as being *Lucernarida*, in which *the reproductive elements are developed in free zoöids, produced by fission from attached Lucernaroids. The umbrella of the generative zoöids is without marginal tentacles, and the polypites are "numerous, modified, forming with the genitalia a dendriform mass depending from the umbrella*."—(Greene.)

The following is a brief summary of the life-history of a member of this extraordinary order (fig. 24), the illustration, however, representing the development of *Chrysaora*, one of the *Pelagidæ*, in which the phenomena are essentially the same. The embryo is a free-swimming, oblong, ciliated body, termed a "planula" (*a*), of a very minute size, and composed of an outer and inner layer enclosing a central cavity. The planula soon becomes pear-shaped, and a depression is formed at its larger end. "Next, the narrower end attaches itself to some submarine body, whilst the depression at the opposite extremity, becoming deeper and deeper, at length communicates with the interior cavity. Thus, a mouth is formed, around which may be seen four small protuberances, the rudiments of tentacula. In the interspaces of these four new tentacles arise; others in quick succession make their appearance, until a circlet of numerous filiform appendages, containing thread-cells, surrounds the distal margin of the 'Hydra-tuba' (*b*), as the young organism at this stage of its career has been termed by Sir J. G. Dalyell. The mouth, in the mean time, from being a mere quadrilateral orifice, grows and lengthens itself so as to constitute a true polypite, occupying the axis of the inverted umbrella or disc, which supports the marginal tentacles. The space between the walls of the polypite and umbrella is divided into longitudinal canals, whose relations to the rest of the organism, and, indeed, the whole structure of *Hydra-tuba*, closely resemble what may be seen in *Lucernaria*."—(Greene, *Manual of Cœlenterata*.) The *Hydra-tuba* thus constitutes the fixed "Lucernaroid," or the "trophosome" of one of the *Rhizostomidæ*. In height it is less than half an inch, but it possesses the power of forming, by gemmation, large colonies, which may remain in this condition for years, the organism itself being incapable of producing the essential elements of generation. Under certain circumstances, however, reproductive zoöids are produced by the following singular process (fig. 24). The *Hydra-tuba* becomes elongated, and becomes marked by a series of grooves or circular indentations, extending transversely across the body from a little below the tentacles to a little above the fixed extremity. At this stage the organism was described as new by Sars, under the name "*Scyphistoma*" (*c*). The annulations or constrictions go on deepening, and become lobed at their margin, till the *Scyphistoma* assumes the aspect of a pile of saucers, arranged one upon another with their con-

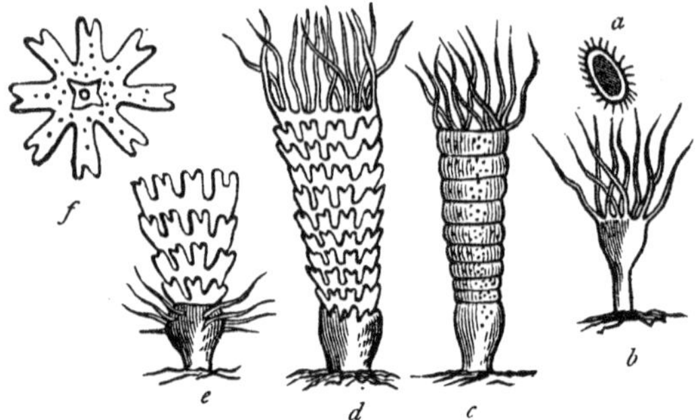

Fig. 24.—Development of Lucernarida (*Chrysaora*). *a* Ciliated embryo or "planula;" *b* Hydra-tuba; *c* Hydra-tuba undergoing fission, or "Scyphistoma;" *d* The fission still further advanced, constituting the "Strobila;" *e* A form still further advanced, in which a fresh circlet of tentacles has been developed near the base; *f* Free-swimming medusoid or "Ephyra," produced by fission from the hydra-tuba.

cave surfaces upwards. This stage was described by Sars under the name of "*Strobila*" (*d*). The tentacular fringe which originally surrounded the margin of the *Hydra-tuba* now disappears, and a new circlet is developed below the annulations, at a point a little above the fixed extrémity of the *Strobila* (*e*). "The disc-like segments above the tentacles gradually fall off, and, swimming freely by the contractions of the lobed margin which each presents, they have been described by Eschscholtz as true *Medusidæ* under the name of *Ephyræ*" (*f*). Each *Ephyra*, however, soon shows its true nature by becoming developed into a free-swimming reproductive body, usually of large size, with umbrella, hooded lithocysts and tentacles, constituting, in fact, a *Steganophthalmate Medusa*. The reproductive zoöid now swims freely by the contractions of its umbrella, and it eats voraciously and increases largely in size. The essential elements of generation are then developed in special cavities in the umbrella, and the fertilised ova, when liberated, appear as free-swimming, ciliated "planulæ," which fix themselves, become *Hydra-tubæ*, and commence again the cycle of phenomena which we have above described.

As regards the *size* of these reproductive zoöids as compared with the organism by which they are given off, it may be mentioned that the umbrella of *Cyanea arctica* has been found in one specimen to be seven feet in diameter, with tentacles more than fifty feet in length, the fixed *Lucernaroid* from which it was produced not being more than half an inch in height.

Fig. 25.—Hidden-eyed Medusæ. Generative zoöid of one of the *Pelagidæ* (*Chrysaora hysoscella*), after Gosse.

As regards the special structure of these gigantic reproductive bodies, considerable differences obtain between the *Rhizostomidæ* and that section of the *Pelagidæ* in which this method of reproduction is employed. In the *Pelagidæ*, namely, the generative zoöids possess a general, though chiefly mimetic, resemblance both to the genuine *Discophora* and to the free-swimming medusiform gonophores of so many of the *Hydrozoa*, and they have the following structure. Each (fig. 25) consists of a bell-shaped, gelatinous disc, the "umbrella," from the roof of which is suspended a large polypite, the lips of which are extended into lobed processes often of considerable length, "the folds of which serve as temporary receptacles for the ova in the earlier stages of their development." The polypite—manubrium or proboscis—is hollowed into a digestive sac, which communicates with a cavity in the roof of the umbrella, from which arise a series of radiating canals, the so-called "chylaqueous canals." These canals, which are never less than eight in number, branch freely and anastomose as they pass towards the periphery of the umbrella, where the entire series is connected by a circular marginal canal. This, in turn, sends tubular processes into the marginal tentacles, which are often of great length. Besides the tentacles, the margin of the umbrella is furnished with a series of peculiar bodies, termed "lithocysts," each of which is protected by a sort of process or hood derived from the ectoderm, and consists essentially of a combined "vesicle" and "pigment-spot," such as have been described as occurring in the *Medusidæ*. These marginal bodies likewise communicate with the chylaqueous canals. The reproductive elements "are lodged in saccular processes of the lower portion of the central cavity, immedi-

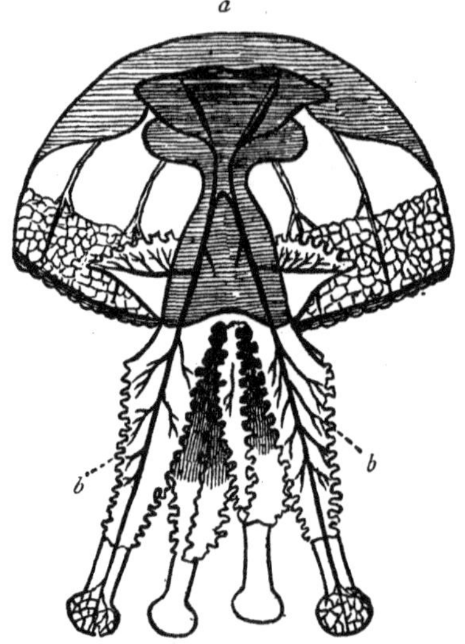

Fig. 26.—Rhizostomidæ. Generative zoöid of *Rhizostoma* (after Owen). *a* Umbrella; *bb* "Stomatodendra," covered with clavate tentacles and minute polypites; *cc* Anastomosing network of canals.

ately above the bases of the radiating canals, and, being usually of some bright colour, form a conspicuous cross shining through the thickness of the disc."—(Greene.)

In the *Rhizostomidæ* the reproductive zoöids (fig. 26) differ from those we have just described as occurring in the first section of the *Pelagidæ*, in not possessing tentacles on the margin of the umbrella, and in having the simple central polypite replaced by a composite dendriform process, which bears numerous polypites, projects far below the umbrella, and is thus described by Professor Huxley:—"In the *Rhizostomidæ* (fig. 26) a complex, tree-like mass, whose branches, the 'stomatodendra,' end in, and are covered by, minute polypites, interspersed with clavate tentacula, is suspended from the middle of the umbrella in a very singular way. The main trunks of the dependent polypiferous tree, in fact, unite above into a thick, flat, quadrate disc, the 'syndendrium,' which is suspended by four stout pillars, the 'dendrostyles,' one springing from each angle, to four corresponding points on the under surface of the umbrella, equidistant from its centre. Under the middle of the umbrella, therefore, is a chamber, whose floor is formed by the quadrate disc, whilst its roof is constituted by the under wall of the central cavity of the umbrella, and its sides are open. The reproductive elements are developed within radiating folded diverticula of the roof of this genital cavity."

It appears, finally, that amongst the old Pulmograde Acalephæ, or amongst what would commonly be called Jelly-fishes, we have the following distinct sets of beings, which resemble each other more or less closely in appearance, but differ in their true nature:—

1. Free medusiform gonophores of various *Corynida*, *Sertularida*, *Campanularida*, and the *Oceanic Hydrozoa.*

2. True *Medusidæ*, entirely resembling the former in anatomical structure, but differing in the fact that their ova do not give rise to a fixed zoöid, but to free-swimming organisms exactly like the parent hydrosoma (*Trachynemidæ* and *Æginidæ*).

3. *Hydrozoa*, which are provided with an "umbrella" (with all the peculiarities belonging to this structure), but which reproduce themselves without the intervention of free generative zoöids produced by fission (*Pelagia*).

4. The free generative zoöids of most of the *Pelagidæ*, with an umbrella and a single polypite, the primitive hydrosoma being fixed and sexless (*Aurelia*, *Cyanea*, &c.)

5. The free generative zoöids of the *Rhizostomidæ*, with an umbrella and a complex central tree bearing many polypites (*Rhizostoma*, *Cephea*, &c.)

Of these five classes of organisms, Nos. 1 and 2 constitute the Gymnophthalmate *Medusæ* of Professor E. Forbes, whilst Nos. 3, 4, and 5 are the Steganophthalmate *Medusæ* of the same naturalist.

SUB-CLASS V. GRAPTOLITIDÆ.—The organisms included at present under this head are all extinct, and they are in many respects so dissimilar, and their structure is so far from being entirely understood, that it is doubtful if any definition can be framed which will include *all* the supposed members of the family. The following definition, however, will include all the most typical Graptolites:—

Hydrosoma compound, occasionally branched, consisting of numerous polypites united by a cœnosarc; the latter being enclosed in a strong tubular polypary, whilst the former were protected by hydrothecæ. In the great majority of Graptolites the hydrosoma was certainly unattached; but in some aberrant forms—doubtfully belonging to the sub-class—there is reason to believe that the hydrosoma was fixed. The polypites are never separated from the cœnosarc by any partition. In many cases the hydrosoma was strengthened by a solid chitinous rod, the "solid axis," somewhat analogous to the chitinous rod recently described by Professor Allman in the singular Polyzoön, *Rhabdopleura.*

From the above definition, it will be seen that the nearest living allies to the Graptolites are the Sertularians. In point of fact, if we do not insist upon the presence of a "solid axis" as part of the definition, the Graptolites differ from the Sertularians in no essential point, save that the hydrosoma is always attached in the latter, and was certainly free in the most typical examples of the former. Indeed, certain forms at present placed among the *Graptolites*—such as *Ptilograpsus* and *Dendrograpsus*—are so similar to some living Sertularians, that it might be well to remove them altogether from the *Graptolitidæ*, and to regard them as extinct representatives of the *Sertularida.*

As regards the value of the "solid axis" as an element in defining Graptolites, we fear that much stress cannot be laid upon its presence or absence. It is true that it is present in all the most characteristic members of the sub-class, but it seems to be certainly absent in some—*e.g.*, in *Retiolites Geinitzianus*, and in all species of *Rastrites*—and there do not seem to be sufficient grounds for excluding these from the *Graptolitidæ* on this account alone.

Taking such a simple Graptolite as *G. sagittarius* (fig. 27, 1) as the type of the sub-class, the hydrosoma is found to consist of the "solid axis," the "common canal," and the "cellules." The entire polypary is corneous and flexible, and the solid axis is a cylindrical fibrous rod, which gives support to the entire organism, and is often prolonged beyond one or both ends of the hydrosoma. The common canal is a tube which encloses the cœnosarc, and gives origin to a series of cellules, these being little cups corresponding to "hydrothecæ," and enclosing the polypites. Not only are the essential details of the structure—with the exception of the solid axis—strictly comparable with that of a Sertularian, but there is a good evidence, as shown by Hall and the author, that the reproductive process was also carried on in a manner similar to what we have seen in the other *Hydroida*—namely, by generative buds or gonophores.

No *Graptolite*, however, has hitherto been certainly proved to have been fixed by a "hydrorhiza," and it is only in certain aberrant forms that there are any traces of a "hydrocaulus."

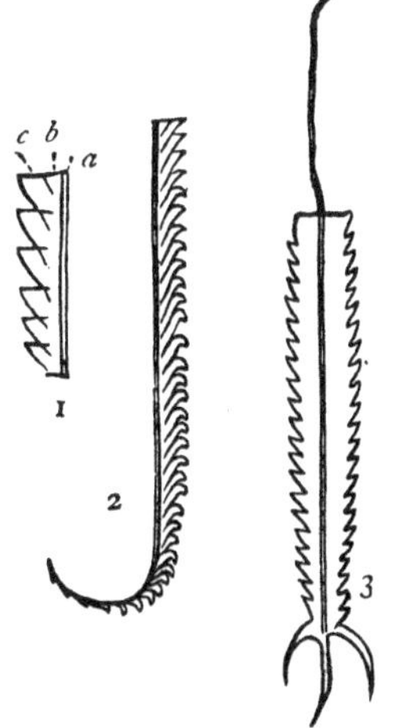

Fig. 27.—Morphology of Graptolites. 1. Portion of *Graptolites sagittarius* enlarged; *a* Solid axis; *b* Common canal; *c* Cellules. 2. Monoprionidian Graptolite (*G. argenteus*). 3. Diprionidian Graptolite (*Diplograpsus pristis*, variety with long basal spines).

Besides the simple forms of Graptolites with a row of cellules on one side (monoprionidian) (fig. 27, 2,) there are others with a row of cellules on each side (diprionidian) (fig. 27, 3). Many other curious modifications are known; but there is only another peculiarity which is worthy of notice here. This is the occurrence in several genera of a basal corneous disc or cup, which is probably the homologue of the "float" or "pneumatophore" of the Physophoridæ. (For distribution of Graptolites see Distribution of Hydrozoa in Time.)

As regards their mode of occurrence, Graptolites are usually found as glistening, pyritous impressions, with a silvery lustre. In some cases, however, they are found in relief.

CHAPTER XII.

DISTRIBUTION OF THE HYDROZOA.

I. Distribution of Hydrozoa in Space.—The genera of *Hydrozoa* have a wide distribution, the mode of reproduction amongst the fixed forms being such as to insure their extension over considerable areas. The various species of *Hydra* are of common occurrence in the fresh waters of Europe. *Cordylophora*, the sole remaining fresh-water genus, has not been found to occur out of the north temperate zone. All the other *Hydrozoa*, without a known exception, are marine in their habits. The fixed forms—viz., the *Corynida*, *Sertularida*, and *Campanularida*—are represented more or less abundantly in almost all seas, extending from the littoral zone to considerable depths. The oceanic *Hydrozoa*, *Calycophoridæ* and

Physophoridæ, are chiefly characteristic of tropical seas; but they are found also in the Mediterranean, and even in seas not far from, or even within, the Arctic circle.

II. Distribution of Hydrozoa in Time.—With the exception of the impression of a *Medusa* said to have been observed by Professor Agassiz in the fine-grained lithographic slate of Solenhofen (Oolite), there are no fossil remains which would be universally conceded to be of a *Hydrozoal* nature. The *Oldhamia* of the Cambrian rocks of Ireland has, indeed, been regarded as belonging to the *Hydrozoa;* but it is believed by Mr Salter to be really a plant. It consists of a main stem with numerous secondary branches, springing from the axis in an umbellate manner, but exhibiting no traces of hydrothecæ.

The occurrence of *Corynida* in a fossil condition can hardly be said to be free from doubt. Remains probably referable to this order have been, however, recently discovered in the Palæozoic Rocks. The oldest of these was described by the author some years ago from the Lower Silurian rocks of Dumfriesshire under the name of *Corynoides.* More lately a form called *Palæocoryne* has been described from the Carboniferous rocks of Scotland.

The *Sertularida* and *Campanularida* are not certainly known to occur in a fossil condition. The fossils called *Dendrograpsus, Callograpsus, Ptilograpsus,* and *Dictyonema,* all at present placed amongst the *Graptolites,* are, however, not improbably truly referable to the *Sertularida.*

There can be little doubt but that the large and singular family of the *Graptolitidæ* should really be looked upon as extinct *Hydrozoa,* though good authorities still place them amongst the *Polyzoa.* As regards their distribution two facts are chiefly noticeable. In the first place, no Graptolite, except the doubtful genus *Dictyonema,* has hitherto been found to occur above the Silurian rocks. The Graptolites may therefore be regarded as characteristic fossils of the Silurian period. Secondly, the diprionidian Graptolites, or those with a row of cellules on each side (genera *Diplograpsus, Climacograpsus,* and *Dicranograpsus*), have never yet been certainly shown to occur above the horizon of the *Lower* Silurian rocks. The common genus *Didymograpsus* (comprising the "twin" Graptolites, fig. 28) is still more characteristic of the Lower Silurian period. In *Didymo-*

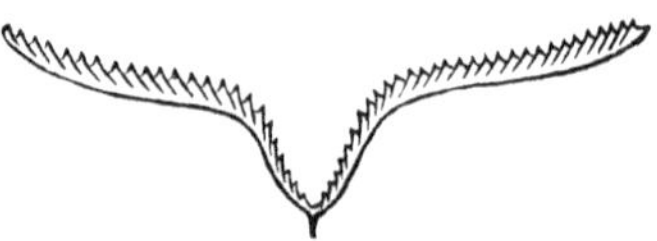

Fig. 28.—*Didymograpsus V-fractus.*

grapsus the polypary consists of two lateral symmetrical branches, with cellules on one side only, springing from a central point or base, which is usually marked by a little spine or "radicle."

CHAPTER XIII.

ACTINOZOA.

1. General Characters of the Actinozoa. 2. Characters of the Zoantharia. 3. Zoantharia Malacodermata. 4. Zoantharia Sclerobasica. 5. Zoantharia Sclerodermata.

Class II. Actinozoa. — The *Actinozoa* are defined as *Cœlenterata with a differentiated digestive sac opening below into the somatic cavity, but separated from the body-walls by an intervening "perivisceral space," which is divided into a series of compartments by vertical partitions, or "mesenteries," to the faces of which the reproductive organs are attached.*

The *Actinozoa* (fig. 30), therefore, differ fundamentally from the *Hydrozoa* in this, that whereas in the latter the digestive cavity is identical with the somatic cavity, in the former there is a distinct digestive sac, which opens, indeed, into the somatic cavity, but is, nevertheless, separated from it by an intervening perivisceral space. As a result of this, the body of a typical *Actinozoön* (fig. 29), exhibits on transverse section two concentric tubes, one formed by the digestive sac, the other by the parietes of the body; whereas the transverse section of a *Hydrozoön* exhibits but a single tube, formed by the walls of the combined digestive and somatic cavity.

Histologically, the tissues of the *Actinozoa* are essentially the same as those of the *Hydrozoa,* consisting of the two fundamental layers, the "ectoderm" and the "endoderm." In the *Actinozoa,* however, there is a much greater tendency to a differentiation of these into specialised structures, and in some members of the class muscular fibres are well developed. The ectoderm, especially, shows a tendency to break up into two layers, which are differentiated in opposite directions from an intermediate zone, and are termed by Huxley the "ecderon" and "enderon," corresponding respectively to the epidermis and derma of man. Cilia are often present, especially in the interior of the somatic cavity, where they serve to promote a circulation of the digestive fluids contained therein. The sole digestive apparatus in the *Actinozoa* consists of a tubular stomach-sac, which communicates freely with the outer world

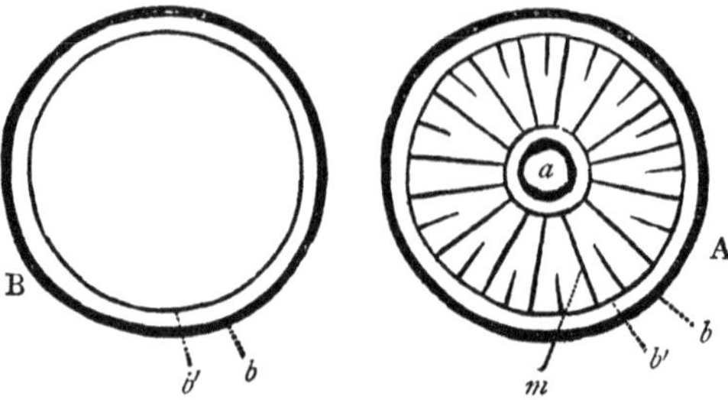

Fig. 29.—A Transverse section of an *Actinozoon*. *a* Digestive sac; *b* Wall of the body; *m* Mesenteries connecting the stomach with the body-walls, and dividing the space between them into a number of vertical compartments. B Transverse section of a *Hydrozoön*, showing the single tube formed by the walls of the body.

by means of the mouth, and opens inferiorly directly into the general body-cavity. In most, the "perivisceral space" between the body-walls and the digestive sac is subdivided into compartments by a series of vertical lamellæ, which are called the "mesenteries" (fig. 30, *b*). Upon the faces of these

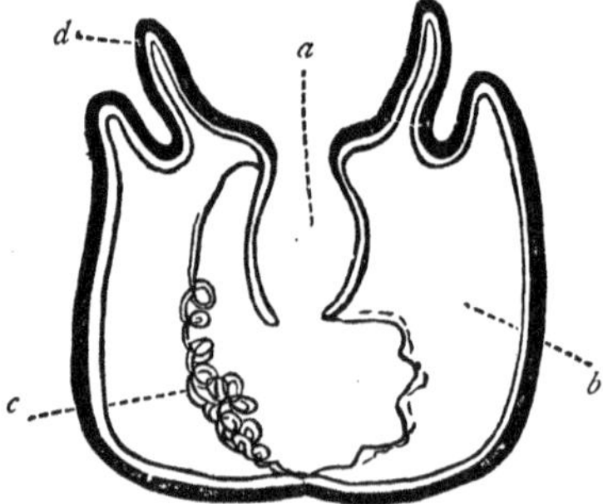

Fig. 30.—Morphology of Actinozoa. Diagrammatic vertical section of *Actinia*. *a* Stomach-sac; *b* Mesentery; *c* Craspedum; *d* Tentacle.

are borne the reproductive organs in the form of band-like ovaria or spermaria.

Thread-cells, often of very complicated structure, are almost universally present, some of the *Ctenophora* having been asserted to be without them; and some of the *Actinozoa* are able to sting very severely.

A nervous system has not yet been proved to exist in any of the *Actinozoa*, except in the *Ctenophora*, and in none are there any traces of a vascular system.

Distinct reproductive organs occur in all the *Actinozoa*, but these are internal, and are never in the form of external processes as in the *Hydrozoa*. Sexual reproduction occurs in all the members of the class, but in many forms gemmation or fission constitutes an equally common mode of increase. Some *Actinozoa*, therefore, such as the common Sea-anemones, are simple organisms; whilst others, such as the reef-building corals, are composite, the act of gemmation or fission giving rise to colonies composed of numerous zoöids united by a cœnosarc. In these cases the separate zoöids are termed "polypes," the term "polypite" being restricted to the *Hydrozoa*. In the simple *Actinozoa*, however, the term "polype" is employed to designate the entire organism. In other words, the "actinosoma," or entire body of any *Actinozoön*, may be composed of a single "polype," or of several such, produced by a process of continuous gemmation or fission, and united by a common connecting structure, or cœnosarc.

Most of the *Actinozoa* are permanently fixed; some, like the Sea-anemones, possess a small amount of locomotive power; and one order, the *Ctenophora*, is composed of highly active, free-swimming organisms. Some of the *Actinozoa* are unprovided with any hard structure or support, as in the Sea-anemones and in all the *Ctenophora;* but a large number secrete a calcareous or horny, or partially calcareous and partially horny, framework or skeleton, which is termed the "coral," or "corallum."

The *Actinozoa* are divided into four orders—viz., the *Zoantharia*, the *Alcyonaria*, the *Rugosa*, and the *Ctenophora;* but the last is sometimes placed amongst the *Hydrozoa*, and it has been recently proposed to remove the *Rugosa* also to the same class.

ORDER I. ZOANTHARIA.—The *Zoantharia* or "Helianthoid Polypes" are defined *by the disposition of their soft parts in multiples of five or six, and by the possession of simple, usually numerous, tentacles. There may be no corallum, or rarely a "sclerobasic" one. Usually there is a "sclerodermic" corallum, in which the septa in each corallite, like the mesenteries, are arranged in multiples of five or six.*

The *Zoantharia* are divided into three sub-orders, the *Zoantharia malacodermata*, the *Z. sclerobasica*, and the *Z. sclerodermata;* according as the corallum is entirely absent or very rudimentary, is "sclerobasic," or is "sclerodermic."

SUB-ORDER I. ZOANTHARIA MALACODERMATA.—In this section of the *Zoantharia* there is either no corallum or a very rudimentary one, in the form of a few scattered spicules. The "actinosoma" is usually composed of but a single polype. (The term "actinosoma" is a very convenient one to express in the *Actinozoa* what "hydrosoma" expresses in the *Hydrozoa*, namely, the entire organism, whether simple or compound.)

There are three families in this section, of which the *Actinidæ* will require a somewhat detailed examination, since they may be taken as typical of the entire class of the *Actinozoa*.

FAMILY I. ACTINIDÆ.—The members of this family are commonly known as Sea-anemones, and are distinguished by having no evident corallum, by being rarely compound, and by having the power of locomotion.

The body of a Sea-anemone (fig. 31, *a*) is a truncated cone, or a short cylinder, termed the "column," and is of a soft, leathery consistence. The two extremities of the column are

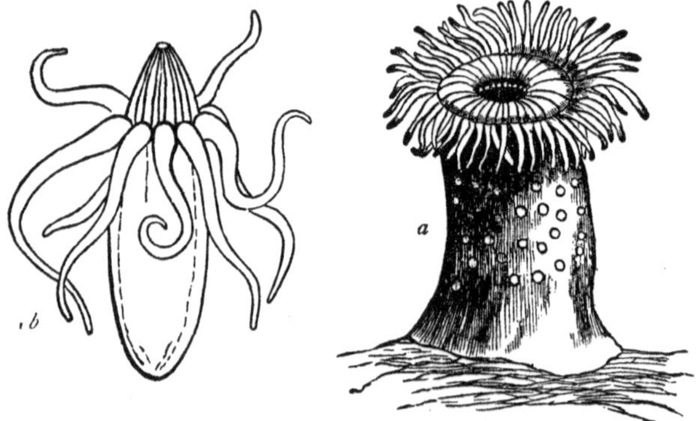

Fig. 31.—Morphology of Actinidæ. *a Actinia rosea*; *b Arachnactis albida*. (After Gosse.)

termed respectively the "base" and the "disc," the former constituting the sucker, whereby the animal attaches itself at will, whilst the mouth is situated in the centre of the latter. In a few cases (*Cerianthus* and *Peachia*) the centre of the base is perforated, but the object of this arrangement is unknown. Between the mouth and the circumference of the disc is a flat space, without appendages of any kind, termed the "peristomial space." Round the circumference of the disc are placed numerous tentacles, usually retractile, arranged in alternating rows, and amounting to as many as 200 in number in the common *Actinia*. The tentacles are tubular prolongations of the ectoderm and endoderm, containing diverticula from the somatic chambers, and sometimes having apertures at their free extremities. The mouth leads directly into the stomach, which is a wide membranous tube, opening by a large aperture into the general body-cavity below, and extending about half-way between the mouth and the base. The wide space between the stomach and column-wall is subdivided into a number of compartments by radiating vertical lamellæ, termed the "primary mesenteries," arising on the one hand from the inner surface of the body-wall, and attached on the other to the external surface of the stomach. As the stomach is considerably shorter that the column, it follows that the inner edges of the primary mesenteries below the stomach are free; and these free edges, curving at first outwards and then downward and inwards, are ultimately attached to the centre of the base. Besides the primary mesenteries, there are other lamellæ which also arise from the body-wall, but which do not reach so far as the outer surface of the stomach, and are called "secondary" and "tertiary" mesenteries, according to their breadth. The reproductive organs are in the form of reddish bands, which contain ova and spermatozoa, and are situated on the faces of the mesenteries. Most of the *Actiniæ* are diœcious—that is to say, the same individual does not develop both ova and spermatozoa. Along the free margins of the mesenteries there also occur certain singular convoluted cords, charged with thread-cells, and termed "craspeda," the function of which is not yet understood. It is believed, however, that the apertures, termed "cinclides," in the column-walls of some of the *Actinidæ*, are for the emission of the craspeda. No traces of a nervous system have as yet been proved to exist in any *Actinia*.

The embryo of the *Actiniæ* is a free-swimming ciliated body, at first rounded, but afterwards somewhat ovate. The rudimentary mouth is soon marked out by a depression at the larger extremity; thread-cells appear as a layer in the ectoderm; a fold is prolonged inwards from the mouth to form the digestive sac; and the primitive tentacles are at first either five or six in number, but usually double themselves rapidly.

FAMILY II. ILYANTHIDÆ.—In this family there is no corallum, and the polypes are single and free, with a rounded or tapering base (fig. 31, *b*). *Ilyanthus* is in all essential respects identical with the ordinary *Actiniæ*, but it is of a pointed or conical shape, the base being much attenuated, though whether its habit of life is free or not, is a matter of some uncertainty.

Arachnactis is certainly free, and, according to Professoı E. Forbes, it can not only swim like a jelly-fish, but "it can convert its posterior extremity into a suctorial disc, and fix itself to bodies in the manner of an *Actinia*." It is by no means certain, however, that *Arachnactis* is a mature form, and there is some reason to suppose that it is merely the young stage of some at present unknown *Actinozoön*.

FAMILY III. ZOANTHIDÆ.—In the *Zoanthidæ* there is a spicular corallum, and the polypes are attached by a fleshy or coriaceous base or cœnosarc. In *Zoanthus* the separate polypes closely resemble small *Actiniæ*, but they are united together at their bases by a thin fleshy cœnosarc.

SUB-ORDER II. ZOANTHARIA SCLEROBASICA.—The members of this sub-order are always composite, and always possess a corallum, but this is "sclerobasic," and there are no spicular tissue-secretions.

It appears advisable to explain here what is understood by the terms "sclerobasic" and "sclerodermic," as applied to corals. The "corallum" is the term which is applied to the hard structures deposited by the tissues of any *Actinozoön*, many of which are so familiarly known as "corals." Usually the corallum is composed of carbonate of lime; but it may be corneous, or partly corneous and partly calcareous. Whatever their composition may be, all coralla may be divided into two sections, termed respectively "sclerobasic" and "sclerodermic," which must be carefully distinguished from one another. The "sclerobasic" corallum, of which the red coral of commerce may be taken as the type, is in reality an exoskeleton, somewhat analogous to the shell of a *Crustacean*, being a true tegumentary secretion. At the same time it is not a *shell* or external envelope, but it forms an axis, upon which the entire actinosoma is spread. The actinosoma, in fact, is *inverted*, and the "sclerobasis" is secreted by the *outer* surface of the ectoderm. The sclerobasic corallum is therefore truly "outside the bases of the polypes and their connecting cœnosarc, which, at the same time, receive support from the hard axis which they serve to conceal."—(Greene.) Upon this view the sclerobasis is termed "foot-secretion" by Mr Dana. In other words, the sclerobasic coral is a hard skeleton which belongs solely to the *cœnosarc* of the actinosoma, and which can therefore be produced by a compound organism only.

The "sclerodermic" corallum, on the other hand, is secreted *within* the bodies of the polypes, apparently by the *inner* layer of the ectoderm—the "enderon" of Huxley—and it is therefore termed "tissue-secretion" by Mr Dana. In the sclerodermic corallum each polype has a complete skeleton of its own, and the entire coral may consist of one such skeleton, or of several such united by the calcareous matter of the cœnosarc.

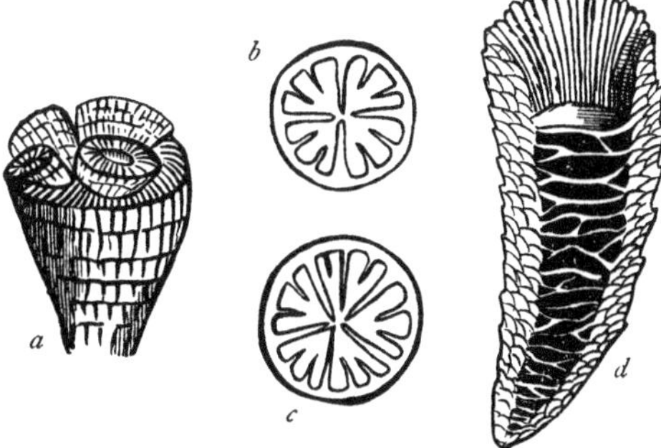

Fig. 32.—Morphology of Corals *a* Cup of *Acervularia ananas*, showing calicular gemmation, enlarged; *b* Diagram of Rugose Coral (*Polycœlia profunda*), showing the quadripartite arrangement of the septa; *c* Diagram of a recent coral, showing the sextuple arrangement of the septa; *d* Vertical section of *Campophyllum flexuosum*, showing tabulæ.

A sclerodermic corallum, therefore, like the animal which produces it, may be simple or composite, according as it is produced by a single polype or by several united by a cœnosarc. It consists, therefore, of a single calcareous cup, or "corallite;" or of several such united by a common calcareous bond or basis, the "cœnenchyma." Taking a single "corallite" (fig. 32, *d*) as the type, we find that it shows its origin and nature plainly in its form. It consists of a cylindrical or conical tube of carbonate of lime, the outer wall of which is called the "theca." The upper part of the space included by the "theca" is vacant, and it is termed the cup or "calice;" but the lower part is subdivided into a series of chambers, or "loculi," by a series of radiating, vertical, calcareous plates, which are called the "septa" (fig. 32, *b*). The septa extend from the inner surface of the theca towards its centre, where they usually unite to form an axial column, called the "columella." Many of the septa, however, do not reach the centre, but stop short at some distance from the columella, often being broken up into upright pillars, called "pali." The parts thus described as essentially composing a corallite in a typical sclerodermic corallum are related in the most obvious manner to the soft structures of the animal by which they are secreted. Thus, the "theca" clearly corresponds to the "column-wall," or the general wall of the body; the "columella," when present, corresponds to "that part of the enderon which forms the floor of the somatic cavity below the digestive sac;" whilst the "septa" correspond to the "mesenteries," and, like them, are called "primary" and "secondary," according as they reach the columella or fall short of it. When there are several corallites, the bond of union between them, the "cœnenchyma," is secreted by the "cœnosarc," to which it corresponds. In many *Actinozoa*, however, the sclerodermic corallum is not present in the typical form above described, but simply in the form of calcareous spicules or nodules scattered through the tissues of the animal. There are, also, members of the class in which both a sclerodermic and a sclerobasic corallum are present, the latter constituting the main skeleton, whilst the former is represented by scattered spicules. The coral tissue itself is known as "sclerenchyma," and it varies considerably in texture, being sometimes extremely compact, and at other times very loosely put together.

From what has been said it will be seen that a sclerobasic corallum can easily be distinguished from a sclerodermic by inspection; the former (fig. 33, *b*) being usually more or less

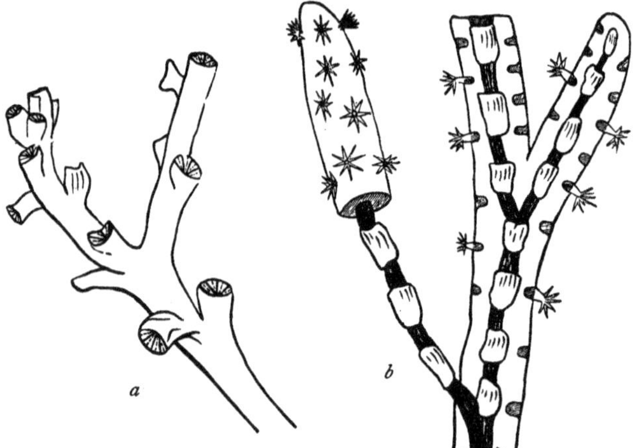

Fig. 33.—Sclerodermic and Sclerobasic Corals. *a* Portion of branch of *Dendrophyllia nigrescens*, a sclerodermic coral (after Dana); *b* Longitudinal section of *Isis hippuris*, a sclerobasic coral, exhibiting the external bark or cœnosarc, with its embedded polypes, supported by the internal axis or skeleton (after Jones).

smooth, and being invariably devoid of the cups or receptacles for the separate polypes, which are always present in the latter (fig. 33, *a*). The more important variations of detail which occur in both classes of corals will be noticed under the different families in which they occur. It only remains to add that doubt has been thrown by eminent zoologists upon the validity of the general distinction between sclerobasic and sclerodermic corals, as above defined.

Returning now to the *Zoantharia Sclerobasica*, we find the sub-order to contain the two families of the *Antipathidæ* and the *Hyalonemadæ* (or *Hyalochætidæ*). Of these the *Antipathidæ* are chiefly noticeable because of their likeness to some of the *Gorgonidæ*, from which, however, they are readily distinguished by the fact that the number of their tentacles is a multiple of six, whereas in the latter it is a multiple of four. *Antipathes* itself possesses a horny sclerobasic corallum, which may be simple or branched, and is covered with numerous small polypes, united together by a cœnosarc, and possessing six tentacles each.

The second family, that of the *Hyalonemadæ*, contains the so-called "Glass-zoophytes," the true nature and position of which has been a subject of much controversy. By Dr Gray the *Hyalonemadæ* are believed to be true *Actinozoa*, and he defines them as follows:—"Social Zoanthoid polypes secreting a central, siliceous, internal, axial coil for their support. The upper half of the coil covered by a uniform cylindrical bark, regularly studded with retractile polypes." The lower portion of the siliceous rope-like axis, which looks exactly like a skein of threads of glass, is sunk in the sand at the bottom of the sea. The upper portion of the *Hyalonema* is often occupied by a cup-shaped sponge, called *Carteria*, which Dr Gray believes to be a parasitic growth. By Professors Löven, Perceval Wright, Wyville Thomson, and others, the sponge *Carteria* is looked upon as the true artificer of the siliceous rope, and the polypes are regarded as parasitic, and as referable to *Palythoa*. This last view, by which *Hyalonema* would be placed amongst the siliceous sponges, appears, upon the whole, to be most probably the correct one. In this case there is no *Actinozoön*, as far as is yet known, which possesses the power of secreting a siliceous skeleton, in this respect presenting a striking contrast to the *Protozoa*.

SUB-ORDER III. ZOANTHARIA SCLERODERMATA.—The members of this sub-order include the great bulk of the coral-producing or "coralligenous" zoophytes of recent seas. They are defined by the possession of a sclerodermic corallum, the parts of which are arranged in multiples of five or six. The actinosoma may be simple, consisting of a single polype, or it may be composite, consisting of several polypes united by a cœnosarc.

The divisions of the sub-order are founded upon the nature of the corallum, for the due comprehension of which it will be necessary to consider some points in connection with these structures somewhat more minutely. As already described, a typical corallite consists of an outer wall or "theca," with a cup or "calice" above, and divided below into numerous chambers or "loculi" by vertical partitions or "septa." Often the larger or "primary" septa coalesce centrally to form a median calcareous rod or "columella." The chief additional structures to be remarked are what are known as "tabulæ," and "dissepiments." The "tabulæ" (fig. 32, *d*) are transverse plates or floors running at right angles to the axis of the corallite, and dividing the theca into so many horizontal compartments or stories, each of which is vertically subdivided by the septa, when these exist. As a rule, however, the septa are absent when there are tabulæ, though the two structures coexist in many extinct corals. The "dissepiments" are incomplete transverse plates, which, "growing from the sides of the septa, interfere, to a greater or less extent, with the perfect continuity of the loculi."—(Greene.) The septa, too, are often furnished with styliform or spine-like processes growing from their sides, which often meet so as to form "transverse props extending across the loculi like the bars of a grate, and termed 'synapticulæ.'"

The *Zoantharia Sclerodermata* are divided into the four following groups, founded upon the characters of the corallum:—

1. *Tabulata.*—Septa rudimentary, or entirely absent; tabulæ well developed, and dividing the visceral chamber into a series of stories.

2. *Perforata.* — Septa well developed; dissepiments rudimentary; no tabulæ. Corallum composed of porous sclerenchyma.

3. *Aporosa.* — Septa well developed, lamellar; no tabulæ. Corallum composed of compact, imperforate sclerenchyma.

4. *Tubulosa.*—Septa indicated by mere striæ; thecæ pyriform, occasionally united by a basal cœnenchyma.

GEMMATION AND FISSION AMONGST CORALS. — As regards the modes in which the composite corals are produced, the following is a summary of Professor Greene's remarks upon this subject. (See *Cœlenterata*, p. 185 *et seq.*) The production of the composite *Actinozoa* is effected either by gemmation or by fission. In the former method three varieties have been distinguished, termed respectively "basal," "parietal," and "calicular" gemmation.

In basal gemmation the mode of increase is by means of a rudimentary cœnosarc, which is put forth by the original polype, and from which the young polype-buds are produced. It "affords very different products according as the cœnosarc remains soft, or deposits a cœnenchyma; appears under the form of stolons, or of stouter connecting stems; or even spreads out in several directions as a continuous horizontal expansion;" in which last case the youngest polypes are, of course, those nearest to the periphery of the mass.

The parietal mode of gemmation is the commonest, and it gives rise chiefly to dendroid, or tree-like, corals. In this method the buds are produced from the sides of the original polype, and they often repeat the process indefinitely.

Calicular gemmation is not known to occur in any recent coral, but it was a common mode of increase amongst extinct forms. In this method "the primitive polype sends up from its oral disc two or more similar buds; these, in their turn, produce other young polypes, and thus the process is repeated until an inverted pyramidal mass of considerable size is produced, all the parts of which rest upon the narrow base of the first budding polype (fig. 32, *a*). Fission in the *Actinozoa* differs from gemmation chiefly in the fact, that the polypes produced fissiparously resemble one another in organisation, and often in size, as soon as they become distinct. In gemmation, on the other hand, the polype-bud consists primarily of a mere process of ectoderm and endoderm, enclosing a cæcal process of the somatic cavity, and a mouth and other structures are at first wanting. Amongst the coralligenous *Actinozoa* fission is usually effected by "oral cleavage," the divisional groove commencing at the oral disc, and deepening to a certain extent, the proximal extremity always remaining undivided. More rarely, fission "is effected by the separation of small portions from the attached base of the primitive organism, whose form and structure they subsequently, by gradual development, tend to assume."

"The coral-structures which result from a repetition of the fissiparous process are of two principal kinds, according as they tend most to increase in a *vertical* or in a *horizontal* direction. In the first of these cases the corallum is *cæspitose*, or tufted, convex on its distal aspect, and resolvable into a succession of short diverging pairs of branches, each resulting from the division of a single corallite." In the second case the coral becomes *lamellar*. "Here the secondary corallites are united throughout their whole height, and disposed in a linear series, the entire mass presenting one continuous theca."

Both these forms of corallum "are liable to become *massive* by the union of several rows or tufts of corallites throughout the whole or a portion of their height. An illustration of this is afforded by the large *gyrate* corallum of *Meandrina*, over the surface of whose spheroidal mass the calicine region of the combined corallites winds in so complex a manner as at once, to suggest that resemblance to the convolutions of the brain which its popular name of Brain-stone Coral has been devised to indicate."

CHAPTER XIV.

ALCYONARIA.

ORDER II. ALCYONARIA.—The second great division of living *Actinozoa* is that of the *Alcyonaria*, defined by the possession of *polypes with eight pinnately-fringed tentacles, the mesenteries and somatic chambers being also some multiple of four. The corallum, when present, is usually sclerobasic, or spicular; if "thecæ" are present, as is rarely the case, there are no septa.*

The *Alcyonaria* or "Asteroid Polypes" differ numerically from the *Zoantharia* in having their soft parts arranged in multiples of *four*, instead of *five or six*, as in the latter. Their tentacles, too, are pinnate, and are not simply rounded. Numerically the *Alcyonaria* agree with the extinct order *Rugosa*, but the latter invariably possess a well-developed sclerodermic corallum, the thecæ of which exhibit either septa or tabulæ, or both combined.

With the exception of the single genus *Haimeia*, the *Alcyonaria* are all composite, their polypes being connected together by a common cœnosarc, "through which permeate prolongations of the somatic cavity of each, forming a sort of canal system, whose several parts freely communicate," and permit of a free circulation of nutrient fluids. As a rule, the entire colony forms a lobate or branched mass. Anatomically the polypes of the *Alcyonaria* do not differ in any essential particular from those of the *Zoantharia;* the numerical distinction being the one by which they are chiefly separated from one another. The *Alcyonaria* are divided into four families, viz., the *Alcyonidæ*, the *Tubiporidæ*, the *Pennatulidæ*, and the *Gorgonidæ*.

FAMILY I. ALCYONIDÆ.—This family is characterised by the possession of a fixed actinosoma, which is provided with a sclerodermic corallum in the form of calcareous spicula embedded in the tissues. The spicules are mostly fusiform in shape, and are generally present both in the polypes themselves and in the connecting cœnosarc; but there is no central solid axis.

Alcyonium may be taken as the type of the family, and it is well known to fishermen under the name of "Dead-men's fingers." It forms spongy-looking, orange-coloured crusts or lobate masses, which are attached to submarine objects, and are covered with little stellate apertures, through which the delicate polypes can be protruded and retracted at will. The polypes communicate with one another by an anastomosing system of aquiferous tubes, and the corallum is in the form of cruciform, calcareous spicula scattered through its substance. In the allied *Sarcodictyon* the actinosoma is creeping and linear.

FAMILY II. TUBIPORIDÆ.—In the *Tubiporidæ*, or "organ-pipe corals," of which *T. musica* is a familiar example, there is a well-developed sclerodermic corallum, with thecæ, but without septa. The corallum is composed of a number of bright-red, tubular, cylindrical thecæ, which are united together externally by horizontal plates or floors, which are termed "epithecæ," and represent external tabulæ. The polypes are usually bright green in colour, and possess eight tentacles each.

FAMILY III. PENNATULIDÆ.—The *Pennatulidæ*, or "Sea-pens," are defined by their free habit, and by the possession of a sclerobasic, rod-like corallum, sometimes associated with sclerodermic spicules.

Pennatula (fig. 34), or the "Cock's-comb," consists of a free cœnosarc, the upper end of which is fringed on both sides with feather-like lateral pinnæ, which bear the polypes; whilst its proximal end is smooth and fleshy, and is probably sunk in the mud of the sea-bottom. This latter portion of the cœnosarc is likewise strengthened by a long, slender, styliform sclerobasis, resembling a rod in shape, whilst spicula occur also in the tentacles and ectoderm. The general colour of *Pennatula* is a deep reddish purple, the proximal extremity of the cœnosarc being orange-yellow. Our British species (*Pennatula phosphorea*) varies from two to four inches in length, and is found on muddy bottoms in tolerably deep water. Its specific name is derived from the fact that it phosphoresces brilliantly when irritated.

In *Virgularia* (fig. 35), which, like *Pennatula*, occurs not uncommonly in British seas, the actinosoma is much longer and more slender than in the preceding, and the polype-bearing fringes are short. The polypes have eight tentacles. The sclerobasis is in the form of a long calcareous rod, like a knitting-needle, and part of it is usually naked. No spicula are found in the tissues of *Virgularia*. In the nearly-allied *Pavonaria* the polype-mass is quadrangular in shape.

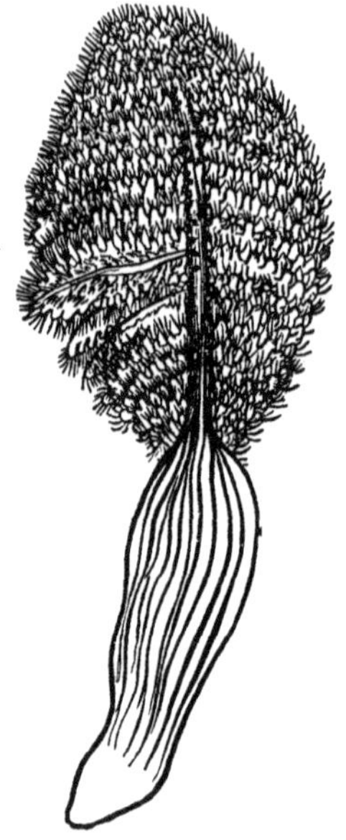

Fig. 34.—*Pennatula phosphorea* (after Johnston).

Fig. 35. — Pennatulidæ. *Virgularia mirabilis*. *a* A portion of the stem in the living condition, enlarged; *b* Portion of the stem in its dead condition.

FAMILY IV. GORGONIDÆ. — In the *Gorgonidæ*, or "Sea-shrubs," there is an arborescent cœnosarc permanently rooted and provided with a grooved, or sulcate, branched sclerobasis, which is sometimes associated with true tissue-secretions, termed "dermo-sclerites."

The sclerobasis of the *Gorgonidæ* varies a good deal in its composition. In some it is corneous, and these have often been confounded with the *Antipathidæ*, amongst the *Zoantharia*. The distinction, however, between them is easy, when it is remembered that the polypes in the *Gorgonidæ* have tentacles in multiples of *four*, whilst in the *Antipathidæ* they are in *sixes*. The sclerobasis, too, in the former is always marked by grooves, whereas in the latter it is always either smooth or spinulous.

In *Isis* (fig. 33, *b*) and *Mopsea* the sclerobasis consists of alternate calcareous and horny segments, branches being developed in the former from the calcareous, and in the latter from the horny segments.

In *Corallium rubrum*, the "red coral" of commerce, the sclerobasis is unarticulate, or unjointed, and is entirely calcareous. It is the most familiar member of the family, and is largely imported for ornamental purposes. Red coral consists of a branched densely calcareous sclerobasis, which is finely grooved upon its surface, and is of a bright-red colour. The corallum is invested by a cœnosarc, also of a red colour, which is studded by the apertures for the polypes, which are white, and possess eight pinnately-fringed tentacles. The entire cœnosarc is channelled out by a number of anastomosing canals, which communicate with the somatic cavities of the polypes, and are said to be in direct communication with the external medium by means of numerous perforations in their walls. The entire canal system is filled with a nutrient fluid, containing corpuscles, and known as the "milk."

CHAPTER XV.

RUGOSA.

Order III. Rugosa.—The members of this order are entirely extinct, and, with the exception of *Holocystis elegans* from the Lower Cretaceous rocks, and a few more modern forms, are not known to occur in deposits younger than the Palæozoic epoch. With the soft parts of the *Rugosa* we are, of course, entirely unacquainted, and the definition of the order must therefore be founded upon the characters of the corallum. The corallum in the *Rugosa* is highly developed, sclerodermic, with true thecæ, and often presenting both septa and tabulæ combined. The septa are in multiples of *four* (fig. 32, *b*), unlike the recent sclerodermic coralla, in which they are in multiples of *five* or *six*. There is, further, no true cœnenchyma. Some of the Rugosa are simple; but others are composite, increasing either by parietal or by calicular gemmation.

Recently it has been shown that some very abnormal Rugose corals were provided with a lid or operculum, closing the mouth of the calice. In the genus *Calceola*, formerly referred to the *Brachiopoda*, and very abundant in certain parts of the

Devonian System, the operculum consisted of a single valve or piece. In *Goniophyllum* four valves were present, and in *Cystiphyllum prismaticum* there were four or more valves in the operculum. It is worthy of notice that some recent corals (species of *Primnoa*, *Paramuricea*, and others) exhibit also a more or less complete operculum. According to Professor Agassiz, the *Rugosa* and the *Tabulate* division of the *Zoantharia* ought not to be considered as belonging to the *Actinozoa*, but should be placed amongst the *Hydrozoa*. This radical change, however, cannot be accepted without the production of very conclusive evidence in its favour. A strong argument against referring the Rugose and Tabulate Corals, as proposed by Agassiz, to the *Hydrozoa*, is their possession in most cases of well-developed *septa*, implying, of course, the existence in the living animal of *mesenteries*, structures which are wholly wanting in the *Hydrozoa*.

Distinctions between the Coralla of the Orders of Actinozoa.—Having now considered all the orders of the Actinozoa in which coralla are developed, it may be as well briefly to review their more striking differences.

In the first place, a sclerobasic corallum may be distinguished by inspection from a sclerodermic corallum by the fact that the latter, unless composed simply of spicules, presents the cups or "thecæ," in which the polypes were contained; the surface of the former being invariably destitute of these receptacles.

A sclerobasic corallum is found in the families *Antipathidæ* and *Hyalonemadæ* (?) amongst the *Zoantharia*, and in the families *Pennatulidæ* and *Gorgonidæ* amongst the *Alcyonaria*; the following being the differences between them:—

1. *Antipathidæ*.—Sclerobasis spinulous or smooth; tentacles and soft parts in multiples of six.

2. *Hyalonemadæ* (?).—Sclerobasis siliceous, composed of numerous threads; tentacles in multiples of five.

3. *Pennatulidæ*.—Sclerobasis sulcate, free; soft parts in multiples of four.

4. *Gorgonidæ*.—Sclerobasis sulcate, attached proximally; soft parts in multiples of four.

Sclerodermic coralla fall under two heads, according as they are simply composed of scattered spicules, or are provided with true *thecæ*.

I. *Spicular* coralla occur in the *Zoantharia Malacodermata* (occasionally), and in the *Alcyonidæ*; and no differences can be stated between the coralla themselves. The animals, however, differ entirely, the soft parts of the former being in multiples of five or six, those of the latter being in multiples of four.

II. A *thecal* sclerodermic corallum occurs in three distinct sections of *Actinozoa*:—1. In the *Zoantharia Sclerodermata*. 2. In the *Tubiporidæ*, amongst the *Alcyonaria*; and, 3. In the *Rugosa*; and the following are the distinctions between them:—

1. *Zoantharia Sclerodermata*.—Septa in multiples of five or six, sometimes absent; tabulæ often present.

2. *Tubiporidæ*.—Septa absent; thecæ united externally by distinct, horizontal "epithecæ."

3. *Rugosa*.—Septa in multiples of four; tabulæ usually present.

CHAPTER XVI.

CTENOPHORA.

ORDER IV. CTENOPHORA.—The *Ctenophora* comprise "*transparent, oceanic, gelatinous Actinozoa, swimming by means of 'ctenophores,' or parallel rows of cilia disposed in comb-like plates. No corallum*."—(Greene.)

The members of this order are all free-swimming organisms, and they are placed by many amongst the *Hydrozoa*, from which, however, they appear to be clearly separated by the possession of a differentiated digestive sac, as well as by their analogies with the *Actinozoa*, and their generally superior degree of organisation.

Pleurobrachia (*Cydippe*) (fig. 36) may be taken as the type of the order, the structure of all being similar to this in essential points. *Pleurobrachia* possesses a transparent, colourless, gelatinous, melon-shaped body, or "actinosoma," in which the two poles of the sphere are termed respectively the "oral" and "apical," and the rest of the body constitutes the "interpolar region." At the oral pole is the transverse mouth, bounded by lateral, slightly protuberant margins. "Eight meridional bands, or 'ctenophores' bearing the comb-like fringes, or characteristic organs of locomotion, traverse at definite intervals the interpolar region, which they divide into an equal number of lune-like lobes, termed the "actinomeres"; but this division of the body does not extend into the immediate vicinity of the poles, before reaching which the ctenophores gradually diminish in diameter, each terminating in a point."—

(Greene.) The normal number of the ctenophores appears to be eight, and each consists of a band of surface elevated transversely into a number of ridges, to each of which a fringe of cilia is attached, so as to form a comb-like plate. The cilia in the middle of these transverse ridges are the longest, and they gradually diminish in length towards the sides, so that the form of each comb is somewhat crescentic. Besides the comb-like groups of vibratile cilia, *Pleurobrachia* is provided with two very long and flexible tentacular processes, which are fringed on one side with smaller cirrhi. These filamentous processes arise each from a sac, situated on one of the lateral actinomeres, within which they can be completely and instantaneously retracted at the will of the animal.

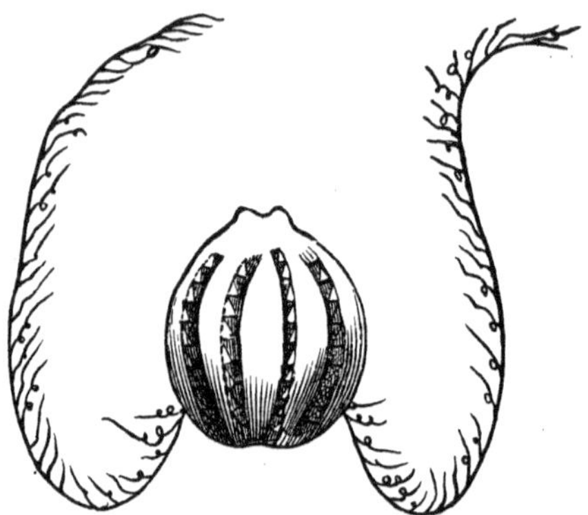

Fig. 36.—Ctenophora. *Pleurobrachia pileus.*

The mouth of *Pleurobrachia* (fig. 37, *a*) opens into a fusiform digestive sac, or stomach (*b*), the lower part of which is provided with brown cells, supposed to discharge the functions of a liver. The stomach opens below into a shorter and wider cavity (*c*), termed the "funnel," from which two canals diverge in the direction of the vertical axis of the organism, to open at the "apical pole." These canals are known as the "apical canals" (*e*), and their apertures as the "apical pores." From the funnel two other pairs of canals are given off. Of these, one pair—known as the "paragastric canals"—turns upwards, one running parallel to the digestive sac on each side (*d*), and "terminating cæcally before quite reaching the oral extremity." The second pair of canals (*i*)—the so-called "radial canals"—branch off from the funnel laterally, each dividing into two, and then again into two, as they proceed towards the periphery of the body. Thus, the two "primary" radial canals produce four "secondary" canals (*k*), and these, in turn, give rise to eight "tertiary" radial canals (*l*), which finally terminate by opening "at right angles into an equal number of longitudinal vessels, the 'ctenophoral canals' (*f*), whose course coincides with that of the eight locomotive bands. These canals end cæcally both at their oral and apical extremities."—(Greene.) The whole of this complex canal-system is lined by a ciliated endoderm, and a constant circulation of the included nutrient fluids is thus maintained.

Immediately within the apical pole is situated a small cyst or vesicle, supposed to be an organ of sense, and termed the

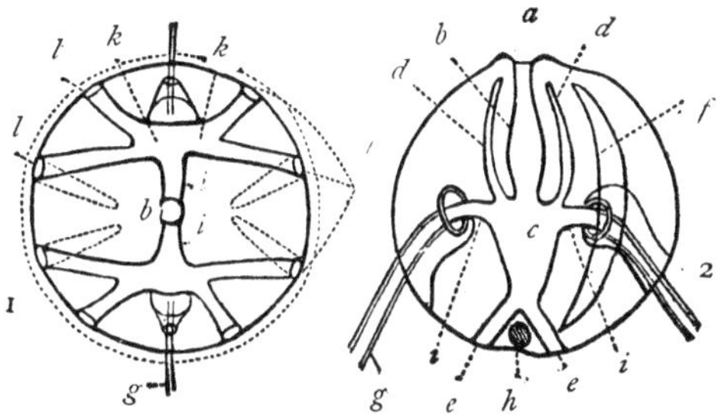

Fig. 37.—Morphology of Ctenophora. 1. Diagrammatic transverse section of *Pleurobrachia*. *b* Digestive cavity; *i i* Primary radial canals; *k k* Secondary radial canals; *l l* Tertiary radial canals; *g* Tentacle.
2. Longitudinal section of *Pleurobrachia*. *a* Mouth; *b* Digestive cavity; *e* Funnel; *d d* Paragastric canals; *e e* Apical canals; *f* Ctenophoral canal; *g* Tentacle; *h* Ctenocyst. (After Greene.)

"ctenocyst" (*h*). In structure the "ctenocyst" consists of a spherical vesicle, lined with a ciliated epithelium, and filled with a clear fluid, which contains mineral particles, probably of carbonate of lime. Resting upon the ctenocyst is a small ganglionic mass, giving origin to a number of delicate filaments, and generally admitted to be a rudimentary form of nervous system. The reproductive organs of *Pleurobrachia* are in the form of folds, containing either ova or spermatozoa, and situated beneath the endodermal lining of the ctenophoral canals, one on each side.

The embryo *Pleurobrachia* is at first rudely cylindrical in form, a belt of cilia passing round the middle of its body. This soon breaks up into two lateral groups, which eventually disappear altogether, "the ctenophores, at first very broad and few in number, at an early period taking on the performance of their special function."—(Greene.)

As regards the homologies between *Actinia* and *Pleurobrachia*, the following may be quoted from Professor Greene:—

"If now a comparison be made between this nutrient system" (the canal-system of the *Ctenophora*) "and that of *Actinia*, the digestive sacs of the two organisms are clearly seen to correspond in form, in relative size, and mode of communication with the somatic cavity. The funnel and apical canals of *Pleurobrachia*, though more distinctly marked out, are the homologues of those parts of the general cavity, which in *Actinia* are central in position, and underlie the free end of the digestive sac. So also the paragastric and radial canals may be likened to those lateral portions of the somatic cavity of *Actinia* which are not included between the mesenteries. Lastly, the ctenophoral canals of *Pleurobrachia* and the somatic chambers of *Actinia* appear to be truly homologous, the chief difference between the two forms being, that while in the latter the body-chambers are wide and separated by very thin partitions, they are in *Pleurobrachia* reduced to the condition of tubes; the mesenteries which intervene becoming very thick and gelatinous, so as to constitute, indeed, the principal bulk of the body." The "apical" canals, again, by which the digestive sac communicates inferiorly with the external medium, may be compared with the perforation which is found in some of the *Actinidæ* (*Cerianthus* and *Peachia*) traversing the axis of the base or foot.

The remaining members of the *Ctenophora* conform in most essential respects with *Pleurobrachia*, the most important differences being found in the canal-system. For purposes of comparison this system may be divided into four portions as follows:—1. The "axial system," consisting of the mouth, stomach, funnel, and apical canals. 2. The "paraxial system," comprising the paragastric canals. 3. The "radial system," comprising the primary, secondary, and tertiary radial canals. 4. The "ctenophoral system," consisting of the tubes which run underneath the locomotive bands.

In *Beroe*, which is in other respects very similar to *Pleurobrachia*, the axial system of canals is the same as we have seen in the latter. The paraxial system, however, consists of *two* pairs of paragastric canals, which, instead of terminating cæcally, open into a circular canal which surrounds the mouth. The ctenophoral canals, likewise, open into the oral vessel, instead of terminating cæcally as in *Pleurobrachia*. Lastly, the radial system is not developed, the ctenophoral canals simply curving round towards their apical extremities, and opening into the funnel directly.

Amongst the *Beroidæ* the mouth extends entirely across the oral extremity of the body; hence they have been termed *Eurystomata*, the term *Stenostomata* being applied collectively to all the other *Ctenophora.*

The *Beroidæ* further differ from *Pleurobrachia* in being destitute of the long tentacular appendages so characteristic of the latter.

In *Cestum*, or "Venus's Girdle," "elongation takes place to an extraordinary extent, at right angles to the direction of the digestive track, a flat, ribbon-shaped body, three or four feet in length, being the result."

DIVISIONS OF THE CTENOPHORA.—The following arrangement of the *Ctenophora* has been adopted by Gegenbaur (see Greene) :—

Order CTENOPHORA.

Sub-order I. *Stenostomata.*

Family I. CALLYMMIDÆ.
Body furnished with a pair of antero-posterior oral lobes, and other smaller lateral appendages. *Tentacles* various, turned towards the mouth.

Family II. CESTIDÆ.
Body ribbon-shaped, extended in a lateral direction, without oral lobes. *Tentacles* two in number, antero-posterior, turned towards the mouth.

Family III. CALLIANIRIDÆ.
Body produced into a pair of wing-like lateral lobes, bearing the ctenophores. *Tentacles* two in number, lateral, turned from the mouth.

Family IV. PLEUROBRACHIADÆ.
Body oval or spheroidal, without oral lobes. *Tentacles* two in number, lateral, turned from the mouth.

Sub-order II. *Eurystomata.*

Family V. BEROIDÆ.
Body oval, elongated, without oral lobes. *Tentacles* absent.

CHAPTER XVII.

DISTRIBUTION OF ACTINOZOA.

1. DISTRIBUTION OF ACTINOZOA IN SPACE. 2. CORAL REEFS. 3. DISTRIBUTION OF ACTINOZOA IN TIME. 4. APPENDIX.

DISTRIBUTION OF ACTINOZOA IN SPACE.—The *Zoantharia malacodermata* appear to have an almost cosmopolitan range, sea-anemones being found on almost every coast; some of the tropical forms attaining a very large size. The *Ctenophora*, too, have an almost world-wide distribution, occurring in all seas from the equator to within the arctic circle. In habit all the *Ctenophora* are pelagic, being found, like the oceanic *Hydrozoa*, swimming near the surface far from land. *Pennatulidæ* and *Gorgonidæ* are found in the seas of the temperate zone, but the latter attain their maximum within the tropics. The Red Coral of commerce (*Corallium rubrum*) is derived from the Mediterranean.

The so-called "reef-building" Corals have their distribution conditioned by the mean winter temperature of the sea, a temperature of not less than 66° being necessary for their existence. The seas, therefore, which possess the necessary temperature may be said to be all comprised within a distance of about 1800 miles of the equator on each side. Within these limits, however, apparently owing to the influence of Arctic currents, no coral-reefs are found on the western coasts of America and Africa. They are found chiefly on the east coast of Africa, the shores of Madagascar, the Red Sea and Persian Gulf, throughout the Indian Ocean and the whole of Polynesia, and around the West Indian Islands and the coast of Florida.

All known *Actinozoa* are marine, no member of the class having hitherto been found in fresh water.

CORAL-REEFS.—A "coral-reef" is a mass of coral, sometimes many hundred miles in length, and it may be two thousand feet or more in thickness, produced by the combined growth of different species of coralligenous *Actinozoa*. As before said, a mean winter temperature of not less than 66° is necessary for their existence, and, therefore, nothing worthy of the name of a "coral-reef" is to be found in seas so far removed from the equator as to possess a lower winter temperature than the above. The headquarters of the reef-building Corals may be said to be around the islands and continents of the Pacific Ocean. According to Darwin, coral-reefs may be divided into three principal forms—viz., Fringing-reefs, Barrier-reefs; and Atolls, distinguished by the following characters:—

1. *Fringing-reefs* (fig. 38, 1).—These are reefs, seldom of great size, which may either surround islands, or skirt the shores of continents. These shore-reefs have no channel of any great depth intervening between them and the land, and the soundings on their seaward margin indicate that they repose upon a gently-sloping surface.

2. *Barrier-reefs* (fig. 38, 2).—These, like the preceding, may either encircle islands, or may skirt continents. They are dis-

tinguished from fringing-reefs by the fact that they occur usually at a much greater distance from land, that there intervenes a channel of deep water between them and the shore, and that soundings taken close to their seaward margin indicate enormous depths. If the barrier-reef surround an island, it is sometimes called an "encircling barrier-reef," and it constitutes with its island what is called a "lagoon-island."

As an example of this class of reefs may be taken the great barrier-reef on the N.E. coast of Australia, the structure of which is on a perfectly colossal scale. This reef runs, with a few breaches in its continuity, for a distance of more than a

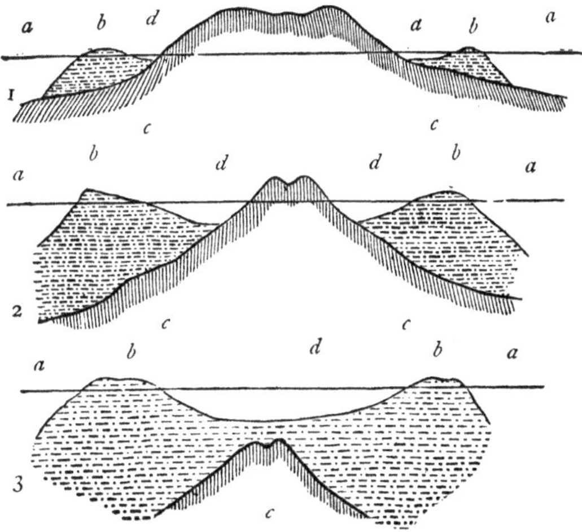

Fig. 38.—Structure of coral-reefs. 1 Fringing-reef; 2 Barrier-reef; 3 Atoll. *a* Sea level; *b* Coral-reef; *c* Primitive land; *d* Portion of sea within the reef, forming a channel or lagoon.

thousand miles, its average distance from the shore being between twenty and thirty miles, and the depth of the inner channel being from ten to sixty fathoms, whilst the sea outside is "profoundly deep" (in some places over 1800 feet).

3. *Atolls* (fig. 38, 3).—These are nearly circular reefs of coral, enclosing a central expanse of water or lagoon. They seldom form complete rings, the reef being usually breached by one or more openings, which are always situated on the leeward side, or on that side which is most completely sheltered from the prevailing winds. In their structure they are identical with "encircling barrier-reefs," and differ from these only

in the fact that the lagoon which they enclose does not contain an island in its centre.

If a coral-reef be observed—say a portion of an encircling barrier-reef—the following are the general phenomena which may be noticed. The general shape of the reef is triangular, presenting a steep and abrupt wall on the seaward side, and having a long and gentle slope towards the land. The outer margin of the reef is exposed to the beating of a tremendous surf, whilst the soundings taken just outside the line of breakers always indicate great depths. The longer inner slope is washed by the calm waters of the inner lagoon or channel. The reef is only very partially composed of living corals, which are found to occupy a mere strip, or zone, along the seaward margin of the reef, whilst all above this, as well as all below, is constituted by dead coral, or "coral-rock."

As to the method in which such a reef is produced, the following facts have been established:—

A. The coral-producing polypes cannot exist at levels higher than extreme low water, exposure to the sun, even for a short period, proving rapidly fatal. It follows from this that no coral-reef can be raised above the level of the sea by the efforts of its builders. The agency whereby reefs are raised above the surface of the sea, is the denuding power of the breakers which constantly fall upon their outer margins. These detach large masses of dead coral, and heap them up in particular places, until an island is gradually produced. The fragments thus accumulated are compacted together by the finer *detritus* of the reef, and are cemented together by the percolation of water holding carbonate of lime in solution. In this way the upper surface of the reef, along a line of greater or less breadth, is more or less completely raised above the level of high water. It is obvious, however, that the reef might be entirely destroyed by a continuation of this process—the sea being quite competent to undo what it had done—unless some counteracting force were brought into play. This counteracting force is found in the vital activity of the living corals which form the seaward margin of the reef, and which, by their growth, prevent the sea from *always* destroying the masses of sediment which it may have thrown up.

B. The coral-producing polypes cannot exist at depths exceeding some 15 to 30 fathoms. It follows from this that no coral-reef can be commenced upon a sea-bottom deeper than about 30 fathoms. The question now arises—In what way have reefs been produced, which, as we have seen, rise out of depths of 300 fathoms or more? This question has been answered by Darwin, who showed that the production of barrier-reefs and atolls was really to be ascribed to a gradual subsidence of the foundations upon which they rest. Thus, if a fringing-reef which surrounds an island is supposed gradually to sink beneath the sea, the upward growth of the corals will neutralise the downward movement of the land, so far, at any rate, that the reef will appear to be stationary, whilst it is really growing upwards. The island, however, as subsidence goes on, will gradually diminish in size, and a channel will be formed between it and the reef. If the depression should be still continued, the island will be reduced to a mere peak in the centre of a lagoon; and the reef, from a "fringing-reef," will have become converted into an "encircling barrier-reef." As the growth of the reef is chiefly vertical, the continued depression will, of course, have produced deep water all round the reef. If the subsidence be continued still further, the central peak will disappear altogether, and the reef will become a more or less complete ring surrounding a central expanse of water; thus becoming converted into an "atoll." The production, therefore, of encircling barrier-reefs and atolls is thus seen to be due to a process of subsidence of the sea-bottom. The existence, however, of fringing-reefs is only possible when the land is either slowly rising, or is stationary; and as a matter of fact, fringing-reefs are often found to be conjoined with upraised strata of post-tertiary age. Atolls and encircling barrier-reefs, on the other hand, are not found in the vicinity of active volcanoes—regions where geology teaches us that the land is either stationary or is undergoing slow upheaval.

C. Different portions of a coral reef are occupied by different kinds of corals. According to Agassiz, the basement of a coral-reef is formed by a zone of massive *Astræans*. These cannot flourish at depths of less than six fathoms of water, and consequently when the surface of the reef has reached this level, the *Astræans* cease to grow. Their place is now taken by *Meandrinas* (Brain-corals) and *Porites;* but these, too, cannot extend above a certain level. Finally, the summit of the reef is formed by an aggregation of less massive corals, such as *Madreporidæ*, *Milleporidæ*, and *Gorgonidæ*.

Distribution of Actinozoa in Time.—With the single exception of the Mollusca, no division of the animal kingdom contributes such important and numerous indications of its past existence as the *Actinozoa.*

In the Palæozoic Rocks the majority of corals belong to the division *Rugosa*, these seeming to have filled the place now taken by the sclerodermic *Zoantharia.* The order *Rugosa* is entirely Palæozoic, with the single exception of the genus *Holocystis* which is represented in the Secondary Rocks by a single species (viz., *H. elegans*, from the lower Greensand) and a few Tertiary forms. In the lower Palæozoic Rocks the *Rugosa* are especially abundant; but in the Permian formation the order is represented by the single genus *Polycœlia*.

The *Zoantharia Sclerodermata*, though attaining their maximum at the present day, nevertheless are well represented in past time, beginning in the Silurian period. One subdivision of this group, the *Tubulosa*, is entirely confined to the Palæozoic Rocks, and another, the *Tabulata*, is chiefly Palæozoic. The *Perforata* and *Aporosa*, on the other hand, are more abundant in the Mesozoic and Kainozoic Epochs.

The *Zoantharia Sclerobasica* are hardly known as fossils, but the Miocene deposits of Piedmont (Middle Tertiary) have yielded a species of *Antipathes*.

The *Zoantharia Malacodermata*, from the soft nature of their bodies, are obviously incapable of leaving any traces of their existence; though we are by no means therefore justified in asserting that they did not exist in past geological epochs.

The *Alcyonaria* are very doubtfully represented in rocks older than the Chalk; the Lower Silurian fossil called *Protovirgularia* being more probably referable to the *Hydrozoa*. One of the *Pennatulidæ* (viz., *Graphularia*) has been found in the London Clay (Eocene), and the same formation has likewise yielded two species of *Gorgonidæ* (*Mopsea* and *Websteria*). The genus *Corallium* has likewise been found in deposits of Miocene age.

The *Ctenophora*, being entirely destitute of any hard structures, are not known at all as occurring in the fossil condition.

Appendix giving a Tabular View of the Divisions of the Zoantharia Sclerodermata and Rugosa (after Milne-Edwards and Jules Haime).

A. The *Zoantharia Sclerodermata* are defined by the possession of a sclerodermic corallum, the parts of which are arranged in multiples of five or six. Septa generally well developed, but not combined, as a rule, with tabulæ.

The following chief divisions of the *Zoantharia Sclerodermata* are, with few alterations, those adopted by the above-mentioned authorities :—

I. Tabulata.—Septa rudimentary or absent; tabulæ well developed, dividing the visceral chamber into a series of stories.

1. *Thecidæ*.—Corallum massive; a dense spurious cœnenchyma formed by the lateral union of the septa; tabulæ numerous.
2. *Favositidæ*.—Septa and corallites distinct; little or no true cœnenchyma.

3. *Seriatoporidæ.*—Corallum arborescent; sclerenchyma abundant and compact; tabulæ few.
4. *Milleporidæ.*—Corallum massive or foliaceous; septa not numerous; sclerenchyma tabular or cellular.

II. PERFORATA.—Septa well developed; no tabulæ; dissepiments rudimentary; sclerenchyma porous.

5. *Eupsammidæ.*—Corallum simple or composite; septa well developed and lamellar; columella spongiose.
6. *Poritidæ.*—Corallum composed of spongy, reticulated sclerenchyma. Septa never lamellar, but consisting wholly of a more or less definite series of trabeculæ; no tabulæ.
7. *Madreporidæ.*—Corallum usually composite; cœnenchyma abundant and spongy; thecæ porous, not distinct from the cœnenchyma; septa distinct, but slightly perforate.

III. APOROSA.—Septa well developed, completely lamellar, and primitively consisting of six elements; no tabulæ; sclerenchyma imperforate.

8. *Fungidæ.*—Corallum simple or compound; thecæ ill developed, and somewhat porous; no dissepiments or tabulæ; synapticulæ numerous.
9. *Astræidæ.*—Corallum simple or compound; no proper cœnenchyma; numerous dissepiments; no synapticulæ. Corallites well defined, and separated from one another by perfect walls.
10. *Oculinidæ.*—Corallum composite; cœnenchyma abundant and compact; dissepiments few in number. Walls of the corallites without perforations, not distinct from the cœnenchyma.
11. *Turbinolidæ.*—Corallum usually simple; no cœnenchyma; septa well developed; no dissepiments, nor synapticulæ.

IV. TUBULOSA.—Septa indicated by mere striæ; thecæ pyriform; corallites sometimes connected by a creeping basal cœnenchyma.

12. *Auloporidæ.*—This being the only family in the *Tubulosa*, its characters are necessarily the same as those of the division itself.

B. ORDER RUGOSA.—Characterised by the possession of a sclerodermic corallum, usually with septa and tabulæ combined, the former being in multiples of four. The corallites are always distinct, and are never united together by a cœnenchyma. The septa are usually incomplete, but are never porous, and never bear synapticulæ. The order is divided into the following four families:—

Family 1. *Stauridæ.*

Corallum simple or composite; septa incomplete, united by lamellar dissepiments; four large primary septa, forming a cross.

Family 2. *Cyathaxonidæ.*

Corallum simple; septa complete; no dissepiments or tabulæ; without four primary septa.

Family 3. *Cyathophyllidæ.*

Corallum simple or composite; septa incomplete; tabulæ generally present.

Family 4. *Cystiphyllidæ.*

Corallum simple, composed chiefly of a vesicular mass, with but slight traces of septa.

ANNULOIDA.

CHAPTER XVIII.

1. General Characters of the Annuloida. 2. General Characters of the Echinodermata.

Sub-kingdom III. Annuloida (= *Echinozoa*, Allman).—This sub-kingdom was proposed by Professor Huxley for the reception of the two groups of the *Echinodermata* and the *Scolecida*, of which the former belonged to the old sub-kingdom *Radiata*, whilst the latter was formerly classed with the *Annulosa*. The same sections have been grouped by Professor Allman together, under the name *Echinozoa;* the *Rotifera*, however, being excluded from this division and classed with the *Annulosa*. By others again, the *Annuloida* are looked upon as a section of the *Annulosa*, and not as a distinct sub-kingdom. Provisionally, however, it seems best to regard the *Annuloida* as one of the primary divisions of the animal kingdom, it being impossible, in the meanwhile, to frame a definition common to it and to the *Annulosa*. The name *Vermes* has sometimes been employed to designate the sub-kingdom *Annuloida*, certain classes being sometimes removed elsewhere, or certain others being added. In its most modern signification, the term *Vermes* may be held as synonymous with *Annuloida*, *minus* the *Echinodermata* and *plus* the whole of the *Anarthropodous* division of the *Annulosa*.

The *Annuloida* are distinguished *by the presence of a distinct nervous system, and the possession of an alimentary canal which is entirely shut off from the general cavity of the body. A peculiar system of canals, usually communicating with the exterior, and termed the "water-vascular" or "aquiferous" system, is present in all; and a true vascular apparatus is sometimes present. In none is the body of the adult composed of definite segments, or provided with "bilaterally disposed successive pairs of appendages."*

The union of the *Echinodermata* with the *Scolecida* in a single sub-kingdom, as proposed by Huxley, must be regarded as a purely provisional arrangement. Many other classifications have been proposed, each with some obvious advantages and some disadvantages. Perhaps the most natural arrangement would be to establish a separate sub-kingdom for the *Echinodermata*, and to group the *Scolecida* with the *Anarthropoda* under the name of *Vermes*. In the confessedly imperfect state of our knowledge, however, it will be as well to retain for the present the sub-kingdom *Annuloida*.

The *Annuloida* are divided into two great classes, the *Echinodermata* and the *Scolecida*.

Class I.—Echinodermata.

The members of this class are known commonly as Sea-urchins, Star-fishes, Brittle-stars, Feather-stars, Sea-cucumbers, &c., and the following are their leading characteristics. They are all animals which, in the adult condition, show a more or less distinctly radiate condition of their parts, especially of those around the mouth; whilst in their embryonic stages they are more or less distinctly bilaterally symmetrical. Whilst radial symmetry in the great majority of cases preponderates in the adult *Echinoderm*, there are, nevertheless, many instances in which the fully-grown animal shows distinct traces of bilateral symmetry. The external envelope of the body ("perisome") is either composed of numerous calcareous plates, articulated together, or of a coriaceous integument, in which calcareous granules and spicules are usually developed. In all adult *Echinoderms* there is a system of tubes, termed the "ambulacral system," which generally subserves locomotion, and usually communicates with the exterior. This water-vascular system surrounds the commencement of the alimentary canal, and in almost all cases gives off secondary vessels in a radiating manner. An alimentary canal is always present, and is completely shut off from the body-cavity. In many, if not in all, both neural and hæmal systems are developed. The nervous system in all the adult *Echinoderms* is a ring-like gangliated cord, which surrounds the œsophagus and sends branches parallel to the radiating ambulacral canals.

The special features of the structure of the *Echinodermata* will be noticed under each order, but it will be as well to give here an abstract of Professor Huxley's description of the process of development in the members of the class. In the great majority, if not in all, of the *Echinodermata* the impregnated ovum is developed into a free-swimming, ciliated, ovoid embryo. Soon the cilia become restricted to one, two, or more bands, which are generally disposed transversely to the long axis of the body, and are in all cases bilaterally symmetrical. The parts of the body which support the cilia are usually developed into protuberances, or processes, which are symmetrically disposed upon the two sides of the body. "The larvæ of *Asteridea* and *Holothuridea* are devoid of any continuous skeleton, but those of *Ophiuridea* and *Echinidea* possess a very remarkable, bilaterally symmetrical, continuous, calcareous skeleton, which extends into and supports the processes of the body." In this stage the larva form of the two orders last mentioned was described by Müller as a distinct animal under the name of *Pluteus*, from its resemblance to a painter's easel. (See fig. 41, 1.)

An alimentary canal soon appears in the larva, forming a curve with an open angle towards the ventral surface of the organism. The parts of the alimentary canal consist of a mouth, gullet, globular stomach, and short intestine, with a distinct anal aperture; the whole being "disposed in a longitudinal and vertical plane, dividing the larval body into two symmetrical halves." Besides the digestive canal, no other organs have hitherto been discovered in these larvæ. In the further process of development, "an involution of the integument takes place upon one side of the dorsal region of the body, so as to give rise to a cæcal tube, which gradually elongates inwards, and eventually reaches a mass of formative matter, or blastema, aggregated upon one side of the stomach. Within this, the end of the tube becomes converted into a circular vessel, from which trunks pass off, radially, through the enlarging blastema. The latter, gradually expanding, gives rise, in the *Echinidea*, the *Asteridea*, the *Ophiuridea*, and the *Crinoidea*, to the body-wall of the adult; the larval body and skeleton (when the latter exists), with more or less of the primitive intestine, being either cast off as a whole, or disappearing, or becoming incorporated with the secondary development, while a new mouth is developed in the centre of the ring formed by the circular vessel. The vessels which radiate from the latter give off diverticula to communicate with the cavities of numerous processes of the body—the so-called feet—which are the chief locomotive organs of the adult. The radiating and circular vessels, with all their appendages, constitute what is known as the 'ambulacral system'; and in *Asterids* and *Echinids* this remarkable system of vessels remains in communication with the exterior of the body by canals, connected with perforated portions of the external skeleton—the so-called 'madreporic canals' and 'tubercles.' In *Ophiurids* the persistence of any such communication of the ambulacral system with the exterior is doubtful, and still more so in *Crinoids*. In *Holothurids* no such communication obtains; the madreporic canals and their tubercles depending freely from the circular canal into the perivisceral cavity."

By Professor Wyville Thomson the larva of the *Echinodermata* is termed the "pseud-embryo," since it leads a perfectly independent existence, and the true *Echinoderm* is usually developed out of a portion only of its substance. The great peculiarity, therefore, in the development of the *Echinodermata* is found in the possession by the larva of provisional organs, which may be either absorbed or cast off, but which are not converted into the corresponding structures of the adult. Thus the *Pluteus* of an Echinoid possesses a mouth and alimentary canal which are not converted into, and in no way correspond with, the mouth and alimentary canal of the adult.

The *Echinodermata* are divided into seven orders—viz., the *Crinoidea*, *Cystoidea*, *Blastoidea*, *Ophiuroidea*, *Asteroidea*, *Echinoidea*, and *Holothuroidea*. Of these, the first is almost extinct and the two next are entirely so; they are really the lowest orders; but their structure will be better understood if the higher orders are considered first.

CHAPTER XIX.

ECHINOIDEA.

ORDER ECHINOIDEA.—The members of this order—commonly known as Sea-urchins—are characterised by the possession of a subglobose, discoidal, or depressed body, encased in a "test" or shell, which is composed of numerous, immovably connected, calcareous plates. The intestine is convoluted, and there is a distinct anus. The mouth is usually armed with calcareous teeth, and is always situated on the inferior surface of the body, but the position of the anal aperture varies. The larva is pluteiform, and has a skeleton.

The "test" of the *Echinoidea* is composed of numerous calcareous plates, firmly united to one another by their edges, and bearing different names according to their position and function. In one or two exceptional cases, though the essential structure of the shell is the same as in the ordinary forms, the plates of the test are so thin, and are so united together, that the entire test becomes flexible and soft. In all recent members of the order the test is composed of twenty rows of these plates, arranged in ten alternating zones, which pass from the one pole of the animal to the other, each zone being composed of two similar rows. Five of these double rows are composed of large plates, which are not perforated by any apertures (fig. 39); the zones formed by these imperforate plates being termed the "inter-ambulacral areas." The other five double rows of plates alternate regularly with the former, and are termed the "ambulacral areas," or "poriferous zones."

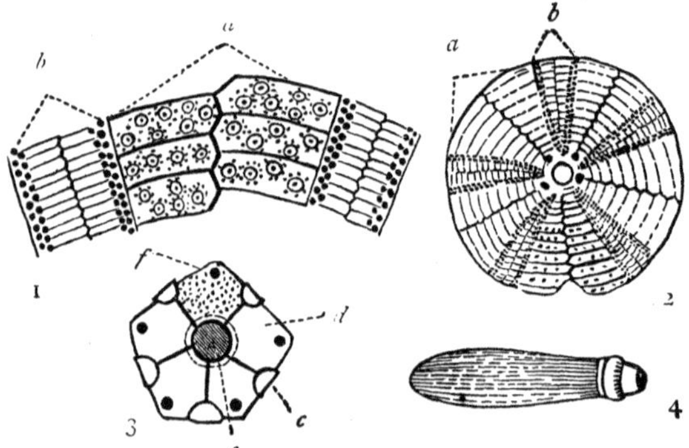

Fig. 39.—Morphology of Echinoidea. 1. Portion of the test of *Galerites hemisphericus* enlarged, showing an inter-ambulacral area (*a*), and an ambulacral area (*b*). 2 *Galerites hemisphericus* viewed from above. *a* Inter-ambulacra; *b* Ambulacra. 3. Genital and ocular disc of *Hemicidaris intermedia* enlarged. *c* Ocular plate; *d* Genital plate; *e* Anal aperture; *f* Madreporiform tubercle. 4. Spine of the same. (After Forbes.) The tubercles are mostly omitted on figs. 2 and 3 for the sake of clearness.

Each of these zones is composed of two rows of small plates, which are perforated by minute apertures for the emission of the "ambulacral tubes," or "tube-feet." Growth of the test is carried on by additions made to the edge of each individual plate, by means of an organised membrane which passes between the sutures, where the plates come into contact with one another. The plates of the test are studded with large tubercles, which are more numerous on the inter-ambulacral areas than on the ambulacral, and are wanting on all the plates which do not belong to either area. These tubercles carry spines (fig. 40), used defensively and in locomotion, which are articulated to their apices by means of a sort of "universal" or "ball-and-socket" joint. Occasionally a small ligamentous band passes between the head of the tubercle and the centre of the concave articular surface of the spine, thus closely resembling the "round ligament" of the hip-joint of man. Besides the main rows of plates just described, forming the so-called "corona," other calcareous pieces go to make up the test of an *Echinus*. The mouth is surrounded by a coriaceous peristomial membrane, which contains a series of small calcareous pieces, known as the "oral plates"; whilst a corresponding series of "anal plates" is found in the membrane surrounding the opposite termination of the alimentary canal. Surrounding the aperture of the anus at the summit of the test is the "apical disc," composed of the so-called genital and

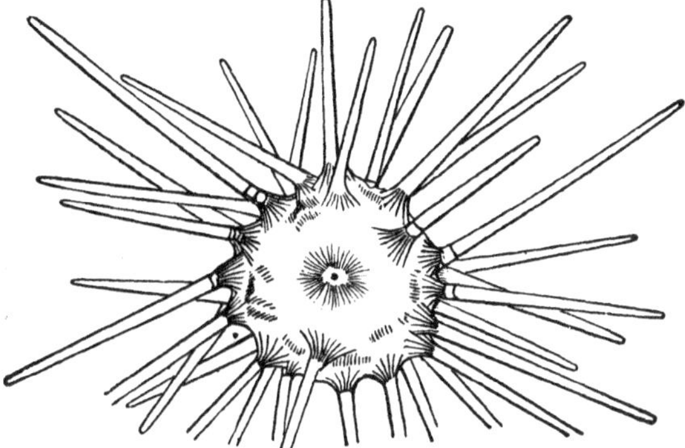

Fig. 40.—*Cidaris papillata*. (After Gosse.)

ocular plates (fig. 39, 3). The "genital plates" are five large plates of a pentagonal form, each of which is perforated by the duct of an ovary or testis. One of the genital plates is larger than the others, and supports a spongy tubercle, perforated by many minute apertures, like the rose of a watering-pot, and termed the "madreporiform tubercle." The genital plates occupy the summits of the interambulacral areas. Wedged in between the genital plates, and occupying the summits of the ambulacral areas, are five smaller, heart-shaped, or pentagonal plates, known as the "ocular plates," each being perforated by a pore for the reception of an "ocellus" or "eye."

Besides the spines, which are sometimes of a very great length, the test often bears curious little appendages, called "pedicellariæ," and often supposed to be parasitic. Each of these consists of a stem, bearing two or three blades or claws, which snap together and close upon foreign objects, like the beak of a bird. Their action appears to be independent of the will of the animal, and their true function is not known; but they may be regarded as peculiarly modified spines.

Locomotion in the *Echinoidea* is effected by means of a singular system of contractile and retractile tubes, which constitute the "ambulacral tubes," or "tube-feet," and are connected with the "ambulacral system" of aquiferous canals (fig. 41). From the perforated "madreporiform tubercle" on the largest of the genital plates, there proceeds a membranous canal, known as the "stone," or "sand canal," whereby water is conveyed from the exterior to a circular tube, surrounding the œsophagus, and constituting the centre of the water-vascular or ambulacral system. The function of the madreporiform tubercle appears to be that of permitting the ingress of water from the exterior, but of excluding any solid particles, which might be injurious. The "circular canal," surrounding the gullet, is situated between the nervous and blood-vascular rings, and gives off five branches—the "radiating canals"—which proceed radially along the "ambulacral areas" in the interior of the shell. In this course they give off numerous short lateral tubes—the "tube-feet"—which pass through the "ambulacral pores" to gain the exterior of the test, and terminate in suctorial discs. Besides the radiating ambulacral canals, there are connected with the circular canal certain vesicles of unknown function, known as the "Polian vesicles" (*ampullæ Polianæ*). The ambulacral tubes, or tube-feet, can be protruded at the will of the animal through the pores which perforate the ambulacral areas, and can be again retracted. By means of these locomotion is effected, the tube-feet being capable of protrusion to a length greater than that of the longest spines of the body. The mechanism by which the tube-feet are protruded and retracted is as follows:—Each tube-foot, shortly after its origin, gives rise to a secondary lateral branch, which terminates in a vesicle. These vesicles or "ampullæ" are provided with circular muscular fibres, by the contraction of which their contained fluid is forced into the tube-feet, which are thus protruded. Retraction of the ambulacral tubes is effected by proper muscular fibres of their own, which expel again the fluid which has been forced into them by the vesicles. According to Owen, the terminal sucker in each tube-foot of the *Echinus* is "supported by a circle of five, or sometimes four, reticulate calcareous plates, which intercept a central foramen, and by a single, delicate, reticulated, perforate plate on the proximal side of the preceding group. The centre of the suctorial disc is perforated by an aperture conducting to the interior of the ambulacral tube-foot." This perforation of the suctorial discs of the ambulacra, though affirmed by Valentin, is denied by Müller; and it is difficult to believe that it would not impair the functions of the feet in the act of protrusion.

The digestive system of the *Echinus* consists of a mouth, armed with five long, calcareous, rod-like teeth, which perforate five triangular pyramids, the whole forming a singular structure, known as "Aristotle's Lantern." The mouth conducts by a pharynx and a tortuous œsophagus to a stomach,

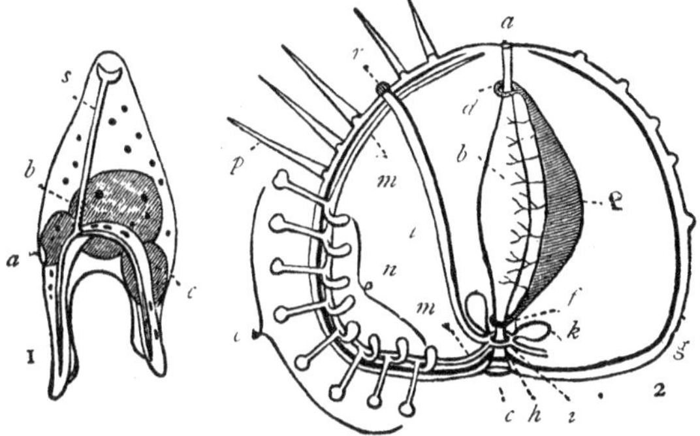

Fig. 41.—Morphology of Echinoidea. 1. Echinid larva. *a* Mouth; *b* Stomach; *c* Intestine; *s* Skeleton.

2. Diagram of Echinus. The spines and the ambulacra are represented over a small portion of the test; the vascular system is cross-shaded; the nervous system is represented by the black line. *a* Anus; *b* Stomach; *c* Mouth; *d* and *f* Vascular rings round the alimentary canal; *e* Heart; *g* Test; *h* Nervous ring round the gullet; *i* Ambulacral ring or "circular canal" round the gullet; *kk* Polian vesicles; *l* Sand canal; *m m* Radiating ambulacral canal; *n* Secondary ambulacral vesicles; *o* Ambulacral tubes, or "tube-feet"; *p* Spines; *r* Madreporiform tubercle.

opening into a convoluted intestine, which winds round the interior of the shell, and terminates in a distinct anus. The mouth is always situated at the base of the test, and may be central, sub-central, or altogether excentric in position. The anus varies considerably in its position, being usually situated within the apical disc, and surrounded by the genital and ocular plates, when the test is said to be "regular." Sometimes, however, the anal aperture is without the apical disc, and is removed to some distance from the genital plates, when the test is said to be "irregular." The convolutions of the alimentary canal are attached to the interior of the test by a delicate mesentery; the surface of which, as well as that of the lining-membrane of the shell, is richly ciliated, and subserves the purposes of respiration.

The proper blood-vascular system (fig. 41, 2) consists of a central, fusiform, contractile vesicle, or heart. This gives off one vessel, which forms a ring round the intestine near the anus, and another which passes downwards, and forms a circle round the gullet, above the "circular canal" of the ambulacral system. From the anal vessel proceed five arterial branches, which run along the ambulacral spaces, and return their blood by five branches, which run alongside of them in an opposite direction. This blood-vascular system has been thought to be homologous with the pseudohæmal system of the *Annelida*, rather than with the true circulatory system of higher animals. By Huxley, however, the pseudohæmal vessels of the Annelides are looked upon as homologous with the water-vascular system of the *Scolecida*, to which the water-vascular or ambulacral system of the Echinoderms is unquestionably comparable. The hæmal system, therefore, of the Echinoderms must be regarded as something not represented amongst the Annelides.

The nervous system consists of a ganglionated circular cord, which surrounds the gullet below, or superficial to, the "circular canal" of the ambulacral system, and which sends five branches along the ambulacral spaces, in company with the radiating ambulacral canals.

There is no distinct respiratory organ, but the function of aëration of the blood appears to be performed partly by the vascular lining of the test and the mesentery, and partly by the secondary ambulacral vesicles. The perivisceral cavity is filled with sea-water, but the mode in which this is admitted, or renovated, is not known with certainty. According to Tiedemann, the water gains access to the interior by means of short, branched processes, which are attached to the extremities of the inter-ambulacral areas round the mouth; but others deny that these are perforated by any apertures. These processes are apparently nothing more than greatly-developed tube-feet, and they are probably homologous with the crown of feathery tentacles surrounding the mouth in the Holothurians; though this is perhaps represented by some very large tube-feet placed just round the mouth.

The sexes are distinct in all the *Echinoidea*, and the reproductive organs are in the form of five membranous sacs, which occupy the inter-ambulacral areas, and open on the exterior by means of the apertures in the genital plates. In the "irregular" Echinoids (such as the "heart-urchins") there are only four genital glands, and, therefore, only four genital plates in the apical disc.

The *Echinoidea* may be divided into the following principal families:—

SYNOPSIS OF THE FAMILIES OF ECHINOIDEA (AFTER POMEL).

SUB-ORDER I. ECHINIDA.—Test composed of no more than twenty rows of plates.

a. Spatiformes.—Mouth excentric, in front; anus behind; anterior ambulacrum obliterated; form obovate. (The so-called "Spantangoid" Sea-urchins.)

1. *Ananchytida.*—With simple ambulacra.
2. *Spantangida.*—With petaloid ambulacra.

b. Lampadiformes.—Mouth central, or nearly so; toothed or toothless; anus more or less posterior, but often mounting high enough to enter into the genital disc; ambulacra similar.

3. *Echinoneida.*—Toothless; ambulacra simple.
4. *Cassidulida.*—Toothless; ambulacra petaloid.
5. *Clypeastrida.*—Toothed; ambulacra petaloid.
6. *Echinoconida.*—Toothed; ambulacra simple.

c. Globiformes.—Mouth central; anus opposite the mouth, surrounded by the genital plates.

7. *Cidarida.*—Ambulacra prolonged on the buccal membrane and destitute of buccal branchiæ.
8. *Echinidæ.*—Ambulacra not prolonged on the buccal membrane, but provided with buccal branchiæ.

SUB-ORDER II. PERISCHOECHINIDÆ OR TESSELATA.—Corona of the test, consisting of more than twenty rows of plates. (The Palæozoic Sea-urchins, *Archæocidaris* and *Palæchinus.*)

CHAPTER XX.

ASTEROIDEA AND OPHIUROIDEA.

ORDER ASTEROIDEA (*Stellerida*).—This order comprises the ordinary star-fishes, and is defined by the following characters:—The body (fig. 42) is star-shaped or pentagonal, and consists of a central body or "disc," surrounded by five or more lobes, or "arms," which radiate from the body, are hollow, and contain prolongations of the viscera. The body is not enclosed in an immovable box, as in the *Echinoidea*, but the integument ("perisome") is coriaceous, and is strengthened by irregular calcareous plates, or studded by calcareous spines. No dental apparatus is present. The mouth is inferior, and central in position; the anus either absent or dorsal. The ambulacral tube-feet are protruded from grooves on the under surface of the rays. The larva is vermiform, and has no pseudembryonic skeleton.

The skeleton of the *Asteroidea* is composed of a vast number of small calcareous plates, or ossicula, united together by the coriaceous perisome, so as to form a species of chain-armour. Besides these, the integument is abundantly supplied with spines, tubercles, and "pedicellariæ." Lastly, the radiating ambulacral vessels run underneath a species of internal skeleton, occupying the axis of each arm, and composed of a great number of bilateral "vertebral ossicles" or calcareous plates, which are movably articulated to one another, and are provided with special muscles by which they can be brought together or drawn apart. The upper surface of a star-fish corresponds to

Fig. 42.—*Cribella oculata*. (After Forbes.)

the combined inter-ambulacral areas of an *Echinus*, and exhibits the aperture of the anus (when present), and the "madreporiform tubercle," which is situated near the angle between two rays. The inferior or ventral surface corresponds to the ambulacral areas of an *Echinus*, and exhibits the mouth and ambulacral grooves.

The mouth is central in position, and is not provided with teeth; it leads, by a short gullet, into a large stomach, from which a pair of sacculated diverticula are prolonged into each ray. A distinct intestine and anus may, or may not, be present; but the anus is sometimes wanting (in the genera, *Astropecten*, *Ctenodiscus*, and *Luidia*).

The ambulacral system is essentially the same as in the *Echinoidea*, and is connected with the exterior by means of the "madreporiform tubercle," or "nucleus," two, three, or more of these being occasionally present. The ambulacral tube-feet are arranged in two or four rows, along grooves in the under surface of the arms.

The blood-vascular system consists, as in the *Echinus*, of two circular vessels, one round the intestine, and one round the gullet, with a dilated tube, or heart, intervening between them. There are no distinct respiratory organs, but the surfaces of the viscera are abundantly supplied with cilia, and doubtless subserve respiration; the sea-water being freely admitted into the general body-cavity by means of numerous contractile ciliated tubes, which project from the dorsal surface of the body, and are perforated at their free extremities (Owen).

The nervous system consists of a gangliated cord, surrounding the mouth, and sending filaments to each of the rays. At the extremity of each ray is a pigment-spot, corresponding to one of the ocelli of an *Echinus*, and, like it, supposed to be a rudimentary organ of vision. The eyes are often surrounded by circles of movable spines, called "eyelids."

The generative organs are in the form of ramified tubes, arranged in pairs in each ray, and emitting their products either into the surrounding medium, by means of efferent ducts which open round the mouth, or into the general body-cavity, by dehiscence, the external medium in this latter case being ultimately reached through the respiratory tubes. In their development, the *Asteroidea* show the same general phenomena as are characteristic of the class; but the larvæ are not provided with any continuous endoskeleton. In some *Asteroids* the larval forms have been described under the name of *Bipinnariæ*, and in these, as in the *Pluteus* of the Echinoids, a large portion of the larva is cast off as useless. In *Bipinnaria asterigera* (Sars) the digestive cavity is a simple sac which sends no prolongations into the rays, and the mouth is inter-radial, instead of being placed in the centre of the ambulacral system. The mouth of the adult is at this stage closed by the soft external skin of the larva.

The general shape of the body varies a good deal in different members of the order. In the common star-fish (*Uraster rubens*) the disc is small, and is furnished with long, finger-like rays, usually five in number. In the *Cribellæ* (fig. 42) the general shape of the body is very much the same. In the *Solasters* the disc is large and well marked, and the rays are from twelve to fifteen in number, and are narrow and short (about half the length of the diameter of the body). In the

Goniasters the body is in the form of a pentagonal disc, flattened on both sides; the true "disc" and rays being only visible on the under surface of the body. In none of the true star-fishes, however, are the arms ever sharply separated from the disc, as in the *Ophiuroidea*, but they are always an immediate continuation of it.

The order *Asteroidea* has been divided by Dr Gray as follows:—

ORDER ASTEROIDEA.

Section a. Ambulacra with four rows of feet.

Family 1. *Asteriadæ.* Dorsal wart simple.

Section b. Amublacra with two rows of feet.

Family 2. *Astropectinidæ.* Back flattish, netted with numerous tubercles, crowned with radiating spines at the tip, called "paxillæ."

Family 3. *Pentacerotidæ.* Body supported by roundish or elongated pieces, covered with a smooth or granular skin, pierced with minute pores between the tubercles.

Family 4. *Asterinidæ.* Body discoidal or pyramidal; sharp-edged; skeleton formed of flattish, imbricate plates; dorsal wart single, rarely double.

ORDER OPHIUROIDEA.—This order comprises the small but familiar group of the "Brittle-stars" and "Sand-stars," often considered as belonging to the *Asteroidea*, to which they are nearly allied. The body in the *Ophiuroidea* (fig. 43) is discoidal, and is covered with granules, spines, or scales, but pedicellariæ are wanting. From the body—which contains all the viscera—proceed long slender arms, which may be simple or branched, but which do not contain any prolongations from the stomach, nor have their under surface excavated into ambulacral grooves. The arms, in fact, are not simple prolongations of the body, as in the *Asteroidea*, but are special appendages, superadded for locomotive and prehensile purposes. Each arm is enclosed by four rows of calcareous plates, one on the dorsal surface, one on the ventral surface, and two lateral. In the centre of each arm is a chain of quadrate ossicles, forming a central axis, and between this axis and the row of ventral plates is placed the ambulacral vessel. Each ossicle of the central chain is composed of two symmetrical halves, but these are immovably articulated together, and are not movable upon one another, as in the *Asteroidea*. The mouth is situated in the centre of the inferior surface of the body, is provided with a masticatory apparatus, and is surrounded by tentacles. It opens directly into a sac-like ciliated stomach, which is not continued into an intestine, the mouth serving as an anal aperture. The stomach is destitute of lateral diverticula. The reproductive organs are situated near the bases of the arms, and open by orifices on the ventral surface of the body or in the interbrachial areas.

It is very questionable whether the ambulacral system in the adult *Ophiuroidea* communicates with the exterior; its place as a locomotive apparatus being taken by the arms. The radial vessels of the ambulacral system are not provided with secondary vesicles or "ampullæ," as they are in the *Echinoidea* and *Asteroidea*, and the lateral "feet" which they give off have no terminal suckers. The madreporiform tubercle is either

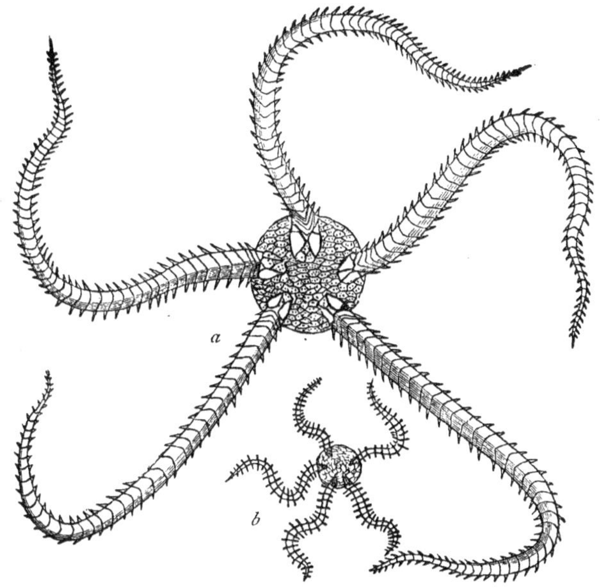

Fig. 43.—Ophiuroidea. *a Ophiura texturata*, the common Sand-star; *b Ophiocoma neglecta*, the grey Brittle-star (after Forbes).

placed on the inferior surface of the body or is partially concealed by one of the plates surrounding the mouth.

The larva of the *Ophiuroidea* is pluteiform, and is furnished with a continuous endoskeleton; and in some, as in *Ophiolepis squamata*, the echinoderm-body appears within the larva, when the latter has attained but a very imperfect degree of development.

In *Euryale* the body is in the form of a sub-globose disc with five obtuse angles, and the arms are prehensile. In *As-*

terophyton, the Medusa-head star, the arms are divided from the base, first dichotomously, and then into many branches. In *Ophiura*, the sand-star, the arms serve for reptation (creeping), and are undivided, often exceeding the diameter of the disc many times in length.

The order *Ophiuroidea* may be divided into two families, as follows:—

Family 1. *Ophiuridea*.
Genital fissures two or four in number. Arms five, always simple.
Family 2. *Asterophydiæ*.
Genital fissures ten in number. Arms five, simple or branched.

CHAPTER XXI.

CRINOIDEA, CYSTOIDEA, AND BLASTOIDEA.

ORDER CRINOIDEA.—The members of this order are *Echinodermata*, in which *the body is fixed, during the whole or a portion of the existence of the animal, to the sea-bottom by means of a longer or shorter, jointed, and flexible stalk.* The body is distinct, composed of articulated calcareous plates, bursiform, or cup-shaped, and provided with solid arms, which are primarily from five to ten in number, are independent of the visceral cavity, and are grooved on their upper surfaces for the ambulacra. (The position of the body being reversed, the *upper* surface is *ventral;* whilst the *dorsal* surface is *inferior*, and gives origin to the pedicle.) The tubular processes, however, which are given off from the radiating ambulacral canals of the *Crinoidea*, unlike those of the *Echinoidea* and *Asteroidea*, are not used in locomotion, but have probably a respiratory function. The mouth is central, and looks upwards, an anal aperture being sometimes present, sometimes absent. The ovaries are situated beneath the skin in the grooves on the ventral surfaces of the arms or pinnules, as are also the ambulacral or respiratory tubes. The arms are furnished with numerous lateral branches or "pinnulæ." The embryo is "free and ciliated, and develops within itself a second larval form, which becomes fixed by a peduncle."—(Huxley.)

Of those *Crinoidea* which are permanently fixed to the sea-bottom by a jointed pedicle, there exist but a few living forms, of which the best known is the *Pentacrinus Caput-Medusæ*. In this type of the *Crinoidea*—largely represented in past geological epochs—the body is composed of a series of calcareous plates, united together so as to form a cup or "calyx," the bottom of which is continued into a "column," or pedicle, composed of a series of calcareous joints or articulations, whereby the animal is fixed to some foreign body. The upper part of the calyx is roofed over by a series of calcareous plates, and is perforated by the apertures of the mouth and anus, the latter being sometimes absent. In the recent species the mouth is central, and there is a distinct anus at one side. The margin of the calyx gives origin to the arms, which are grooved on their upper (or ventral) surfaces for the ambulacra. In the living *Crinoids* the ambulacral grooves are continued along the upper surface of the calyx to the mouth. In the Palæozoic *Crinoids* there is only a single opening on the upper surface of the calyx, which is sometimes central and sometimes lateral, and which serves both as a mouth and anus. In many cases this aperture is level with the surface of the calyx, but in many species it is placed at the summit of a long projecting tube, which is termed the "proboscis." The ambulacral grooves in the Palæozoic *Crinoids* are found on the ventral surfaces of the arms, as in the living species; but instead of being continued over the surface of the body to the mouth, they stop short at the bases of the arms, where they gain access to the interior of the calyx by a series of special apertures.—(Billings.)

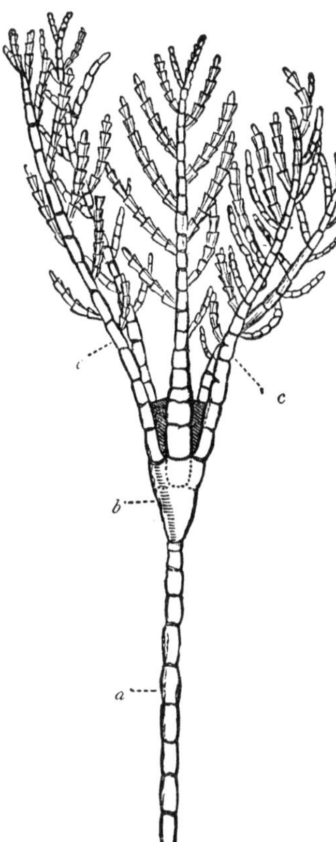

Fig. 44.—Crinoidea: *Rhizocrinus Lofotensis*, a living Crinoid (after Wyville Thomson), four times the natural size. *a* Stem. *b* Calyx. *c c* Arms.

More recently a stalked Crinoid has been discovered in the Atlantic and North Sea, and has been described under the name of *Rhizocrinus Lofotensis* (fig. 44). The chief interest of this form is the fact that it belongs to a group of the *Crinoidea* hitherto believed to be exclusively confined to the Mesozoic Rocks—viz., the *Apiocrinidæ* or "Pear-encrinites." In fact, *Rhizocrinus* is very closely allied to the Cretaceous genus *Bourgueticrinus*, and it may even be doubted if it is generically separable from it. The late remarkable researches into the life of the deeper parts of the ocean have brought to light several

Fig. 45.—Crinoidea. *Comatula rosacea*, the Feather-star; *a* Free adult; *b* Fixed young. (After Forbes.)

new Crinoids, which will doubtless, when fully investigated, still further fill up the interval between the living and extinct *Crinoidea*.

In the second type of the *Crinoidea*—represented in our seas by the *Comatula* (fig. 45), or Feather-star—the animal is not permanently fixed, but is only attached by a stalk when young (fig. 45, *b*), in which condition it was described as a distinct species, under the name of *Pentacrinus Europæus*. In its adult condition, however, the *Comatula* is free, and consists of a pentagonal disc, which gives origin to ten slender arms, which are fringed with many marginal pinnulæ or "cirri." The mouth and anus are on the ventral surface of the disc, which in this case is again the inferior surface, since the animal creeps about by means of its pinnated arms. The arms, in fact, of *Comatula* appear to be purely locomotive in function, and to be never employed as prehensile organs. The animal lives upon very minute organisms, drawn into the mouth by the action of the cilia lining the alimentary canal. The dorsal cirri, however, are employed to moor the animal temporarily to solid objects. The mouth is central in position, and the anus, which in some species forms a tubular projection, is situated on one side. Both the arms and the lateral pinnulæ are grooved on their ventral surfaces for the ambulacral vessels; and the pinnules also serve for the support of the reproductive organs. It is extremely doubtful if the ambulacral system, in the adult, has any communication with the exterior. The function, in fact, of the water-vascular system appears to be wholly respiratory, locomotion being entirely effected by means of the arms. The alimentary canal is confined entirely to the disc, and the stomach sends no diverticula along the arms as it does in the *Asteroidea*. The larva or pseudembryo of *Comatula rosacea* is a small ovate organism, with four transverse ciliated bands, a key-hole-shaped mouth, and a small vent and rudimentary intestine, the whole showing no traces of radiation. The young Crinoid is produced within the pseudembryo, and develops a fresh mouth, anus, and stomach for itself; the first being originally oro-anal in function, and being placed in the centre of the ambulacral system.

Several species of *Comatula* are known, and the genus appears to be cosmopolitan in its distribution. The genus *Ophicrinus* has been formed for species said to possess no more than five undivided arms.

ORDER CYSTOIDEA (*Cystidea*).—The members of this order are all extinct,* and are entirely confined to the Palæozoic period. The body (fig. 46) was more or less spherical, and was protected by an external skeleton, composed of numerous polygonal calcareous plates, accurately fitted together, and enclosing all the viscera of the animal. The body was in most cases permanently attached to the sea-bottom by means of a jointed calcareous "column," or pedicle, but this was much shorter than in the majority of *Crinoids*. Upon the

* Recently Professor Löven has described a singular Australian Echinoderm as being most closely allied to, if not truly referable to, the order *Cystoidea*. He has named this curious form *Hyponome Sarsi*, and believes it to be nearly related to the Cystidean genus *Agelacrinites*.

upper surface of the body were two, sometimes three apertures, the functions of which have been a matter of considerable controversy. One of these is lateral in position, is defended by a series of small valvular plates, and is believed by some to be the mouth, whilst by others it is asserted to have been an ovarian aperture. The view advocated by Mr Billings, that this aperture was the mouth, appears from every point of view to be most probably correct; or rather it was oro-anal, as was also the proboscis of the Palæocrinoids. The second opening is central in position, and is believed by Mr Billings to be the "ambulacral orifice," as it is always in the centre of the arms when these are present. The third aperture is only occasionally present, and doubtless discharged the functions of an anus.

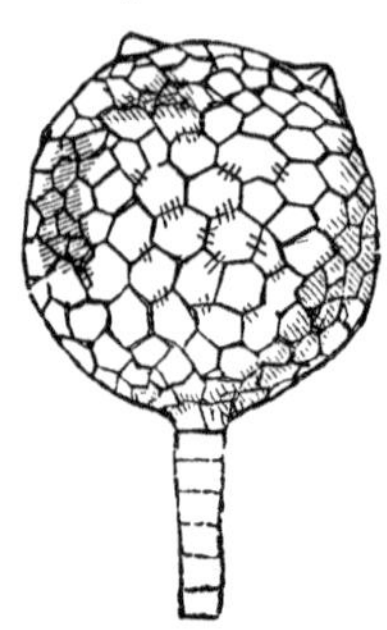

Fig. 46.—Cystidea. *Echinosphærites aurantium*, a Cystidean from the Bala Limestone (Lower Silurian).

In some *Cystoidea* there were no arms, properly speaking, but only small pinnulæ. In a second section true arms were present, but these were bent backwards, and were immovably soldered down to the body. In one single species (*Comarocystites punctatus*, Billings) the development has gone further, the arms being free, and provided with lateral pinnulæ, as in the true *Crinoids*.

Many Cystideans are likewise provided with a system of pores, or fissures, penetrating the plates of the body, and usually arranged in definite groups. These groups are termed "pectinated rhombs," but their exact function is doubtful. By Mr Billings, however, they are believed, and apparently with good reason, to have admitted water to the body-cavity, and to have thereby subserved a respiratory function.

Order Blastoidea.—The members of this order, like those of the preceding, are all extinct, and are entirely confined to the Palæozoic period. The body was fixed to the bottom of the sea by means of a short jointed pedicle; it was globular or oval in shape, and composed of solid polygonal calcareous plates, firmly united together, and arranged in five inter-ambulacral and as many ambulacral areas. (These ambulacral areas are termed by M'Coy "pseud-ambulacra," upon the belief that they were not pierced for tube-feet, but that they carried a double row of little jointed tentacles or arms.) The ambulacral areas are petaloid in shape, having a deep furrow down the centre, and striated transversely. They converge to the mouth, which is superior and central in position, and is surrounded by five ovarian apertures. No arms are present.

The *Blastoidea* are known more familiarly under the name of *Pentremites*, and they occur most commonly in the Carboniferous Rocks.

CHAPTER XXII.

HOLOTHUROIDEA.

ORDER HOLOTHUROIDEA.—The members of this order are commonly known by the name of "sea-cucumbers," "trepangs," or "bèches-de-mer," and are the most highly organised of all the *Echinodermata*. The body is elongated and vermiform, or rarely slug-shaped, and is not provided with a distinct test, but is enclosed in a coriaceous skin, sometimes containing scattered calcareous granules or spicules. The ambulacral tube-feet, when present, are usually disposed in five rows, which divide the body into an equal number of longitudinal segments or lobes. The mouth is surrounded by a circlet of feathery tentacles, containing prolongations from the central ring of the water-vascular system; and an anus is situated at the opposite extremity of the body. There is a long, convoluted intestine. A special respiratory, or water-vascular, system is usually developed, in the form of a system of arborescent tubes, which admit water from the exterior. The larva is vermiform, and has no skeleton (fig. 47). At a certain period of their existence, the young Holothurians are barrel-shaped, with transverse rings of cilia. They rotate rapidly on their long axis, and have at this stage been described as a distinct genus under the name of *Auricularia*.

In the *Holothuriæ* proper, locomotion is chiefly effected by means of rows of ambulacral tube-feet, or by alternate extension and contraction of the worm-like body; but in the *Synaptidæ* there are no ambulacra, but only the central circular canal of the ambulacral system, and the animal moves by means of anchor-shaped spicula, which are scattered in the integument. When developed, the ambulacral system consists of a "circular canal," surrounding the mouth, bearing one or more "Polian vesicles," and giving off branches to the tentacula; and of five "radiating canals" which run down the interspaces between the great longitudinal muscles. These radiating canals give off the tube-feet and their secondary vesicles, just as in the *Echinus*. In the typical forms there are five rows of tube-feet, but these may be scattered over the whole body, or may be restricted to the ventral surface. There is also a "sand-canal," which arises from the circular canal, and is terminated by a madreporiform tubercle; but this, instead of opening on the exterior, hangs down freely in the perivisceral cavity. The fluid, therefore, with which the ambulacral system is filled, is derived from the perivisceral cavity, and not from the exterior, as is usually the case.

The mouth in *Holothuria* is situated anteriorly, and is surrounded by a beautiful fringe of branched, retractile tentacles (fig. 47), which arise from a ring of calcareous plates, and into which are sent prolongations from the circum-oral ring of the ambulacral system. The mouth opens into a pharynx, which conducts to a stomach. The intestine is long and convoluted, and opens into a terminal dilatation, termed the "cloaca," which serves both as an anus and as an aperture for the

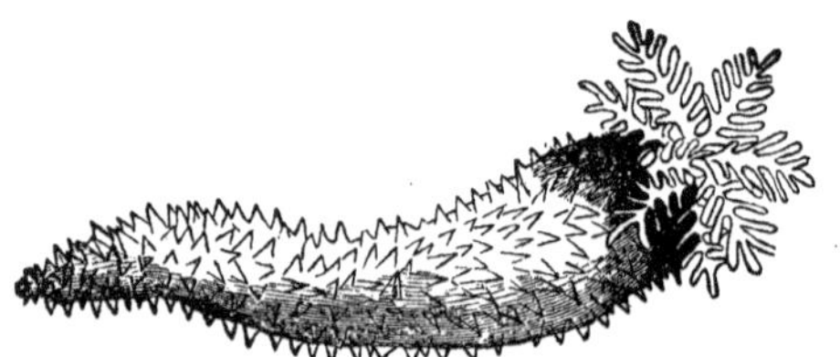

Fig. 47.—Holothuroidea. *Thyone papillosa*. (After Forbes.)

admission of sea-water to the respiratory tubes. From the cloaca arise two branched and arborescent tubes, the terminations of which are probably cæcal. These run up towards the anterior extremity of the body, and together constitute the so-called "respiratory tree." They are highly contractile, and they perform the function of respiratory organs, sea-water being admitted to them from the cloaca. The nervous system consists of a cord, surrounding the gullet, and giving off five branches, which run alongside of the radiating ambulacral canals. The generative organs are in the form of long, ramified, cæcal tubes, which open externally by a common aperture, situated near the mouth. There is thus no trace of that radial symmetry which is observed in the arrangement of the reproductive organs in the other orders of the *Echinodermata*. The vascular system consists of two main vessels—one dorsal, and the other ventral—connected with a circum-œsophageal ring.

The skin in the *Holothuriæ* is highly contractile, and the body is provided with powerful longitudinal and circular muscles, in compensation for the absence of any rigid integumentary skeleton. Many of the Sea-cucumbers, in fact, are endowed with such high contractility that they can eject their internal organs entirely, if injured or alarmed.

In *Synapta* there is no ambulacral system of tube-feet, nor respiratory tree. Locomotion is effected by means of little, anchor-shaped, calcareous spicules, placed upon little papillæ of the integument. Respiration is effected in the abdominal cavity, into which the water is admitted by five openings between the tentacles.

The order *Holothuroidea* may be divided into the following two families:—

Family I. *Holothuridæ.*

Body free, cylindrical, with a coriaceous integument containing scattered calcareous particles. An ambulacral system always, and a respiratory tree usually, present.

Family II. *Synaptidæ.*

Body free, covered with a coriaceous, sometimes soft, integument, containing minute, anchor-shaped spicules, by means of which the animal moves. The ambulacral system rudimentary, not giving rise to tube-feet, and not connected with locomotion. A respiratory tree sometimes present, sometimes absent.

CHAPTER XXIII.

DISTRIBUTION OF ECHINODERMATA IN SPACE AND TIME.

Distribution of Echinodermata in space.—The *Crinoidea* are represented by very few forms in recent seas, and these have a very local distribution. The *Comatulæ* are the commonest, and species have been found in most seas. The *Pentacrinus Caput-Medusæ* is exclusively confined, as far as is known, to the Caribbean Sea. *Rhizocrinus Lofotensis* has been dredged on the coast of Norway, and a form believed to be the same has been found in the Gulf of Mexico.

The *Asteroidea*, *Ophiuroidea*, and *Echinoidea* are represented in almost all seas, whether in tropical or temperate zones, some occurring very far north. The *Holothuroidea* have their metropolis in the Pacific Ocean, occurring abundantly on the coral-reefs of the Polynesian Archipelago. One species is collected in large numbers, and is exported to China, where it is regarded as a great delicacy.

DISTRIBUTION OF ECHINODERMATA IN TIME.—Numerous remains of *Echinodermata* occur in most sedimentary rocks, beginning with the base of the Lower Silurian Rocks, and extending up to the recent period. The two orders *Cystoidea* and *Blastoidea*, which are the most lowly organised of the entire class, are exclusively Palæozoic; and the *Crinoidea* are mostly referable to the same epoch. The more highly organised *Asteroidea* and *Ophiuroidea* commenced to be represented in the Silurian period; but the *Echinoidea*, with a single exception, have no representative earlier than the Carboniferous Rocks. The following exhibits the geological distribution of the different orders of the *Echinodermata* in somewhat greater detail:—

1. CRINOIDEA.*—The *Crinoidea* attained their maximum in the Palæozoic period, from which time they have gradually diminished down to the present day. As has already been described, the Palæozoic *Crinoidea* differ in some important particulars from those which succeeded them. The order is well represented in the Silurian, Devonian, and Carboniferous Rocks, but especially in the latter; many Carboniferous limestones (crinoidal limestones and entrochal marbles) being almost entirely made up of the columns and separate joints of Crinoids. In the Secondary Rocks Crinoids are still abundant. In the Trias the beautiful "Stone-lily" (*Encrinus liliiformis*) is peculiar to its middle division (Muschelkalk). In the Jurassic period occur many species of *Apiocrinus* (Pear-encrinite), *Pentacrinus*, and *Extracrinus*. The Chalk also abounds in Crinoids, amongst which is a remarkable unattached form (the Tortoise-encrinite or *Marsupites*).

Of the non-pediculate *Crinoidea*, which are a decided advance upon the stalked forms, there are few traces; but remains of *Comatula* have been discovered in the lithographic slate of Solenhofen (Oolite) and in the Chalk.

2. BLASTOIDEA.—The *Blastoidea*, or Pentremites, are entirely

* As regards the calyx of the fossil *Crinoidea*, the following terms are employed to designate its different parts. The base of the cup, or calyx, is termed the "pelvis," and it is made up of five, four, or sometimes three plates, which are termed the "basals." To the "basals" succeed two or three rows of plates, which are termed respectively the "primary radials," "secondary radials," and "tertiary radials," according to their distance from the basals. The axillary radials, which are the furthest removed, give origin to the arms, and are occasionally called the "scapulæ" (for this reason), whilst the primary and secondary radials are called the "costæ."

Palæozoic, and attain their maximum in the Carboniferous Rocks, some beds of which in America are known as the Pentremite Limestone, from the abundance of these organisms. They are, however, also found in the Silurian and Devonian Rocks.

3. CYSTIDEA.—These, like the preceding, are entirely Palæozoic; but they are, as far as is yet known, exclusively confined to the Upper Cambrian and Silurian Rocks, being especially characteristic of the horizon of the Bala Limestone. Forms supposed to be Cystideans have been described from the Devonian Rocks, but their true nature is doubtful. The oldest known Echinoderms are two extremely simple Cystideans (*Trochocystites* and *Eocystites*) which have been discovered in the primordial zone of North America.

4. ASTEROIDEA.—These have a very long range in time, extending from the Lower Silurian period up to the present day. In the Silurian Rocks the genera *Palæaster*, *Stenaster*, *Palæodiscus*, and *Petraster* are among the more important, the greater number of forms being Upper Silurian. The next period in which star-fishes abound is the Oolitic (Mesozoic); the more important genera being *Uraster*, *Luidia*, *Astropecten*, *Plumaster*, and *Goniaster*, some of which have survived to the present day. Many star-fishes occur, also, in the Cretaceous Rocks, the genera *Oreaster*, *Goniodiscus*, and *Astrogonium* being among the more noticeable. In the Tertiary Rocks few star-fishes are known to occur, but *Goniaster* and *Astropecten* are represented in the London Clay (Eocene).

5. OPHIUROIDEA.—The "brittle-stars" are represented in the Silurian Rocks by the single genus *Protaster*. In the Oolitic, Cretaceous, and Tertiary Rocks several genera of *Ophiuroidea* are known; some being extinct, whilst others (such as *Ophioderma*, *Ophiolepis*, and *Ophiocoma*) still survive at the present day.

6. ECHINOIDEA.—This order is represented in the Palæozoic Rocks by a single aberrant family; but it is numerously represented in the Mesozoic and Kainozoic periods.

For the Palæozoic *Echinoidea* the formation of a separate sub-order has been proposed by Professor M'Coy under the name of *Perischoechinidæ*, since they differ in some fundamental points from all the other known members of the order. The test is composed of *more than twenty* rows of calcareous plates, divided into five ambulacral and five inter-ambulacral areas. The five ambulacra are continuous from pole to pole, and are surmounted dorsally by the ocular plates. The five inter-ambulacra are composed, each, of three, five, or more rows of plates, and are surmounted dorsally by the ovarian plates. The two genera *Archæocidaris* and *Palæchinus* comprise all the known forms of the family, the former being entirely confined to the Carboniferous Limestone, whilst the latter occurs also in the Upper Silurians.

The Secondary and Tertiary *Echinoidea* resemble those now living in being composed of not more than twenty rows of calcareous plates. The Oolitic and Cretaceous Rocks are especially rich in forms belonging to this order, many genera being peculiar; but the number of forms is too great to permit of any selection.

7. HOLOTHUROIDEA.—This order, comprising, as it does, soft-bodied animals, can hardly be said to be known as occurring in the fossil condition. Some calcareous plates and spicules, supposed to belong to a *Holothurid*, have, however, been described as occurring in the Secondary Rocks, and the shield of *Psolus* has been found in Post-tertiary deposits in Bute.

CHAPTER XXIV.

SCOLECIDA.

CLASS II. SCOLECIDA.—This class was proposed by Professor Huxley for the reception of the remaining members of the *Annuloida*, comprising the *Rotifera*, the *Turbellaria*, the *Trematoda*, the *Tæniada*, the *Nematoidea*, the *Acanthocephala*, and the *Gordiacea*. Of these the *Rotifera* stand alone, whilst the *Turbellaria*, *Trematoda*, and *Tæniada* constitute the old division of the *Platyelmia* (Flat Worms), and the *Nematoidea*, *Acanthocephala*, and *Gordiacea* make up the old *Nematelmia* (Round Worms or Thread-worms). For some purposes these old divisions are sufficiently convenient to be retained, though they are of little scientific value. The term *Entozoa* has acquired such a general currency that it is necessarily employed occasionally, but it has been used in such widely different senses by different writers, that it would be almost better to discard it altogether. It certainly cannot be used as synonymous with *Scolecida*, many of these not being parasitic at all. It will, therefore, be employed here, in a restricted sense, to designate those orders of the *Scolecida* which are internal parasites, comprising the *Trematoda*, *Tæniada*, *Nematoidea* (in part), *Acanthocephala*, and *Gordiacea*. The *Turbellaria* and *Rotifera*, with a section of the *Nematoidea*, lead a free existence, and are not parasitic within other animals.

The *Scolecida* are defined by the possession of a "water-vascular system," consisting of a "remarkable set of vessels which communicate with the exterior by one or more apertures situated upon the surface of the body, and branch out, more or less extensively, into its substance."—(Huxley.) No proper vascular apparatus is present, and the nervous system (when present) "consists of one or two closely approximated ganglia." The habits and mode of life of the different members of the *Scolecida* are so different, that no other character, save the above, can be predicated which would be common to the entire class, and would not be shared by some other allied division.

DIVISION I. PLATYELMIA.—This section includes those *Scolecida* which possess a more or less flattened body, usually somewhat ovate in shape, and not exhibiting anything like distinct segmentation. The division includes two parasitic orders—the *Tæniada* and the *Trematoda*; and one non-parasitic order—viz., the *Turbellaria*. A sub-order, however, of this last, the *Nemertidæ*, does not conform to the above definition; but their other characters are such as to forbid their separation.

ORDER I. TÆNIADA (*Cestoidea*).—This order comprises the internal parasites, called Tape-worms (Cestoid worms), and the old order of the "Cystic worms" (Cystica); the latter being now known to be merely immature forms of the Tape-worms.

In their mature condition, the *Tæniada* (see fig. 48) are always found inhabiting the alimentary canal of some warm-blooded vertebrate animal; and they are distinguished by their great length, and by being composed of a number of flattened joints or articulations. These joints are not, however, an example of true segmentation, nor do they really constitute the Tape-worm; the true animal being found in the small, rounded, anterior extremity, the so-called "head," or "nurse," whilst the joints are simply hermaphrodite, generative segments, which the "head" throws off by a process of gemmation. The "head" (fig. 48, 3), which constitutes the real Tape-worm, is a minute, rounded body, which is furnished with a circlet of hooks or suckers, or both, whereby the parasite is enabled to maintain its hold upon the mucous membrane of the intestines of its host. No digestive organs of any kind are present, not even a mouth; and the nutrition of the animal is entirely effected by imbibition. The nervous system consists of two small ganglia, which send filaments backwards; but there is considerable obscurity on this point, and it has been asserted that the nervous system is entirely wanting, or that there is only a single ganglion. The "water-vascular system" consists of a series of long vessels which run down each side of the body, communicating with one another at each articulation by means of a transverse vessel, and opening in the last joint into a contractile vesicle. It thus appears that all the joints are organically connected together. Whilst the "head" constitutes the real animal, it, nevertheless, contains no reproductive organs, and these are developed in the joints or segments (fig. 48, 4), which are produced from the head posteriorly by budding. After the first joint, each new segment is intercalated between the head and the segment, or segments, already formed; so that the joints nearest the head are those latest formed, and those furthest from the head are the most mature. Each segment, when mature, contains both male and female organs of generation, and is, therefore, sexually perfect. To such a single segment the term "proglottis" is applied, from its resemblance in shape to the tip of the tongue. The ovary is a branched tube, which occupies the greater part of the proglottis, and opens, along with the efferent duct of the male organ, at a common papilla, which is perforated by an aperture, termed the "generative pore." The position of this pore varies, being placed in the centre of one of the lateral margins of the proglottis in the common Tape-worm (*Tænia solium*), but being situated upon the flat surface of the segment in the rarer *Bothriocephalus latus*. These two elements—namely, the minute head, with its hooklets and suckers, and the aggregate of the joints, or proglottides—together compose what is commonly called a "Tape-worm," such as is found in the alimentary canal of man, and of many animals. The length of this composite organism varies from a few inches to several yards.

Singular as is the composition of the mature Tape-worm, still more extraordinary are the phenomena observed in its development, of which the following is a brief account:—

"Proglottides," or the sexually mature segments of a Tape-worm, are only produced within the alimentary canal of man, or of some other warm-blooded vertebrate. The development of the ova which are contained in the proglottides, cannot, however, be carried out in this situation; hence the comparative harmlessness of this parasite, and hence the name of "solitary worm," which is sometimes applied to it. For the production of an embryo, it is necessary that the ovum should be swallowed by some animal other than the one inhabited by the mature Tape-worm. If this does not take place, the fecundated ovum is absolutely unable to develop itself. To secure this, however, the dispersion of the ova is provided for by the expulsion of the ripe proglottides from the bowel, all their contained ova having been previously fertilised. After their discharge from the body, the proglottides decompose, and the ova are liberated (fig. 48, 1), when they are found to be covered by a capsule which protects them from all ordinary mechanical, and even chemical, agencies, which might prove injurious to them. In this stage, the embryo is often so far developed within the ovum that its head may be recognised by its possession of three pairs of siliceous hooklets. For further development, it is now necessary that the ovum be swallowed by some warm-blooded vertebrate, and should thus gain access to its alimentary canal. When this takes place, the protective capsule or covering of the microscopically minute ovum is ruptured, either mechanically during mastication, or chemically by the action of the gastric juice; and the embryo is thus liberated. The liberated embryo is now called a "proscolex," and consists of a minute vesicle, which is provided with three pairs of siliceous spines, fitted for boring through the tissues of its host. Armed with these, the proscolex perforates the wall of the stomach, and may either penetrate some contiguous organ, or may gain access to some blood-vessel, and be conveyed by the blood to some part of the body, the liver being the one most likely.

Having by one of these methods reached a suitable resting-place, the proscolex now proceeds to surround itself with a cyst, and to develop a vesicle, containing fluid, from its posterior extremity, when it is called a "scolex" (fig. 48, 2). In some of the *Tæniada* the scolices are called "hydatids," and it is these, also, which constituted the old order of the "Cystic Worms." When thus encysted within the tissues of an animal, the "scolex" consists simply of a tænioid head, with a circlet of hooklets and four "oscula" or suckers, united by a contracted neck to a vesicular body. It contains no reproductive organs, or, indeed, organs of any kind, and cannot attain any further stage of development, unless it be swallowed and be taken for the second time into the alimentary canal of a warm-blooded vertebrate. It may increase, and produce fresh scolices, but this takes place simply by a process of gemmation. In some cases, however, a very partial and limited development does actually take place in the scolex prior to this change of abode, but this is an exceptional occurrence. In these cases the "neck" of the scolex becomes partially segmented, so that it comes to resemble an imperfectly developed *Tænia*, and is called a "strobila-embryo." The series of changes, however, whereby the scolex is converted into the "strobila," or adult tape-worm, cannot be carried out unless the scolex gain access to the alimentary canal of a warm-blooded vertebrate. In this case, the scolex attaches itself to the mucous membrane of the intestinal tube by means of its cephalic hooklets (when these are present) and suckers. The caudal vesicle now drops off, and the scolex is thus converted into the "head" of the tape-worm. Gemmation then com-

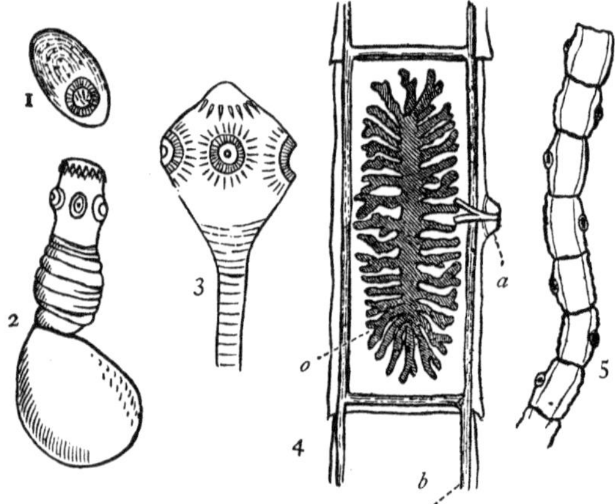

Fig. 48.—Morphology of Tæniada. 1. Ovum containing the embryo in its leathery case. 2. *Cysticercus longicollis.* 3. "Head" of adult *Tænia solium* enlarged, showing the hooklets and cephalic suckers. 4. A single generative joint, or proglottis, magnified, showing the dendritic ovary (*o*), the generative pore (*a*), and the water-vascular canals (*b*). 5. A portion of a Tape-worm (strobila), showing the alternate arrangement of the generative pores.

mences from its posterior extremity, the first segments being immature. As the first-formed joints, however, are pushed further from the head by the constant intercalation of fresh articulations, they become sexually mature, thus constituting the "proglottides" of the adult Tape-worm with which the cycle began. To the entire organism, with its "head" and its mature and immature joints ("proglottides"), the term "strobila" is now applied.

In the development, therefore, of the Tape-worm we have to remember the following stages:—

1. The *ovum*, set free from a generative joint, or proglottis.

2. The *proscolex*, or the minute embryo which is liberated from the *ovum*, when this latter has been swallowed by any warm-blooded vertebrate.

3. The *scolex*, or the more advanced, but still sexually imperfect, embryo, into which the *proscolex* develops, when it has encysted itself within the tissues of its host. (Under this head come the so-called "Cystic Worms.")

4. The *strobila*, or adult Tape-worm, into which the *scolex* develops itself, when received into the alimentary canal of a warm-blooded vertebrate. The strobila is constituted by the "head," and by a number of immature and mature generative segments or joints, termed the "proglottides."

The subject will, perhaps, be more clearly understood by following the development of one of the common Tape-worms of man—viz., the *Tænia solium*. Commencing with an individual who is already suffering from the presence of this parasite, one of the most distressing symptoms of the case is found to be the escape of the joints of the animal from the bowel. These joints are the ripe "proglottides," containing the fecundated ova. When the ova—which are microscopic in size—are liberated by the decomposition of the proglottis, they may gain access to water, or be blown about by the wind. In many ways, it is easy to understand how one of them may be swallowed by a pig. When this occurs, a "proscolex" is liberated from the ovum, and bores its way through the walls of the stomach, to become a "scolex." It now takes up its abode, generally in the muscles, in which position it was originally described as a cystic worm under the name of *Cysticercus cellulosæ*, constituting what is commonly known as the "measles" of the pig. In this state the scolex will continue for an indefinite period; but if a portion of "measly" pork be eaten by a man, then the scolex will develop itself into a tape-worm. The scolex fixes itself to the mucous membrane of the intestine, throws off its caudal vesicle, and commences to produce "proglottides" instead, becoming, thus, the "strobila" of the *Tænia solium*, with which we originally started. The other common tape-worm of man—viz., the *Tænia mediocanellata*—is derived in an exactly similar manner from the "measles" of the ox. The young, however, of another of the Tape-worms of man (viz., the *Bothriocephalus latus*) is said not to be "cystic." In like manner, the tape-worm of the cat (*Tænia crassicollis*) is the mature form of the cystic worm of the mouse (*Cysticercus fasciolaris*); the tape-worm of the fox (*Tænia pisiformis*) is derived from the cystic worms of hares and rabbits (*Cysticercus pisiformis*); and the tape-worm of the dog (*Tænia serrata*) is the developed form of the *Cœnurus cerebralis* of the sheep, the cystic worm which causes the "staggers" in the latter animal.

Besides tape-worms, however, man is liable to be affected with "scolices," which are the larvæ of the tape-worms of other animals. Thus, what are professionally called "hydatids" in the human subject, are really the scolices of the tape-worm of the dog. The disease is indicated by the presence of the so-called "hydatid-tumour," which consists of a strong membranous cyst—the "hydatid" proper—situated in some solid organ, most commonly the liver, and filled with a watery fluid. To the interior of the cyst are attached numerous minute scolices, many others also floating freely in the contained fluid. These "Echinococci," as they are called, do not differ in structure from other scolices, consisting of a head, provided with four suckers and a circlet of recurved hooklets, a vesicular body, and an intermediate contracted portion or neck. The Echinococci multiply within the hydatid cyst by gemmation, but they develop no reproductive organs. If, however, an Echinococcus should gain access to the alimentary canal of a dog, it then becomes the tape-worm peculiar to that animal—the *Tænia echinococcus*.

CHAPTER XXV.

TREMATODA AND TURBELLARIA.

ORDER TREMATODA.—This order includes a group of animals, which, like the preceding, are parasitic, and are commonly known as "suctorial worms," or "Flukes." They inhabit various situations in different animals—mostly in birds and fishes—and they are usually flattened or roundish in shape. The body is provided with one or more suctorial pores for adhesion. An intestinal canal, with one exception, is always present, but this is simply hollowed out of the substance of the body, and does not lie in a free space, or "perivisceral cavity." The intestinal canal is often much branched, and possesses but a single external opening, which serves alike as an oral and an anal aperture, and is usually placed at the bottom of an anterior suctorial disc. The sexes are united in the same individual. A "water-vascular system" is always present, and is sometimes "divided into two portions, one with contractile and non-ciliated walls, the other with non-contractile and ciliated walls."—(Huxley.)

The Trematode Worms are all hermaphrodite, and they pass through a series of changes in their development somewhat analogous to those observed in the *Tæniada*. This subject, however, is still involved in great obscurity, and it is too complicated to admit of description in this place. The larvæ are often tailed, but never possess cephalic hooklets, and are never "cystic."

From the absence of a perivisceral cavity, the *Trematoda* were formed by Cuvier into a separate division of *Entozoa*, under the name of *Vers Intestinaux parenchymateux*, along with the *Tæniada* and *Acanthocephala*, in which no alimentary canal is present. By Owen, for the same reason, they are included in a distinct class, under the name of *Sterelmintha*.

The *Distoma hepaticum* (fig. 49) may be taken as the type of the *Trematoda*. It is the common "Liver-fluke" of the sheep,

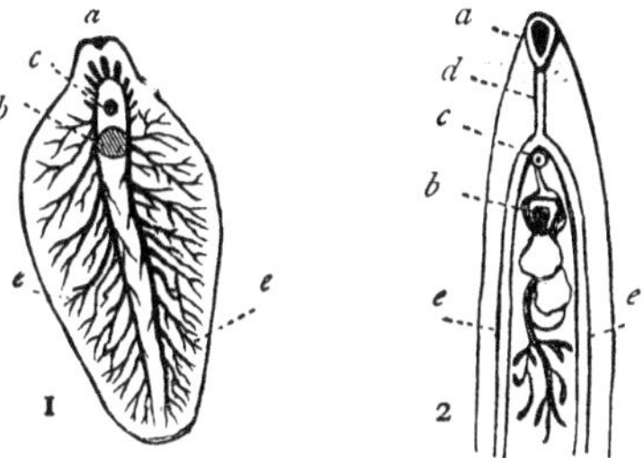

Fig. 49.—Trematoda. 1. *Distoma hepaticum*, the "Liver-fluke," showing the branched alimentary canal. 2. Anterior extremity of *Distoma lanceolatum*. *a* Anterior sucker; *b* Posterior sucker; *c* Generative pore; *d* Œsophagus; *e* Alimentary canal. (After Owen.)

and inhabits the gall-bladder or biliary ducts, giving rise to the disease known as the "rot." In form it is ovate, and flattened on its two sides, and it presents two suctorial discs, the anterior of which is perforated by the aperture of the mouth, whilst the posterior is impervious. Between the suckers is the "genital pore," at which the efferent ducts of the reproductive organs open on the exterior. A branched water-vascular system is present, and opens posteriorly by a small aperture. The alimentary canal bifurcates shortly behind the mouth, the two divisions thus produced giving off numerous lateral diverticula, and terminating posteriorly in blind extremities. The nervous system consists of a ring round the gullet, giving off filaments both forwards and backwards. The larvæ of *Distoma* are tailed or "cercariform," and are found in the interior of fresh-water snails.

In *Distoma lanceolatum* (fig. 49, 2) the intestine has not the ramose, complex character of that of *D. hepaticum*. On the other hand, the alimentary canal, after its bifurcation, is continued on each side of the body to the posterior extremity without giving off any branches on the way, and it terminates simply in blind extremities.

Diplostomum, in its essential characters, does not differ much from *Distoma;* but it is found living gregariously in the vitreous humour and lens of the eyes of certain fresh-water fishes, such as the common Perch.

Other members of the order infest the intestines of birds and Batrachians, the gills of fishes, or the paunch of Ruminants.

ORDER TURBELLARIA. — The members of this order are almost all aquatic, and are all non-parasitic; thus differing entirely from the animals which compose the two preceding orders. Their external surface is always and permanently ciliated, and they never possess either suctorial discs or a circlet of cephalic hooklets. A "water-vascular system" is always present, opening externally by one or more apertures, or appearing to be entirely closed in the adult (*Nemertida*). As in the *Trematoda*, the alimentary canal is imbedded in the parenchyma of the body, and, except in the *Nemertida*, there is no "perivisceral cavity." The intestine is either straight or branched, and a distinct anal aperture may, or may not, be present. The nervous system consists of ganglia situated in the fore-part of the body, united to one another by transverse cords, and sending filaments backwards.

The *Turbellaria* are divided into two sections, termed respectively the *Planarida* and the *Nemertida*.

SUB-ORDER I. PLANARIDA.—The Planarians (fig. 50) are mostly ovoid or elliptical in shape, flattened, and soft-bodied. They are for the most part aquatic in their habits, occurring in fresh water, or on the sea-shore, but occasionally found in moist earth. The integument is abundantly provided with vibratile cilia, which subserve locomotion, and it also contains numerous cells which have been compared to the "cnidæ," or nettle-cells, of the *Cœlenterata*. There is always a considerable portion of the body situated in front of the mouth, constituting the so-called "præ-oral region," or "prostomium"; and this is often modified into a singular protrusible and retractile organ, called the "proboscis," the exact use of which is not known. The mouth opens into a muscular pharynx, which is often evertible; and the intestine may be either straight or branched, but always terminates cæcally behind, and is never provided with an anal aperture. The "water-vascular system" communicates with the exterior by two or more contractile apertures. The nervous system consists of two ganglia, situated in front of the mouth, united by a commissure, and giving off filaments in various directions. Pigment-spots, or rudimentary eyes, from two to sixteen in number, are often present, and are always placed in the præ-oral region of the body. The male and female organs are united in the same individual, and the process of reproduction may be either sexual, by means of true ova, or non-sexual, by internal gemmation or transverse fission.

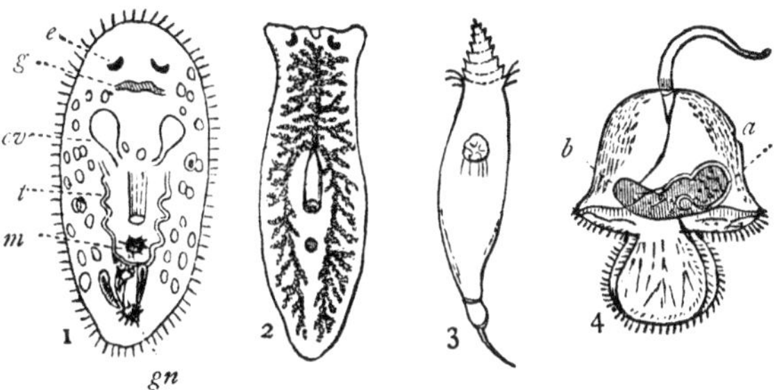

Fig. 50.—Morphology of Turbellaria. 1. *Planaria torva* (Muller); *m* Mouth; *g* Nerve-ganglion; *e* Eyes; *ov* Ovary; *t* Testis; *gn* Genital opening. 2. *Planaria lactea*, showing the branched (dendrocœl) intestine. 3. Microscopic larva of *Alaurina*, a marine Turbellarian. 4. *Pilidium*, the "pseudembryo" of a Nemertid; *a* The alimentary canal; *b* Rudiment of the Nemertid.

The *Planarians* have been divided into two sections, as follows:—

Section A. RHABDOCŒLA.—Intestine straight, not branched. Body elongated, rounded, or oval.

Section B. DENDROCŒLA.—Intestine branched or arborescent. Body flat and broad.

SUB-ORDER II. NEMERTIDA.—The *Nemertida*, or "Ribbon-worms," agree in most essential respects with the *Planarida*. They are distinguished, however, by their elongated, vermiform shape, by the presence of a distinct anus, by the possession of a distinct perivisceral cavity, by the absence of an external aperture to the water-vascular system of the adult, and by the fact that the sexes, with one or two exceptions, are distinct. The *Nemertida* further differ from the other *Platyelmia* in possessing a pseudohæmal system in addition to, and distinct from, the water-vascular system.

Reproduction takes place by the formation of true ova, by internal gemmation, or by transverse fission. In *Nemertes*, however, the egg gives rise to a larva, from which the adult is developed in a manner closely analogous to that described as characteristic of the *Echinodermata*. The larval form of *Nemertes* was described by Johannes Müller, under the name of *Pilidium* (fig. 50, 4). It is "a small hemet-shaped larva, with a long flagellum attached like a plumelto the summit of the helmet, the edges and side-lobes of which are richly ciliated. A simple alimentary canal opens upon the under surface of the body between the lobes. In this condition the larva swims about freely; but, after a while, a mass of formative matter appears on one side of the alimentary canal, and, elongating gradually, takes on a worm-like figure. Eventually it grows round the alimentary canal, and, appropriating it, detaches itself from the *Pilidium* as a Nemertid—provided with the characteristic proboscis, and the other organs of that group of *Turbellaria*."—(Huxley.)

CHAPTER XXVI.

NEMATELMIA.

1. Acanthocephala. 2. Gordiacea. 3. Nematoda.

Division II. Nematelmia.—This section may be considered as comprising those Scolecids in which the body has an elongated and cylindrical shape. Strictly speaking, it should include the *Nemertida*, but the division is not founded upon anatomical characters, and is employed here simply for convenience. Most of the *Nematelmia* possess an annulated integument; but there is no true segmentation, and there are rarely any locomotive appendages attached to the body. The majority are unisexual, and parasitic during the whole or a part of their existence. Three orders are comprised in this division—viz., the *Acanthocephala*, the *Gordiacea*, and the *Nematoda*.

Order I. Acanthocephala.—The *Acanthocephala* are entirely parasitic, vermiform in shape, and devoid of any mouth or alimentary canal. They are provided with a kind of snout or proboscis armed with recurved hooks, which is continued backwards into a bandlike structure (*ligamentum suspensorium*), to which the reproductive organs are attached. "Immediately beneath the integument lies a series of reticulated canals containing a clear fluid, and it is difficult to see with what these can correspond if not with some modification of the water-vascular system."—(Huxley.) This system of water-vascular canals, however, does not communicate, so far as is known, in any way with the exterior. At the base of the proboscis is placed a single nervous ganglion, which gives off radiating filaments in all directions.

Besides the presence of a water-vascular system and the absence of any alimentary canal, another point of affinity between the *Acanthocephala* and the *Tæniada* has recently been established by the discovery that the adult worm is developed within a hooked embryo, from which it is secondarily produced.

The "Thorn-headed worms" include some of the most formidable parasites with which we are acquainted. The *Echinorhynchus* (fig. 51) is found in the intestinal canal of many vertebrate animals, especially of birds and fishes.

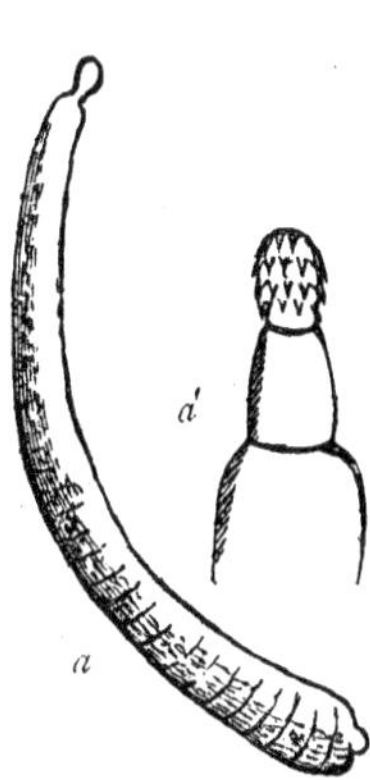

Fig. 51. — Acanthocephala; *a. Echinorhynchus gigas*, nat. size; *a'* The head of the same magnified.

ORDER II. GORDIACEA.—The *Gordiacea*, or "Hair-worms," are thread-like parasites, which in the earlier stages of their existence inhabit the bodies of various insects, chiefly of beetles and grasshoppers. They possess a mouth and alimentary canal, but they are not provided with a distinct anal aperture. In *Gordius* itself the gullet is said to open directly into the body-cavity; but it is more probable that this is an error, and that there is a complete intestine opening posteriorly into a cloaca. The sexes are distinct, and they leave the bodies of the insects which they infest in order to breed; subsequently depositing their ova in long chains, either in water or in some moist situation. At the time of its migration the mouth of the adult *Gordius* appears to be obliterated, and the anterior portion of the alimentary canal becomes atrophied.

In form the *Gordiacea* are singularly like hairs, and they often attain a length many times greater than that of the insect which harbours them.

ORDER III. NEMATODA (or *Nematoidea*).—The *Nematoda*—"Thread-worms" or "Round-worms"—are of an elongated and cylindrical shape; and are often, though by no means always, parasitic in the interior of other animals. They possess a distinct mouth and an alimentary canal which is freely suspended in an abdominal cavity, and terminates posteriorly in a distinct anus. They also possess a system of canals, in some cases contractile, which open externally near the anterior part of the body, on the ventral surface, or by lateral pores, and are probably homologous with the water-vascular system of the *Tæniada* and *Trematoda*. The sexes are distinct, and the males are usually less frequently met with, and of smaller size, than the females. The nervous system is mostly well developed, and is in the form of a ganglionic ring, surrounding the œsophagus, and sending filaments backwards.

As before said, most of the *Nematoda* are internal parasites, inhabiting the alimentary canal, the pulmonary tubes, or the areolar tissue, in man and in many other vertebrate animals; but a large section of the order are of a permanently free habit of existence.

The most familiar examples of the parasitic *Nematoda* are the *Ascaris lumbricoides*, the little *Oxyuris*, the *Trichina*, and the Guinea-worm.

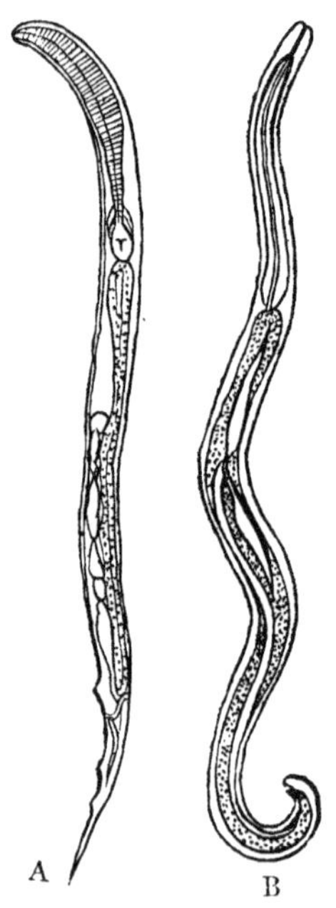

Fig. 52.—Nematoda. A *Anguillula aceti*; B *Dorylaimus stagnalis*. Magnified.

The *Ascaris lumbricoides*, or common Round-worm, inhabits the intestine of man, often attaining a length of several inches. The ova are probably expelled with the fæces, and the embryo is developed within the ovum prior to its rupture. When fully formed, the embryo is about one-hundredth of an inch in length, and its development is not exactly known, though it appears to be directly transferred from river or pond water to the alimentary canal of some vertebrate animal.

The *Oxyuris vermicularis*, or "Small Thread-worm," is a gregarious worm, which inhabits the rectum, especially of children. It is the smallest of the intestinal worms of man, its average length not being more than a quarter of an inch, but the females are much bigger than the males.

The *Trichina spiralis* is a singular Nematoid, which gives rise to a painful and very generally fatal train of symptoms, somewhat resembling rheumatic fever, and known as Trichiniasis. The

Trichina is known in two different conditions, sexually immature or mature. In its sexually immature condition it inhabits the muscles, usually of the pig, in vast numbers, each worm being coiled up in a little capsule or cyst. In this condition the worm is incapable of further development, and may remain, apparently for an indefinite period, without change, and without seeming to produce any injurious results to the animal affected. If, however, a portion of trichinatous muscle be eaten by a warm-blooded vertebrate, and so introduced into the alimentary canal, an immediate development of young *Trichinæ* is the result. The immature worms escape from their enveloping cysts, grow larger, develop sexual organs, and give birth to numerous progeny, which they produce viviparously. The young *Trichinæ* thus produced perforate the walls of the alimentary canal, and, after working their way amongst the muscles, become encysted. If the animal in which these changes go on has sufficient vitality to bear up under the severe symptoms which are produced by the migration of the *Trichinæ*, he is now safe; since they cannot become sexually mature, or develop themselves further, until again transferred to the alimentary canal of some other animal.

The Guinea-worm (*Dracunculus* or *Filaria medinensis*) is a Nematode worm, which inhabits, during one stage of its existence, the cellular tissue of the human body, generally attacking the legs, and often attaining a length of several feet. All known specimens of this parasite are impregnated females, containing a large number of young. The worm remains imbedded in the body, in a more or less quiescent condition, for a year or more, at the end of which time it seeks the surface, in order to get rid of its young. No external aperture to the genital organs has hitherto been proved to exist, and it seems possible that the young are produced within the body of the parent by a process of internal gemmation. The young *Filaria* consists of a vermiform body, terminating in a hair-like tail; and when set free from the parent, its further development probably takes place in water, when it is believed to be converted into one of the "Tank-worms" so common in India. In this condition it is possible, as some believe, that sexual organs are developed, and that the females are impregnated. The worm is believed to gain access to the body of bathers, when still extremely minute. According to Dr Bastian, however, it appears probable that the Guinea-worm "is a parasite only accidentally, and that it and its parents were originally free Nematoids."

The second section of the *Nematoda* comprises worms, which are not at any time parasitic, but which are permanently free. These "free Nematoids" (fig. 52) constitute the family of the *Anguillulidæ*, of which about two hundred species have been already described, mostly inhabiting fresh water or the shores of the sea. They resemble the parasitic Nematoids in all the essential features of their anatomy, but they differ in often possessing pigment-spots, or rudimentary eyes, in being mostly provided with a terminal sucker, and in bringing forth comparatively few ova at a time; the dangers to which the young are exposed being much less than in the parasitic forms. Amongst the more familiar Nematoids are the Vinegar Eel (*Anguillula aceti*, fig. 52, A) and the *Tylenchus* (or *Vibrio*) *tritici*, which produces a sort of excrescence or gall upon the ear of wheat, causing the disease known to farmers as the "Purples," or "Ear Cockle."

The parasitic and free Nematoids are connected together by an *Ascaris* (*A. nigrovenosa*), which in succeeding generations is alternately free and parasitic. This *Ascaris* has long been known as inhabiting the lungs of the frog, but it has been shown by Mecznikow that "the young of this animal become real, free Nematoids; for, after passing from the intestine of the frog into damp earth or mud, they grow rapidly, and actually develop in the course of a few days, whilst still in this external medium, into sexually mature animals. Young, differing somewhat in external characters from their parents, are soon produced by them, and these attain merely a certain stage of development whilst in the moist earth, arriving at sexual maturity only after they have become parasites, and are ensconced in the lung of the frog."—(Bastian.) This extraordinary history is rendered still more remarkable, if it should be proved that the young of the parasitic forms of this *Ascaris* are produced by a process of parthenogenesis; and this seems to be highly probable, since none of the individuals which are found as parasites are males, but are universally females.

CHAPTER XXVII.

ROTIFERA.

SUB-CLASS ROTIFERA (*Rotatoria*).—The *Rotifera*, or "Wheel-animalcules," constitute a very natural group, the exact position of which has been a good deal disputed, and is still doubtful. They are looked upon here as a distinct division of the *Scolecida*, following Huxley; but they are very frequently placed with the *Annelida* amongst the lower division of the *Annulosa* (*Anarthropoda*).

The *Rotifera* are *Annuloida of a minute size, never parasitic, inhabiting water, and usually provided with an anterior ciliated disc, capable of inversion and eversion. In the females there is a distinct mouth, intestinal canal, and anus. A nervous system is also present, consisting of ganglia, situated near the anterior extremity of the body, and sending filaments backwards. A water-vascular system is also present.*

Most of the *Rotifera* are entirely invisible to the naked eye, and they are all extremely minute, none of them attaining a greater length than 1-36th of an inch. Nevertheless, as remarked by Mr Gosse, "so elegant are their outlines, so brilliantly translucent their texture, so complex and yet so patent their organisation, so curious their locomotive wheels, so unique their apparatus for mastication, so graceful, so vigorous, so fleet, and so marked with apparent intelligence their movements, so various their forms and types of structure," that they form one of the most interesting departments of zoological and microscopical study. They are all aquatic in their habits, and in the great majority of cases are free-swimming animals, some, however, being permanently fixed, as is the case with *Stephanoceros*, *Melicerta* (fig. 53, B), and *Floscularia*. They are usually simple, but are occasionally composite, forming colonies, as in *Megalotrocha*. As a rule, the male and female *Rotifera* differ greatly from one another, the males being smaller than the females, destitute of any masticatory or digestive apparatus, and more or less closely resembling the young form of the species. The most characteristic organ in the great majority of the *Rotifera* is the so-called "wheel-organ," or "trochal disc," which is always situated at the cephalic or distal end of the body, and consists of a retractile disc surrounded by a circlet of cilia, which, when in action, vibrate so rapidly as to produce the illusory impression that the entire disc is rotating. The disc, which carries the cilia, is capable of eversion and inversion, and may be circular, reniform, bilobed, four-lobed, or divided into several lobes. It serves the purpose of locomotion in the free-swimming forms, acting somewhat like the propeller of a screw-steamer, and in all it serves to produce currents in the water, which convey the food to the mouth.

In *Chætonotus*, and one or two other forms, there is no true wheel-organ, capable of protrusion and retraction, but the cilia are variously disposed over the surface of the body. The *Chætonoti* or Hairy-backed Animalcules have no jaws, and have the ventral surface of the body clothed with cilia. They have usually been placed in the *Turbellaria*, but there seem to be good reasons for regarding them as an aberrant group of *Rotatoria.*

The proximal extremity of the body in the free forms terminates in a caudal process, or "foot," sometimes telescopic, which ends in a suctorial disc, or in a pair of diverging "toes," which act as a pair of forceps (fig. 53, A).

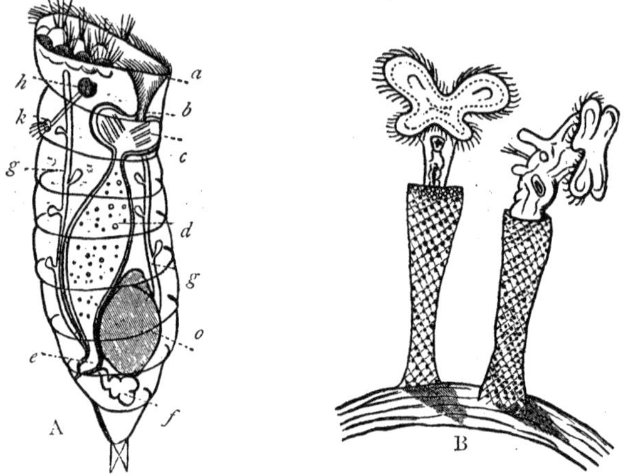

Fig. 53.—Rotifera. A Diagrammatic representation of *Hydatina senta* (generalised from Pritchard). *a* Depression in the ciliated disc leading to the digestive canal; *b* Mouth; *c* Pharyngeal bulb or mastax, with the masticatory apparatus; *d* Stomach; *e* Cloaca; *f* Contractile bladder; *g g* Respiratory or water-vascular tubes; *h* Nerve-ganglion giving filament to ciliated pit (*k*); *o* Ovary. B *Melicerta ringens.* (After Gosse.)

The mouth usually opens into a pharynx, or "buccal funnel," which is generally provided with a muscular coat, constituting the "mastax" or "pharyngeal bulb," and which generally contains a very complicated masticatory apparatus.* The parts of this apparatus are horny, and are believed by Mr Gosse to be homologous with the parts of the mouth in Insects. In the females of almost all known species of *Rotifera* the intestinal canal is a more or less simple tube, extending

* The lower jaws, or "incus," consist of a fixed portion, the "fulcrum," to which are attached two movable blades—the "rami." The upper jaws, or "mallei," consist each of a handle, or "manubrium," to which is hinged a toothed blade, or "uncus."

through a well-developed perivisceral cavity, and terminating posteriorly in a dilatation, or "cloaca," which forms the common outlet for the digestive, generative, and water-vascular systems.

In both sexes there is a well-developed water-vascular system, usually consisting of the following parts:—In the hinder part of the body, close to the cloaca, and opening into it, is a sac or vesicle, which is termed the "contractile bladder," and exhibits rhythmical contractions and dilatations. From the contractile bladder proceed two tubes—"the respiratory tubes"—which pass forwards along the sides of the body, and terminate anteriorly in a manner not quite ascertained. Attached to the sides of the respiratory tubes, in all the larger *Rotifera*, is a series of ovate or pyriform vesicles, each of which is furnished internally with a single central cilium, which is fixed to the free end of the vesicle. It is asserted, however, that these ciliated vesicles communicate internally with the perivisceral cavity with its contained corpusculated fluid. The exact function of this water-vascular system is not known, but it is most probably respiratory and excretory. Dr Leydig believes that water enters the perivisceral cavity by endosmose, where it mingles with the absorbed products of digestion, to form the so-called "chylaqueous fluid"; and that the effete fluid is excreted by the respiratory tubes, and ultimately discharged into the cloaca by the contractile bladder. Taking this view of the subject, Mr Gosse believes that "the respiratory tubes represent the kidneys, and that the bladder is a true urinary bladder"; and consequently that "the respiratory and urinary functions are in the closest relation with one another." This observer, further, finds a decided analogy between the above system in the *Rotifera* and the long and tortuous renal tubes of the *Insecta*, to which class he believes the *Rotifera* to be most nearly allied. No central organ of the circulation or heart and no organs of respiration are present, but the perivisceral cavity is filled with a corpusculated fluid.

The nervous system of the *Rotifera* constitutes a bilobate cerebral mass, "which for its proportionate volume may compare with the brain of the highest vertebrates." It is placed anteriorly, and usually on the dorsal aspect of the body, and the eye—in the shape of a red pigment spot or spots—is invariably situated like a wart upon it. Other sense-organs, probably tactile, are often present in the form of two knobs surmounted by tufts or bristles, placed at the back of the head. The ovaries constitute conspicuous organs in the female *Rotifera*, but in summer the young Rotifers appear to be produced by the females without having access to the males.

The muscular system of the *Rotifera* is well developed, consisting of bands which produce the various movements of the body and foot, whilst others act upon the various viscera, and others effect the movements of the jaws.

The typical group of the *Rotifera* is that of the *Notommatina* (*Hydatinea* of Ehrenberg). In this group the animals are all permanently free, and are never combined into colonies, while the integument is flexible, and the body is never encased in a tube.

Stephanoceros and *Floscularia*, on the other hand, are fixed, and are enclosed in a gelatinous tube which is secreted by the animal. *Melicerta* (fig. 53, B) inhabits a tubular case, which the animal forms for itself by means of a special organ for the purpose; whilst *Polyarthra* and *Triarthra* are protected by a stiff shell, or "lorica."

In *Triarthra* there are twelve ensiform fins, jointed to the body by distinct shelly tubercles, and moved by powerful muscles. These natatory organs are considered by Mr Gosse to be homologous with the articulated limbs of the *Arthropoda*.

In *Asplanchna*, whilst the masticatory organs, gullet, and stomach are well developed, there is no intestine, the stomach "hanging like a globe in the centre of the body-cavity," but not communicating with the body-cavity.

Affinities of Rotifera.—In their external appearance the *Rotifera* approximate closely to the *Infusoria*, but the organisation of the former presents a very striking advance when compared with that of the latter. Thus, in the *Infusoria* there is no differentiated body-cavity, bounded by distinct walls, and the alimentary canal is imperfect, the digestive sac simply opening inferiorly into the diffluent sarcode of the centre of the body. Further, there are no traces of a nervous system, and the contractile vesicles, if looked upon as representing the water-vascular system, are a very rudimentary form of this apparatus. In the *Rotifera*, on the other hand, the alimentary canal forms a complete tube, having an oral and an anal aperture, and not communicating with the surrounding perivisceral cavity; and there is a well-developed nervous system, and a highly complex water-vascular system. A real affinity is found to subsist, however, between the *Rotifera* and the *Planarida*; both possessing external cilia, a nervous system, and a well-developed water-vascular apparatus, the characters of which are not dissimilar in the two groups. In the *Planarida*, however, the sexes are united in the same individual, and there is no anal aperture; whereas in the *Rotifera* the sexes are distinct, and there is a distinct anus. To the true *Arthropoda*, as already pointed out, the *Rotifera* shows some points of affinity, but these are hardly sufficiently numerous or decided to warrant the removal of the group from the *Annuloida* to the *Annulosa*.

ANNULOSA.

CHAPTER XXVIII.

ANNULOSA.

1. General Characters of Annulosa. 2. General Characters of Anarthropoda. 3. Class Gephyrea. 4. General Characters of the Class Annelida.

Sub-kingdom Annulosa.—The members of this sub-kingdom are distinguished by the possession of *a body which is composed of numerous segments, or "somites," arranged along a longitudinal axis. A nervous system is always present, and consists of a double chain of ganglia, running along the ventral surface of the body, and traversed anteriorly by the œsophagus* (fig. 54). *The limbs (when present) are turned towards the neural aspect of the body.*

The sub-kingdom *Annulosa* may be divided into two primary divisions, according as the body is provided with articulated appendages, or not; these divisions being termed respectively

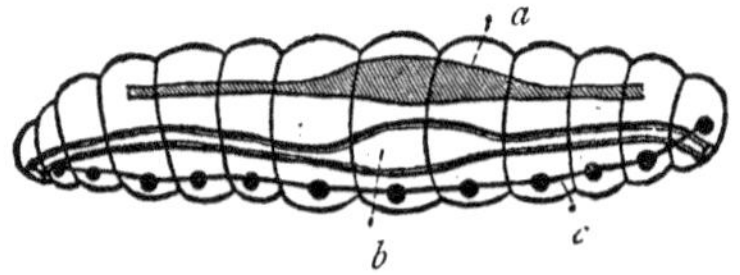

Fig. 54.—Diagram of an Annulose Animal. *a* Blood-vascular or hæmal system; *b* Digestive system; *c* Neural system.

the *Arthropoda* and *Anarthropoda.* The first of these comprises Crustaceans, Spiders, Scorpions, Centipedes, and Insects; whilst the latter includes the Spoon-worms, Leeches, Earthworms, Tube-worms, and Sand-worms.

Division I. Anarthropoda.—In this division of the *Annulosa the locomotive appendages are never distinctly jointed or articulated to the body.* In this division are included three classes, viz.:—the *Gephyrea*, the *Annelida*, and the *Chætognatha.*

Class I. Gephyrea (= *Sipunculoidea*).—This class includes certain worm-like animals in which the body is sometimes obviously annulated, sometimes not; but there are no ambulacral tubes nor foot-tubercles, though there are sometimes bristles concerned in locomotion. The nervous system consists of an œsophageal nerve-collar and a cord placed along the ventral surface of the body.

The *Sipunculus* and its allies (fig. 55) make up this class, and from their affinity to the worm-like Holothurians they have often been placed amongst the *Echinodermata*. They are not, however, provided with an ambulacral system, the integument is not capable of secreting calcareous matter, and there are no traces of any radiate arrangement of the nervous system.

The *Sipunculus* is a worm which is found burrowing in the sand of the coasts of most of our European seas, or which inhabits the cast-away shells of dead univalve Molluscs. The different species differ much in length, varying from half an inch to a foot or more. The body is cylindrical, covered by a delicate cuticle, beneath which is a thick, muscular, and highly

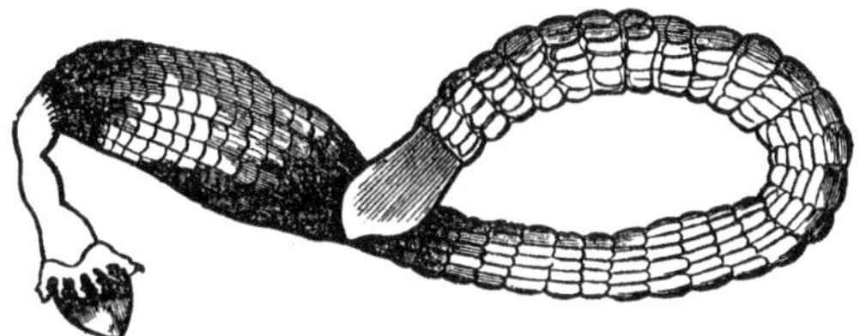

Fig. 55.—Gephyrea. *Syrinx nudus.* (After Forbes.)

contractile coat. The anterior portion of the body forms a retractile trunk or proboscis, at the extremity of which is the mouth surrounded by a circlet of simple tentacles. The alimentary canal is proportionately of great length, and is much convoluted. Upon reaching the posterior extremity of the body it is reflected forwards, and it terminates in a distinct anus, which is placed anteriorly near the junction of the body with the proboscis. The sexes are distinct in all the *Gephyrea*, and the young pass through a metamorphosis.

In *Echiurus*, which is found on the coasts of the North Sea, the body is provided posteriorly with zones of horny bristles; and in the *Sternaspis* of the Adriatic similar zones of bristles are found anteriorly as well as posteriorly. In the *Echiuridæ*, also, there are branched tubes connected with the termination of the intestine, which are doubtless homologous with the "respiratory tree" of the Holothurians. In *Bonellia* there is a proboscis formed by a folded fleshy plate, susceptible of great elongation, and forked at its extremity. The vent is at the opposite end of the body, and the intestine is very long, and folded several times.

The British species of the class are grouped by Professor E. Forbes as follows:—

Fam. I. *Sipunculacea*, having a retractile proboscis, at the base of which the anus is placed, and round the extremity of which is seen a circlet of tentacles.

Fam. II. *Priapulacea*, having a retractile proboscis but no tentacula, and having the anus placed at the extremity of a long, filiform, caudal appendage.

Fam. III. *Thalassemacea*, having a proboscis to which a long fleshy appendage is attached. There are no oral tentacula, and the anus is placed at the posterior extremity of the body.

CLASS II. ANNELIDA (= *Annulata*).—The *Annelida* are distinguished from the preceding by the possession of distinct external segmentation; the nervous system is composed of a ventral, double, gangliated cord, with an œsophageal collar and præ-œsophageal ganglion.

This class comprises elongated, worm-like animals, in which the integument is always soft, and the body is more or less distinctly segmented, each segment usually corresponding with a single pair of ganglia in the ventral cord. All the segments are similar to one another except those at the anterior and posterior extremities of the body. Each segment may also be provided with a pair of lateral appendages, but these are never articulated to the body, and are never so modified in the region of the head as to be converted into masticatory organs.

In the higher *Annelida* each segment (fig. 56) consists of two

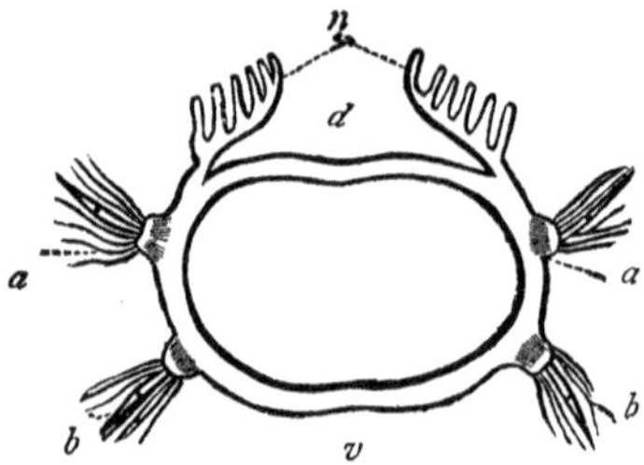

Fig. 56.—Diagrammatic transverse section of an Annelide. *d* Dorsal arc; *v* Ventra arc; *n* Branchiæ; *a* Notopodium or dorsal oar; *b* Neuropodium or ventral oar, both carrying setæ and a jointed cirrhus.

arches, termed, from their position, respectively the "dorsal arc" and the "ventral arc"; and each bears two lateral processes, or "foot-tubercles" (*parapodia*), one on each side. Each "foot-tubercle" is double, being composed of an upper process, called the "notopodium," or "dorsal oar," and a lower process, termed the "neuropodium," or "ventral oar." The foot-tubercles, likewise, support bristles, or "setæ," and a soft, cylindrical appendage, which is termed the "cirrhus" (fig. 56).

The number of the segments varies much, being as many as 400 in *Eunice gigantea;* and, generally, there is not a distinct head which is separable from the succeeding rings of the body. When such a distinct head appears to be present, it is not comparable with the head of the *Arthropoda*, but is really a greatly modified præ-oral region, or "prostomium," as is shown by the position of the mouth.

The digestive system of the Annelides consists of a mouth, sometimes armed with horny jaws, a gullet, stomach, intestine, and a distinct anus. Except in the *Hirudinea*, the alimentary canal is suspended in a capacious perivisceral space, divided into compartments by more or less complete partitions. The alimentary canal is, with one exception, not convoluted, and extends straight from the mouth to the anus; but lateral diverticula are often present.

As regards the vascular system, "no Annelide ever possesses a heart comparable to the heart of a Crustacean or Insect; but a system of vessels, with more or less extensively contractile walls, containing a clear fluid, usually red or green in colour, and in some cases only corpusculated, is very generally developed, and sends prolongations into the respiratory organs, when such exist."—(Huxley.) This system has been termed the "pseudo-hæmal system," and its vessels are considered by Professor Huxley as being "extreme modifications of organs homologous with the water-vessels of the *Scolecida*"; since the perivisceral cavity, with its contained corpusculated fluid (chylaqueous fluid), is believed by M. de Quatrefages to be the true homologue of the vascular system of Crustacea and insects. The pseudo-hæmal system, therefore, of the Annelides is to be regarded as essentially respiratory in function. The pseudo-hæmal vessels are sometimes wanting, and in these cases respiration appears to be effected by the cilia lining the perivisceral cavity.

Respiration is effected by the general surface of the body, by saccular involutions of the integument, or by distinct external gills, or branchiæ.

The nervous system consists of a double, ventral, gangliated cord, which is traversed anteriorly by the œsophagus; the "præ-œsophageal," or "cerebral," ganglia being connected by lateral cords or commissures with the "post-œsophageal" ganglia. Pigment-spots, or "ocelli," are present in many, generally upon the proboscis, sometimes in each segment, or on the branchiæ, or on the tail; and the head often supports two or more feelers which differ from the "antennæ" of Insects and Crustacea in not being jointed.

The sexes in the *Annelida* are sometimes distinct, and sometimes united in the same individual. The embryos are almost universally ciliated, and even in the adult cilia are almost always, if not always, present, in both of which respects this class differs from the *Arthropoda*.

The *Annelida* may be divided into two sections, characterised by the presence or absence of external respiratory organs or branchiæ. The Abranchiate section comprises the Leeches and the Earth-worms; whilst the Branchiate division includes the Tube-worms (*Tubicola*) and the Sand-worms (*Errantia*). The *Annelida* are also often divided into two sections, called *Chætophora* and *Discophora*, according as locomotion is effected by chitinous setæ (Earth-worms, Tube-worms, and Sand-worms) or by suctorial discs (Leeches).

CHAPTER XXIX.

ORDERS OF ANNELIDA.

ORDER I. HIRUDINEA (*Discophora* or *Suctoria*).—This order includes the Leeches, and is characterised by the possession of a locomotive and adhesive sucker, posteriorly or at both extremities, and by the absence of bristles and foot-tubercles. The sexes are united in the same individual, and the young do not pass through any metamorphosis.

The Leeches are aquatic, vermiform animals, mostly inhabiting fresh water, though a few species are marine. Locomotion is effected either by swimming by means of a serpentine bending of the body, or by means of one or two suctorial discs. In those forms in which there is only a single sucker (posterior), the head or anterior extremity of the body can be converted into a suctorial disc. The body is ringed, as many as one hundred annulations being present in the common Leech; but it is not divided into distinct somites, and, with one exception, there are no lateral appendages of any kind. The mouth is sometimes edentulous, but it is usually armed with teeth. The alimentary canal is short, and is united to the skin by means of a spongy vascular tissue. The pseudo-hæmal system consists principally of four great longitudinal trunks, connected by lateral vessels, and devoid of any special dilatations.* Respiration appears to be partly effected by means of a number of sacs, which are formed simply by an involution of the integument, and which open externally by minute apertures, termed "stigmata." In the common Leech there are about seventeen of these vesicles on each side of the body, their openings being placed on the abdominal surface. These saccular involutions of the integument certainly secrete the mucus with which the body of the animal is lubricated; and it is believed by some that their function is solely excretory, and that they answer to the kidneys of higher animals. In this case respiration must be effected by the general surface of the body; but there is no reason why the same organs should not perform both functions, since a close relationship subsists between the two. These sacculi are generally known as the "segmental organs," and in most of the *Hirudinea* they are closed internally, and only open externally by the "stigmata." In some of the *Hirudinea*, however, the "segmental organs" agree with those of the great majority of the Annelides in not only opening externally, but in also communicating internally with the perivisceral cavity. The segmental organs of the Leeches differ further from those of the other *Annelida* in not being in any way connected with the process of reproduction.

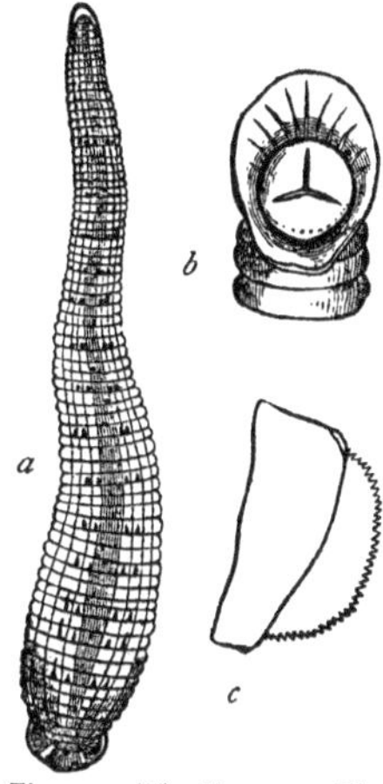

Fig. 57.—Hirudinea. *a* The Medicinal Leech (*Sanguisuga officinalis*), natural size; *b* Anterior extremity of the same magnified, showing the sucker and triradiate jaws; *c* One of the jaws detached, showing the semicircular toothed margin.

The nervous system consists of a præ-œsophageal ganglion, which gives branches to a number of simple eyes, or ocelli, which are placed on the head, and which is united by lateral œsophageal cords to the ventral gangliated chain.

The sexes are united in the same individual, but the Leeches are nevertheless incapable of self-fertilisation.

* If *Branchiobdella* be regarded as a true Leech, then the absence of gills is not universal in this order, for it possesses branchiæ. It has two suckers, and is parasitic on the Torpedo and on the gills of the Cray-fish.

Reproduction, also, is always effected by means of the sexes, and never by fission or gemmation.

The common Horse-leech is only provided with a few blunt teeth; but the Medicinal Leech (*Sanguisuga officinalis*, fig. 57) has its mouth furnished with three crescentic jaws, the convex surfaces of which are serrated with minute teeth. This species is chiefly imported from Hungary, Bohemia, and Russia. In *Sanguisuga medicinalis*, also used in medicine, the abdomen has numerous black spots. In both species the oral and caudal extremities are narrowed before dilating into the suckers, and the body has from ninety to one hundred rings. The marine *Pontobdellæ* have the body tuberculated, and attach themselves to the bodies of fishes, especially skates. The anterior sucker is separated in these from the body by a distinct constriction or neck. In the little fresh-water *Clepsinæ* the anterior sucker is wanting, and there is a proboscidiform mouth. They are found attached to the stems of water-plants.

ORDER II. OLIGOCHÆTA (*Terricola*).—The members of this order, comprising the Earth-worms (*Lumbricidæ*) and the Water-worms (*Naïdidæ*), are distinguished by the fact that their locomotive appendages are in the form of chitinous setæ or bristles, attached in rows to the sides and ventral surface of the body. They are all hermaphrodite. The *Oligochæta* are divided into the two groups of the *Terricolæ* or Earthworms, and the *Limicolæ* or Mud-worms and Water-worms (*Sænuridæ* and *Naïdidæ*).

In the common Earth-worm (*Lumbricus*) the body is cylindrical, attenuated at both extremities, and carrying in the adult a thickened zone, which occupies from six to nine rings in the anterior part of the body, is connected with reproduction, and is termed the "clitellum," or "saddle." Locomotion is effected by eight rows of short bristles or setæ, four of which are placed laterally and four on the ventral surface of the body; these representing the foot-tubercles of the higher Annelides. The mouth is edentulous, and opens into a short œsophagus, which leads to a muscular crop, or "pro-ventriculus," succeeded by a second muscular dilatation, or gizzard. The intestine is continued straight to the anus, and is constricted in its course by numerous transverse septa, springing from the walls of the perivisceral cavity. The perivisceral cavity (as in all the *Oligochæta*) is lined by a cellular membrane, which is continuous with a yellow cellular layer covering the intestine and large vessels, and which casts off its cells into the perivisceral fluid. The pseudo-hæmal system is well developed; and there exists, in even greater numbers, the same series of lateral sacculi or "segmental organs" which we have seen in the Leeches, and which have either a respiratory or a renal function. In all the *Oligochæta* the segmental organs communicate internally with the perivisceral cavity as well as externally with the outer medium. A portion of the segmental organs is ciliated, and in all cases the segmental organs of certain of the segments have the special function of acting as efferent ducts for the generative organs.

Of the little *Naïdidæ*, the most familiar is the *Tubifex rivulorum*, which is of common occurrence in the mud of ponds and streams. It is from half an inch to one inch and a half in length, and of a bright-red colour. The pseudo-hæmal system is provided with two contractile cavities or hearts; and there is present the same system of lateral tubes, opening externally by pores, as occurs in the Earth-worms.

The *Naïdidæ* are chiefly noticeable on account of the singular process of non-sexual reproduction which they present before they attain sexual maturity. In this process the *Naïs* throws out a bud between two rings, at a point generally near the middle of the body. Not only is this bud developed into a fresh individual, but the two portions of the parent marked out by the budding point likewise became developed into separate individuals. The portion of the parent in front of the bud develops a tail, whilst the portion behind the bud develops a head. Prior to the detachment of the bud, other secondary buds are formed from the same segment, each in front of the one already produced; and in this way, before separation takes place, a chain of organically connected individuals is produced, all of which are nourished by the anterior portion of the primitive worm. Besides their non-sexual reproduction, the *Naïdidæ* possess generative organs when adult, and exhibit true sexual reproduction. With the development of the generative organs, a new segment is added to the body, and certain other modifications take place; so that the process of attaining sexual maturity is actually attended with a species of metamorphosis.

ORDER III. TUBICOLA (*Cephalobranchiata*).—The Annelides which are included in this order inhabit tubes, which may be calcareous, and secreted by the animal itself, or may be composed of grains of sand or pieces of broken shell, cemented together by a glutinous secretion from the body. The body-rings are mostly provided with fasciculi of bristles set upon lateral foot-tubercles or parapodia, by means of which the animal is enabled to draw itself in and out of its tube. The alimentary canal is loosely attached to the integument. The

Tubicola are unisexual, and the young pass through a metamorphosis.

When the tube of a Tubicolar Annelide is a true calcareous secretion from the body of the animal, it is, nevertheless, readily distinguished from the shell of the Mollusca, by the fact that there is no organic connection of any kind between the animal and its tube.

The pseudo-hæmal system has its usual arrangement, and the contained fluid is usually red in colour, but is olive-green in *Sabella*. The respiratory organs are in the form of filamentous branchiæ, attached to, or near, the head, generally in two lateral tufts, arranged in a funnel-shaped or spiral form. Each filament is fringed with vibrating cilia, and the tufts are richly supplied with fluid from the pseudo-hæmal system. There is no special apparatus required to drive the blood back to the heart, but this is effected by the contractile power of the gills themselves. From the position of the branchiæ upon, or near, the head, the *Tubicola* are often known as the "cephalobranchiate" Annelides (fig. 58).

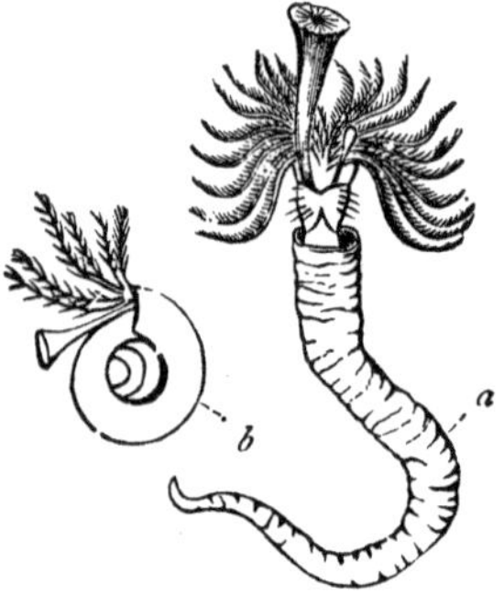

Fig. 58.—Tubicola. *a Serpula contortuplicata*, showing the branchiæ and operculum; *b Spirorbis communis*.

Reproduction in the *Tubicola* is generally sexual, the sexes being in different individuals; but spontaneous fission has also been observed. As regards their development, the process has been thus described, as it occurs in *Terebella*:—The embryo, which is at first a free-swimming, ciliated body, "lengthens, and the cilia, which were at first generally diffused, become confined to a cincture behind the head, a transverse ventral band near the tail, and a small circle round that part. The head is distinguished by two red eye-specks; new segments are successively added, one behind the other, and always in front of the anal one; but as yet the embryo is apodal. The tubercles and setæ are next developed in the same order, and a free-swimming or "errant" Annelide ensues. Finally, the cilia of the buccal rings are lost, the young *Terebella* reposes, and envelops itself in a mucous tube."—(Owen.) As the young tubicolar Annelide is thus free, or "errant," before it becomes finally enveloped in a tube, it is generally believed that the *Tubicola* should be looked upon as really higher than the next order of *Annelida*—viz. the *Errantia*. It appears, however, more probable that the stationary condition of the adult *Tubicola* should rather be regarded as an instance of "retrograde development."

The most familiar of the Tubicola is the *Serpula* (fig. 58, *a*), the contorted and winding calcareous tubes of which must be known to almost every one as occurring on shells or stones on the sea-shore. One of the cephalic cirrhi in *Serpula* is much developed, and carries at its extremity a conical plug, or operculum, whereby the mouth of the tube is closed, when the animal is retracted within it. The operculum of *Serpula* has a more than ordinary interest in the fact that it is the only instance in the *Annelida* in which calcareous matter is deposited within the integument. In *Spirorbis* (fig. 58, *b*) the shelly tube is coiled into a flat spiral, one side of which is fixed to some solid object. It is of extremely common occurrence on the fronds of sea-weed and on other submarine objects.

Equally familiar with *Serpula* is *Terebella*, the animal of which is included in a tube composed of sand and fragments of shell, cemented together by a glutinous secretion. In the *Sabellidæ* the tube is composed of granules of sand or mud. In *Pectinaria* the tube is free, membranous, or papyraceous, and in the form of a reversed cone of considerable length.

ORDER IV. ERRANTIA (*Nereidea*).—This order comprises *free* Annelides,* which possess setigerous foot-tubercles. The respiratory organs are generally in the form of tufts of external branchiæ, arranged along the back or the sides of the body. They are unisexual, and the young pass through a metamorphosis. This order includes most of the animals which are commonly known as Sand-worms and Sea-worms, together with the familiar Sea-mice.

The integument is soft, and the body is very distinctly divided into a great number of rings or segments, each of which, in the typical forms, possesses the following structure. The segment consists of two arches, a lower or "ventral arc," and an upper or "dorsal arc," with a "foot-tubercle" on each side. Each foot-tubercle consists of an upper process, or "notopodium," and a lower process, or "neuropodium," each of which carries a tuft of bristles, or "setæ," and a species of tentacle termed the "cirrhus" (fig. 56).

The anterior extremity of the body is usually so modified as to be distinctly recognisable as the head, and is provided

* Fritz Müller describes an errant Annelide belonging to the *Amphinomidæ* as living parasitically within the shell of the common Barnacle (*Lepas*), showing that the members of this group may sometimes lose their free habit.

with eyes, and with two or more feelers, which are not jointed, and are, therefore, not comparable with the antennæ of Crustacea and Insects. The mouth is placed on the inferior surface of the head, and is often furnished with one or more pairs of horny jaws, working laterally. The pharynx is muscular, and forms a sort of proboscis, being provided with special muscles, by means of which it can be everted and again retracted. In most there is no distinction between stomach and intestine, and the epithelium of the alimentary canal, like that of the preceding orders, is ciliated. The perivisceral cavity is filled with a colourless corpusculated fluid—the "chylaqueous fluid"—which "performs one of the functions of an internal skeleton, acting as the fulcrum or base of resistance to the cutaneous muscles, the power of voluntary motion being lost when the fluid is let out."—(Owen.)

The pseudo-hæmal system is well developed, and consists essentially of a long dorsal vessel, and a similar ventral one, connected by transverse branches, and furnished at the bases of the branchiæ with pulsating dilatations. The contained fluid is mostly red, but is yellow in *Aphrodite* and *Polynoe*.

Respiration is carried on by means of a series of external branchiæ or gills, arranged in tufts upon the sides of the body on its dorsal aspect, along the middle of the body only, or along its entire length. From the position of the branchiæ, the members of this order are often spoken of as the "Dorsibranchiate" (or more properly "Notobranchiate") Annelides. The "segmental organs," with few exceptions, communicate with the perivisceral cavity internally, and in certain segments they are always specialised to act as efferent ducts for the reproductive organs.

In the Sea-mouse (*Aphrodite*, fig. 60, B), the back is covered with a double row of membranous imbricated plates, which are called "elytra," or "squamæ," and respiration is effected by the periodical elevation and depression of these plates, whereby water is alternately admitted into, and expelled from, a space beneath them. This space is separated by a membrane from the perivisceral cavity below, and contains the gills in the form of small fleshy crests. The pharynx is thick and muscular, and can be everted like a proboscis, and the intestine has a number of lateral branched cæca.

The nervous system in the *Errantia* has its typical form, consisting of a double gangliated ventral cord, two ganglia of which are appropriated to each segment. The præ-œsophageal, or cerebral, ganglia are of large size, and send filaments to the ocelli and feelers.

The sexes in the *Errantia* are in different individuals, and reproduction is usually sexual, though in some cases gemmation is known to occur. The process of gemmation is carried on by a single segment, and so long as it continues, the budding individual remains sexually immature, though the young thus produced develop generative organs. Thus, there is in these cases a kind of alternation of generations, or rather an alternation of generation and gemmation; the oviparous individuals producing eggs from which the gemmiparous individuals are born; these, in their turn, but by a non-sexual process, producing the oviparous individuals.

The embryo usually appears, on its liberation from the ovum, as a free-swimming, ciliated body, possessing a mouth, intestine, and anus. The cilia are primarily diffused, but become aggregated so as to form a single median belt, or two bands, one about each extremity. The head, with its feelers and eye-

Fig. 59.—"Errant" Annelide. *Nereis*, showing the "head" with its appendages, and the setigerous parapodia.

specks, appears at one extremity, whilst the segments of the body begin to be formed at the other. Each segment is developed in four parts, the two principal ones forming half-rings, united by shorter side-pieces, from which the setigerous foot-tubercles are developed. The ciliated band or bands finally disappears, and new rings are rapidly added by intercalation between the head and the segments already formed.

Amongst the best known of the *Errantia* is the common Lob-worm (*Arenicola piscatorum*, fig. 60, C), which is used by fishermen for bait. The Lob-worm lives in deep canals which it hollows out in the sand of the sea-shore, literally eating its way as it proceeds, and passing the sand through the alimentary canal, so as to extract from it any nutriment which it may contain. It possesses a large head, without eyes or jaws, and with a short proboscis. There are thirteen pairs of branchiæ, placed on each side in the middle of the body.

In the *Nereidæ*, or "Sea-centipedes," the body is greatly

elongated, and consists of a great number of similar segments, with rudimentary branchiæ. The head is distinct, and carries eyes and feelers, whilst the mouth is furnished with a large proboscis, and often with two horny jaws (fig. 59). In the *Eunicea* the branchiæ are usually well developed and of large size, and the mouth is armed with seven, eight, or nine horny jaws. *Eunice gigantea* attains sometimes a length of over four feet, and may consist of more than four hundred rings.

DISTRIBUTION OF ANNELIDA IN TIME.—Of the *Annelida* the only orders which are known to have left any traces of their existence in past time are the *Tubicola* and the *Errantia;* of which the former are known by their investing tubes, whilst

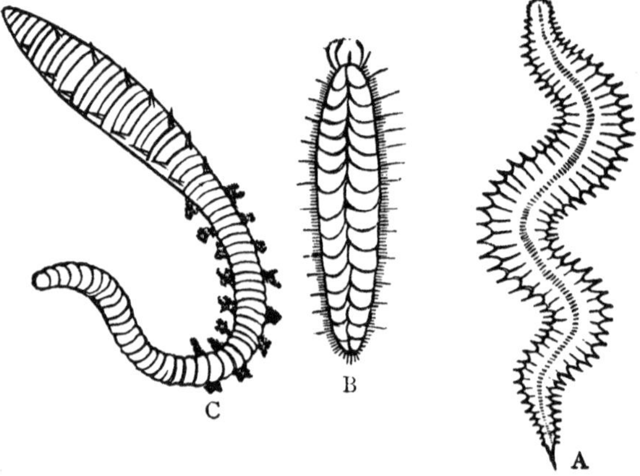

Fig. 60.—Errant Annelides. A. Hairy-bait (*Nephthys*); B. Sea-mouse (*Aphrodite*); C. Lob-worm (*Arenicola*). (After Gosse.)

the latter are only recognised by the tracks which they left upon ancient sea-bottoms, or by their burrows in sand or mud. These tracks and burrows of Annelides are found commonly in rocks of almost all ages from the Cambrian period upwards. Those tracks which have been caused simply by the passage of the worm over the surface of the mud are termed by Mr Salter *Helminthites*, whilst the burrows are called *Scolites* (or *Scolithus*).

Tubicolar Annelides are known to occur from the Silurian Rocks upwards. The well-known Silurian fossil, *Tentaculites*, is generally believed to belong to this order, but it is referred by M. Barrande to the *Pteropoda* (*Mollusca*). *Cornulites* and

Trachyderma, however, are undoubted Silurian *Tubicola*. The *Microconchus carbonarius* is a little spiral Tubicolar Annelide, nearly allied to the *Spirorbis* (fig. 58, *b*) of our seas, which is not uncommonly found in strata belonging to the Carboniferous period; and the genus *Spirorbis* itself is represented even in the Silurian period.

TABULAR VIEW OF THE ANNELIDA.

Division A. ABRANCHIATA.—No external organs of respiration.

Order I. *Hirudinea.*—No bristles or foot-tubercles: locomotion by means of a suctorial disc at one or both extremities. Ill. Gen. *Hirudo, Clepsine, Pontobdella.*

Order II. *Oligochæta.*—Locomotion by means of rows of stiff bristles, or "setæ;" no foot-tubercles. Ill. Gen. *Lumbricus, Naïs, Tubifex.*

Division B. BRANCHIATA.—Respiratory organs in the form of external branchiæ.

Order III. *Tubicola.*—Body protected by a calcareous or arenaceous tube. Branchiæ attached to, or near, the head (*Cephalobranchiata*). Ill. Gen. *Serpula, Terebella, Sabella.*

Order IV. *Errantia.*—Animal free, with setigerous foot-tubercles. Branchiæ in tufts, attached on the sides of the body, in the middle of dorsal region only, or along its entire length (*Dorsibranchiata*). Ill. Gen. *Arenicola* (Lob-worm), *Nereis* (Sea-centipede), *Aphrodite* (Sea-mouse).

CLASS III. CHÆTOGNATHA (Huxley).—The remaining class of the *Anarthropoda* has been recently constituted by Professor Huxley under the name of *Chætognatha*, for the reception of the single genus *Sagitta*, which had been formerly placed amongst the *Annelida*. By Professor Rolleston, however, the *Chætognatha* are placed in the division *Nematelmia* of the *Annuloida*, in the immediate neighbourhood of the *Nematoidea.*

The *Sagittæ* are singular marine animals, transparent, and elongated in form, and usually not more than an inch in length. The following are the characters ascribed to the class by Huxley:—

"The head is provided with several, usually six, sets of strong, bilaterally symmetrical oral setæ, two of which, long and claw-like, lie at the sides of the mouth; while the other four sets are short, and lie on that part of the snout which is produced in front of the oral aperture. The posterior part of the body is fringed on each side by a delicate striated fin-like membrane, which seems to be an expansion of the cuticle. In some species the body is beset with fine setæ. The intestine is a simple, straight tube, extending from the mouth to the anus; the latter opens on the ventral surface, just in front of the hinder extremity. A single oval ganglion lies in the abdomen, and sends, forwards and backwards, two pairs of lateral cords. The lateral cords unite in front of and above the mouth into a hexagonal ganglion. This gives off two branches which dilate at their extremities into the spheroidal ganglia, on which the darkly pigmented imperfect eyes rest. The ovaries, saccular organs, lie on each side of the intestine and open on either side of the vent; *receptacula seminis* are present. Behind the anus, the cavity of the tapering caudal part of the body is partitioned into two compartments; on the lateral parietes of these, cellular masses are developed which become detached, and, floating freely in the compartment, develop into spermatozoa. These escape by spout-like lateral ducts, the dilated bases of which perform the part of *vesiculæ seminales*. The embryos are not ciliated, and undergo no metamorphosis."—(See *Introduction to the Classification of Animals*, p. 52.)

CHAPTER XXX.

ARTHROPODA.

DIVISION II. ARTHROPODA, OR ARTICULATA.—The remaining members of the sub-kingdom *Annulosa* are distinguished by the possession of *jointed appendages, articulated to the body;* and they form the second primary division—often called by the name *Articulata*. As this name, however, has been employed in a wider sense than is understood by it here, it is, perhaps, best to adopt the more modern term *Arthropoda*.

The members of this division, comprising the *Crustacea* (Lobsters, Crabs, &c.), the *Arachnida* (Spiders and Scorpions), the *Myriapoda* (Centipedes), and the *Insecta*, are distinguished as follows:—

The body (fig. 54) is composed of a series of segments, arranged along a longitudinal axis; each segment, or "somite," occasionally, and some always, being provided with articulated appendages. Both the segmented body and the articulated limbs are more or less completely protected by a chitinous exoskeleton, formed by a hardening of the cuticle. The appendages are hollow, and the muscles are prolonged into their interior. The nervous system in all, at any rate in the embryonic condition, consists of a double chain of ganglia, placed along the ventral surface of the body, united by longitudinal commissures, and traversed anteriorly by the œsophagus. The hæmal system, when differentiated, is placed dorsally, and consists of a contractile cavity, or heart, provided with valvular apertures, and communicating with a perivisceral cavity, containing corpusculated blood. Respiration is effected by the general surface of the body, by gills, by pulmonary sacs, or by tubular involutions of the integument, termed "tracheæ." In no member of the division are vibratile cilia known to be developed. According to Professor Huxley, an additional constant character of the *Arthropoda* is to be found in the structure of the head, which is typically composed of six segments, and never contains less than four.

The *Arthropoda* are divided into four great classes—viz., the *Crustacea*, the *Arachnida*, the *Myriapoda*, and the *Insecta;* which are roughly distinguished as follows :—

1. CRUSTACEA.—*Respiration by means of gills, or by the general surface of the body. Two pairs of antennæ. Locomotive appendages more than eight in number, borne by the segments of the thorax, and usually of the abdomen also.*

2. ARACHNIDA.—*Respiration by pulmonary vesicles, by tracheæ, or by the general surface of the body. Head and thorax united into a cephalothorax. Antennæ (as such) absent. Legs eight. Abdomen without articulated appendages.*

3. MYRIAPODA.—*Respiration by tracheæ; head distinct; remainder of the body composed of nearly similar somites. One pair of antennæ. Legs numerous.*

4. INSECTA.—*Respiration by tracheæ. Head, thorax, and abdomen distinct. One pair of antennæ. Three pairs of legs, borne on the thorax. Abdomen destitute of limbs. Generally two pairs of wings on the thorax.*

CHAPTER XXXI.

CRUSTACEA.

CLASS I. CRUSTACEA.—The members of this class are commonly known as Crabs, Lobsters, Shrimps, King-crabs, Barnacles, Acorn-shells, &c. They are nearly allied to the succeeding order of the *Arachnida* (Spiders and Scorpions); but may usually be distinguished by the possession of articulated appendages upon the abdominal segments, by the possession of two pairs of antennæ, and by the presence of branchiæ.

"In the *Crustacea* the body is distinguishable into a variable number of 'somites,' or definite segments, each of which may be, and some of which always are, provided with a single pair of articulated appendages. . . . In most Crustacea, and probably in all, one or more pairs of appendages are so modified as to subserve manducation. A pair of ganglia is primitively developed in each somite, and the gullet passes between two successive pairs of ganglia, as in the *Annelida*.

"No trace of a water-vascular system, nor of any vascular system similar to that of the *Annelida*, is to be found in any Crustacean. All *Crustacea* which possess definite respiratory organs have branchiæ or outward processes of the wall of the body, adapted for respiring air by means of water; the terrestrial *Isopoda*, some of which exhibit a curious rudimentary representation of a tracheal system, forming no real exception to this rule. When they are provided with a circulatory organ, it is situated on the opposite side of the alimentary canal to the principal chain of ganglia of the nervous system; and communicates by valvular apertures with the surrounding venous sinus—the so-called 'pericardium.'"—(Huxley.)

In addition to these characters, the body in the *Crustacea* is always protected by a chitinous or sub-calcareous exoskeleton, or "crust," and the number of pairs of articulated limbs is generally from five to seven. They all pass through a series of metamorphoses before attaining their adult condition, and every part that is found in an embryonic form, even though only temporarily developed, may be represented in a permanent condition in some member of a lower order.

The classification of the Crustacea is extremely complicated, and hardly any two writers adhere to the same arrangement. The tabular view which follows embodies the arrangement which appears to be most generally adopted, and the diagnostic characters of each order will be briefly given, a more detailed description being reserved for the more important divisions of the class. Before proceeding further, however, it will be as well to give a description of the morphology of a typical *Crustacean*, selecting the Lobster as being as good an example as any.

The body of a typical *Crustacean* may be divided into three regions—a *head*, a *thorax*, and an *abdomen*, each of which is composed of a certain number of somites, though opinions differ both as to the number of segments in each region, and as to their number collectively. By the majority of writers the body is looked upon as being typically composed of *twenty-one* segments, of which seven belong to the head, seven to the thorax, and seven to the abdomen. In many *Crustacea*, however, the segments of the head and thorax are welded together into a single mass, called the "cephalothorax;" in which case the body shows only two distinct divisions, of which the cephalothorax claims fourteen segments, whilst the remaining seven are allotted to the abdomen. By Professor Huxley, on the other hand, the terminal joint of the abdomen, termed the "telson," is regarded as an *appendage*, and not as a somite. Upon this view, the body of a typical Crustacean will consist of *twenty* segments only. Professor Huxley, further, differs from the above-mentioned view in the allotment of the somites, and he divides the body into six cephalic, eight thoracic, and six abdominal somites.* Fritz Müller and Claus deny that the eyes are limbs, or that there is an ocular segment. The telson, on the other hand, is regarded by the former as a true somite, chiefly because the intestine usually opens in this piece.

Whilst the normal number of segments in the body of any Crustacean may thus be regarded as being twenty-one, or twenty, there occur cases in which this number is exceeded, and others in which the number of somites is apparently less. In these latter cases, however, the apparent diminution in the number of segments is really due to some having been fused together, as is shown by the number of appendages, since each pair of appendages indicates a separate somite. In other cases, however, in which the number of somites is really less than the normal, this is due to an arrest of development. According to Milne-Edwards:—

"In the embryo these segments are formed in succession from before backwards, so that, when their evolution is checked, the later, rather than the earlier, rings are those which are wanting; and, in fact, it is generally easy to see in those specimens of full-grown Crustaceous animals whose bodies present fewer than twenty-one segments, that the anomaly depends on the absence of a certain number of the most posterior rings of the body." According to Dana, however, the abortion of segments, with their appendages, almost always takes place at the posterior end of the cephalothorax.

In no single example can a general view be obtained of the different segments and their appendages in the *Crustacea*. "Indeed, the only segment that may be said to be persistent, is that which supports the mandibles, for the eyes may be

* In reality the five hindmost segments of the eight somites here allotted to the thorax, should alone be regarded as constituting the *abdomen* proper, —that is, the region corresponding to the "abdomen" of insects and Arachnida. The six somites allotted above to the *abdomen* belong to what is strictly called the "*post-abdomen*" of the *Crustacea*.

wanting, and the antennæ, though less liable to changes than the remaining appendages, are nevertheless subject to very extraordinary modifications, and have to perform functions equally various. Being essentially and typically organs of touch, hearing, and perhaps of smell, in the highest Decapods, they become converted into burrowing organs in the *Scyllaridæ*, organs of prehension in the *Merostomata*, claspers for the male in the *Cyclopoidea*, and organs of attachment in the *Cirripedia*. Not to multiply instances, we have presented to us in the *Crustacea* probably the best zoological illustration of a class, constructed on a common type, retaining its general characteristics, but capable of endless modification of its parts, so as to suit the extreme requirements of every separate species."—(H. Woodward.)

Taking the common Lobster (fig. 74) as a good and readily obtainable type of the *Crustacea*, the body is at once seen to be composed of two parts, familiarly called the "head" and the "tail," the latter being jointed and flexible. The so-called "head" is really composed of both the head, properly so called, and the thorax, which have coalesced so as to form a single mass, technically called the "cephalothorax." The so-called "tail," on the other hand, is truly the "abdomen." The various appendages of the animal are arranged along the lower surface of the body, and consist of the feelers, jaws, claws, legs, &c. The entire body, with the articulated appendages, is enclosed in a strong chitinous "shell," or exoskeleton, and the cephalothorax is covered by a great cephalic shield or plate, which is termed the "carapace."

Each segment of the body may be regarded as essentially composed of a convex upper plate, termed the "tergum," which is closed below by a flatter plate, called the "sternum," the line where the two unite being produced downwards and outwards, into a plate, which is called the "pleuron," or "pleura" (fig. 62, 2).

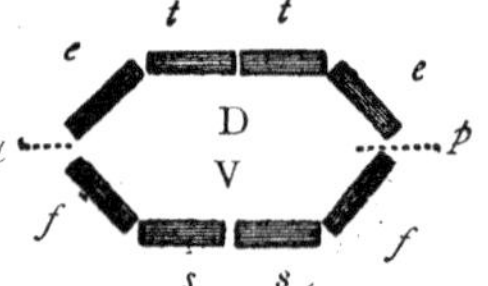

Fig. 61.—Theoretical figure illustrating the composition of the tegumentary skeleton of the Crustacea (after Milne-Edwards). D Dorsal arc; *t t* Tergal pieces; *e e* Epimeral pieces; V Ventral arc; *s s* Sternal pieces; *f f* Episternal pieces; *p p* Insertion of the extremities.

Strictly speaking, the composition of the typical somite is considerably more complex, each of the primary arcs of the somite being really composed of four pieces. The tergal arc is composed of two central pieces, one on each side of the middle line of the body, united together, and constituting the "tergum" proper. The superior arc is completed by two lateral pieces, one on each side of the tergum, which are termed the "epimera." In like manner the ventral or sternal arc is composed of a central plate, composed of two pieces united together in the middle line, and constituting the "sternum" proper; the arc being completed by two lateral pieces, termed the "episterna." These plates are usually more or less completely anchylosed together, and the true structure of the somite in these cases is often shown by what are called "apodemata." These are septa which proceed inwards from the internal surface of the somite, penetrating more or less deeply between the various organs enclosed by the ring, and always proceeding from the line of junction of the different pieces of the segment (fig. 61).

It must be borne in mind that though the so-called "head" —that is to say, the "cephalothorax"—of the Lobster is produced by an amalgamation of the various somites of the head and thorax, this is not the case with the great shield which covers this portion of the body. This shield—the so-called "cephalic buckler," or "carapace"—is not produced by the union of the tergal arcs of the various cephalic and thoracic segments, as would at first sight appear to be the case. On the contrary, the "carapace" in the higher *Crustacea* is produced by an enormous development of the tergal pieces, or of the "epimera" of one or two of the cephalic segments: the tergal arcs of the remaining somites being overlapped by the carapace and remaining undeveloped.

Examining the somites from behind forwards (for simplicity's sake), the last segment comes to be first described. This is the so-called "telson," which forms the last articulation of the abdomen, and never bears any appendages. For this reason, many authorities do not regard it as a somite, properly speaking, but simply as an azygos appendage—that is to say, as an appendage without a fellow. In the next segment (the last but one, or the last, of the abdomen, according to the view which is taken of the "telson"), there is a pair of natatory appendages, called "swimmerets." Each swimmeret (fig. 62, 2) consists of a basal joint, which articulates with the sternum, and is called the "protopodite" or propodite, and of two diverging joints, which are attached to the former; the outer of these being called the "exopodite," and the inner the "endopodite." In this particular segment, the exopodite and endopodite are greatly expanded, so as to form powerful paddles, and the exopodite is divided into two by a transverse joint. In the succeeding somites of the abdomen—with the exception of the first, in which there is some modification—the appendages are in the form of swimmerets, essentially the same as those attached to the penultimate segment, and differing only in the fact, that the exopodite and endopodite are much narrower, and the former is undivided (fig. 62, 2). The last thoracic somite—immediately in front of the abdomen—carries a pair of the walking or ambulatory legs, each consisting of a short basal piece, or "propodite," and of a long jointed "endopodite," the "exopodite" not being developed. The

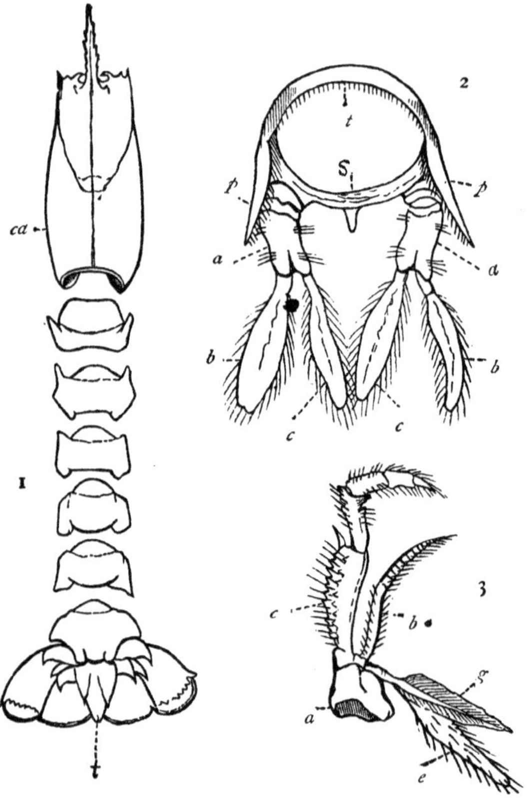

Fig. 62.—Morphology of Lobster. 1. Lobster with all the appendages, except the terminal swimmerets, removed, and the abdominal somites separated from one another. *ca* Carapace; *t* Telson. 2. The third abdominal somite separated. *t* Tergum; *s* Sternum; *p* Pleuron; *a* Protopodite; *b* Exopodite; *c* Endopodite. 3. One of the last pair of foot-jaws or maxillipedes. *e* Epipodite; *g* Gill; the other letters as before.

next thoracic segment carries another pair of ambulatory limbs, quite similar to the last, except for the fact that the protopodite bears a process which serves to keep the gills apart, and is termed the "epipodite." The succeeding segment supports a pair of limbs similar to the last in all respects, except that its extremities, instead of being simply pointed, are converted into nipping claws, or "chelæ." The next segment of the thorax carries a pair of chelate limbs, just like the preceding, and the next is furnished with appendages, which are essentially the same in structure, but are much larger, constituting the great claws. The next two segments of the thorax, and the segment in front of these (by some looked upon as belonging to the head, by others as referable to the thorax), bear each a pair of modified limbs, which are termed "maxillipedes," or "foot-jaws." These are simply limbs with the ordinary structure of protopodite, exopodite, endopodite, and epipodite, but modified to serve as instruments of mastication, the hindmost pair being less altered than the two anterior pairs (fig. 62, 3). The next two somites carry appendages, which are in the form of jaws, and are termed respectively the first and second pairs of "maxillæ." Each consists of the parts aforementioned, but the epipodite of the first pair of maxillæ is rudimentary, whilst that of the second pair is large, and is shaped like a spoon. It is termed the "scaphognathite," and its function is to cause a current of water to traverse the gill-chamber by constantly baling water out of it. The next segment carries the biting jaws, or "mandibles;" each of which consists of a large protopodite, and a small endopodite, which is termed the "palp," whilst the exopodite is undeveloped. The aperture of the mouth is situated between the bases of the mandibles, bounded behind by a forked process, called the "labium," or "metastoma," and in front by a single plate, called the "labrum" (upper lip). The next segment bears the long antennæ, or feelers (fig. 74, *ga*), each consisting of a short protopodite, and a long, jointed, and segmented endopodite, with a very rudimentary exopodite. In front of the great antennæ are the next pair of appendages, termed the "antennules," or smaller antennæ (fig. 74, *a*), each composed of a protopodite, and a segmented endopodite and exopodite, which are nearly of equal size. Finally, attached to the first segment of the head are the eyes, each of which is borne upon an eye-stalk formed by the protopodite. The gill-chamber is formed by a great prolongation downwards of the pleura of the thoracic segments, and the gills are attached to the bases of the legs.

As regards the digestive system of the *Crustacea*, the alimentary canal is, with few exceptions, continued straight from the mouth to the aperture of the anus. There are no salivary glands, but a large and well-developed liver is usually present.

A heart is generally, but not always, present. In most of the lower forms it is a long vasiform tube, very like the "dorsal vessel" of Insects. The exact course of the circulation has been differently stated by different writers, but the following appear to be the facts of the case: In some of the lower forms (*e.g.*, *Copepoda*) there are no arterial vessels, and the venous blood returned from the body is collected into a venous sinus —the so-called "pericardium," which surrounds the heart and opens into it by valvular apertures. In the higher forms, the heart gives off a number of arteries by which the blood is driven to all parts of the body and to the gills. The arteries do not terminate in a system of capillary vessels, but in a series of irregular lacunæ occupying all the interstices between the different organs of the body. From this interstitial lacunar system arise the venous trunks, which are generally dilated into more or less extensive sinuses. Whether the whole of the venous blood is submitted to the action of the gills, or whether the blood sent to the gills is derived mainly from the heart, is a matter of question; but the former is the more probable view. Be this as it may, the blood is invariably returned to a large venous sinus which surrounds the heart, and opens into it by a number of valvular apertures. It follows from this description, that the heart of the *Crustacea* is mainly, if not altogether, a *systemic* heart, being concerned chiefly, if not entirely, in driving the aerated blood to all parts of the body.

Distinct respiratory and circulatory organs may be altogether wanting; but, as a rule, distinct branchiæ are present. The exact form and structure of the gills differ in different cases, but their leading modifications will be alluded to in treating of the different orders.

Tabular View of the Divisions of the Crustacea.

Sub-class I. Epizoa (*Haustellata*).

Order 1. *Ichthyophthira*.
" 2. *Rhizocephala*.

Sub-class II. Cirripedia.

Order 3. *Thoracica*. { Balanidæ. Verrucidæ. Lepadidæ.

" 4. *Abdominalia*.
" 5. *Apoda*.

Sub-class III. ENTOMOSTRACA.

Order 6. *Ostracoda.* } *Legion*, Lophyropoda.
„ 7. *Copepoda.* }
„ 8. *Cladocera.* }
„ 9. *Phyllopoda.* } *Legion*, Branchiopoda.
„ 10. *Trilobita.* }
„ 11. *Merostomata.*

Sub-class IV. MALACOSTRACA.

Division A. EDRIOPHTHALMATA.

Order 12. *Læmodipoda.*
„ 13. *Isopoda.*
„ 14. *Amphipoda.*

Division B. PODOPHTHALMATA.

Order 15. *Stomapoda.*
„ 16. *Decapoda.*
Tribe *a. Macrura.*
„ *b. Anomura.*
„ *c. Brachyura.*

CHAPTER XXXII.

EPIZOA AND CIRRIPEDIA.

SUB-CLASS I. EPIZOA (*Haustellata*).—The members of this sub-class of the *Crustacea* are in the adult state parasitic upon the bodies of fishes, and are usually deformed; but in the young condition they are locomotive, and are furnished with antennæ and eyes. The mouth is suctorial, and the limbs are terminated by suckers, hooks, or bristles. There are no differentiated respiratory organs, but respiration is performed by the surface of the body. The males are rudimentary, and are much smaller than the females, which are usually furnished with external ovisacs. The *Epizoa* are closely allied to the *Copepoda*, and may, indeed, be regarded as parasitic Copepods, having the mouth modified so as to form a suctorial tube or beak, resulting from the elongation of the labrum and labium. Within this are almost always two stylets or lancet-shaped mandibles, used in piercing. The feet are often deformed by age, or wanting, but are primitively natatory. Not only does their developmental history bear out this view, but cases are known (in some *Lernæa*) in which the males do not undergo retrograde metamorphosis, but remain permanently in the condition of free Copepods.

This division includes the single order *Ichthyophthira*, the characters of which are therefore the same as those of the subclass, comprising various parasites upon fishes belonging to the genera *Lernæa*, *Achtheres*, *Peniculus*, &c.

ORDER I. ICHTHYOPHTHIRA.—The members of this order are attached in the adult condition to the skin, eyes, or gills of fishes, and when mature possess an elongated body, having a more or less distinct head, and in the females usually a pair of long, cylindrical ovisacs, depending from the extremity of the abdomen. Some adhere by a suctorial mouth, or by cephalic processes (*Cephaluna*); others are attached by a suctorial disc, developed at the extremities of the last pair of thoracic limbs, which are united together (*Brachiuna*); whilst in others (*Onchuna*) attachment is effected by hooks at the free extremities of the first pair of thoracic limbs.—(Owen.)

The males are usually not attached, but adhere to the females, of which, from their much smaller size, they appear to be mere parasites. The chief anatomical peculiarities of the female are the following:—The head is provided usually with a pair of jointed antennæ, and the body is divided into a cephalothorax and abdomen. The alimentary canal consists of a mouth, gullet, and intestine, terminating posteriorly in a distinct anus. The nervous system consists of a double ventral cord.

The embryo (fig. 63, *a*) is free-swimming, and is provided with visual organs and locomotive appendages. The two

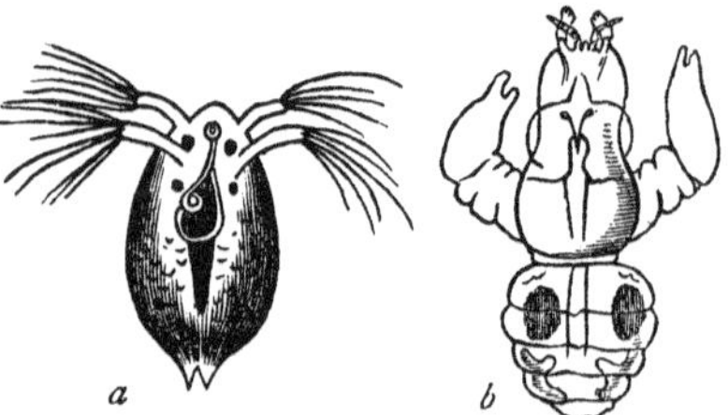

Fig. 63.—Ichthyophthira. *a* Free-swimming larva of *Achtheres percarum*, in its first stage; *b* Adult male of the same. Enlarged. (After Owen.)

sexes are now alike, and the conversion of the active embryo, or larva, into the swollen and deformed adult, must be regarded as an instance of "retrograde metamorphosis." In *Achtheres* *percarum* (fig. 63), the primitive form of the young is a "Nauplius;"* but a wholly different larva, resembling the *Cyclops* in shape, but with fewer limbs and somites, is prepared within the Nauplius-skin, and is liberated by the rupture of the same.

ORDER II. RHIZOCEPHALA.—The name *Rhizocephala* has been proposed for another group of *Crustacea*, which are fixed, parasitic, and greatly deformed when adult; but which are locomotive when young, and which are most nearly allied to the *Cirripedia*. The larvæ are "Naupliiform," with an ovate unsegmented body, an unpaired median eye, and a dorsal shield or carapace. The abdomen terminates in a movable caudal fork, and there is neither mouth nor alimentary canal. In their second stage (as so-called "pupæ"), the young of the *Rhizocephala* are enclosed in a bivalve shell, the foremost pair of limbs constitute peculiar organs of adhesion ("prehensile antennæ" of Darwin), the two following pairs of limbs are cast off, and six pairs of powerful biramose natatory feet are formed on the abdomen. There is still no mouth. The "pupæ" now attach themselves to the abdomen of Crabs, *Porcellanæ*, and Hermit-crabs; they remain astomatous; "they lose all their limbs completely, and appear as sausage-like, sack-shaped, or discoidal excrescences of their host, filled with ova; from the point of attachment closed tubes, ramified like roots, sink into the interior of the host, twisting round its intestine, or becoming diffused amongst the sac-like tubes of its liver. The only manifestations of life which persist in these *non plus ultras* in the series of retrogressively metamorphosed Crustacea are powerful contractions of the roots, and an alternate expansion and contraction of the body, in consequence of which water flows into the brood-cavity, and is again expelled through a wide orifice."—(Fritz Müller.) The branched roots of the *Rhizocephala* appear to be the homologues of the "cement-ducts" of the *Cirripedia*, and to be, therefore, really the antennæ.

SUB-CLASS II. CIRRIPEDIA.—This sub-class includes, amongst others, the common Acorn-shells and the Barnacles or Goose-mussels. All the *Cirripedia* are distinguished by the fact, that in the adult condition they are permanently fixed to some solid object by the anterior extremity of the greatly metamorphosed head; the first three cephalic segments being much developed, and enclosing the rest of the body. The larva is

* The name of "Nauplius" was given by O. F. Müller to the unsegmented ovate larva of the lower *Crustacea*, with a median frontal eye, but without a true carapace; and this name may be conveniently employed to designate all the larval forms which agree in these characters.

free and locomotive, and the subsequent attachment, and conversion into the fixed adult, is effected by means of a peculiar secretion, or cement, which is discharged through the antennæ of the larva, and is produced by a special cement-gland, which is really a portion of the ovary. In the *Cirripedia*, therefore, the head of the adult is permanently fixed to some solid object, and the visceral cavity is protected by an articulated calcareous shell, or by a coriaceous envelope. The posterior extremity of the animal is free, and can be protruded at will through the orifice of the shell. This extremity consists of the abdomen, and of six pairs of forked, ciliated limbs, which are attached to the thorax, and serve to provide the animal with food. The two more important types of the *Cirripedia* are the Acorn-shells (*Balanidæ*) and the Barnacles (*Lepadidæ*). In the former the animal is sessile, the larval antennæ, through which the cement exudes, being imbedded in the centre of the membranous or calcareous "basis" of the shell. In the latter the animal is stalked, and consists of a "peduncle" and a "capitulum." The peduncle consists of the anterior extremity of the body, with the larval antennæ usually cemented to some foreign body. The capitulum is supported upon the peduncle, and consists of a case composed of several calcareous plates, united by a membrane, enclosing the remainder of the animal.

Before giving a more detailed description of this singular and important sub-class, the following definition, as given by Owen, may be advantageously appended :—

"*Body*, chitinous, or chitino-testaceous, sub-articulated, mostly symmetrical, with aborted antennæ and eyes. *Mouth*, prominent, composed of a labrum, palpi, two mandibles, and two pairs of maxillæ. *Thorax*, attached to the sternal internal surface of the carapace, with six pairs of multiarticulate, biramous, setigerous limbs. *Abdomen*, rudimentary. Vascular system diffused; white blood. Branchiæ, when present, attached to the inferior lateral part of the surface. Most are hermaphrodite; a few have minute, rudimentary, male individuals, parasitically attached to the females. *Penis*, proboscidiform, multiarticulate, attached to the hinder end of the abdomen. No oviducts. *Metamorphosis* and *metagenesis*, resulting in a permanent parasitic attachment of the fully-developed female or hermaphrodite individual."

As regards the development of the *Cirripedia*, the larva is at first a "Nauplius," with an unsegmented pyriform body, a median eye, and a dorsal shield or buckler. The abdomen is produced beneath the anus into a long forked caudal appendage, and there is a long spine over the anus. A mouth, intestine, and vent are present. After several moults the young Nauplii become "pupæ" (fig. 64). The dorsal shield is folded so as to form a bivalve shell; the anterior limbs (antennæ) are transformed into prehensile organs; the two following pairs of limbs are cast off; and six pairs of strong, biramose, natatory feet are developed upon the abdomen. A pair of composite eyes are present, but there is no mouth. Finally, the pupæ fix themselves by the prehensile antennæ to rocks, drift-wood, ships, Sponges, Cetaceans, Turtles, Crustaceans, or even

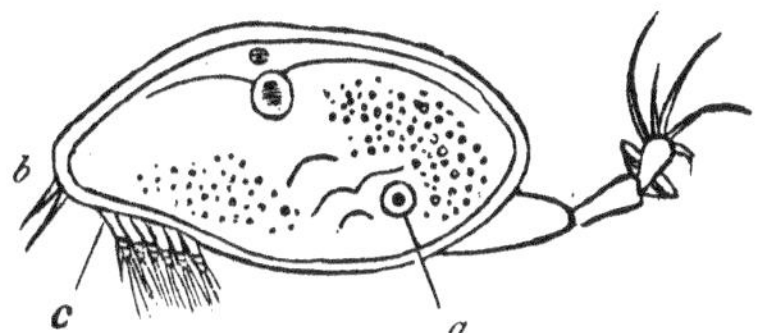

Fig. 64.—Locomotive "pupa" of *Balanus*. *a* Eye; *b* caudal bristles; *c* Setigerous limbs.

Jelly-fish. The prehensile antennæ are glued down permanently by the secretion of a peculiar cement-gland. The carapace becomes, as a rule, the seat of definite calcifications, by which it is converted into a multivalve calcareous "test;" the mouth is developed, and the six pairs of natatory feet are converted into long jointed "cirri," by which food is conveyed to the mouth. "The 'cement-ducts' can be traced as far as the third or 'disc-segment' of the antennæ. There the cement seems to transude and fasten down the disc; soon both antennæ are surrounded by a common border of cement, which gradually increases in extent after the metamorphosis. In the *Lepas fascicularis* the cement is poured forth in sufficient quantities to form, itself, the substance to which the peduncle of the adult barnacle adheres, and for a cluster of which barnacles it constitutes a central vesicular float."—(Owen.) The cement-gland, as shown by Darwin, is "part of, and continuous with, the branching ovaria," and the cement-ducts open through the prehensile antennæ.

The form of the adult, as already said, differs considerably, but the two most important types are those presented respectively by the Sessile and by the Pedunculated *Cirripedia*.

In the common Acorn-shells (*Balani*, fig. 65, *a*) the anterior portion of the head is not elongated, but is fixed to the centre of a basal, membranous, or shelly plate, termed the "basis," which adheres by its external surface to some solid body. Above the basis rises a more or less limpet-shaped, or conical, shell, which is open at the top, but is capable of being completely closed by a pyramidal lid, or "operculum." Both the shell itself and the operculum are composed of calcareous plates usually differing from one another in shape, and distinguished by special names. Within the shell the animal is fixed, head downwards. The thoracic segments, six in number, bear six pairs of limbs, each of which consists of a jointed protopodite and a much segmented exopodite and endopodite, both of which are ciliated, and constitute the so-called "cirri," from which the name of the sub-class is derived. These twenty-four cirri—the "glass hand" of the *Balanus*—are in incessant action, being protruded from the opening of the shell, and again retracted within it, constantly producing currents of water, and thus bringing food to the animal. There are no specialised respiratory organs in the family of the *Balanidæ*. *Balani* sometimes attain a very considerable size, and *Balanus psittacus* is largely eaten on the coast of Chili.

In the Barnacles (*Lepadidæ*, fig. 65, *b*) the anterior extremity of the animal is enormously elongated, forming with

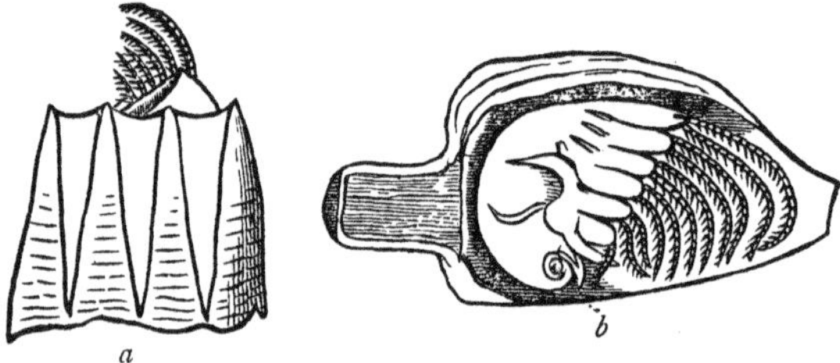

Fig. 65.—Morphology of Cirripedia. *a* Sessile Cirripede or Balanoid, *Balanus sulcatus*. *b* Pedunculate Cirripede or Lepadoid, *Lepas anatifera*.

the prehensile antennæ, the cement-ducts, and their exudation, a long stalk or peduncle, whereby the animal is attached to some solid object. At its free extremity the peduncle bears the "capitulum," which corresponds to the shell of the Balanoids, and is composed of various calcareous plates, united together by a membrane, moved upon one another by appropriate muscles, and protecting in their interior the body of the animal with its appendages. The thorax and limbs resemble those of the *Balanus;* but "slender appendages, which from their position and connections are homologous with the branchiæ of the higher *Crustacea*, are attached to, or near to, the bases of a greater or less number of the thoracic feet, and extend in an opposite direction outside the visceral sac."—(Owen.)

All the *Balanidæ* are hermaphrodite, and this is also the

case with most of the *Lepadidæ*, but some extraordinary exceptions occur in this latter order. Thus, in some species of *Scalpellum* the individual forming the ordinary shell is female, and each female has two males lodged in transverse depressions within the shell. These males "are very singular bodies; they are sac-formed, with four bead-like, rudimental valves at their upper ends; they have a conspicuous internal eye; they are absolutely destitute of a mouth, or stomach, or anus; the cirri are rudimental and furnished with straight spines, serving apparently to protect the entrance of the sac; the whole animal is attached like the ordinary Cirripede, first by the prehensile antennæ, and afterwards by the cementing substance. The whole animal may be said to consist of one great sperm-receptacle, charged with spermatozoa; as soon as these are discharged, the animal dies."

"A far more singular fact remains to be told; *Scalpellum vulgare* is, like ordinary Cirripedes, hermaphrodite, but the male organs are somewhat less developed than is usual; and, as if in compensation, several shortlived males are almost invariably attached to the occludent margin of both scuta. . . . I have called these beings *complemental males*, to signify that they are complemental to an hermaphrodite, and that they do not pair like ordinary males with simple females."—(Darwin.)

Divisions of Cirripedia.—(After Darwin.)

Order I. Thoracica.

Carapace, either a capitulum on a pedicle, or an operculated shell with a basis. *Body*, formed of six thoracic segments, generally furnished with six pairs of limbs; abdomen rudimentary, but often bearing caudal appendages. *Mouth*, with labrum not capable of independent movements. *Larva*, firstly one-eyed, with three pairs of legs; lastly two-eyed, with six pairs of legs.

Fam. 1. *Balanidæ*.

Sessile, without a peduncle; scuta and terga (forming the *operculum*) provided with depressor muscles; the rest of the valves immovably united together.

Fam. 2. *Verrucidæ*.

Sessile. Shell asymmetrical, with scuta and terga, which are movable, but not furnished with a depressor muscle.

Fam. 3. *Lepadidæ*.

Pedunculated. *Peduncle* flexible, provided with muscles. Scuta and terga, when present, not furnished with a depressor muscle. Other valves, when present, not united into a single immovable case.

Order II. Abdominalia.

Carapace flask-shaped; body formed of one cephalic, seven thoracic, and three abdominal segments, the latter bearing three pairs of cirri, but the thoracic segments being without limbs. *Mouth*, with the labrum greatly produced, and capable of independent movements. *Larva*, firstly egg-shaped, without external limbs, or an eye; lastly binocular, without thoracic limbs, but with abdominal appendages.

Genus. *Cryptophialus.*

ORDER III. APODA.

Carapace, reduced to two separate threads, serving for attachment. *Body* consisting of one cephalic, seven thoracic, and three abdominal segments, all destitute of cirri. Mouth suctorial.

Genus. *Proteolepas.*

CHAPTER XXXIII.

SUB-CLASS ENTOMOSTRACA.

SUB-CLASS III. ENTOMOSTRACA.—The term *Entomostraca* has been variously employed, and few authorities include exactly the same groups of the *Crustacea* under this name. By most the division is simply defined as including all those *Crustacea* in which the segments of the thorax and abdomen, taken together, are more or fewer than fourteen in number—the parasitic *Epizoa* and the *Cirripedia* being excluded. By Professor Rupert Jones the following definition of the *Entomostraca* has been given:—

"Animal aquatic, covered with a shell, or carapace, of a horny consistency, formed of one or more pieces, in some genera resembling a cuirass or buckler, and in others a bivalve shell, which completely or in great part envelops the body and limbs of the animal. In other genera the animal is invested with a multivalve carapace, like jointed plate-armour; the branchiæ are attached either to the feet or to the organs of mastication; the limbs are jointed, and more or less setiferous. The animals, for the most part, undergo a regular moulting or change of shell, as they grow; in some cases this amounts to a species of transformation."

The *Entomostraca* are divided into two great divisions, or "legions," the *Lophyropoda* and the *Branchiopoda*, with which the order *Merostomata* may be conveniently considered.

DIVISION A. LOPHYROPODA.—The members of this division possess few branchiæ, and these are attached to the appendages of the mouth. The feet are few in number, and mainly subserve locomotion; the carapace is in the form either of a shield protecting the cephalothorax, or of a bivalve shell enclosing the entire body. The mouth is not suctorial, but is furnished with organs of mastication.

This division comprises the two orders *Ostracoda* and *Copepoda*.

ORDER I. OSTRACODA.—Small Crustaceans having the entire body enclosed in a shell or carapace, which is composed of two valves united along the back by a membrane. The branchiæ are attached to the posterior jaws, and there are only two or three pairs of feet, which subserve locomotion, but are not adapted for swimming. A distinct heart is sometimes present (*Cypridina*), but is more usually wanting (*Cypris* and *Cythere*).

Little is known of the development of the *Ostracoda*, but the young of *Cypris* are said to be "shell-bearing Nauplius forms" (Claus), possessing only the three anterior pairs of limbs, but protected by a bivalve shell. As in other Nauplii, the third pair of limbs, though now locomotive, are ultimately transformed into the mandibles. They pass through several stages, with complete moults, before arriving at sexual maturity. The *Cytherides*, on the other hand, have at birth the two pairs of antennæ and two pairs of jaws, with three pairs of rudimentary abdominal limbs.

The order includes the *Cyprides* (fig. 66, *a*), which are of almost universal occurrence in fresh water. The common

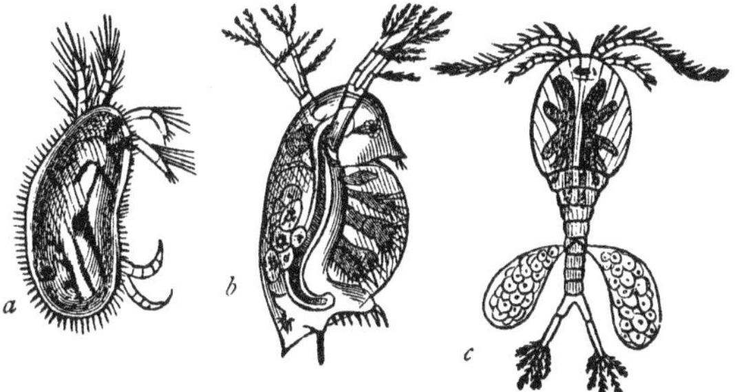

Fig. 66.—Fresh-water Entomostraca. *a Cypris tris-striata*; *b Daphnia pulex*; *c Cyclops quadricornis*.

Cypris is completely protected from its enemies by a bivalve carapace, which it can open and shut at will, and out of which it can protrude its feet. Locomotion is mainly effected by means of a pair of caudal appendages. The *Cypris* is extremely prolific, and a single impregnation appears to last the female for its entire lifetime. It appears, also, that the young females, produced in this way, are capable for some generations of producing fresh individuals without the influence of a male (parthenogenesis).

ORDER II. COPEPODA.—Small Crustaceans, having the head and thorax covered by a carapace, and furnished with five pairs of natatory feet. Usually there are two caudal locomotive appendages. A distinct heart is sometimes absent (as in the *Cyclopidæ*), but is sometimes present. Both marine and fresh-water Copepods are known.

The larvæ of the Copepods are Naupliiform, with unpaired eyes, three pairs of limbs (the future antennæ and mandibles), and two terminal setæ. Next the maxillæ are produced, and then three other pairs of limbs (the foot-jaws and the two front pairs of natatory feet). At the next moult, the larva assumes the *Cyclops* form, but has at first much fewer limbs and somites.

In the *Cyclops* (fig. 66, *c*), which is one of the commonest of the "Water-fleas," the cephalothorax is protected superiorly by a carapace, and the abdominal somites are conspicuous. In front of the head is situated a single large eye, behind which are the great antennæ and the antennules. The feet are five pairs in number, each consisting of a protopodite and a segmented exopodite and endopodite, usually furnished with hairs and forming an efficient swimming apparatus. The young pass through a metamorphosis, and are not capable of reproducing the species until after the third moult or change of skin. The female *Cyclops* carries externally two ovisacs, in which the ova remain till they are hatched. A single congress with the male is apparently sufficient to fertilise the female for life.

The *Copepoda*, or Oar-footed Crustaceans, are all of small size, and are of common occurrence in fresh water in all parts of Europe. By good authorities the *Ichthyophthira* are regarded as merely *Copepoda* peculiarly modified to suit a life of parasitism.

DIVISION B. BRANCHIOPODA.—The Crustaceans included in this division have many branchiæ, and these are attached to the legs, which are often numerous, and are formed for swimming. In other cases the legs themselves are flattened out so as to form branchiæ. The body is either naked, or is protected by a carapace, which may enclose either the entire body, or the head and thorax only. The mouth is provided with organs of mastication.

The *Branchiopoda* comprise the *Cladocera*, the *Phyllopoda*, and probably the *Trilobita*, though this order departs in many respects from the above definition. The *Merostomata* may be considered along with these, though these, too, are in many respects peculiar.

ORDER I. CLADOCERA.—The members of this order are

small Crustaceans, which have a distinct head, and have the whole of the remainder of the body enclosed within a bivalve carapace, similar to that of the *Ostracoda*. The feet are few in number (usually four, five, or six pairs), and are mostly respiratory, carrying the branchiæ. Two pairs of antennæ are present, the larger pair being of large size, branched, and acting as natatory organs. The *Cladocera* quit the egg with the full number of limbs proper to the adult.

In the *Daphnia pulex* (fig. 66, *b*), or "branched-horned Water-flea," which occurs commonly in our ponds, the body is enclosed in a bivalve shell, which is not furnished with a hinge posteriorly, and which opens anteriorly for the protrusion of the feet. The head is distinct, not enclosed in the carapace, and carrying a single eye. The mouth is situated on the under surface of the head, and is provided with two mandibles and a pair of maxillæ. The gills are in the form of plates, attached to the five pairs of thoracic legs. The males are very few in number, compared with the females, and a single congress is all that is required to fertilise the female for life. Not only is this the case, but the young females produced from the original fecundated female appear to be able to bring forth young without having access to a male. In this way the influence of a single fecundation appears to be transmitted through several generations. Two kinds of eggs occur in *Daphnia*. In the first of these, or "summer eggs," the ova (from ten to fifty in number) are deposited in an open space between the valves, and are retained there till the young are ready to be hatched. In the second of these, or "winter eggs," the ova (generally two in number) are placed in a peculiar receptacle, which is formed on the back of the carapace, and is called the "ephippium" or saddle. After a time the ephippium is cast off, and floats about till spring, when its contained eggs are hatched by the warmer temperature of the water.

ORDER II. PHYLLOPODA.—*Crustacea*, mostly of small size, the carapace protecting the head and thorax, or the body entirely naked. Feet numerous, never less than eight pairs, mostly foliaceous or leaf-like, branchial in function. The eyes sometimes confluent, sometimes distinct and sub-pedunculate. There are two horny mandibles without palps, and the first pair of feet are oar-like, with setiform terminal appendages. The remaining feet are branchial, and adapted for swimming. The Phyllopods undergo a metamorphosis, the youngest forms being "Nauplii." In *Nebalia*, however, which is the only marine Phyllopod, "Zoea-stages" are superadded as well.

The *Phyllopoda* are chiefly interesting from their affinity to the extinct Trilobites. In the typical genera *Limnadia* and *Apus* the body is protected by a carapace, which is bivalve in the former and shield-like in the latter. In *Limnadia* the carapace covers the greater part of the body, and opens along the ventral margin. There are from 18 to 30 pairs of membranaceous and respiratory feet. In *Apus* the carapace is

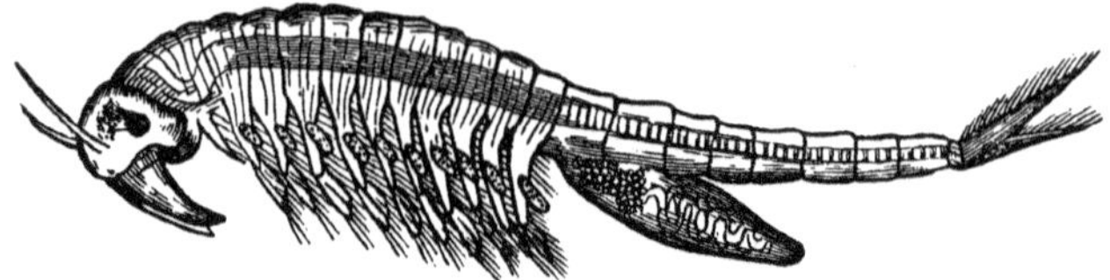

Fig. 67.—Phyllopoda. Fairy Shrimp (*Chirocephalus diaphanus*)—after Baird.

clypeiform and covers a portion of the abdomen; and there are sixty pair of feet, of which all but the first pair are foliaceous. *Apus* is gregarious, fresh-water in habit, and often found in great numbers in pools and ditches in Europe. The different species of *Branchipus* have the body unprotected by any carapace, and are found in ponds and swamps in various parts of the world. The various "Brine-shrimps" (*Artemia*) are found inhabiting the brine-pans in salt-works, or occur in salt-lakes in both hemispheres, being especially abundant in Great Salt Lake in Utah.

ORDER III. TRILOBITA.—This order is entirely extinct, none of its members having survived the close of the Palæozoic period. It is probable that the Trilobites should be placed near the *Phyllopoda;* but their exact position is uncertain, as, with one exception, no traces of any appendages of any kind, except the labrum, have hitherto been discovered in any Trilobite.

The body of a Trilobite (fig. 68) was covered with a "crust," or exoskeleton, which shows more or less markedly a division into three longitudinal lobes, from the presence of which the name of the order is derived. The shell is composed of a cephalic shield, a certain number of free and movable thoracic rings, and a caudal shield, or "pygidium," the rings of which are more or less completely anchylosed. On the under surface of the body nothing has hitherto been discovered, except the "hypostome," or "labrum," which was a plate placed in front of the mouth. No traces of ambulatory or natatory limbs, of branchiæ, or of antennæ, have ever been discovered. The eyes, when present, are compound, and usually sessile, but are sometimes supported upon projecting processes It has generally been supposed that the body of the Trilobite occupied the medium lobe of the crust, commencing with the "glabella" in front, and terminating with the "pygidium" behind, whilst the axial lobes protected a series of delicate respiratory feet; but this view is doubted by many authorities, and the question is one which we have at present no means of deciding. Quite recently, however, a specimen of a Trilobite has been discovered in which it is said that the bases of the legs were distinctly recognisable. The specimen in question was an *Asaphus;* but the great number and excellent preservation of Trilobites, as a general rule, render it highly probable that in most cases the limbs were destitute of a chitinous exoskeleton, and were therefore incapable of being preserved in a fossil state. According to Spence Bate, "the young of the Trilobites are of the Nauplius form."

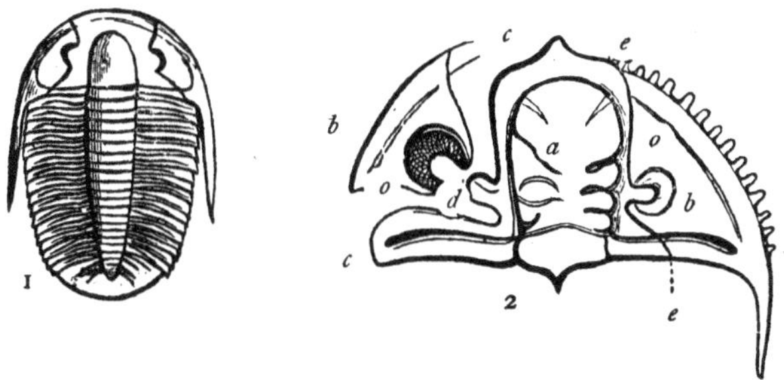

Fig. 68.—Morphology of Trilobites. 1. *Angelina Sedgwickii;* 2. Diagram of the cephalic shield of a Trilobite (after Salter). *a* Glabella; *b b* Free cheeks, bearing the eyes (*o o*); *c c* Fixed cheek, including the eye-lobe (*d*); *e e* Facial suture.

The cephalic shield of a typical Trilobite is more or less completely semicircular (fig. 68, 2), and is composed of a central and of two lateral pieces, of which the two latter may, or may not, be united together in front of the former.

The median portion is usually elevated above the remainder of the cephalic shield, and is called the "glabella;" it protected the region of the stomach, and is usually divided into from three to four lobes by lateral grooves. At each side of the glabella, and continuous with it, is a small semicircular area, called the "fixed cheek." The glabella, with the "fixed cheeks" is separated from the lateral portions of the cephalic shield—termed the "movable" or "free cheeks"—by a peculiar suture or line of division, which is known as the "facial suture," and is quite unknown amongst recent *Crustacea,* except for a faint indication in the *Limulus,* and more or less doubtful traces in certain other forms. The movable cheeks bear the eyes, which are generally crescentic or reniform in shape, are rarely pedunculated, and consist of an aggregation of facets covered by a thin cornea. The facial sutures may join one another in front of the glabella—in which case the free cheeks will form a single piece—or they may cut the anterior margin of the shield separately—in which case the free cheeks will be discontinuous. The posterior angles of the free cheeks are often produced into long spines.

Behind the cephalic shield comes the thorax, composed of a variable number of segments, which are not soldered together, but are capable of free motion upon one another, so as to allow the animal to roll itself up after the manner of a wood-louse, or hedgehog. The thorax is usually strongly trilobed, and each thorax-ring shows the same trilobation, being composed of a central, more or less strongly convex, portion, called the "axis," and of two flatter side-lobes, called the "pleuræ."

The "pygidium," or "tail," is usually trilobed also, and, like the thorax, consists of a median axis and of a marginal limb, the composition of the whole out of anchylosed segments being shown by the existence of axial and pleural grooves.

ORDER IV. MEROSTOMATA.—The members of this order are *Crustacea*, often of gigantic size, in which the mouth is furnished with mandibles and maxillæ, the terminations of which become walking or swimming feet, and organs of prehension.

This order comprises the recent King Crabs, and the extinct *Pterygoti* and *Eurypteri*.

SUB-ORDER I. XIPHOSURA.—"Crustacea having the anterior segments welded together to form a broad convex buckler, upon the dorsal surface of which are placed the compound eyes and ocelli; the former sub-centrally, the latter in the centre in front. The mouth is furnished with a small labrum, a rudimentary metastoma and six pairs of appendages. Posterior segments of the body more or less free, and bearing upon their ventral surfaces a series of broad lamellar appendages; the telson, or terminal segment, ensiform."—(Henry Woodward.)

The *Xiphosura* include no other recent forms than the *Limuli* (King Crabs, or Horse-shoe Crabs) (fig. 69). They are distinguished by the possession of *six pairs of chelate limbs, placed round the mouth, having their bases spinous and officiating as jaws.* Six other pairs of foliaceous appendages are attached to the abdomen, and the last five of these carry branchiæ. The body, which is often of great size, when viewed from above exhibits a division into three portions:—(1) An anterior semicircular shield, which carries two compound and two simple eyes; (2) a posterior, irregularly hexagonal shield, which covers the abdomen; and (3) a long, sword-like telson, articulated to the dorsal buckler, and giving the name to the sub-order.

The chief features, therefore, which characterise the *Limulus* are as follows:—1. The possession of six pairs of appendages which are placed round the mouth, have their bases spinous, act as jaws, and have their free extremities developed into claws; 2. The possession of six abdominal pairs of appendages, expanded for swimming, and carrying the gills; 3. The possession of a semicircular buckler, covering the cephalothorax, and carrying the eyes upon its upper surface; 4. The possession of a second buckler, or "operculum," covering the abdomen;

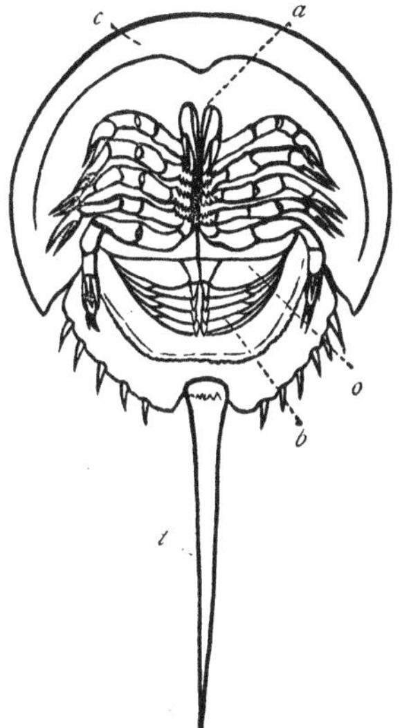

Fig. 69.—Xiphosura. *Limulus polyphemus*, viewed from below. *c* The cephalic shield carrying the sessile eyes upon its upper surface; *o* "Operculum," covering the reproductive organs; *b* Branchial plates; *a* First pair of antennæ (antennules) ending in chelæ. Below these is the aperture of the mouth surrounded by the spiny bases of the remaining five pairs of appendages, which are regarded by Woodward as being respectively, from before backwards, the great antennæ, the mandibles, the first maxillæ, the second maxillæ, and a pair of maxillipedes. All have their extremities chelate.

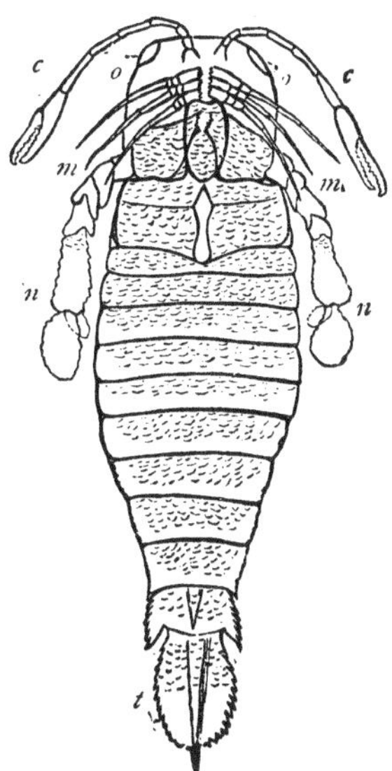

Fig. 70.—Eurypterida. *Pterygotus Anglicus*, restored (after H. Woodward). *c c* Chelate antennæ; *o o* Eyes, situated at the anterior margin of the carapace; *m m* The mandibles, the first and second maxillæ; *n n* The maxillipedes; the basal margins of these are serrated, and are drawn as if seen through the metastoma or post-oral plate, which serves as a lower lip. Immediately behind this is seen the operculum or thoracic plate which covers the two anterior thoracic somites. Behind this are five thoracic and five abdominal somites, and lastly there is the telson (*t*).

5. The presence of a long, sword-shaped telson, or tail-spine, articulated to the dorsal shield.

The larval *Limulus* does not possess the ensiform post-anal spine of the adult, and is further stated to show a decided resemblance to the Trilobites.

The King-crabs are found in the Indian and Japanese Seas, on the coasts of North America, and in the Antilles. They sometimes attain a large size, and both the eggs and the flesh are eaten by the Malays.

SUB-ORDER 2. EURYPTERIDA.—"Crustacea with numerous, free, thoracico-abdominal segments, the first and second (?) of which bear one or more broad lamellar appendages upon their ventral surface, the remaining segments being devoid of appendages; anterior rings united into a carapace, bearing a pair of larval eyes (*ocelli*) near the centre, and a pair of large, marginal, or sub-central eyes; the mouth furnished with a broad post-oral plate, or *metastoma*, and five pairs of movable appendages, the posterior of which form great swimming-feet; the telson, or terminal segment, extremely variable in form; the integument characteristically sculptured."—(Henry Woodward.)

The *Eurypterida* are all extinct, and are entirely confined to the Palæozoic period. Many of them attained to a comparatively gigantic size; *Pterygotus Anglicus* (fig. 70) being supposed to have reached a length of probably six feet. In their characters they present many larval features; resembling the larvæ of the *Decapoda* especially in the fact that all the free somites of the abdomen (except the two anterior ones) were totally devoid of appendages.

CHAPTER XXXIV.

MALACOSTRACA.

SUB-CLASS IV. MALACOSTRACA.—The *Crustacea* of this sub-class are distinguished by the possession of a generally *definite* number of body-segments; seven somites going to make up the thorax, and an equal number entering into the composition of the abdomen (counting, that is, the telson as a somite). The *Malacostraca* are divided into two primary divisions, termed respectively the *Edriophthalmata* and the *Podophthalmata* according as the eyes are sessile, or are supported upon eye-stalks.

DIVISION A. EDRIOPHTHALMATA.—This division comprises those *Malacostraca* in which the eyes are sessile, and the body is mostly not protected by a carapace. It comprises the three orders, *Læmodipoda*, *Isopoda*, and *Amphipoda*. The eyes are generally compound, but sometimes simple, and are placed on the sides of the head. The head is almost always distinct from the body, and the mandibles are often furnished with a palp. Typically there are seven pairs of feet in the adult, hence this division is called *Tetradecapoda* by Agassiz. In certain Isopods (*Tanais*) alone is there a carapace.

ORDER I. LÆMODIPODA.—The *Læmodipoda* are small Crustaceans, which are distinguished amongst the *Edriophthalmata* by the rudimentary condition of the abdomen. The first thoracic segment is amalgamated with the head, and the limbs of this segment appear to be inserted beneath the head, or, as it were, beneath the throat (fig. 71); hence the name given to the order. The respiratory organs are in the form of two or three pairs of membranous vesicles attached to the segments of the thorax, or to the bases of the legs. The last pair of feet are either inserted at the end of the last somite, or are followed by not more than one or two small segments. There are four setaceous antennæ, and the mandibles are without palps. The body is generally linear, of eight or nine joints, but is sometimes oval. The feet are hooked. The *Læmodipoda* are all marine, and one section of the order comprises parasitic Crustaceans, of which the Whale-louse (*Cyamus Ceti*) is the most familiar. The entire order is now generally regarded as being merely a section of the *Amphipoda*.

Fig. 71.—Læmodipoda. *Caprella phasma.*

ORDER II. AMPHIPODA.—The members of this order resemble those of the preceding in the nature of the respiratory organs, which consist of membranous vesicles attached to the bases of the thoracic limbs. The first thoracic segment, however, is distinct from the head, and the abdomen is well developed, and is composed of seven segments. There are seven pairs of thoracic limbs, directed partly forwards, and partly backwards, the name of the order being derived from this circumstance. As in the *Læmodipoda*, the heart has the form of a long tube extending through the six segments following the head, and having the blood admitted to its interior by three pairs of valvular fissures. The three posterior pairs of abdominal limbs are bent backwards, and form, with the telson, a natatory or saltatorial tail. The young Amphipod acquires its full number of segments and limbs before its liberation from the egg; and as a rule the young undergo little or no metamorphosis in reaching maturity.

All the *Amphipoda* are small, the "Sand-hopper" (*Talitrus locusta*, fig. 72) and the "fresh-water Shrimp" (*Gammarus pulex*) being two of the commonest forms. The Sand-hoppers and Gammari swim on their side when in the water, and the former leap with great activity on land.

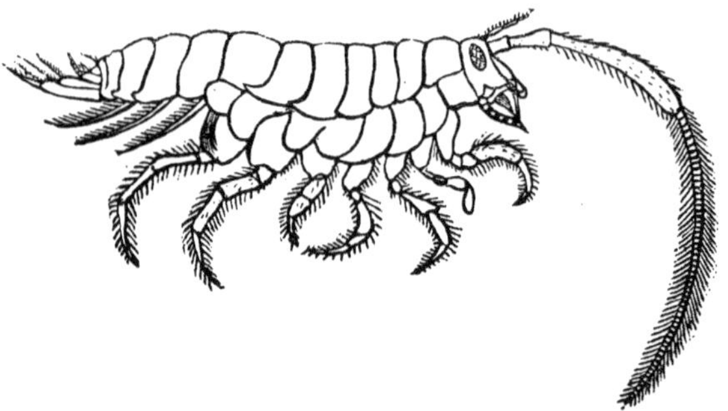

Fig. 72.—Amphipoda. The Sand-hopper, *Talitrus locusta*, enlarged.

ORDER III. ISOPODA.—In this order the head is always distinct from the segment bearing the first pair of feet. The respiratory organs are not thoracic, as in the two preceding orders, but are attached to the inferior surface of the abdomen, and consist of branchiæ, which in the terrestrial species are protected by plates which fold over them. The thorax is composed of seven segments, bearing seven pairs of limbs, which, in the females, have marginal plates attached to their bases, and serving to protect the ova. The number of segments in the abdomen varies, but is never more than seven. The eyes are two in number, formed of a collection of simple eyes, or sometimes truly compound. The heart is sometimes an elongated tube, with three pairs of fissures (as in the *Amphipoda*), sometimes short or spherical, removed towards the abdomen, and with more or fewer fissures than the above. The young Isopod is developed within a larval membrane, destitute of appendages. After a time this membrane bursts, and liberates the young, which resembles the adult in most respects, but possesses only six instead of seven pairs of limbs. Of the members of this order, many are aquatic in their habits, and are often parasitic, but others are terrestrial.

By Milne-Edwards the *Isopoda* are divided into three sections, termed respectively, from their habits, the *Natatorial*, *Sedentary*, and *Cursorial* Isopods. In the *Natatorial Isopoda* the extremity of the abdomen and the last pair of abdominal legs are expanded so as to form a swimming-tail. Some of this section are parasitic upon various fishes (*Cymothoa*), whilst others are found in the sea (*Sphæroma*). In the *Sedentary Isopoda* the animals are all parasitic, with short, incurved, hooked feet. This section includes the single family of the *Bopyridæ*, all the species of which live parasitically either in the gill-chambers, or attached to the ventral surface, of certain of the Decapod *Crustacea*, such as the Shrimps (*Crangones*) and the *Palæmones*.

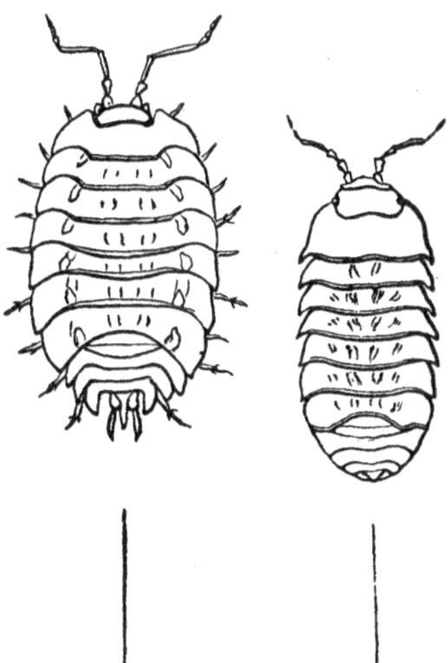

Fig. 73.—Isopoda. Wood-lice (*Oniscus*).

The *Cursorial*, or running, *Isopods* mostly live upon the land, and are therefore destitute of swimming-feet. The most familiar examples of this section are the common Wood-lice (*Oniscus*). Here, also, belongs the little *Limnoria terebrans*, so well known for the destruction which it produces by boring into the wood-work of piers and other structures placed in the sea. Other well-known Isopods are the Water-slaters (*Asellus*) of fresh waters, the Rock-slaters (*Ligia*) of almost all coasts, the Box-slaters (*Idothea*), the Shield-slaters (*Cassidina*), and the Cheliferous Slaters (*Tanais*). These last are remarkable as being the only Isopods in which there is a carapace. The lateral parts of the carapace are highly vascular, and respiration is effected by these, and not by the abdominal feet.

Many Isopods undergo an extensive metamorphosis. "In some Fish-lice (*Cymothoa*) the young are lively swimmers, and the adults are stiff, heavy, stupid fellows, whose short clinging feet are capable of little movement." In the *Bopyridæ* the adult females are usually blind, the antennæ are rudimentary, and the abdominal appendages from natatory become respiratory organs. The males, on the other hand, are dwarfed, and sometimes lose all the abdominal appendages, and all traces of segmentation; until we get forms which, like *Cryptoniscus planarioides*, "would be regarded as a Flat-worm rather than an Isopod, if its eggs and young did not betray its Crustacean nature."—(Fritz Müller).

DIVISION B. PODOPHTHALMATA.—The members of this division have compound eyes supported upon movable stalks or peduncles, and the body is always protected by a cephalothoracic carapace. Most of the *Podophthalma* pass through Zoea-stages in their development. It comprises the two orders *Stomapoda* and *Decapoda*, of which the latter includes all the highest and most familiar examples of the class *Crustacea*.

ORDER I. STOMAPODA.—In this order there are generally from six to eight pairs of legs, and the branchiæ, when present, are not enclosed in a cavity beneath the thorax, but are either suspended beneath the abdomen, or, more rarely, are attached to the thoracic legs. The shell, also, is thin, and often membranous. From all the preceding orders the *Stomapoda* are, of course, distinguished by the possession of pedunculate eyes. The development of the *Stomapoda* would appear to be by means of "Zoeæ."

All the Stomapods are marine, and the Locust Shrimp (*Squilla mantis*) may be taken as a good example of the order. In this Crustacean the carapace is small, and the posterior half of the thorax is unprotected. Several of the anterior appendages are developed into powerfully prehensile and hooked feet. The branchiæ are attached to the first five pairs of abdominal feet. The three posterior thoracic and the abdominal appendages are in the form of "swimmerets," and the tail is expanded into a powerful fin. Besides the Locust Shrimps, the order includes the Glass Shrimps (*Erichthys*) and their allies, and the Opossum Shrimps (*Mysis*).

ORDER II. DECAPODA.—The members of this order are the most highly organised of all the *Crustacea*, as well as being those which are most familiarly known, the Lobsters, Crabs, Shrimps, &c., being comprised under this head. For the most part they are aquatic in their habits, and they are usually protected by strong, resisting shells. There is always a complicated set of "gnathites," or appendages modified for masticatory purposes, surrounding the mouth. The ambulatory feet are made up of five pairs of legs (hence the name of the order), the first pair—and often some other pairs behind this—being "chelate," or having their extremities developed into nipping-claws. The branchiæ are pyramidal, and are contained in cavities at the sides of the thorax. The carapace is large, covering the head and thorax, and the anterior part of the abdomen. The heart of the *Decapoda* is in the form of a more or less quadrate sac, furnished with three pairs of valvular openings. As regards the development of the Decapods enormous differences obtain, even amongst forms very closely allied to one another.

The *Decapoda* are divided into three tribes, termed respectively the *Macrura*, *Anomura*, and *Brachyura*, and characterised by the nature of the abdomen.

TRIBE A. MACRURA.—The "long-tailed" Decapods included in this tribe are distinguished by the possession of a well-developed abdomen, often longer than the cephalothorax, the posterior extremity of which forms a powerful natatory organ or caudal fin. As regards the development of the *Macrura*, most appear at first in the form of "Zoeæ;* but there is little metamorphosis in the common Lobster, and there is said to be none in the Cray-fish (*Astacus fluviatilis*) and in one of the Land-crabs (*Gecarcinus*). Fritz Müller, again, has shown that the primitive form of one of the Shrimps (*Peneus*) is that of a "nauplius." This section comprises the Lobster, Cray-fish, Shrimp, Prawn, &c., of which the Lobster may be taken as the type.

In the Lobster (fig. 74) the somites of the head and thorax are amalgamated into a single mass, the "cephalothorax," covered by a carapace or shield, which is developed from "the lateral or epimeral elements of the fourth cephalic ring, which meet along the back, and give way preparatory to the moult. The tergal elements of the thoracic rings are not developed in either Crabs or Lobsters; when these rings are exposed by lifting up the cephalothoracic shield, the epimeral parts alone are seen, converging obliquely towards one another, but not joined at their apices."—(Owen.)

The first segment of the head bears the compound eyes, which are supported upon long and movable eye-stalks or peduncles. Behind these come two pairs of jointed tactile organs, the larger called the "great antennæ" (fig. 74 *ga*), the smaller the "antennules" (*a*). The mouth is situated on the under surface of the front of the head, and is provided from before backwards with an upper lip ("labrum"), two "mandibles," two pairs of "maxillæ," three pairs of "maxillipedes"

* The young Decapod, in most cases, leaves the egg in a larval form so different to the adult that it was originally described as a distinct animal under the name of *Zoea*. In this stage (fig. 76) the thoracic segments with the five pairs of legs proper to the adult are either wanting or are quite rudimentary. The abdomen and tail are without appendages, and the latter is composed of a single piece. The foot-jaws are in the form of natatory forked feet, and the mandible has no palp. Lastly, there are no branchiæ, and respiration is carried on by the lateral parts of the carapace. The "Zoea" is separated from the "nauplius" by having a segmented body, large paired eyes (sometimes with a median eye), and a carapace. The form proper to the adult is not attained until after several moults, constituting a genuine metamorphosis, though one which is effected by very gradual stages.

or "foot-jaws," and a bifid lower lip, or "metastoma." The five remaining segments of the thorax carry the five pairs of ambulatory legs, of which the first (fig. 74, 1) constitute the great claws, or "chelæ;" the next two pairs (2 and 3) are also

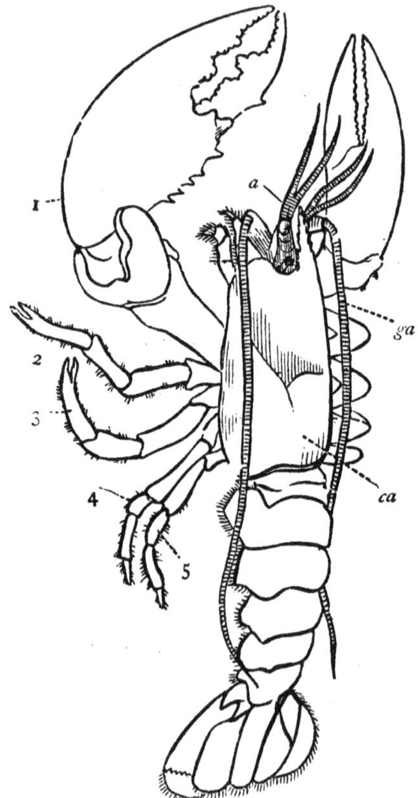

Fig. 74.—Macrura. Common Lobster (*Homarus vulgaris*). 1. First pair of legs, constituting the great chelæ or nipping-claws; 2. Second pair of legs, also chelate; 3. Third pair of legs, also chelate; 4 and 5. Last two pairs of ambulatory legs, with simply pointed extremities; *a* Antennules; *ga* Great antennæ; *ca* Carapace.

chelate, though much smaller; and the last two pairs are terminated by simply pointed extremities. The segments of the abdomen carry each a pair of natatory limbs, or "swimmerets," the last pair being greatly expanded, and constituting, with the "telson," a powerful caudal fin. Most posteriorly of all is the post-anal plate, or "telson," which may be looked upon either as an azygos appendage, or as a terminal segment which has no lateral appendages.

The mouth leads by a short œsophagus into a globose stomach, in the cardiac portion of which is a calcareous apparatus for triturating the food, which is commonly called the "lady in the lobster." The intestine is continued backwards from the stomach without convolutions, and the anal aperture is situated just in front of the telson. There is also a well-developed liver, consisting of two lobes which open by separate ducts into the intestine.

The heart is situated dorsally, and consists of a single polygonal contractile sac, which opens by valvular apertures into a surrounding venous sinus, inappropriately called the "pericardium." The heart is filled with oxygenated blood derived from the gills, and propels the aerated blood through every part of the body. The gills (fig. 62, 3, *g*) are pyramidal bodies attached to the bases of the legs, and protected by the sides of the carapace. They consist each of a central stem supporting numerous laminæ, and they are richly supplied with blood, but are not ciliated. The water which occupies the gill-chambers is renovated partly by the movements of the legs, and partly by the expanded epipodite of the second pair of maxillæ, which constantly spoons out the water from the front of the branchial chamber, and thus causes an entry of fresh water by the posterior aperture of the cavity.

The nervous system is of the normal "homogangliate" type, consisting of a longitudinal series of ganglia of different sizes, united by commissural cords, and placed along the ventral surface of the body. The organs of sense consist of the two compound eyes, the two pairs of antennæ, and two auditory sacs.

The sexes are invariably distinct, and the generative products are conveyed to the exterior by efferent ducts, which open at the base of one of the pairs of thoracic legs. The ovum is "meroblastic," a portion only of the vitellus undergoing segmentation. The neural side of the body—that is to say, the ventral surface—appears on the surface of the ovum, so that the embryo is built up from below, and the umbilicus is situated posteriorly.

TRIBE B. ANOMURA.—The Decapods which belong to this tribe are distinguished by the condition of the abdomen, which is neither so well developed as in the *Macrura*, nor so rudimentary as in Crabs. Further, the abdomen does not terminate posteriorly in a caudal fin, as in the Lobster. The development in the *Anomura* appears invariably to take place through Zoea-forms.

The most familiar of the *Anomura* are the Hermit-crabs (*Paguridæ*). In the common Hermit-crab (*Pagurus Bernhardus*) the abdomen is quite soft, and is merely enclosed in a membrane, so that the animal is compelled to protect itself by adopting the empty shell of some Mollusc, such as the common Whelk, which it changes at will, when too small. The Hermit is provided with a terminal caudal sucker, and with two or three pairs of rudimentary feet developed upon the abdomen, by means of which he retains his position within his borrowed dwelling. The abdominal appendages, however, are mostly unsymmetrical. The carapace is not strong, but the claws are

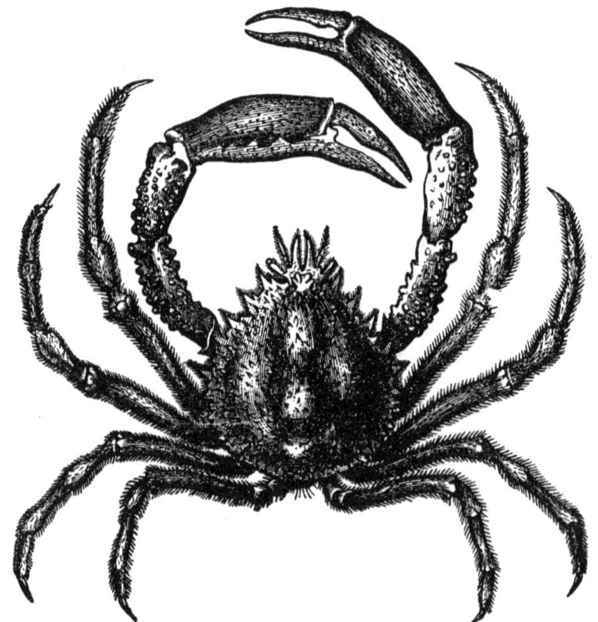

Fig. 75.—Brachyura. The Spiny Spider-Crab (*Maia squinado*).

well developed, one being always larger than the other. Other forms of the *Anomura* are the Sponge-crabs (*Dromia*), the Crab-lobsters (*Porcellanæ*), and the Tree-crabs (*Birgus*).

TRIBE C. BRACHYURA.—The "short-tailed" Decapods, or Crabs, are distinguished from the two preceding tribes by the rudimentary condition of the abdomen, which is very short, and is tucked up beneath the cephalothorax, the latter being disproportionately large. The extremity of the abdomen is not provided with any appendage, and it is merely employed by the female to carry the ova. The Crabs (fig. 75) are mostly furnished with ambulatory limbs, and are rarely formed for swimming, most of them being littoral in their habits, and some even living inland.

In all the essential points of their anatomy the Crabs do not differ from the Lobster and the other *Macrura;* but they are decidedly higher in their organisation. This is especially seen in the disposition of the nervous system, the ventral ganglia in the Crab being concentrated into a single large ganglion, from which nervous filaments are sent to all parts of the body. In the Land-crabs (*Gecarcinus*) respiration is by branchiæ, but there is almost always an aperture behind the carapace for the admission of air. They are distributed over the warm countries of the Old and New Worlds, as well as Australia. They are essentially terrestrial in their habits, and migrate in large bodies to the sea, in order to lay their eggs. Besides the true *Gecarcini*, members of other very different families live more or less constantly on dry land, and have air admitted directly into the branchial chamber. Amongst these are the Calling-crabs (*Gelasimus*), the Frog-crabs (*Ranina*), and the Sand-crabs (*Ocypoda*).

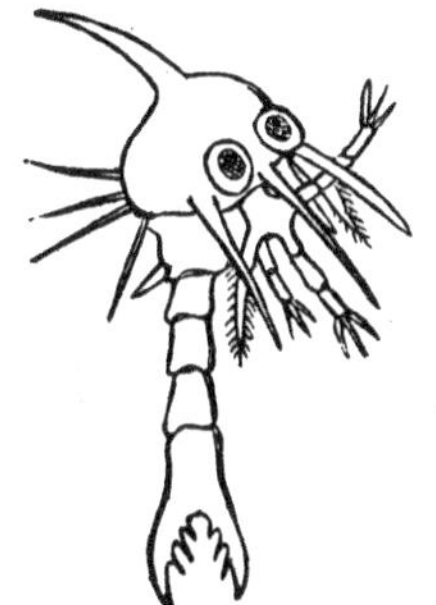

Fig. 76.—Larva (*Zoea*) of Crab (*Pirimela denticulata*), magnified—after Kinahan.

Reproduction in the Crabs is the same as in the *Macrura*, but the larva is exceedingly unlike the adult, and approximates closely to the type of the *Macrura*, another proof that the *Brachyura* stand higher in the Crustacean scale. The larval Crab was originally described as a distinct animal, under the name of *Zoea* (fig. 76), presenting in this condition a long and well-developed abdomen. It is only after several successive moults that the young Crab assumes its characteristic Brachyurous form, and acquires by gradual changes the features which distinguish the adult. The *Zoeæ* of the Crabs are usually distinguished by the possession of long spines developed from the carapace. When first liberated from the egg, the Zoea is enveloped in a larval skin or membrane, which is shed in a few hours.

CHAPTER XXXV.

DISTRIBUTION OF THE CRUSTACEA.

DISTRIBUTION OF CRUSTACEA IN SPACE.—The following general principles have been laid down by Milne-Edwards with regard to the geographical distribution of the *Crustacea*:—

1. The different forms and modes of organisation of the *Crustacea* are more varied and numerous in proportion as we pass from the polar regions towards the equator.

2. The number of different species is not only greater, but the number of types is greater in warm regions as compared with cold.

3. The higher *Crustacea* are either entirely wanting or are sparingly represented in the colder regions of the globe, but increase rapidly in relative numbers as the equator is approached.

4. The size attained by the *Crustacea* is greater on the average in warm regions than in colder climates.

5. The special points of structure which are characteristic of the different groups of *Crustacea* are more strongly manifested in the warmer regions of the globe.

6. There exists a decided relation between the temperature of any given region and the character of its Crustacean fauna; similar *generic* forms being usually found occupying regions of the same climatal character.

DISTRIBUTION OF CRUSTACEA IN TIME.—The class *Crustacea* is largely represented in past time, ranging from the Cambrian Rocks up to the present day. The oldest families of the *Crustacea* are the *Trilobita* and the *Eurypterida*, both of which are exclusively Palæozoic, and died out at the close of the Carboniferous epoch. It is worthy of notice how larval are the characters of these ancient groups when compared with their modern successors. Of the remaining orders the *Cirripedia*, *Ostracoda*, and *Phyllopoda* are the three which are most largely represented.

1. *Cirripedia.*—The Cirripedes are hardly known as Palæozoic fossils, but valves of a singular member of this order (*Turrilepas*) have been found in the Silurian Rocks of Scotland. With this exception, the Cirripedes are entirely confined in past time to the Secondary and Tertiary epochs. The *Balanidæ* are the most common, commencing, as far as is yet known, in the Eocene period, and attaining their maximum in recent seas. The *Verrucidæ* commence in the Chalk, and the

Lepadidæ begin still lower, in the Jurassic Rocks, and attain their maximum of development in the Cretaceous epoch.

2. *Ostracoda.*—Small Ostracode *Crustacea* are extremely abundant as fossils in many formations, and extend from the Lower Silurian period up to the present day.

3. *Phyllopoda.*—Remains of Crustaceans supposed to belong to this order are found in the Palæozoic Rocks. *Hymenocaris* is found in the Upper Cambrian, *Caryocaris* in the Lower Silurian, *Ceratiocaris* in the Upper Silurian, and *Dithyrocaris* in the Carboniferous Limestone. All these forms, with other similar ones, are believed to be most closely allied to the recent *Apus* and *Nebalia*.

4. *Trilobita.*—The Trilobites are exclusively Palæozoic fossils. In the Upper Cambrian Rocks—the so-called "primordial zone"—there occurs a singular group of Trilobites—the so-called primordial Trilobites—distinguished by the possession of many larval characters. In the Lower and Upper Silurian Rocks the Trilobites attain their maximum of development. They are still well represented in the Devonian Rocks; but they die out completely before the close of the Carboniferous epoch, being represented in the Mountain Limestone by three genera only (*Phillipsia*, *Brachymetopus*, and *Griffithides*).

5. *Eurypterida.*—These, like the last, are entirely Palæozoic, attaining their maximum in the Upper Silurian and Devonian formations, and dying out in the Carboniferous Rocks. *Pterygotus*, *Eurypterus*, and *Slimonia* are the most characteristic genera.

6. *Xiphosura.*—The genus *Limulus* commenced, as far as is yet known, in the Permian period, and has survived up to the present day. Its first appearance, therefore, was just at the close of the Palæozoic epoch. The two remaining genera, which constitute with *Limulus* this sub-order (viz., *Belinurus* and *Prestwichia*), are Palæozoic, and are not known to occur out of the Carboniferous Rocks.

7. *Isopoda.*—The earliest known Isopod is the *Prosoponiscus* of the Permian Rocks.

8. *Stomapoda.*—This order is doubtfully represented in the Carboniferous Rocks.

9. *Decapoda.*—The Decapods, with the exception of a single doubtful form from the Carboniferous Rocks, are not known to have existed at all during the Palæozoic period; but they are well represented, in all their three tribes, in the Secondary and Tertiary epochs, attaining their maximum at the present day. The London Clay (Eocene) is especially rich in the remains of *Macrura* and *Brachyura*.

CHAPTER XXXVI.

ARACHNIDA.

CLASS II. ARACHNIDA.—The *Arachnida*—including the Spiders, Scorpions, Mites, &c.—possess almost all the essential characters of the *Crustacea*, to which they are very closely allied. Thus, the body is divided into a variable number of somites, some of which are always provided with articulated appendages. A pair of ganglia is primitively developed in each somite, and the neural system is placed ventrally. The heart, when present, is always situated on the opposite side of the alimentary canal to the chain of ganglia. The respiratory organs, however, whenever these are differentiated, are never in the form of branchiæ as in the *Crustacea*, but are in the form either of pulmonary vesicles or sacs, or of ramified tubes, formed by an involution of the integument, and fitted for breathing air directly. Further, there are never "more than four pairs of locomotive limbs, and the somites of the abdomen, even when these are well developed, are never provided with limbs;" the reverse being the case amongst the *Crustacea.* Lastly, "in the higher *Arachnida*, as in the higher *Crustacea*, the body is composed of twenty somites, six of which are allotted to the head; but, in the former class, one of the two normal pairs of antennæ is never developed, and the eyes are always sessile; while, in the higher *Crustacea*, the eyes are mounted upon movable peduncles, and both pairs of antennæ are developed."—(Huxley.)

The head in the *Arachnida* is always amalgamated with the thorax, to form a "cephalothorax;" the integument is usually chitinous, and the locomotive limbs are mostly similar in form to those of insects, and are usually terminated by two hooks.

In many of the *Arachnida* the integument remains soft over the entire body; in others, as in the majority of Spiders, the abdomen remains soft and flexible, whilst the cephalothorax is more or less hard and chitinous; in the Scorpions, again, the integument over the whole body forms a strong chitinous shell.

The typical somite of the *Arachnida* is constituted upon exactly the same plan as that of the *Crustacea*, consisting essentially of a dorsal and ventral arc; the former composed of a central piece, or "tergum," and of two lateral pieces, or "epimera;" whilst the latter is made up of a median "sternum" and of two lateral "episterna."

As regards the composition of the cephalothorax of Spiders,

"the tergal elements of the coalesced segments are wanting, and the back of the thorax is protected by the elongation, convergence, and central confluence of the epimeral pieces; the sternal elements have coalesced into the broad plate in the centre of the origins of the ambulatory legs, from which it is separated by the episternal elements. . . . . The non-development of the tergal elements explains the absence of wings."—(Owen.)

The mouth is situated, in all the *Arachnida*, in the anterior segment of the body, and is surrounded by suctorial or masticatory appendages. In the higher *Arachnida*, the mouth is provided from before backwards with the following appendages (fig. 77, 4). 1. A pair of "mandibles," used for prehension. 2. A pair of "maxillæ," each of which is provided with a long jointed appendage, the "maxillary palp." 3. A lower lip, or "labium." In the Scorpion, an upper lip, or "labrum," is also present.

In the Spiders (fig. 77, 4) each mandible terminates in a sharp movable hook, which possesses an aperture at its extremity communicating by a canal with a gland, which is placed in the preceding joint of the mandible, and secretes a poisonous fluid. The maxillary palps in the Spiders are long, jointed appendages, terminated in the females by pointed claws, but frequently swollen, and carrying a special sexual apparatus in the males.

In the Scorpions (fig. 77, 1) the mandibles are short, and terminate in strong pincers, or "cheliceræ." The maxillary palpi are also greatly developed, and constitute powerful grasping-claws, or "chelæ." In the genus *Galeodes*, the mandibles, like those of the Scorpion, constitute "cheliceræ," though comparatively much larger and longer; but the maxillary palps are not developed into "chelæ."

With regard to antennæ, these organs, *as such*, do not exist in the *Arachnida*. It is generally believed, however, that the *mandibles* of the *Arachnida* are truly homologues, not of the parts which bear the same name in the other *Arthropoda*, but of the *antennæ*. The antennæ, therefore, of the Spiders, are converted into prehensile and offensive weapons; whilst in the Scorpions, as in the King-crabs, they are developed into nipping-claws, or chelæ.

In the lower *Arachnida*, the organs of the mouth, though essentially the same as in the higher forms, are enveloped in a sheath, formed by the labium and maxillæ, whilst the mandibles are often joined together so as to constitute a species of lancet.

The mouth opens into a pharynx, which is of remarkably small calibre in the true spiders, all of which live simply on the juices of their prey. The intestinal canal is usually short

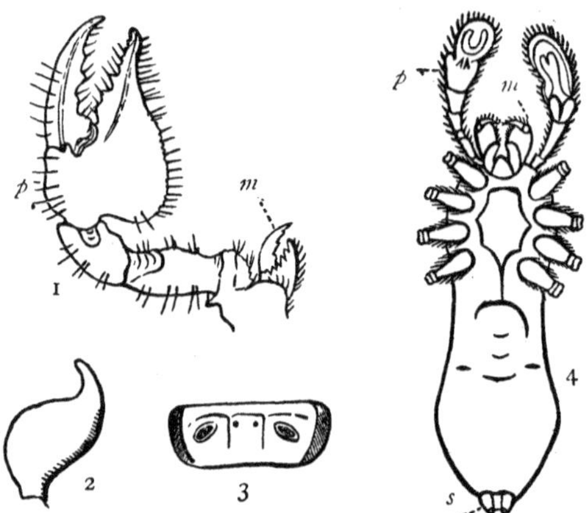

Fig. 77.—Morphology of Arachnida. 1. Organs of the mouth in the Scorpion, on one side; *m* Mandibles (antennæ) converted into chelæ, and called the cheliceræ; *p* Maxillary palpi greatly developed, and forming strong chelæ. 2. Telson of the Scorpion. 3. One of the abdominal segments of the Scorpion, showing the "stigmata," or apertures of the pulmonary sacs. 4. *Tegenaria domestica*, the common Spider (male), viewed from below; *s* Spinnerets; *m* Mandibles with their perforated hooks—below the mandibles are the maxillæ, and between the bases of these is the labium; *p* The maxillary palpi with their enlarged tumid extremities.

and straight, no convolutions intervening between the mouth and the aperture of the anus. Often, however, lateral cæca are appended to the alimentary tube. Salivary glands are also present, as well as ramified tubes, supposed to perform the functions of a kidney, and to correspond to the "Malpighian vessels" of Insects.

The circulation in the *Arachnida* is maintained by a dorsal heart, which is situated above the alimentary canal. Usually the heart is greatly elongated, and resembles the "dorsal vessel" of the *Insecta*. In the lower *Arachnida*, however, there is no central organ of the circulation, and there are no differentiated blood-vessels. All the *Arachnida* breathe the air directly, and the respiratory function is performed by the general surface of the body (as in the lowest members of the class), or by ramified air-tubes, termed "tracheæ," or by distinct pulmonary chambers or sacs; or, lastly, by a combination of tracheæ and pulmonary vesicles. The "tracheæ" consist of ramified or fasciculated tubes, opening upon the surface of the body by distinct apertures, called "stigmata." The walls of the tube are generally prevented from collapsing by means of a chitinous fibre or filament, which is coiled up into a spiral, and is situated beneath their epithelial lining. The pulmonary sacs are simple involutions of the integument, abundantly supplied with blood; the vascular surface thus formed being increased in area by the development of a number of close-set membranous lamellæ, or vascular plates, which project into the interior of the cavity. Like the tracheæ, the pulmonary sacs communicate with the exterior by minute apertures, or "stigmata" (fig. 77, 3).

The nervous system is of the normal articulate type, but is often much concentrated. In the Spiders there is a cephalic or "cerebral" ganglion, a large thoracic ganglion, and in some instances a small abdominal ganglion. In some of the lower forms the articulate type of nervous system is lost, and there is merely a ganglionic mass which is traversed by the gullet. In none of the *Arachnida* are compound eyes present, and in none are the eyes supported upon foot-stalks. The organs of vision, when present, are in the form of from two to eight simple eyes, or "ocelli."

In all the *Arachnida*, with the exception of the *Tardigrada*, the sexes are distinct. The great majority of the *Arachnida* are oviparous, and in most cases the larvæ are like the adult in all except in size. In some cases, however (*Acarina*), the larvæ have only six legs, and do not attain the proper four pairs of legs until after some moults.

The *Arachnida* may be divided into two great sections or sub-classes—viz., the *Trachearia*, in which respiration is effected by the general surface of the body, or by tracheæ, and there are never more than four ocelli; and the *Pulmonaria*, in which respiration is effected by pulmonary sacs, either alone or combined with tracheæ, and there are six or more eyes.

CHAPTER XXXVII.

DIVISIONS OF THE ARACHNIDA.

DIVISION A. TRACHEARIA.—*Respiration cutaneous, or by tracheæ. Eyes never more than four in number.*

The *Trachearia* comprise three orders—viz., the *Podosomata*, the *Acarina* or *Monomerosomata*, and the *Adelarthrosomata*.

ORDER I. PODOSOMATA (*Pantopoda*).—The members of this order, sometimes called "Sea-spiders," have been placed alternately amongst the *Arachnida* and the *Crustacea*, their true position being rendered doubtful by the fact that, though marine in their habits, they possess no differentiated respiratory organs. They possess, however, no more than four pairs of legs, and would therefore appear to be properly referable to the *Arachnida*. The commoner forms of the *Podosomata* (such as *Nymphon* and *Pycnogonum*) may be found on the sea-coast at low water, crawling about amongst marine plants or hiding beneath stones. Some species of the latter genus are asserted to be parasitic upon fishes and other marine animals, but the common British species (*P. littorale*) is free when adult, and does not appear to be parasitic at any stage of its existence (fig. 78, *a*). The legs consist of four pairs, sometimes greatly exceeding the body in length, and sometimes containing cæcal prolongations of the digestive cavity for a portion of their length. The mouth is provided with a pair of "cheliceræ," or chelate mandibles, and with two well-developed maxillary palpi, behind which in the female are a pair of false legs which carry the ova. The abdomen is rudimentary. Though there are no respiratory organs, there is a distinct heart. The sexes are in different individuals.

ORDER II. ACARINA or MONOMEROSOMATA.—The members of this order possess an unsegmented abdomen which is fused with the cephalothorax into a single mass. Respiration is effected by tracheæ. Most of the *Acarina* are parasitic, and the most familiar are the Mites and Ticks.

Family 1. *Linguatulina* or *Pentastomida*.—The members of this family are singular vermiform animals, found as parasites in the frontal sinuses and lungs of some Vertebrates. In their adult condition they possess no external organs except two pairs of hooks, representing limbs, placed near the mouth. They thus closely approximate to the *Tæniada*, beside which they have been generally placed. In the young condition, however, they possess four articulated legs, and even in the adult state the characters of the nervous system are higher than those of the *Scolecida*. There are no differentiated organs of respiration, and there are no circulatory organs, but the sexes are distinct.

Family 2. *Macrobiotidæ* (*Tardigrada* or *Arctisca*).—The "Sloth" or "Bear animalcules," which compose this family, are microscopic animals, very much like *Rotifers*, found in damp moss and in the gutters of houses. The nervous system consists of four ganglia, and there is a suctorial mouth, with rudimentary jaws or stilets. The abdomen is undeveloped, and there are four pairs of rudimentary legs. They exhibit no traces of either circulatory or respiratory organs, and the sexes are united in the same individual.

Family 3. *Acarida.*—This family includes the Mites, Ticks, and Water-mites, some of which are parasitic, whilst others are free, and some are even aquatic in their habits. The mouth is formed for suction. There is no definite line of demarcation between the unsegmented abdomen and the cephalothorax.

In the true *Acari* (fig. 78, *b*), of which the Cheese-mite may be taken as an example, there are four pairs of legs, adapted for walking, and the mouth is provided with distinct mandibles. Besides the Cheese-mite (*A. domesticus*), another well known species is the *Acarus destructor*, which feeds upon various zoological specimens, and is very annoying to the naturalist. In the *Sarcoptes scabiei*—the cause of the skin-disease known as the "itch"—the two anterior pairs of legs are provided with suckers, and the two posterior are terminated by bristles; the mouth, also, is furnished with bristles. In the Ticks (*Ixodes*) the mouth is provided with a beak, or "rostrum," which enables them to pierce the skin, and retain their hold firmly. In the *Hydrachnidæ* (fig. 78, *c*), or Water-mites, the head is fur-

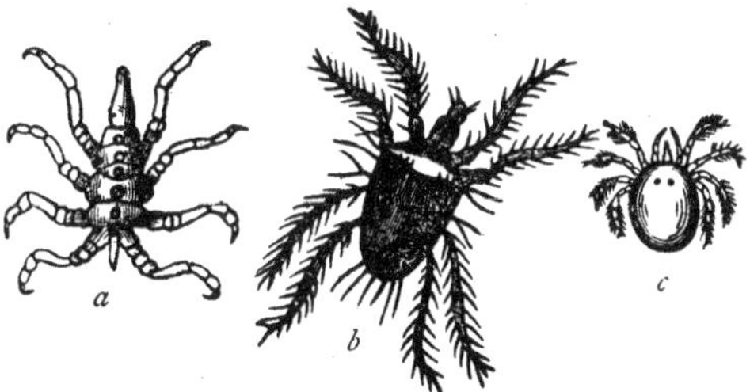

Fig. 78.—Arachnida. *a* ***Pycnogonum littorale;*** *b* ***Tetranychus telarius,*** one of the "Sociable" mites; *c* ***Hydrachna globulus,*** one of the "Water-mites."

nished with two or four ocelli, and there are four pairs of hairy natatory legs. They are parasitic, during at least a portion of their existence, upon Water-beetles and other aquatic insects. They pass through a metamorphosis, the larva being hexapod, or having only three pairs of legs. The Garden-mites (*Trombididæ*) and Spider-mites (*Ganasidæ*) live upon plants; the Wood-mites (*Oribatidæ*) and Harvest-ticks (*Leptidæ*) are to be found amongst moss and herbage, or creeping upon trees or stones; whilst the true Ticks (*Ixodidæ*) attach themselves parasitically by means of their suctorial mouth to the bodies of various Mammals, such as sheep, oxen, dogs, &c. Several Mites (*Thalassarachna*, *Pontarachna*, &c.) have been found to inhabit salt water, and several species of *Trombididæ* live habitually between tide-marks.

Another member of the *Acarina* is the curious little *Demodex folliculorum*, which is found in the sebaceous follicles of man, especially in the neighbourhood of the nose. It is probable that very few, if any, individuals are exempt from this harmless parasite.

ORDER III. ADELARTHROSOMATA.—The members of this order, comprising the Harvest-spiders, the Book-scorpions, &c., are distinguished from the preceding by the possession of an abdomen, which is more or less distinctly segmented, but generally exhibits no line of separation from the cephalothorax, the two regions being of equal breadth and conjoined together. The mouth is furnished with masticatory appendages, and respiration is effected by tracheæ, which open on the lower surface of the body by two or four stigmata.

Family 1. *Phalangidæ*.—The well known "Harvest-spiders" belong to this family. They are characterised by the great length of the legs, and by the filiform maxillary palpi, terminated by simple hooks.

Family 2. *Pseudoscorpionidæ* (*Cheliferidæ*).—The "Book-scorpion" (*Chelifer*) is a common little animal in old books. It is distinguished by the fact that the maxillary palpi are of large size, and are converted into nipping-claws or chelæ, thus giving the animal the appearance of a Scorpion in miniature.

Family 3. *Solpugidæ*.—In this family the abdomen is not only very distinctly segmented, but is also clearly separated from the abdomen. The mandibles in *Galeodes*, which is the type of the group, are chelate, but the maxillary palpi constitute long feet.

DIVISION B. PULMONARIA.—*Respiration by pulmonary sacs alone, or by pulmonary sacs conjoined with tracheæ. Eyes six or more in number. Abdomen usually distinct from the cephalothorax.*

This division comprises the higher *Arachnida*, such as the Scorpions, and the majority of what are commonly known as Spiders; the former constituting the order of the *Pedipalpi*, the latter that of the *Araneida* or *Dimerosomata*.

ORDER I. PEDIPALPI.—In this order are the true Scorpions, together with certain other animals which are in some respects intermediate between the Scorpions and the true Spiders. The members of this order are distinguished by the fact that the abdomen in all is distinctly segmented, but is not separated from the cephalothorax by a well-marked constriction. They agree in this character with the *Adelarthrosomata;* hence the two are sometimes united into a single order (*Arthrogastra*), but they are separated by the nature of the respiratory organs, the latter breathing by tracheæ, and not by pulmonary sacs.

Family 1. *Scorpionidæ.*—The Scorpions are amongst the best known of the *Arachnida*, as well as being amongst the largest. They are distinguished by their long, distinctly segmented abdomen, terminating in a hooked claw (figs. 77, 79). This claw, which is really a modified "telson," is the chief offensive weapon of the Scorpion, and is perforated at its point by the duct of a poison-gland which is situated at its base. The abdomen is composed of twelve somites, but there is no evident line of demarcation between this region and the cephalothorax. The thoracic segments carry four pairs of ambulatory feet. There are six, eight, or twelve simple eyes. The maxillary palpi are

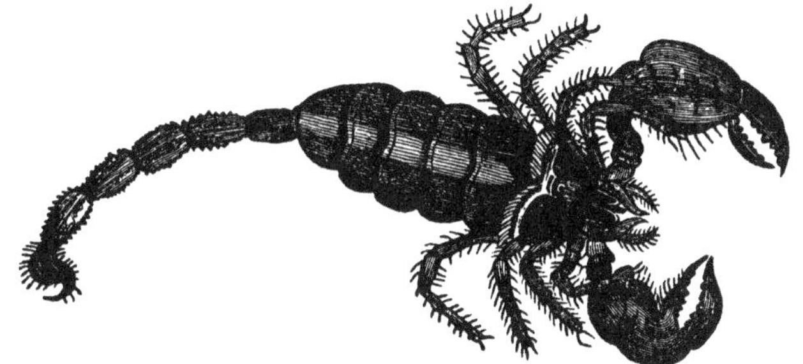

Fig. 79.—Scorpion (reduced).

greatly developed, and constitute strong nipping-claws, or "chelæ" (figs. 77, 79). The mandibles (antennæ) also form claws, or "chelicerae." The respiratory organs are in the form of pulmonary sacs, four on each side, opening upon the under surface of the abdomen by as many stigmata, each of which is surrounded by a raised margin, or "peritrema" (fig. 77, 3).

The Scorpions are mostly inhabitants of warm regions, and their sting, though much exaggerated, is of a very severe nature. They live under stones or in dark crevices, and run swiftly, carrying the tail curved over the back. They feed on Insects, which they hold in the chelate palpi and sting to death.

Family 2. *Thelyphonidæ.*—The members of this family in

external appearance closely resemble the true Spiders, from which they are separated by the possession of a segmented abdomen, and long spinose palpi, and by the absence of spinnerets. They are distinguished from the *Scorpionidæ* by the amalgamation of the head and thorax into a single mass, which is clearly separated from the abdomen by a slight constriction, as well as by the fact that the maxillary palpi terminate in movable claws instead of chelæ. Further, the extremity of the abdomen is not furnished with a terminal hook, or "sting."

ORDER II. ARANEIDA or SPHÆROGASTRA.—This order includes the true Spiders, which are characterised by the amalgamation of the cephalic and thoracic segments into a single mass, and by the generally soft, unsegmented abdomen, attached to the cephalothorax by a constricted portion, or peduncle. Respiration is effected by pulmonary sacs usually in combination with tracheæ. (Hence the name *Pulmotrachearia*, sometimes applied to the order.) The number of the pulmonary sacs is smaller in the true Spiders than in the Scorpions, being either two or four, opening by as many stigmata upon the under surface of the abdomen.

The head bears from six to eight simple eyes; the mandibles are simply hooked, and are perforated by the duct of a gland which secretes a poisonous fluid; and the maxillary palpi are never chelate.

Spiders (fig. 80) are all predaceous animals, and many of them possess the power of constructing webs for the capture

Fig. 80.—Araneida. *Theridion riparium* (female).

of their prey or for lining their abodes. For the production of the web, Spiders are furnished with special glands, situated at the extremity of the abdomen. The secretion of these glands is a viscid fluid, which hardens rapidly on exposure to air, and which is cast into its proper, thread-like shape, by being passed through what are called the "spinnerets." These are little conical or cylindrical organs, four or six in number, situated below the extremity of the abdomen. The excretory ducts of the glands open into the spinnerets, each of which has its apex perforated by a great number of minute tubes, through which the secretion of the glands has to pass before reaching the air. Many spiders, however, do not construct any web, unless it be for their own habitations, but hunt their prey for themselves.

As regards the reproductive process in the Spiders, it appears certain that the act of copulation, so to speak, is performed by the males by means of the maxillary palpi, the extremities of which are specially modified for this purpose. The testes are abdominal, but the semen appears to be stored up in the enlarged extremities of the maxillary palps, which thus perform the part of the vesiculæ seminales. "The most careful observations, repeated by the most attentive and experienced entomologists, have led to the conviction that the ova are fertilised by the alternate introduction into the vulva of the appendages of the two palpi of the male. Treviranus's supposition that these acts are merely preliminary stimuli, has received no confirmation, and is rejected by Dugés, Westwood, and Blackwall; and with good reason, as the detection of the spermatozoa in the palpal vesicles has shown. . . . Dugés offers the very probable suggestion that the male himself may apply the dilated cavities of the palpi to the abdominal aperture (of the testes), and receive from the vasa deferentia the fertilising fluid, preparatory to the union. . . . Certain it is that an explanation of this singular condition of the male apparatus, in which the intromittent organ is transferred to the remote and outstretched palp, is afforded by the insatiable proneness to slay and devour in the females of these most predaceous of articulated animals."—(Owen.)

The Spiders are oviparous, and the young pass through no metamorphosis; but they cast their skins, or moult, repeatedly, before they attain the size of the adult.

DISTRIBUTION OF ARACHNIDA IN TIME.—The *Arachnida* are only very rarely found in a fossil condition. As far as is yet known, both the Scorpions and the true Spiders appear to have their commencement in the Carboniferous epoch, the former being represented by the celebrated *Cyclophthalmus senior* from the Coal-measures of Bohemia. Spiders are also known to occur in the Jurassic Rocks (Solenhofen Slates) and in the Tertiary period.

CHAPTER XXXVIII.

MYRIAPODA.

CLASS III. MYRIAPODA.—The *Myriapoda* are defined as *articulate animals in which the head is distinct, and the remainder of the body is divided into nearly similar segments, the thorax exhibiting no clear line of demarcation from the abdomen. There is one pair of antennæ, and the number of the legs is always more than eight pairs. Respiration is by tracheæ.*

In this class—comprising the Centipedes (fig. 81) and the Millipedes—the integument is chitinous, the body is divided into a number of somites provided with articulated appendages, and the nervous and circulatory organs are constructed upon a plan similar to what we have seen in *Crustacea* and *Arachnida*. The head is invariably distinct, and there is no marked line of demarcation between the segments of the thorax and those of the abdomen. The body, except in *Pauropus*, always consists of more than twenty somites, and those which correspond to the abdomen in the *Arachnida* and *Insecta* are always provided with locomotive limbs. "The head consists of at least five, and probably of six, coalescent and modified somites; and some of the anterior segments of the body are, in many genera, coalescent, and have their appendages specially modified to subserve prehension."—(Huxley.) *Pauropus* has only nine pairs of legs; but with this exception, eleven pairs of legs is the smallest number known in the order.

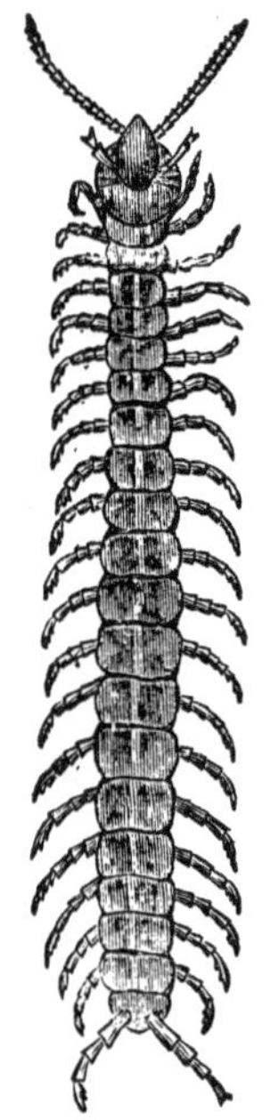

Fig. 81.—Centipede (*Scolopendra*).

The respiratory organs, with one exception, agree with those of the *Insecta* and of many of the *Arachnida* in being "tracheæ"—that is to say, tubes, which open upon the surface of the body by minute apertures, or "stigmata," and the walls of which are strengthened by a spirally-coiled filament of chitine. The tracheæ may or may not anastomose with one another as they do in Insects.

The somites, with the exception of the head and the last abdominal segment, are usually undistinguishable from one another, and each bears a single pair of limbs. In some cases, however, each segment appears to be provided with two pairs of appendages (fig. 82). This is really due to the coalescence of the somites in pairs, each apparent segment being in reality composed of two amalgamated somites. This is shown, not only by the bigeminal limbs, but also by the arrangement of the stigmata, which in the normal forms occur on every alternate ring only, whereas in these aberrant forms they are found upon every ring.

The head always bears a pair of jointed antennæ, resembling those of many Insects, and behind the antennæ there is generally a variable number of simple sessile eyes. In one species (*Scutigera*) compound facetted eyes are present; and in *Pauropus* the antennæ are bifid, and carry many-jointed appendages, thus differing wholly from the antennæ of Insects, and presenting a decided approximation to the *Crustacea*.

The young in some cases, on escaping from the egg, possess nearly all the characters of the parents, except that the number of somites, and consequently of limbs, is always less, and increases at every change of skin ("moult" or "ecdysis"). In most cases, there is a species of metamorphosis, the embryo being at first either devoid of locomotive appendages, or possessed of no more than three pairs of legs, thus resembling the true hexapod Insects. It is believed, however, that the legs of these hexapod larvæ do not correspond homologically with the three pairs of legs proper to adult Insects. In these cases the number of legs proper to the adult is not obtained until after several moults, the entire process being stated to occupy in some species as much as two years, before maturity is reached.

The *Myriapoda* are divided into three orders—viz., the *Chilopoda*, the *Chilognatha*, and the *Pauropoda*.

ORDER I. CHILOPODA.—This order comprises the well-known carnivorous Centipedes and their allies, and is characterised by the number of legs being rarely indefinitely great (usually from 15 to 20 pairs), by the composition of the antennæ out of not less than 14 joints (14 to 40 or more), and by the structure of the masticating organs. These consist of a pair of mandibles with small palpi, a labium, and two pairs of "maxillipedes," or foot-jaws, of which the second is hooked, and is perforated for the discharge of a poisonous fluid. There is not more than one pair of legs to each somite, and the last two limbs are often directed backwards in the axis of the body, so as to form a kind of tail. The body in all the *Chilopoda* is flattened, and the generative organs open at the posterior end of the body.

Scolopendra (fig. 81), *Lithobius*, and *Geophilus* are common

European genera of this order. The ordinary Centipedes of this country are perfectly harmless, but those of tropical regions sometimes attain a length of a foot, or more, and these are capable of inflicting very severe, and even dangerous, bites.

ORDER II. CHILOGNATHA.—This order comprises the vegetable-eating Millipedes (*Iulidæ*) and the Gallyworms (*Polydesmus*). The order is characterised by the great number of legs,—each segment, except the six or seven anterior ones, bearing two pairs; by the composition of the antennæ out of six or seven joints; and by the structure of the masticating organs, which consist of a pair of mandibles without palps, covered by a lower lip, composed of the confluent maxillæ. The generative apertures are placed in the anterior portion of the body.

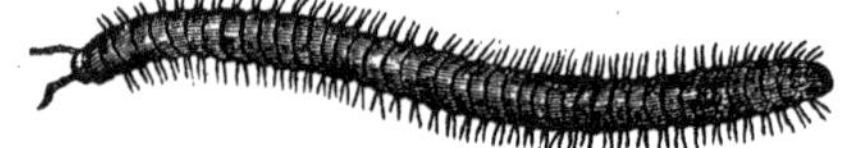

Fig. 82.—Millipede (*Iulus*).

In the common Millipede (*Iulus*) the body is composed of from forty to fifty segments, each of which bears two pairs of minute, thread-like legs. The *Iuli* of this country are of small size, but an American species attains a length of more than half a foot.

ORDER III. PAUROPODA.—In this order is only an extraordinary little *Myriapod*, described by Sir John Lubbock under the name of *Pauropus*. The body is only one-twentieth of an inch in length, and consists of ten somites, furnished with scattered setæ. There are only nine pairs of legs, of which one pair is carried by the 3d segment, whilst the 4th, 5th, 6th, and 7th segments carry each two pairs of legs, and may therefore be regarded as really double. The head is composed of two segments, and is not provided with jaw-feet. The antennæ are five-jointed, bifid, with three long multi-articulate appendages. The body is white and colourless, and there are no tracheæ, so that respiration must be effected entirely by the skin. *Pauropus* is found amongst decaying leaves in damp situations, and species have been described both from Britain and America. It is separated from the *Chilopoda* by its small number of legs, the absence of foot-jaws, and the composition of the antennæ out of no more than five joints.

DISTRIBUTION OF MYRIAPODA IN TIME.—About twenty species of *Myriapoda* are known as fossils, the oldest example of the order having been found in the Carboniferous epoch.

From rocks of this age several species of Chilognathous Myriapods have been discovered. They belong to the genera *Xylobius* and *Archiulus*, and have been placed in a special family under the name of *Archiulidæ*. The occurrence of air-breathing articulate animals (both *Arachnida* and *Myriapoda*) in the Carboniferous period is noticeable, as being contemporaneous with the earliest known terrestrial Molluscs.

CHAPTER XXXIX.

INSECTA.

GENERAL CHARACTERS OF THE INSECTA.

CLASS IV. INSECTA.—The *Insecta* are defined as *articulate animals in which the head, thorax, and abdomen are distinct; there are three pairs of legs borne on the thorax; the abdomen is destitute of legs; a single pair of antennæ is present; mostly, there are two pairs of wings on the thorax. Respiration is effected by tracheæ.*

In the *Insecta* the body is divided into a variable number of definite segments, or somites, some of which are furnished with jointed appendages, and the nervous and circulatory systems are constructed upon essentially the same plan as in the *Crustacea*, *Arachnida*, and *Myriapoda*. The head, thorax, and abdomen are distinct (fig. 83), and the total number of somites in the body never exceeds twenty. "Of these, five certainly, and six probably, constitute the head, which possesses a pair of antennæ, a pair of mandibles, and two pairs of maxillæ, the hinder pair of which are coalescent, and form the 'labium.' Three, or perhaps, in some cases, more, somites unite and become specially modified to form the thorax, to which the three pairs of locomotive limbs, characteristic of perfect Insects, are attached. Two additional pairs of locomotive organs, the wings, are developed, in most insects, from the tergal walls of the second and third thoracic somites. No locomotive limbs are ever developed from the abdomen of the adult insect, but the ventral portions of the abdominal somites, from the eighth backwards, are often metamorphosed into apparatuses ancillary to the generative function."—(Huxley.)

The integument of the *Insecta*, in the mature condition, is more or less hardened by the deposition of chitine, and usually forms a resisting exoskeleton, to which the muscles are attached. The segments of the head are amalgamated into a single piece, which bears a pair of jointed feelers or antennæ, a pair of eyes, usually compound, and the appendages of the mouth. The segments of the thorax are also amalgamated into a single piece; but this, nevertheless, admits of separation into its constituent three somites (fig. 83). These are termed respectively, from before backwards, the "prothorax," "mesothorax," and "metathorax," and each bears a pair of jointed legs. In the great majority of Insects, the dorsal arches of the mesothorax and metathorax give origin each to a pair of wings.

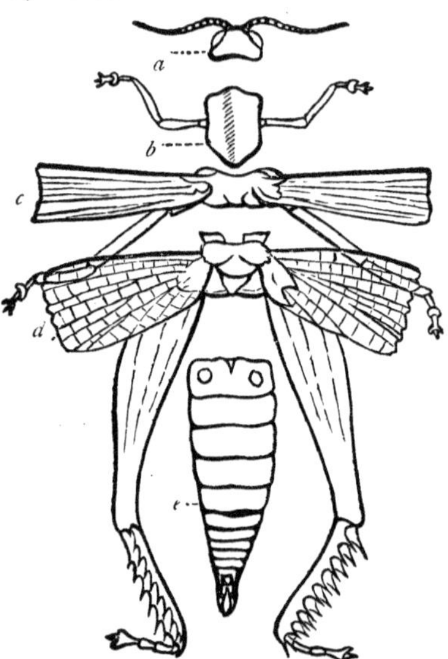

Fig. 83.—Diagram of Insect. *a* Head, carrying the eyes and antennæ; *b* Prothorax, carrying the first pair of legs; *c* Mesothorax, carrying the second pair of legs and first pair of wings; *d* Metathorax, with the third pair of legs and the second pair of wings; *e* Abdemen, without limbs, but having terminal appendages subservient to reproduction.

Each leg consists of from six to nine joints. The first of these, which is attached to the sternal surface of the thorax, is called the "coxa," and is succeeded by a short joint, termed the "trochanter." The trochanter is followed by a joint, often of large size, called the "femur," and this has articulated to it the "tarsus," which may be composed of from two to five joints.

The wings of Insects are membranous "flattened vesicles, sustained by slender but firm hollow tubes, called 'nervures,' along which branches of the tracheæ and channels of the circulation are continued."—(Owen.) According to Newport, the wings of Insects are "expanded portions of the common integument of the sides of the meso- and meta-thorax, occasioned by the enlargement and extension of numerous tracheæ and the accompanying passages for the circulatory fluids, and their motions are intimately connected with the function of respiration." In the *Coleoptera* (Beetles) the anterior pair of wings become hardened by the deposition of chitine, so as to form two protective cases for the hinder membranous wings. In this condition the anterior wings are known as the "elytra," or "wing-cases." In some of the *Hemiptera* this change only affects the inner portions of the anterior wings, the apices of which remain membranous, and to these the term "hemelytra" is applied. In the *Diptera* the posterior pair of wings are rudimentary, and are converted into two capitate filaments, called "halteres," or "balancers." In the *Strepsiptera* the anterior pair of wings are rudimentary, and are converted into twisted filaments.

The primitive number of somites in the abdomen of insects is said to be eleven (*Orthoptera*), but nine is the number ordinarily present; and though these are distinct in most larvæ, it is seldom that more than seven or eight are recognisable in the adult. The abdominal somites are usually more or less freely movable upon one another, and never carry locomotive limbs. The extremity of the abdomen is, however, not infrequently furnished with appendages, which are connected with the generative function, and not infrequently serve as offensive and defensive weapons. Of this nature are the ovipositors of Ichneumons and other Insects, and the sting of Bees and Wasps. In the Earwig (*Forficula*) these caudal appendages form a pair of forceps; whilst in many Insects they are in the form of bristles, by which powerful leaps can be effected, as is seen in the Springtails (*Poduræ*). In some insects (as the Mole-cricket and Cockroach), the 9th or 10th abdominal segment carries jointed antenniform appendages, which, though perhaps partially or even primarily generative in function, are certainly organs of sense, being connected with smell or hearing.

The organs about the mouth in Insects are collectively termed the "trophi," or "instrumenta cibaria." Two principal types require consideration—namely, the masticatory and the suctorial—both types being sometimes modified, and occasionally combined.

In the Masticatory Insects, such as the Beetles (fig. 84, 1), the trophi consist of the following parts, from before backwards:—(1.) An upper lip, or "labrum," attached below the front of the head. (2.) A pair of biting-jaws, or "mandibles." (3.) A pair of chewing-jaws, or "maxillæ," provided with one or more pairs of "maxillary palps," or sensory and tactile filaments. (4.) A lower lip, or "labium," composed of a second coalescent pair of maxillæ, and also bearing a pair of palpi, the "labial palps." The primitive form of the labium, that, namely, of a second pair of maxillæ, is more or less perfectly retained by the *Orthoptera* and some of the *Neuroptera*. The lower or basal portion of the labium is called the "mentum" or chin, whilst the upper portion is more flexible, and is termed the "ligula." The upper portion of the ligula is often developed into a kind of tongue, which is very distinct in some Insects, and is termed the "lingua."

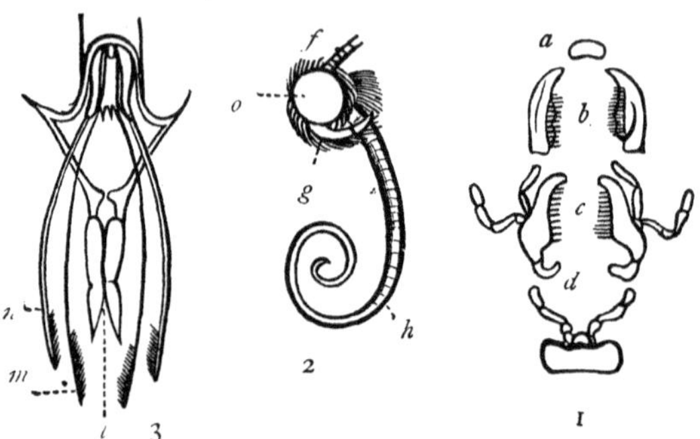

Fig. 84.—Organs of the mouth in Insects. 1. Trophi of a masticating Insect (Beetle): *a* Labrum or upper lip; *b* Mandibles; *c* Maxillæ with their palpi; *d* Labium or lower lip with its palpi. 2. Mouth of a Butterfly: *o* Eye; *f* Base of antennæ; *g* Labial palp; *h* Spiral trunk or "antlia." 3. Mouth of a Hemipterous Insect (*Nepa cinerea*): *l* Labium: *m* Maxillæ; *n* Mandibles.

In the typical suctorial mouth, as seen in the Butterflies (fig. 84, 2), the following is the arrangement of parts:—The labrum and the mandibles are now quite rudimentary; the first pair of maxillæ is greatly elongated, each maxilla forming a half-tube. These maxillæ adhere together by their inner surfaces, and thus form a spiral "trunk," or "antlia" (inappropriately called the "proboscis"), by which the juices of flowers are sucked up. Each maxilla, besides the half-tube on one side, contains also a tube in its interior; consequently on a transverse section the trunk is found really to consist of three canals, one in the anterior of each maxilla, and the third formed between them by their apposition. To the base of the trunk are attached the maxillary palpi, which are extremely small. Behind the trunk is a small labium, composed of the united second pair of maxillæ. The "labial palpi" are greatly developed, and form two hairy cushions, between which the trunk is coiled up when not in use.

In the Bee there exists an intermediate condition of parts, the mouth being fitted partly for biting and partly for suction. The labrum and mandibles are well developed, and retain their usual form. The maxillæ and the labium are greatly elongated; the former being apposed to the lengthened tongue in such a manner as to form a tubular trunk, which cannot be rolled up, as in the Butterflies, but is capable of efficient suction. The labial palpi are also greatly elongated.

In the *Hemiptera*, the "trophi" consist of four lancet-shaped needles, which are the modified mandibles and maxillæ, enclosed in a tubular sheath formed by the elongated labium (fig. 84, 3). Lastly, in the *Diptera*—as in the common House-fly—there is an elongated labium, which is channelled on its upper surface for the reception of the mandibles and maxillæ, these being modified into bristles or lancets.

The mouth in the Masticating Insects leads by a pharynx and œsophagus into a membranous, usually folded, stomach—the "crop," or "ingluvies"—from which the food is transmitted to a second muscular stomach, called the "gizzard" (fig. 85). The gizzard is adapted for crushing the food, often having plates or teeth of chitine developed in its walls, and is succeeded by the true digestive cavity, called the "chylific stomach" (*ventriculus chylopoieticus*). From this an intestine of variable length proceeds, its terminal portion, or rectum, opening into a dilatation which is common to the ducts of the generative organs, and is termed the "cloaca." The œsophagus is furnished with salivary glands of varying size and complexity, and is provided in some of the Suctorial Insects with a dilatation called the "sucking stomach." Behind the pyloric aperture of the stomach, with very few exceptions, are a variable number of cæcal, convoluted tubes (fig. 85, *e*), which open into the intestine, and are called the "Malpighian tubes."

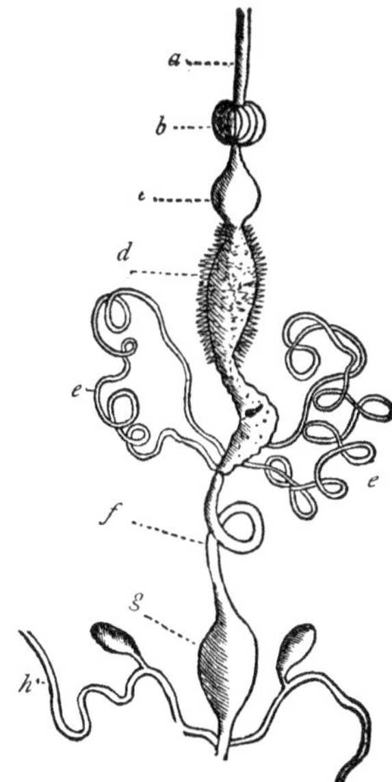

Fig. 85.—Digestive system of a Beetle (*Carabus auratus*). *a* Œsophagus; *b* Crop; *c* Gizzard; *d* Chylific stomach; *e* Malpihgian tubes; *f* Intestine; *g* Cloaca; *h* Supposed renal vessels.

These are often looked upon as representing the liver, but are by some believed to have a renal function. If the Malpighian vessels truly perform the functions of a liver—as their position would appear to prove—then the kidneys will be represented by a series of cæcal tubes which are only occasionally present, and which open into the rectum, close to the cloaca. There are no absorbent vessels, and the products of digestion simply transude through the walls of the alimentary canal into the sinuses or irregular cavities which exist between the abdominal organs. The apparatus of digestion does not differ essentially from the above in any of the Insects; but the alimentary canal is, generally speaking, considerably lengthened in the herbivorous species.

There is no definite and regular course of the circulation in the Insects. The propulsive organ of the circulation is a long contractile cavity, situated in the back and termed the "dorsal vessel." This is composed of a number of sacs (ordinarily eight), opening into one another by valvular apertures, which allow of a current in one direction only—viz., towards the head. The blood is collected from the irregular venous sinuses which are formed by the lacunæ and interstices between the tissues, and enters the dorsal vessel from behind; it is then driven forwards, and is expelled at the anterior extremity of the body.

Respiration is effected by means of "tracheæ," or branched tubes, which commence at the surface of the body by lateral apertures called "stigmata," or "spiracles," and ramify through every part of the animal. In structure the tracheæ are membranous, but their walls are strengthened by a chitinous filament, which is rolled up into a continuous spiral coil. In the aquatic larvæ of many insects, and in one adult insect, branches of the tracheæ are sent to temporary outgrowths which are termed "tracheal gills," and in which the blood is oxygenated. In all, however, except the single insect above mentioned, these temporary external appendages fall off when maturity is attained. The wings, also, whilst acting as locomotive organs, doubtless subserve respiration, the nervures being hollow tubes enclosing tracheæ.

The nervous system in Insects, though often concentrated into special masses, consists essentially of a chain of ganglia, placed ventrally, and united together by a series of double cords or commissures. The cephalic or "præ-œsophageal" ganglia are of large size, and distribute filaments to the eyes and antennæ. The post-œsophageal ganglia are united to the preceding by cords which form a collar round the gullet, and they supply the nerves to the mouth, whilst the next three ganglia furnish the nerves to the legs and wings. In larvæ, thirteen pairs of ganglia, one to each segment, can be recognised. In the imago, however, of the *Coleoptera*, several of these primitive ganglia have coalesced, so that this number is considerably reduced.

The organs of sense are the eyes and antennæ. The eyes in Insects are usually "compound," and are composed of a number of hexagonal lenses, united together, and each supplied with a separate nervous filament. Besides these, simple eyes —"ocelli," or "stemmata"—are often present, or, in rare cases, may be the sole organs of vision. In structure these resemble the single elements of the compound eyes. In a few cases the eyes are placed at the extremities of stalks or peduncles, but in no case are these peduncles movably articulated to the head, as is the case in the Podophthalmous Crustaceans. The antennæ are movable, jointed filaments, attached usually close to the eyes, and varying much in shape in different Insects. They doubtless discharge the functions of tactile organs, but are probably the organ of other more recondite senses in addition.

The sexes in Insects are in different individuals, and most are oviparous. Generally speaking, the young insect is very different in external characters from the adult, and it requires to pass through a series of changes, which constitute the "metamorphosis," before attaining maturity. In some Insects, however, there appears to be no metamorphosis, and in some the changes which take place are not so striking or so complete as in others. By the absence of metamorphosis, or by the degree of its completeness when present, Insects are divided into sections, called respectively *Ametabola*, *Hemimetabola*, and *Holometabola*, which, though not, perhaps, of a very high scientific value, are nevertheless very convenient in practice.

Section 1. *Ametabolic Insects.*—These pass through no metamorphosis, and also, in the mature condition, are destitute of wings. The young of these insects (*Aptera*) on escaping from the ovum resemble their parents in all respects except in size; and though they may change their skins frequently, they undergo no alteration before reaching the perfect condition, except that they grow larger.

Section 2. *Hemimetabolic Insects.*—In the insects belonging to this section there is a metamorphosis consisting of three stages. The young on escaping from the ovum is termed the "larva;" when it reaches its second stage it is called the "pupa," or "nymph;" and in its third stage, as a perfect insect, it is called the "imago." In the *Hemimetabola*, the "larva," though of course much smaller than the adult, or "imago," differs from it in little else except in the absence of wings. It is active and locomotive, and is generally very like the adult in external appearance. The "pupa," again, is a little larger than the larva, but really differs from it in nothing else than in the fact that the rudiments of wings have now appeared, in the form of lobes enclosed in cases. The "pupa" is still active and locomotive, and the term "nymph" is usually applied to it. The pupa is converted into the perfect insect, or "imago," by the liberation of the wings, no other change being requisite for this purpose. From the comparatively small amount of difference between these three stages, and from the active condition of the pupa, this kind of metamorphosis is said to be "incomplete."

In some members of this section however — such as the Dragon-flies—the larva and pupa are aquatic, whereas the imago leads an aerial life. In these cases there is necessarily a considerable difference between the larva and the adult; but the larva and pupa are closely alike, and the latter is active.

Section 3. *Holometabolic Insects.* — These — comprising the Butterflies, Moths, Beetles, &c.—pass through three stages which differ greatly from one another in appearance, the metamorphosis, therefore, being said to be "complete." In these insects (fig. 86) the "larva" is vermiform, segmented, and usually provided with locomotive feet, which do not correspond with those of the adult, though these latter are usually present as well (fig. 86). In some cases the larva is destitute of legs, or is "apodal." The larva is also provided with masticatory organs, and usually eats voraciously. In this stage of the metamorphosis the larvæ constitute what are usually called "caterpillars" and "grubs." Having remained in this condition for a longer or shorter length of time, and having undergone repeated changes of skin, or "moults," necessitated by its rapid growth, the larva passes into the second stage, and becomes a "pupa." The insect is now perfectly

Fig. 86.—Metamorphosis of the Magpie-moth (*Phalæna grossulariata*).

quiescent, unless touched or otherwise irritated; is incapable of changing its place; and is often attached to some foreign object. This constitutes what—in the case of the *Lepidoptera*—is generally known as the "chrysalis," or "aurelia" (fig. 93). The body of the pupa is usually covered by a chitinous pellicle, which closely invests the animal. In some cases (*e g.*, in many Dipterous insects) no traces can be detected in the pupa of the future insect; but in the *Lepidoptera* the thorax and abdomen are distinctly recognisable in the pupæ; whilst in others (*e.g.*, *Hymenoptera*) the parts of the pupa are merely covered by a membrane and are quite distinct. In some cases the pupa is further protected within the dried skin of the larva; and in other cases the larva—immediately before entering upon the pupa-stage—spins, by means of special organs for the purpose, a protective case, which surrounds the chrysalis, and is termed the "cocoon."

Having remained for a variable time in the quiescent pupa-stage, and having undergone the necessary development, the insect now frees itself from the envelope which obscured it, and appears as the perfect adult, or "imago," characterised by the possession of wings.

SEXES OF INSECTS.—The great majority of Insects, as is the case with most of the higher animals, consist of male and female individuals; but there occur some striking exceptions to this rule, as sèen in the Social Insects. In those organised communities which are formed by Bees, Ants, and Termites, by far the greater number of the individuals which compose the colony are either undeveloped females, or are of no fully developed sex. This is the case with the workers amongst Bees, and the workers and soldiers amongst Ants and Termites. And these sterile individuals, or "neuters," as they are commonly called, are not necessarily all alike in structure and external appearance. Amongst the Bees all the neuters resemble one another; but amongst Ants and Termites they are often divided into "castes," which have different functions to perform in the general polity, and differ from one another greatly in their characters.

In all the above-mentioned insects the males are relieved from the performance of any of the duties of life except that of propagating the species; and the females—which are generally solitary in each community—fulfil no other function save that of laying eggs. All the other duties which are necessary for the existence of the community are performed by the workers, or neuters.

The organs of the two sexes are in no case united in the

same individual, or, in other words, there are no hermaphrodite insects. (In some very abnormal cases amongst Bees hermaphrodite individuals have been observed.) As has been noticed, however, before, asexual reproduction is by no means unknown amongst the *Insecta*, and the attendant phenomena are often of extreme interest.—(See Introduction.)

The great majority of insects, during their adult condition, are terrestrial or aerial in their habits, but in many cases, even of these, the larvæ are aquatic. Many other insects live habitually during all stages of their existence in fresh water. A few insects inhabit salt water (either the sea itself or inland salt waters) during the whole or a portion of their existence. (This is the case with two or three Beetles of the families *Hydrophilidæ* and *Dytiscidæ*, some Hemipterous Insects, and the larvæ of various *Diptera*.) Lastly, many insects live parasitically upon the bodies of Birds or Mammals, or upon other Insects.

CHAPTER XL.

DIVISIONS OF INSECTA.

THE class *Insecta* includes such an enormous number of species, genera, and families, that it would be impossible to treat of these satisfactorily otherwise than in a treatise especially devoted to Entomology. Here it will be sufficient to give simply the differential characters of the different orders, drawing attention occasionally to any of the more important points in connection with any given family.

As already said, the *Insecta* are divided into three divisions, termed *Ametabola*, *Hemimetabola*, and *Holometabola*, according as they attain the adult condition without passing through a metamorphosis, or have an incomplete or complete metamorphosis. The Insects which come under the first head (viz. *Ametabola*) are not furnished with wings in the adult condition, and the three orders which compose this section are commonly grouped together under the name *Aptera*. By some, however, this division is entirely rejected, and the three orders in question are placed amongst the *Hemimetabola*, or even grouped with the *Myriapoda*.

SUB-CLASS I. AMETABOLA.—*Young not passing through a metamorphosis, and differing from the adult in size only. Imago destitute of wings; eyes simple, sometimes wanting.*

ORDER I. ANOPLURA.—Minute *Aptera*, in which the mouth is formed for suction; and there are two simple eyes, or none.

This order comprises insects which are commonly parasitic upon man and other animals, and are known as Lice (*Pediculi*). The common Louse is furnished with a simple eye, or ocellus, on each side of a distinctly differentiated head, the under surface of which bears a suctorial mouth. There is little distinction between the thorax and abdomen, but the segments of the former carry three pairs of legs. The legs are short, with short claws or with two opposing hooks, affording a very firm hold. The body is flattened and nearly transparent, composed of eleven or twelve distinct segments, and showing the stigmata very plainly. The young pass through no metamorphosis, and their multiplication is extremely rapid. Most, if not all, Mammals are infested by Lice, each having generally its own peculiar species, and sometimes having two or three. Three species are said to belong to Man—viz., *Pediculus humanus*, *P. capitis*, and *P. pubis*.

ORDER II. MALLOPHAGA.—Minute *Aptera*, in which the mouth is formed for biting, and is furnished with mandibles and maxillæ.

The members of this order are commonly known as "Bird-lice," being parasitic, sometimes upon Mammals, but mostly upon Birds. They strongly resemble the *Pediculi*, but the mouth is formed for biting, to suit their mode of life—since they do not live upon the juices of their hosts, but upon the more delicate tegumentary appendages.

ORDER III. THYSANURA.—*Apterous* insects, usually with a masticatory mouth, and having the extremity of the abdomen furnished with locomotive appendages.

The most familiar members of this order are the *Poduræ*, or "Spring-tails," which are characterised by the possession of a forked caudal appendage, by the extension of which considerable leaps can be effected. In the nearly allied *Lepismæ*, locomotion is assisted by caudal bristles. In both, the body is covered with hairs or scales, the structure of the latter being often very beautiful.

SUB-CLASS II. HEMIMETABOLA.—*Metamorphosis incomplete; the larva differing from the imago chiefly in the absence of wings, and in size; pupa usually active, or, if quiescent, capable of movement.**

* The *Coccidæ*, amongst the *Hemiptera*, undergo a complete metamorphosis. In certain of the *Hemiptera* and *Orthoptera* the adult is apterous, and in these cases there cannot be said to be any metamorphosis, since

ORDER IV. HEMIPTERA.—Mouth suctorial, beak-shaped, consisting of a jointed rostrum, composed of the elongated

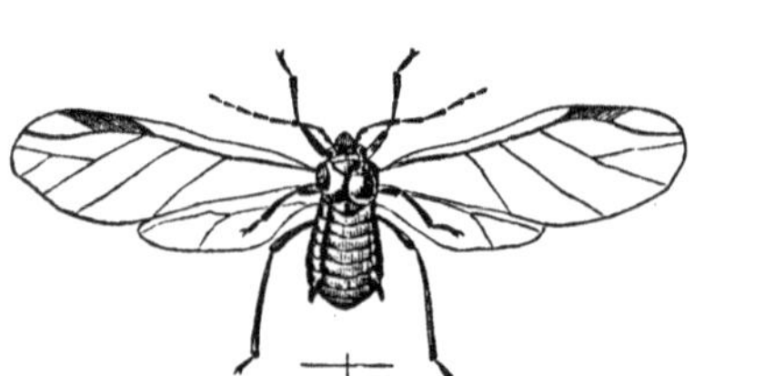

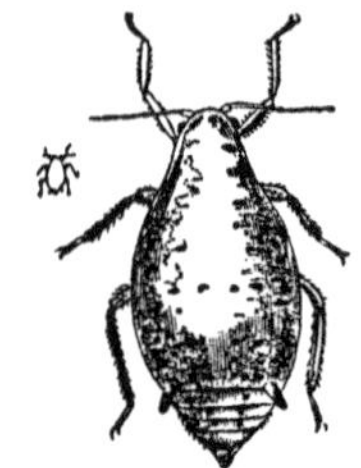

Fig. 87.—Hemiptera. Bean Aphis (*Aphis fabæ*), winged male and wingless female.

labium, which forms a jointed, tubular sheath for the bristle-shaped styliform mandibles and maxillæ. Eyes compound, usually with ocelli as well. Two pairs of wings in most; sometimes wanting.

The *Hemiptera* live upon the juices of plants or animals, which they are enabled to obtain by means of the suctorial rostrum. Amongst the more familiar examples of this order are the Plant-lice (*Aphides*, fig. 87), the Field-bug (*Pentatoma*), the Boat-fly (*Notonecta*), the Cochineal Insects (*Cocci*), and the Cicadas. The order is divided into the following two sub-orders:—

Sub-order a. Homoptera.—The anterior pair of wings of the same texture throughout (membranous); the mouth turned backwards, so that the beak springs from the back of the head. The wings fold over one another when the insect is at rest. The three segments of the thorax are united in a mass, and the pro-thorax is generally shorter than the meso-thorax. There

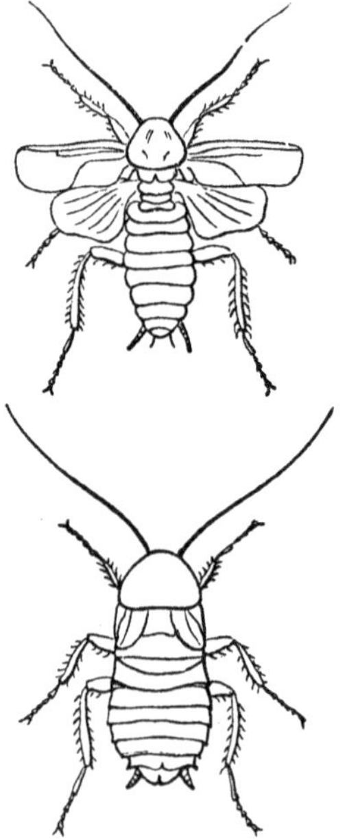

Fig. 88.—Orthoptera The common Cockroach (*Blatta orientalis*) male and female.

the larvæ differ from the adult only in size, in having fewer joints to the antennæ, and in having a smaller number of facets in each of the compound eyes.

are ocelli between the compound eyes, and the antennæ are small and composed of few joints. The females have an ovipositor of three toothed blades. In this section are the Aphides, the Scale Insects (*Coccidæ*), the Cicadas, the Lantern-flies (*Fulgora*), &c.

Sub-order b. Heteroptera.—Anterior wings membranous near their apices, but chitinous towards the base (hemelytra); the rostrum springing from the front of the head. The inner margins of the wings are straight or contiguous. The antennæ are moderate in size, and composed of a few large joints. The pro-thorax is the largest segment of the thorax. They are divided into the two groups of the *Hydrocorisæ* (Water-bugs) and *Geocorisæ* (Land-bugs), according as they are aquatic or mainly terrestrial in their habits.

ORDER V. ORTHOPTERA.—Mouth masticatory; wings four, sometimes wanting; tne anterior pair mostly smaller than the posterior, semi-coriaceous or leathery, usually with numerous nervures, the interspaces between which are filled with many transverse reticulations; sometimes overlapping horizontally (Cockroach), sometimes meeting like the roof of a house (Grasshoppers). Posterior wings usually having their front portion of a different texture from their hinder portion, this latter being almost always more transparent, and when not in use, folded longitudinally like a fan. Posterior wings mostly wanting in the females of the *Blattidæ*. Antennæ usually filiform. Metamorphosis semi-incomplete (sometimes, however, the adult is apterous, when it becomes almost impossible to distinguish the larva, pupa, and imago).

This order includes the Crickets (*Achetina*), Grasshoppers (*Gryllina*), Locusts (*Locustina*), Cockroaches (*Blattina*, fig. 88), &c. Some of them are formed for running (*cursorial*), all the legs being nearly equal in size; whilst in others the first pair of legs are greatly developed, and form powerful raptorial organs, as in the *Mantis*. In others, again, as in the Grasshoppers and Crickets, the hindmost pair of legs are greatly elongated, so as to give a considerable power of leaping to them. All the *Orthoptera* are extremely voracious, and the ravages caused by locusts in hot countries are well known to all. The most destructive of the Locusts is the Migratory Locust (*Acrydium migratorium*, fig. 89) of Africa and Southern Asia. This formidable species is celebrated for the destruction which it causes in certain seasons in the countries in which it occurs, a destruction against which human art has hitherto proved wholly powerless. Vast hordes of this species descend suddenly upon a district, and in a few hours destroy all crops, and devour the leaves of all the trees, migrating as suddenly as they came, so soon as the ground is utterly bare.

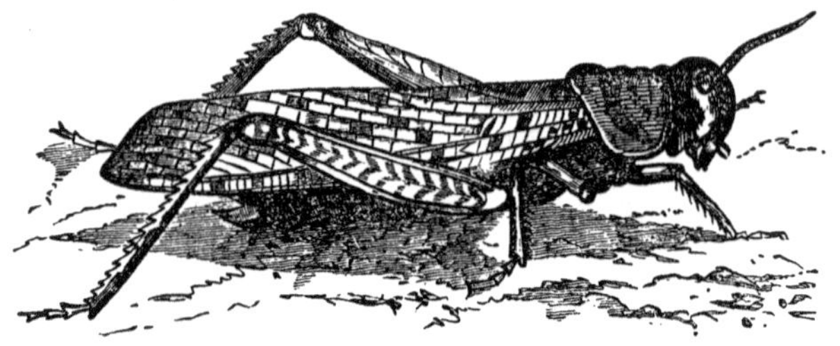

Fig. 89.—Migratory Locust (*Acrydium migratorium*).

ORDER VI. NEUROPTERA.—Mouth usually masticatory; wings four in number, all membranous, generally nearly equal in size, traversed by numerous delicate nervures, having a longitudinal and transverse direction, and giving them a reticulated, lace-like aspect. Metamorphosis generally incomplete, rarely complete. The larva active, hexapod, rarely with pro-legs.

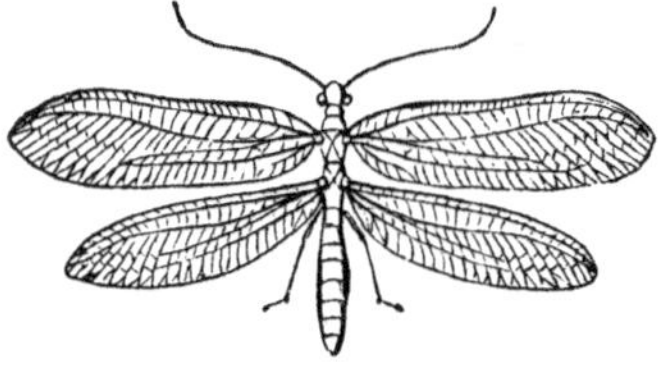

Fig. 90.—Neuroptera. Aphis-lion (*Hemerobiidæ*), imago, larva, and eggs.

This order includes the Dragon-flies (*Libellulidæ*), Caddis-flies (*Phryganeidæ*) May-flies (*Ephemeridæ*),* the Ant-lion (*Myrmeleo*), Termites, &c. The last of these—namely, the Termites or White Ants—are social, and live in communities, and their habits are so singular that a short description of them will not be out of place here. They are mostly inhabitants of hot countries, where they are commonly known as "White Ants;" but it must be borne in mind that they have nothing to do with the insects commonly called Ants, which belong, indeed, to a different order (*Hymenoptera*). The following account is

* By some the Dragon-flies (*Libellulidæ*), the May-flies (*Ephemeridæ*), and the Termites (*Termitidæ*), are placed in the *Orthoptera*, under the common name of *Pseudo-neuroptera*; whilst the Caddis-flies (*Phryganeidæ*) form a separate order under the name of *Trichoptera*.

taken from Mr Bates's work on the Amazons, where there is an excellent description of the habits of these remarkable insects.

Termites are small, soft-bodied insects, which live in large communities, as do the true Ants. They differ, however, from the Ants in the fact that the workers are individuals of no fully developed sex, whereas amongst the latter they are undeveloped females. Further, the neuters of the Termites are always composed of two distinct classes or "castes"—the workers and the soldiers. Lastly, the Ants undergo a quiescent pupa-stage; whereas the young Termites, on their emergence from the egg, do not differ from the adult in any respect except in size.

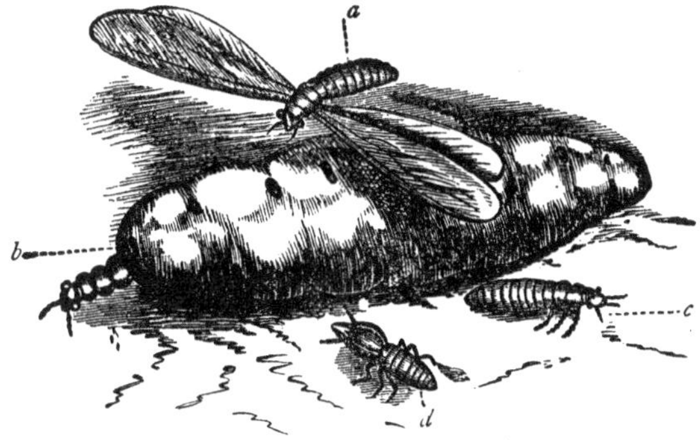

Fig. 91.—Termites (*Termes bellicosus*); *a* King, before the wings are cast off; *b* Queen, with the abdomen distended with eggs; *c* Worker; *d* Soldier.

Each species of Termites consists of several distinct orders or castes, which live together, and constitute populous, organised communities. They inhabit structures known as "Termitaria," consisting of mounds or hillocks, some of which are "five feet high, and are formed of particles of earth worked into a material as hard as stone." The Termitarium has no external aperture for ingress or egress, as far as can be seen, the entrance being placed at some distance, and connected with the central building by means of covered ways and galleries. Each Termitarium is composed of "a vast number of chambers and irregular intercommunicating galleries, built up with particles of earth or vegetable matter, cemented together with the saliva of the insects." Many of "the very large hillocks are the work of many distinct species, each of which uses materials differently compacted, and keeps to its own portion of the tumulus."

A family of Termites consists of a king and queen, of the workers, and of the soldiers. The royal couple are the parents of the colony, and "are always kept together, closely guarded by a detachment of workers, in a large chamber in the very heart of the hive, surrounded by much stronger walls than the other cells. They are both wingless, and immensely larger than the workers and soldiers. The queen, when in her chamber, is always found in a gravid condition, her abdomen enormously distended with eggs, which, as fast as they come forth, are conveyed by a relay of workers in their mouths from the royal chamber to the minor cells dispersed through the hive."

At the beginning of the rainy season a number of *winged* males and females are produced, which, when they arrive at maturity, leave the hive, and fly abroad. They then shed their wings (a special provision for this existing in a natural seam running across the root of the wing and dividing the nervures); they pair, and then become the kings and queens of future colonies.

The workers and the soldiers are distinct from the moment of their emergence from the egg, and they do not acquire their special characteristics in consequence of any difference of food or treatment. Both are wingless, and they differ solely in the armature of the head. The duties of the workers are to "build, make covered roads, nurse the young brood from the egg upwards, take care of the king and queen, who are the progenitors of the whole colony, and secure the exit of the males and females when they acquire wings and fly out to pair and disseminate the race." The duties of the soldiers are to defend the community from all attacks which may be made upon its peace, for which purpose the mandibles are greatly developed.

It may well be admitted, that in such organised communities as those of the Termites, we have the highest development of Insect-life yet known to us. The principle of the division of labour is carried out to its fullest extent—much further, indeed, than is possible amongst human beings,—since the perfection of the greater number of the individuals which compose the community—as organisms—is sacrificed in order to secure the fulfilment of the duties which are necessary for the existence and welfare of the whole. Even the task of perpetuating the species, and of giving origin to fresh colonies, is entirely left to one class of the community, the defence and protection of which is the special object and care of the remainder. No higher development could well be imagined amongst creatures devoid of the higher psychical endowments; and it is worthy of note that at least three distinct and independent families of Insects have attained to this stage—namely, the Termites, the Bees, and the true Ants.

SUB-CLASS III. HOLOMETABOLA.—*Metamorphosis complete; the larva, pupa, and imago differing greatly from one another in external appearance. The larva vermiform, and the pupa quiescent.*

ORDER VII. APHANIPTERA.—Wings rudimentary, in the form of plates, situated on the meso-thorax and meta-thorax. Mouth suctorial. Metamorphosis complete.

This order comprises the Fleas (*Pulicidæ*), most of which are parasitic upon different animals. The larva of the common Flea is an apodal grub, which in about twelve days spins a cocoon for itself, and becomes a quiescent pupa, from which the imago emerges in about a fortnight more. Besides the common Flea (*Pulex irritans*), another well-known and considerably more troublesome member of this order is the Chigoe (*Pulex penetrans*) of the West Indies and South America.

ORDER VIII. DIPTERA.—The anterior pair of wings alone developed; the posterior pair of wings rudimentary, represented by a pair of clubbed filaments, called "halteres," or "balancers" (fig. 92). In a few the wings are altogether wanting. Mouth suctorial. The metamorphosis is complete, the larvæ being generally destitute of feet; but in some cases (*e.g.*, the gnats) the pupæ are aquatic and are actively locomotive. In most cases, however, the pupæ are quiescent.

The proboscis in the *Diptera* consists of a tubular labium, enclosing the other parts of the mouth, and is placed on the under surface of the head. Ocelli are present in addition to the compound eyes. The wings are generally horizontal and transparent, the nervures not very numerous, and for the most part longitudinally disposed. The antennæ are generally small and three-jointed, sometimes many-jointed (*Tipulidæ*), or feathery (*Culicidæ*). The larva is soft and fleshy, with a soft indistinct head, usually apodal, never with thoracic legs, and rarely with pro-legs. The larval skin mostly forms a hardened case for the pupa, but the larvæ sometimes cast their skin when becoming pupæ, or even spin cocoons. In some the eggs are hatched within the body of the mother, so that the insect appears first in the larval state; and in *Pupipara* not only is this the case, but the larvæ continue to reside within the mother until they become pupæ.

The *Diptera* constitute one of the largest of the orders of the *Insecta;* the House-flies and Flesh-flies (*Musca*), Gnats (*Culex*), Forest-flies (*Hippobosca*), Crane-flies (*Tipulidæ*), and Gad-flies (*Tabanidæ*), constituting good examples.

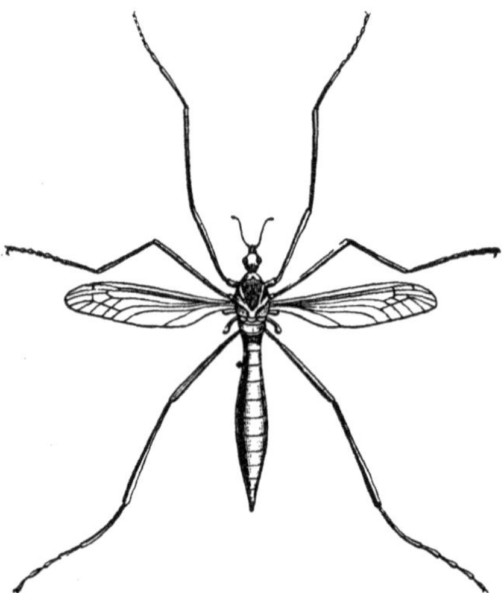

Fig. 92.—Diptera. Crane-fly (*Tipula oleracea*).

ORDER IX. LEPIDOPTERA.—Mouth suctorial, consisting of a spiral trunk or "antlia," composed of the greatly-elongated maxillæ, and protected, when not in use, by the cushion-shaped hairy labial palpi. Maxillæ forming two sub-cylindrical tubes, united together by inosculating hooks, and constituting an intermediate tube by their junction. Maxillary palpi minute; labrum and mandibles rudimentary. Head, thorax, and abdomen more or less covered with hair. Wings, four in number, covered with modified hairs or scales; wanting in the females of a few species. Nervures not very numerous, mostly longitudinal. Antennæ almost always distinct, and composed of numerous minute joints.

This well-known and most beautiful of all the orders of Insects comprises the Butterflies (fig. 93) and the Moths (fig. 94); the former being diurnal in their habits, the latter mostly crepuscular or nocturnal.

The larvæ of *Lepidoptera* (fig. 94), commonly called "caterpillars," are vermiform in shape, normally composed of thirteen segments, the anterior portion forming a distinct horny head, with antennæ, jaws, and usually simple eyes. The mouth of the caterpillar, unlike that of the perfect insect, is formed for mastication. The labium also is provided with a tubular organ —the "spinneret"—which communicates with two internal glands, the functions of which are to furnish the silk, whereby the animal constructs its ordinary abode or spins its cocoon. The three segments behind the head correspond with the prothorax, meso-thorax, and meta-thorax of the perfect insect, and each carries a pair of jointed walking-legs. Besides these thoracic legs, there is a variable number, (generally ten) of soft fleshy legs, which are borne by the segments of the abdomen, and are known as "pro-legs." Each is usually furnished with a crown of small horny hooks, and they are never attached to the 4th, 5th, 10th, and 11th abdominal segments.

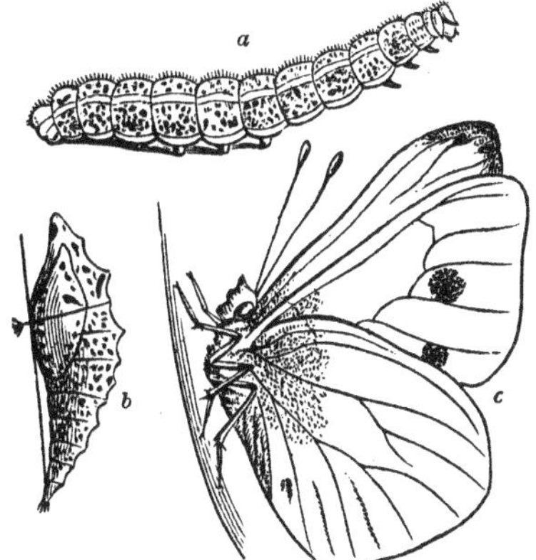

Fig. 93.—Large White Cabbage Butterfly (*Pontia brassicæ*). *a* Larva or caterpillar; *b* Pupa or chrysalis; *c* Imago or perfect insect.

In the Diurnal *Lepidoptera*, or Butterflies proper (fig. 93), the antennæ are knobbed; the wings are usually held erect when the insect is in a state of repose; the larvæ have six thoracic legs, and ten pro-legs; and the pupæ are always naked, attached by the posterior extremity or head downwards, and usually angular.

In the Crepuscular *Lepidoptera*, including those forms which are active during the twilight, the antennæ are fusiform, or grow gradually thicker from the base to the apex; the wings are horizontal or little inclined when the insect is at rest; the posterior wings have their front margins furnished with a rigid spine ("retinaculum") which is received into a hook on the under surface of the anterior wings; and the pupæ are never angular.

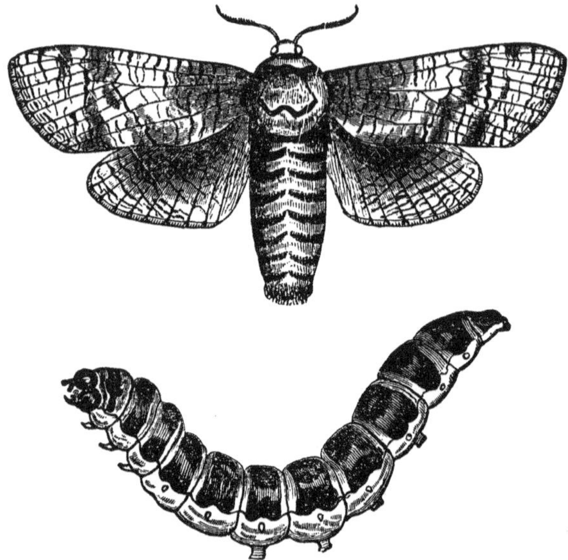

Fig. 94.—Goat-moth (*Cossus ligniperda*) and Caterpillar.

The Nocturnal *Lepidoptera* have the antennæ setaceous, or diminishing gradually from the base to the apex, often serrated or pectinated (fig. 94); the wings in repose are horizontal or deflexed, and the hind-wings are furnished with a "retinaculum," as in the preceding section; the pupæ are mostly smooth, sometimes spiny, and often enclosed in a cocoon.

ORDER X. HYMENOPTERA.—Wings four, membranous, with few nervures; sometimes absent. Mouth always provided with biting-jaws, or mandibles; the maxillæ and labium often converted into a suctorial organ. Females having the extremity of the abdomen mostly furnished with an ovipositor (*terebra* or *aculeus*), consisting chiefly of three elongated processes, of which two serve as a sheath for the third. Besides the compound eyes, there are usually three ocelli placed on the top of the head. The antennæ are generally filiform or setaceous. The metamorphosis is complete, but the various parts of the pupa are visible through the delicate enclosing membrane. The larvæ are sometimes provided with feet, and live on vegetable food (as in the *Tenthredinidæ*, fig. 95); but they are mostly footless, without a distinct head, and fed by the adult.

The *Hymenoptera* form a very extensive order, comprising the Bees, Wasps, Ants, Ichneumons, Saw-flies (fig. 95), &c.

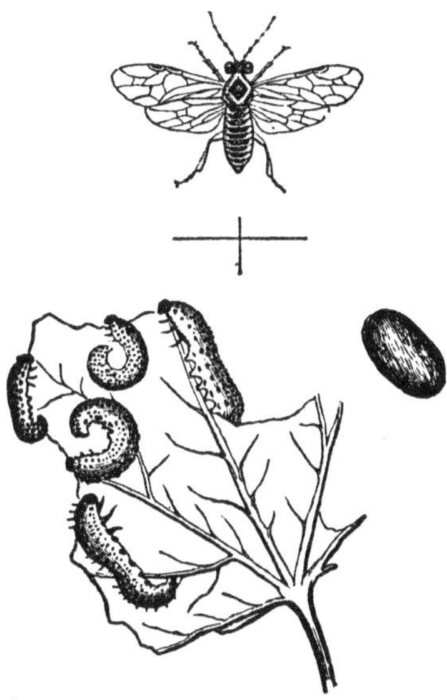

Fig. 95.—Gooseberry Saw-fly (*Tenthredo grossulariæ*), larva, pupa, and imago.

The ovipositor, which is very generally present in the females of this order, is sometimes a boring organ (*terebra*), or in other cases a "sting" (*aculeus*).

Amongst the *Hymenoptera* we find social communities, in many respects resembling those of the Termites, of which a description has already been given. The societies of Bees and Ants are well known, and merit a short description.

The social Bees, of which the common Honey-bee (*Apis mellifica*) is so familiar an example, form organised communities, consisting of three classes of individuals—the males, females, and neuters. As a rule, each community consists of a single female—"the queen"—and of the neuters, or "workers." The impregnation of the female is effected by the production of males, or "drones," during the summer. After impregnation has been effected, the drones, as being then useless, are destroyed by the workers. The eggs produced by the fecundated queen are mostly intended to give origin to neuters, to which end they are placed in the ordinary cells. The ova which are to give origin to females—the "queens" of future colonies—are placed in cells of a peculiar construction, and the larvæ are fed by the workers with a special food. The ova which are to produce males are likewise placed in cells, which are slightly larger than those allotted to the workers. It is asserted, however, that this is not the sole or true cause of the production of the males; but that the ova which are intended to produce drones are not fertilised by the female with the semen which she has stored up in her spermatheca, and are therefore produced by a process of Parthenogenesis. That the males are produced parthenogenetically in some, at any rate, of the *Hymenoptera*, appears to have been placed beyond a reasonable doubt by the researches by Von Siebold.—(See Introduction.)

In the Humble-bees (*Bombidæ*), and in the Wasps (*Vespidæ*), we have societies essentially the same as in the Honey-bee. In a large community of Wasps, or "vespiary," there may be several hundred females, of which few survive the winter, and live to found fresh colonies next spring. The number of males is about equal to that of the females, but, unlike the drones of the Bees, the males work actively and defend the nest. As amongst the Bees, solitary species are not uncommon.

The Ants (*Formicidæ*) likewise form communities, consisting of males, females, and neuters (fig. 96). The males and

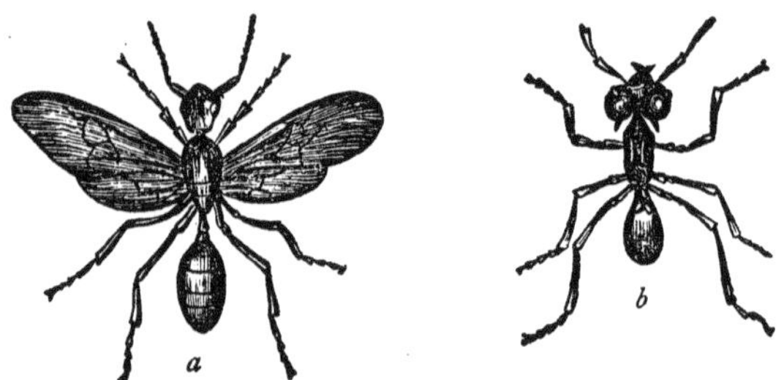

Fig. 96.—Red Ant (*Myrmica rufa*). *a* Winged male; *b* Wingless neuter. Much magnified.

females, as we have seen in the case of the Termites, are winged, and are produced in great numbers at a particular period of the year. They then quit the nest and pair, after which the males die. The females then lose their wings and fall to the ground, when they become the queens of fresh societies. In some Ants, as in the Termites, the neuters are divided into two classes—the workers and the soldiers—of which the former perform all the duties necessary for the preservation of the society except defending the nest, this being left to the soldiers. In other cases, as many as three distinct orders or "castes" of neuters may be present in the same nest.

Amongst the more singular of the habits and instincts of Ants two may be mentioned—the instinct of making slaves, and that of milking, so to speak, the little Plant-lice (*Aphides*). As regards the first of these, it is found that certain Ants possess the extraordinary instinct of capturing the pupæ of other species of Ants, and bringing them up as slaves. The relations between the master and the slaves vary a good deal in different species. In the case of *Formica rufescens*, for instance, the masters are entirely dependent upon their slaves; the males and females do nothing except reproducing the species, and the neuters perform no other labour except that of capturing fresh slaves. The masters are in this case unable even to feed themselves, and their existence is maintained entirely by the devotion of the slaves. In *Formica sanguinea*, on the other hand, the number of slaves is much less, and both masters and slaves occupy themselves in performing most of the duties necessary for the community. The masters, however, go alone when on slave-making expeditions; and in case of a migration, the masters carry the slaves in their mouths.

A second singular fact in the history of Ants is found in the relations which subsist between them and the *Aphides*, or Plant-lice. The *Aphides* secrete, or rather excrete, a peculiar viscid and sweet liquid, by means of a gland which is situated towards the extremity of the abdomen, and communicates with the exterior by two tubular filaments. Ants are extremely fond of this excretion, and it is a well-established fact that the *Aphides* allow themselves to be *milked*, as it were, by the Ants. For this purpose the Ant touches and caresses the abdomen of the Aphis with its antennæ, whereupon the latter voluntarily exudes a drop of the coveted fluid. Ordinarily the Ants seek the Aphides upon plants; but it is asserted that in some cases they *keep* Aphides, much in the same way as human beings keep cows—though this is probably partly imaginary.

ORDER XI. STREPSIPTERA.—Females without wings or feet, parasitic. Males possessing the posterior pair of wings, which are large, membranous, and folded longitudinally like a fan.

The anterior pair of wings rudimentary, represented by a pair of singular twisted organs. Jaws abortive.

The *Strepsiptera* constitute a small order, which includse certain parasites of minute size, found on Bees and other *Hymenoptera*. The female is a soft vermiform grub, without feet, but with a horny head, which it protrudes from between the abdominal segments of its host. The larvæ are active, and possess six feet; whilst the males (fig. 97) are winged, and fly about with great activity.

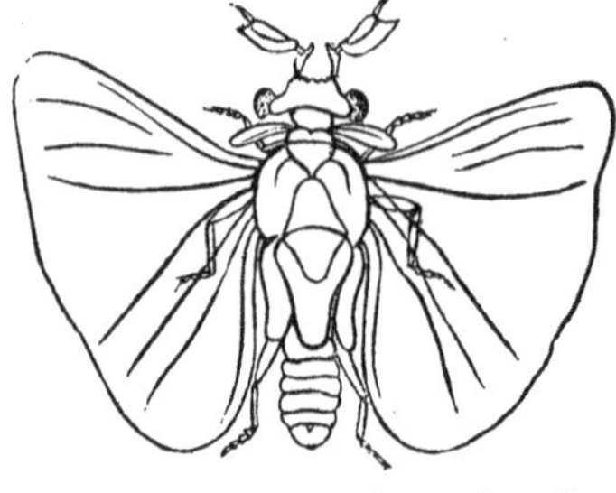

Fig. 97.—Strepsiptera. *Stylops Spencii*, greatly magnified (after Westwood).

ORDER XII. COLEOPTERA.—Mouth masticatory, furnished with an upper lip or labrum, two mandibles, two maxillæ, with maxillary palpi (generally four-jointed), and a movable lower lip or labium, with two jointed labial palpi. The four wings are usually present, and the anterior pair are not adapted for flight, but are hardened by chitine, so as to form protective cases (elytra) for the posterior wings (fig. 98). The inner margins of the

Fig. 98.—Coleoptera. Common Cockchafer (*Melolontha vulgaris*).

elytra are generally straight, and when in contact they form a longitudinal suture. The posterior wings are membranous, and when not in use, are folded transversely beneath the elytra. (Amongst deviations from this state of parts may be mentioned the occasional absence or rudimentary condition of the hinder wings, the soldering together of the elytra, the soft and yielding condition of the elytra, or the absence of both elytra and wings). The eyes are always compound, generally circular, oval, or reniform, but sometimes completely divided. The antennæ are extremely variable in form, generally of eleven joints, sometimes of fewer, rarely of twelve. The thorax is composed of a pro-meso- and meta-thorax, but when the elytra are closed, only the pro-thorax and a little plate ("scutellum") belonging to the meso-thorax are visible. The tarsus is generally composed of five joints, sometimes fewer, never more, and its last joint is usually furnished with two hooked claws.

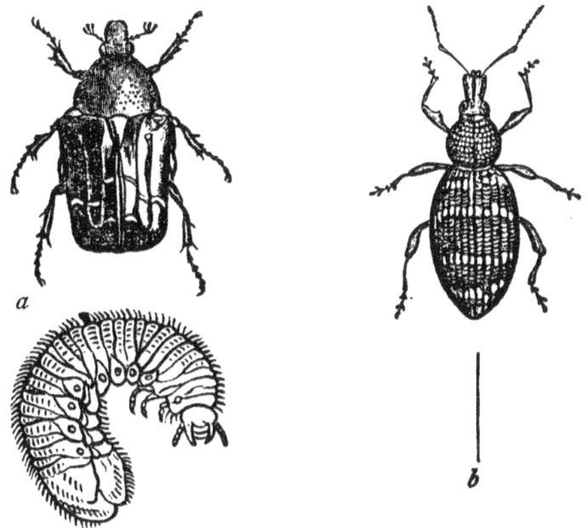

Fig. 99.—*a* Rose-chafer (*Cetonia aurata*) and larva; *b* Vine-weevil (*Curculio sulcatus*).

The larvæ of *Coleoptera* are generally composed of thirteen segments, including the head. The body is generally soft and fleshy, the head horny, and the mouth adapted for mastication. The antennæ are small, usually of three or four joints, with ocelli at their base. They have three pairs of legs attached to the thorax, and sometimes anal pro-legs or fleshy tubercles. The pupa is sometimes enclosed in a cocoon, and the parts of the perfect insect are always distinctly recognisable in the pupa.

The order *Coleoptera* includes all those insects commonly known as "Beetles," and comprises an enormous number of genera and species. They are remarkable, as a general rule, for their hard polished integument, their glittering, often metallic colours, and their voracious habits. They are grouped by Latreille in the following four sections:—

1. *Pentamera.*—Tarsus five-jointed.

2. *Heteromera.*—Tarsus of two anterior pairs of legs five-jointed, of the posterior pair four-jointed.

3. *Tetramera.*—Tarsus four-jointed.

4. *Trimera.*—Tarsus three-jointed.

Distribution of Insecta in Time.—The earliest known insects have been discovered in the Devonian Rocks of America, and consist of the remains of *Neuroptera.** Others, as might have been anticipated, have been found in the Coal-measures. In the Secondary Rocks remains of insects have been found abundantly in certain beds of the Oolitic and Liassic formations. In some Tertiary strata *Lepidoptera* and other insects have been found in a good state of preservation. Amber, which is a fossil resin, has long been known to contain many insects in its interior (in certain specimens); and all of these appear to belong to extinct species, though amber, geologically speaking, is not an ancient product.

* The Devonian *Neuroptera* of North America are most closely allied to the *Ephemeridæ;* but one form is in many respects transitional between the orders *Orthoptera* and *Neuroptera.* Insects belonging to the *Neuroptera* (viz., *Miamia* and *Hemeristia*), to the *Orthoptera* (*Blattina*), and to the *Hemiptera* (*Eugereon*), have also been described from the Carboniferous Rocks of North America.

MOLLUSCA.

CHAPTER XLI.

SUB-KINGDOM MOLLUSCA.

SUB-KINGDOM MOLLUSCA.—The *Mollusca* may be defined as including soft-bodied animals, which are usually provided with an exoskeleton. The intestinal canal is bounded by its own proper walls, and is completely shut off from the perivisceral cavity. The alimentary canal is situated between the hæmal system, which lies dorsally, and the neural system, which is situated towards the ventral aspect of the body. The nervous system (fig. 100) in its highest development consists of three principal ganglia, which are reduced to one in the lower forms. Usually there is a distinct propulsive organ by which the circulation is carried on, but this is occasionally absent. Distinct respiratory organs may or may not be present. Reproduction is sexual, though gemmation is also occasionally superadded. The higher *Mollusca* are all simple animals, but many of the lower forms are capable of forming colonies by continuous gemmation.

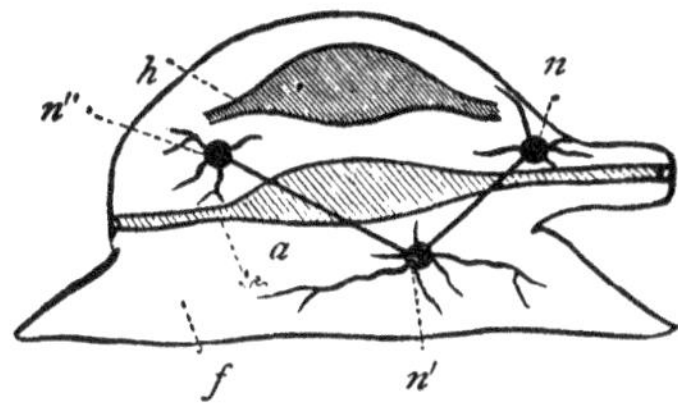

Fig. 100.—Diagram of a Mollusc. *a* Alimentary canal; *h* Heart; *f* Foot; *n* Cerebral ganglion; *n'* Pedal ganglion; *n''* Parieto-splanchnic ganglion.

The digestive system in all the *Mollusca* consists of a mouth, gullet, stomach, intestine, and anus—though in some of the *Brachiopoda*, and in a few other forms, the intestine ends cæcally. In some the mouth is surrounded by ciliated tentacles (*Polyzoa*, fig. 102); in others it is furnished with two ciliated arms (*Brachiopoda*, fig. 105); in the bivalves (*Lamellibranchiata*) it is mostly furnished with four membranous processes or palpi (fig. 106); in others it is provided with a complicated apparatus of teeth (*Gasteropoda*, fig. 108, and *Ptero poda*); and, lastly, the *Cephalopoda* have, in addition, horny or calcareous mandibles, forming a kind of beak. Well-developed salivary glands are usually present; the liver in the higher forms is of large size, and pours its secretion either into the stomach or into the commencement of the intestine; and a renal organ has been detected in most of the *Mollusca* proper. There is no distinct absorbent system, but the products of digestion pass by exosmose into the general abdominal cavity, and thence into the larger veins, which are sometimes pierced by numerous round holes for this purpose.

The blood is colourless, or nearly so. In the *Polyzoa* the circulation is carried on by ciliary action, and there is no distinct propulsive organ, or definite course of the circulating fluid. In the *Tunicata* the heart is a simple tube, open at both ends, and the course of the circulation is periodically reversed. In the *Brachiopoda* the course of the circulation is not definitely ascertained, and it is doubtful if a true heart is present in all. In the higher *Mollusca* a distinct heart is always present, and consists of an auricle which receives the aerated blood from the breathing-organ, and a muscular ventricle which propels it through the systemic vessels. That a system of capillaries in many cases intervenes between the arteries and veins, appears from recent researches to be probable. In all cases the heart of the *Mollusca* is systemic, distributing the aerated blood to the body, and in no case is it respiratory, propelling the non-aerated blood to the breathing-organ.

In the *Polyzoa* there is no differentiated respiratory organ, and the function of respiration is discharged mainly by the oral crown of ciliated tentacles. In the *Tunicata* respiration is effected by means of the pharyngeal or branchial sac; and in the *Brachiopoda* by the oral arms, and possibly, to some extent, by an "atrial" or "water-vascular" system, furnished with contractile dilatations. In the higher *Mollusca* a distinct breathing-organ is always present, a portion of the mantle being specialised for this purpose. In the *Lamellibranchiata*, and the branchiate *Gasteropoda*, the breathing-organs are in the form of lamellar and pectinate gills; and the same is the case with the *Cephalopoda*. In the pulmonate *Gasteropoda*, in which respiration is aerial, a pulmonary sac or air-chamber is produced by the folding of a portion of the mantle, over the interior of which the pulmonary vessels are distributed. The chamber thus formed communicates with the exterior by a round aperture which can be opened or closed at will; and the renovation of the effete air within the sac appears to be effected mainly or entirely by simple diffusion.

The nervous system varies considerably in its development. In the *Polyzoa*, *Tunicata*, and *Brachiopoda*—which collectively constitute the *Molluscoida*—the nervous system consists of a single ganglion, or of a principal pair with accessory ganglia, placed between the oral and anal apertures, or on the ventral surface of the body. The true Molluscan type (fig. 100), however, of nervous system is constituted by the presence of three pairs of ganglia, connected with one another by commissures, but distributed in a characteristically scattered manner (heterogangliate type). One of these ganglia is situated above the œsophagus, and is called the "supra-œsophageal" or "cerebral" ganglion. A second is placed below the œsophagus; and is termed the "infra-œsophageal" or "pedal" ganglion (from its supplying the nerves to the "foot"). The third pair is the most persistent, and is termed the "branchial" or "parieto-splanchnic" ganglion.

Organs of sight exist in some of the lower, and in the majority of the higher, *Mollusca*. In the *Cephalopoda*, and in some of the *Gasteropoda* (e.g. *Strombidæ*), the eyes are of a very high type of organisation. In the *Lamellibranchiata* the adults are either destitute of organs of vision, or possess numerous simple eyes ("ocelli") placed along the margins of the mantle-lobes. Similar ocelli are also found in some of the *Tunicata*, placed between the oral tentacles. Organs of hearing exist in the more highly organised *Mollusca*, especially in the *Gasteropoda* and *Cephalopoda*, and supposed olfactory organs occur in some of the latter.

Reproduction amongst the *Mollusca* is almost invariably sexual, but it is by continuous gemmation that the colonies of the *Polyzoa*, and the social and compound *Tunicata*, are produced, and the "statoblasts" of the former offer a good example of non-sexual reproduction. The sexes are usually distinct, but are in many cases united in the same individual. In many forms the ova are arranged in rows, so as to form a strap or ribbon-shaped structure, termed the "nidamental ribbon."

As implied by their scientific name, the *Mollusca* are mostly soft-bodied animals; but their popular name of "Shell-fish" expresses the fact, that the presence of a shell, protecting the soft body, is likewise a very characteristic feature in the sub-kingdom. At the same time, a shell is not universally present, and many of the *Mollusca* are either permanently naked, or possess nothing that would be ordinarily looked upon as a shell. When there is either no shell at all, or merely a rudimentary shell enclosed in the mantle, the Mollusc is said to be "naked." The shell of the "testaceous" *Mollusca* is very closely related to the respiratory organs; "indeed it may be regarded as a *pneumoskeleton*, being essentially a calcified portion of the mantle, of which the breathing-organ is at most a specialised part. . . . In its most reduced form the shell is only a hollow cone or plate, protecting the breathing-organ and heart, as in *Limax*, *Testacella*, and *Carinaria*. Its peculiar features always relate to the condition of the breathing-organ, and in *Terebratula* and *Pelonaia* it becomes identified with the gill. In the *Nudibranchs* the vascular mantle performs, wholly or in part, the respiratory office. In the *Cephalopods* the shell becomes complicated by the addition of a distinct, internal, chambered portion (*phragmacone*), which is properly a *visceral* skeleton."—(Woodward.) In a great many of the *Mollusca* proper the shell consists of but a single piece, and they are called "univalves." In many others the shell consists of two separate plates or "valves," and these are called "bivalves." In others, again, as in the *Chiton*, the shell consists of more than two pieces, and is said to be "multivalve." Most, however, of the multivalve shells of older writers are in reality referable to the *Cirripedia*.

All the testaceous *Mollusca* (except the Argonaut), and most of the "naked" forms, acquire a rudimentary shell before their liberation from the ovum. In the latter this rudimentary shell is cast off as the embryo grows, but in the former it becomes the "nucleus" of the adult shell. In the bivalves the embryonic shell or "nucleus" is situated at the beak or "umbo" of each valve, and is often very unlike the remainder of the shell.

In composition the shell of the *Mollusca* consists of carbonate of lime—usually having the atomic arrangement of calcite—with a small proportion of animal matter. In the *Pholadidæ*, however, the calcareous matter exists in the allotropic condition of arragonite, which is very much harder than calcite. As regards their texture, three principal varieties of shells may be distinguished—viz., the "porcellanous," the "nacreous," and the "fibrous." In the "nacreous" or pearly shells, as seen in "mother-of-pearl," the shell has a peculiar lustre, due to the minute undulations of the edges of alternate layers of carbonate of lime and membrane. The "fibrous" shells are composed of successive layers of prismatic cells. The "porcellanous" shell has a more complicated structure, and is composed of three layers or strata, each of which is made up of very numerous plates, "like cards placed on edge." The direction in which these vertical plates are placed, is sometimes transverse in the central layer, and lengthwise in the two others; or longitudinal in the middle, and transverse in the outer and inner strata.

All living shells have an outer layer of animal matter, which is known as the "epidermis," or "periostracum." This is sometimes of extreme tenuity, but is sometimes very thick, the latter being especially the case with those shells which are found in fresh water.

In many of the spiral univalves, as the animal grows it withdraws itself from the upper portion of the shell, often partitioning off the space thus left vacant. In many instances the portion thus abandoned falls off, and the shell becomes "truncated," or "decollated;" this being the normal condition in fully-grown examples of some shells.

In the great majority of univalves the shell is coiled into a spiral, the direction of which is right-handed, but in some cases the spiral is left-handed, and the shell is said to be "reversed," or "sinistral." The reversed shell may occur as the normal condition of the species, or it may occur simply as a variety of a form which is normally right-handed, or "dextral."

The sub-kingdom *Mollusca* is divided into two great divisions, termed respectively the *Molluscoida*, and the *Mollusca proper*. In the former of these the nervous system consists of a single ganglion or principal pair of ganglia, and there is either no circulatory organ or an imperfect heart. In the latter the nervous system consists of three principal pairs of ganglia, and there is a well-developed heart, consisting of at least two chambers.

MOLLUSCOIDA.

CHAPTER XLII.

POLYZOA.

DIVISION A. MOLLUSCOIDA.—*Nervous system consisting of a single ganglion, or of a principal pair with accessory ganglia; no distinct organ of the circulation, or an imperfect heart.*

This division includes three classes—viz., the *Polyzoa*, the *Tunicata*, and the *Brachiopoda.*

CLASS I. POLYZOA (*Bryozoa*).—The members of this class are defined as follows:—"Alimentary canal suspended in a double-walled sac, from which it may be partially protruded by a process of evagination, and into which it may be again retracted by invagination. Mouth surrounded by a circle or crescent of hollow, ciliated tentacles; animals always forming composite colonies."—(Allman.)

All the *Polyzoa* live in an associated form in colonies or "polyzoaria," which are sometimes foliaceous (fig. 102, 1), sometimes branched and plant-like, sometimes encrusting, and very rarely are free. Each "polyzoarium" consists of an assemblage of distinct but similar zoöids arising by continuous gemmation from a single primordial individual. The colonies thus produced are in very many respects closely similar to those of many of the Hydroid Polypes, with which, indeed, the *Polyzoa* were for a long time classed. The "polyzoarium," however, of a *Polyzoön* differs from the polypidom of a composite Hydroid in the *general* fact that the separate cells of the former do not communicate with one another otherwise than by the continuity of the external integument; whereas the zoöids of the latter are united by an organic connecting medium, or "cœnosarc," from which they take their origin. On this point Mr Busk observes:—

"It has been before said that the *Polyzoa* are always associated into compound growths, made up of a congeries of individuals, which, though distinct, yet retain some degree of intercommunication, comparable in kind perhaps, though not in degree, to what obtains in many of the compound Ascidians. That this community exists is proved by the otherwise inexplicable circumstance that the polyzoaria in many instances present elements common to the whole growth, and not belonging specially to any individual. The chief bond of connection would appear to reside partly in the continuity of the external integument, and partly also, in all probability, in a slow interchange of the vital fluid with which the cavities of the cells are charged."

In one sub-order of the *Polyzoa* (*Ctenostomata*), the polyzoarium consists of a series of cells arising from a common tube, but this exception does not affect the value of the above *general* distinction between the *Polyzoa* and the *Hydroida*.

A second point of difference is found in the invariably corneous (or chitinous) texture of the polypidoms of the *Hydroida*, whereas those of the *Polyzoa* may be corneous or fleshy, but are in the majority of instances more or less highly charged with carbonate of lime.

The homomorphism, however, which subsists between the *Polyzoa* and the *Hydroida* is shown most decisively not to be a true affinity, when the structure of the individual zoöids is examined. The polypite of a Hydroid Zoophyte, as we have already seen, possesses no alimentary canal distinct from the general cavity of the body; there are no traces of a nervous system, and the reproductive organs are in the form of external processes of the body-wall. In the zoöid of all the *Polyzoa* (fig. 101, 2), on the other hand, there is a distinct alimentary canal, completely shut off from the somatic cavity; a nervous system is present, and the reproductive organs are contained within the body.

The following are the more important differences in the terminology employed to designate the various parts of the compound growths of the *Polyzoa* and the *Hydrozoa*. In the *Hydroida* the entire colony is called the "hydrosoma," and its investing layer, when present, is called the "polypary," or "polypidom;" whilst the individuals composing the hydrosoma are called the the "polypites," and the cups in which these are in some cases contained are called "hydrothecæ." In the *Polyzoa* the entire colony—or its entire dermal system—is called the "polyzoarium" or "cœnœcium;" the separate zoöids are called "polypides;" and the little chambers in which each is contained are called the "cells."

It will be seen, therefore, that the term *polypite* is restricted to the zoöid of a compound *Hydrozoön*, or to the entire hydrosoma of a simple member of the class. The term *polype* is applied to a simple *Actinozoön*, or to the zoöids of a compound actinosoma. Lastly, the term *polypide* is exclusively employed to designate the zoöid of one of the *Polyzoa*.

The construction of a typical polypide of a *Polyzoön* is thus described by Professor Allman (fig. 101, 2):—

"Let us imagine an alimentary canal, consisting of œsophagus, stomach, and intestine, to be furnished at its origin with long ciliated tentacula, and to have a single nervous ganglion placed upon one side of the œsophagus. Let us now suppose this canal to be bent back upon itself towards the side of the ganglion, so as to approximate the termination to the origin. Let us further imagine the digestive tube thus constituted to be suspended in a fluid contained in a membranous sac with two openings, one for the mouth and the other for the vent, the tentacula alone being external to the sac. Let us still further suppose the alimentary tube, by means of a system of muscles, to admit of being retracted or protruded according to the will of the animal; the retraction being accompanied by an invagination of the sac, so as partially or entirely to include the oral tentacles within it; and if to these characters we add the presence of true sexual organs in the form of ovary and testis, occupying some portion of the interior of the sac, and the negative character of the absence of all vestige of a heart, we shall have, perhaps, as correct an idea—apart from all considerations of homology or derivation from an archetype—as can be conveyed of the essential structure of a *Polyzoön* in its simplest and most generalised condition.

"To give, however, more actuality to our ideal *Polyzoön*, we may bear in mind that the immediately investing sac has the power, in almost every case, of secreting from its external surface a secondary investment of very various constitution in the different groups; and we may, moreover, conceive of the entire animal with its digestive tube, tentacula, ganglion, muscles, generative organs, circumambient fluid, and investing sacs, repeating itself by gemmation, and thus producing one or more precisely similar systems, holding a definite position relatively to one another, while all continue organically united, and we shall then have the actual condition presented by the *Polyzoa* in their fully-developed state."

The vast majority of the *Polyzoa* are fixed, but this is not universally the case. Thus the singular fresh-water *Cristatella* is free and locomotive, creeping about by means of a flattened discoid base, not unlike the foot of the *Gasteropoda*.

The two fundamental structures of the "cœnœcium" of a *Polyzoön*—viz., the immediately investing sac, and its secondary investment, are sometimes termed the "endoderm" and

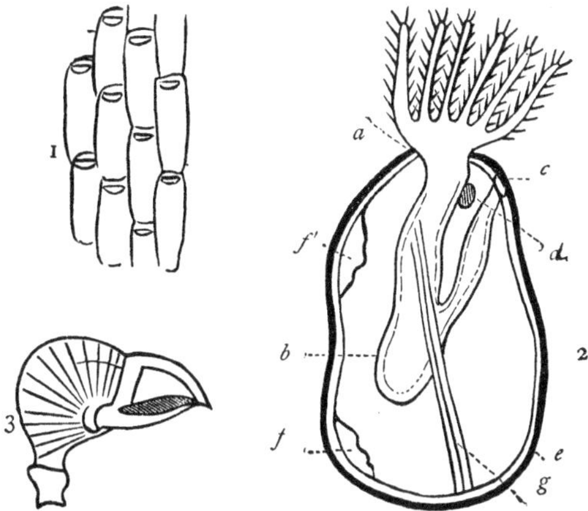

Fig. 101.—Morphology of Polyzoa. 1. Portion of the cœnœcium of *Flustra truncata*, magnified. 2. Diagram of a Polyzoön (after Allman): *a* Region of the mouth surrounded by tentacles; *b* Alimentary canal; *c* Anus; *d* Nervous ganglion; *e* Investing sac (ectocyst); *f* Testis; *f'* Ovary; *g* Retractor muscle. 3. Bird's-head process, or "avicularium," of a Polyzoön.

"ectoderm;" but as these terms are employed in describing the *Hydrozoa*, it is better to make use of the terms "endocyst" and "ectocyst," proposed by Dr Allman.

The "ectocyst," or external investment of the cœnœcium, is usually a brown, pergamentaceous, probably chitinous, but often highly calcareous, membrane; and it is by the ectocyst that the "cells" are formed. In *Cristatella*, alone of the *Polyzoa*, there is no ectocyst, and in *Lophopus* (fig. 102, 3) the ectocyst is gelatinous in its consistence. In many cases the ectocyst is provided with singular appendages, supposed to be weapons of offence and defence, termed "avicularia" (fig. 101, 3) and "vibracula." The avicularia, or "bird's-head processes," differ a good deal in shape, but consist essentially of "a movable mandible and a cup furnished with a horny beak, with which the point of the mandible is capable of being brought into apposition."—(Busk.) In shape the avicularia often closely resemble the head of a bird, and they are in many respects comparable with the "pedicellariæ" of the *Echinodermata*.* In the "vibracula," the place of the mandible of the

* There is great reason, however, as shown by Huxley, to regard the

avicularium is taken by a bristle, or seta, which is capable of extensive movement.

The endocyst is always soft, contractile, and membranous. It lines the interior of the cells formed by the ectocyst, and is reflected backwards at the mouth of the cell, so as to be invaginated, or inverted into itself; and it finally terminates by being attached to the base of the circlet of tentacles. This invagination of the endocyst is more or less permanently present in all the fresh-water *Polyzoa*. A portion of the inner surface of the endocyst, if not the whole, is furnished with vibratile cilia.

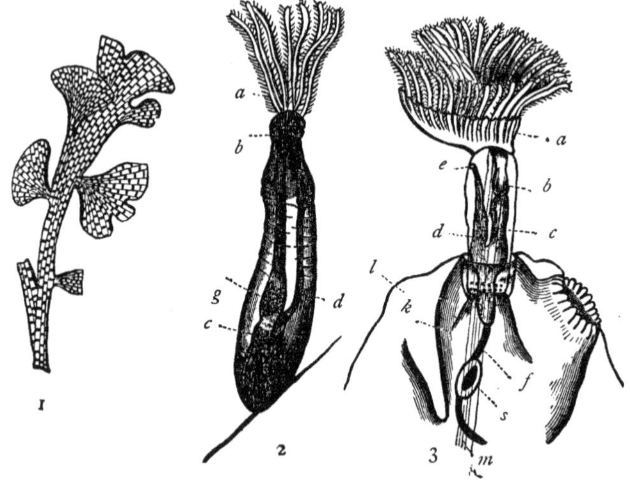

Fig. 102.—1. Fragment of *Flustra truncata*, one of the Sea-mats, natural size. 2. A single polypide of *Valkeria*, magnified, showing the orbicular crown of tentacles 3. A polypide of *Lophopus crystallinus*, a fresh-water Polyzoön, highly magnified, showing the horse-shoe-shaped crown of tentacles: *a* Tentacular crown; *b* Gullet; *c* Stomach; *d* Intestine; *e* Anus; *g* Gizzard; *k* Endocyst; *l* Ectocyst; *f* Funiculus.

The mouth of each polypide is surrounded by a crown of tubular, non-retractile tentacles, which have their sides ciliated, and are arranged sometimes in a circle and sometimes in a crescent. In the fresh-water *Polyzoa* the tentacles are united towards their bases by a funnel-shaped membrane, known as the "calyx." The tentacles are borne upon a kind of disc, or

avicularia, not as mere appendages or organs of any kind, but as peculiarly-modified *zoöids*, having many singular points of affinity with the *Brachiopoda*. The *avicularia*, like the *pedicellariæ* of the *Echinodermata*, continue their movements long after the death of the animal.

stage, which is termed by Professor Allman the "lophophore." In the majority of *Polyzoa*—including almost all the marine species—the lophophore is circular (fig. 102, 2); but in most of the fresh-water forms it has its neural side extended into two long arms, so that the entire lophophore becomes crescentic or "horse-shoe-shaped" (fig. 102, 3); hence this section is sometimes collectively termed the "Hippocrepian" *Polyzoa*. In all the *Polyzoa* in which this crescentic condition of the lophophore exists, there is also a singular valve-like organ which arches over the mouth, and is termed the "epistome." The only marine forms in which the lophophore is bilateral are *Pedicellina* and *Rhabdopleura;* the only fresh-water species in which the lophophore is orbicular are *Paludicella* and *Urnatella*.

The mouth conducts by an œsophagus into a dilated stomach. In some cases a pharynx may be present, and in others there is in front of the stomach a muscular proventriculus, or gizzard. From the stomach proceeds the intestine, which shortly turns forward to open by a distinct anus close to the mouth. As the nervous ganglion is situated on that side of the mouth towards which the intestine turns in order to reach its termination, the intestine is said to have a "neural flexure," and this relation is constant throughout the entire class.

Respiration in the *Polyzoa* appears to be carried on by the ciliated tentacles, and by the "perigastric space," which is filled with a clear fluid, containing solid particles in suspension. A kind of circulation is kept up in this "perigastric fluid" by means of the cilia lining the inner surface of the endocyst. Beyond this there is nothing that could be called a circulation, and there are no distinct circulatory organs of any kind.

The nervous system in all the *Polyzoa* consists of a single small ganglion (fig. 101, 2), placed upon one side of the œsophagus, between it and the anal aperture. Besides the single ganglion which belongs to each polypide, there is also in many, if not in all, of the *Polyzoa*, a "colonial nervous system;" that is to say, there is a well-developed nervous system, which unites together the various zoöids composing the colony, and brings them into relation with one another. It is probably in virtue of this system that the *avicularia* are enabled to continue their movements, and retain their irritability after the death of the polypides.

The muscular system is well developed, and consists of various muscular bands, with special functions attaching to each. The most important fasciculi are the retractor muscles (fig. 101, 2, *g*), which retract the upper portion of the polypide within the cell. These muscles arise from the inner surface of the endocyst near the bottom of the cell, and are inserted into the upper part of the œsophagus. The polypide, when retracted, is again exserted, chiefly by the action of the "parietal muscles," which are in the form of circular bundles running transversely round the cell.

As far as is known, all the *Polyzoa* are hermaphrodite, each polypide containing an ovary and testis (fig. 101, 2). The ovary is situated near the summit of the cell, and is attached to the inner surface of the endocyst. The testis is situated at the bottom of the cell, and a curious cylindrical appendage, called the "funiculus," usually passes from it to the fundus of the stomach. There are no efferent ducts to the reproductive organs; and the products of generation—*i.e.*, the spermatozoa and ova—are discharged into the perigastric space, where fecundation takes place; but it is not certainly known how the impregnated ova escape into the external medium.

As already mentioned, continuous gemmation occurs in all the *Polyzoa*, the fresh zoöids thus produced remaining attached to the organism from which they were budded forth, and thus giving rise to a compound growth.

A form of discontinuous gemmation, however, occurs in many of the *Polyzoa*, in which certain singular bodies, called "statoblasts," are developed in the interior of the polypide. The statoblasts are found in certain seasons lying loose in the perigastric cavity. In form "they may be generally described as lenticular bodies, varying, according to the species, from an orbicular to an elongated-oval figure, and enclosed in a horny shell, which consists of two concavo-convex discs united by their margins, where they are further strengthened by a ring which runs round the entire margin, and is of different structure from the discs. . . . When the statoblasts are placed under circumstances favouring their development, they open by the separation from one another of the two faces, and there then escapes from them a young *Polyzoön*, already in an advanced stage of development, and in all essential respects resembling the adult individual in whose cell the statoblasts were produced."—(Allman.) The statoblasts are formed as buds upon the "funiculus"—the cord already alluded to as extending from the testis to the stomach—upon which they may usually be seen in different stages of growth. They do not appear to be set free from the perigastric space prior to the death of the adult, and when liberated they are enabled to float near the surface of the water, in consequence of the cells of the marginal ring, or "annulus," being spongy and filled with air. They must be looked upon as "*gemmæ* peculiarly encysted, and destined to remain for a period in a quiescent or pupa-like state."—(Allman.)

As regards the development of the *Polyzoa*, the embryo upon its emergence from the ovum presents itself as a ciliated, free-swimming, sac-like body, from which the polypide is subsequently produced by a process of gemmation. In the singular *Rhabdopleura* the primitive bud is enclosed between two fleshy lobes or valve-like plates, attached along their dorsal margin, and giving exit in front to the rudimentary lophophore. As the development proceeds, these plates cease to keep pace in their growth with the rest of the bud; till ultimately they appear as a peculiar shield-like organ on the hæmal side of the lophophore. These lobes have been compared by Dr Allman with the mantle-lobes of the *Lamellibranchiata*.

DIVISIONS OF THE POLYZOA.—The *Polyzoa* are divided into two divisions or orders—the *Phylactolæmata* (fig. 102, 3), distinguished by the possession of a bilateral horse-shoe-shaped lophophore, and of an "epistome" arching over the mouth; and the *Gymnolæmata* (fig. 102, 2), in which the lophophore is orbicular, and there is no epistome.

TABLE OF THE DIVISIONS OF THE POLYZOA.

ORDER I. PHYLACTOLÆMATA.

Lophophore bilateral; mouth with an epistome.

Sub-order 1. *Lophopea* (fresh-water).

Arms of lophophore free or obsolete; consistence horny, sub-calcareous.

Sub-order 2. *Pedicellinea* (marine).

Arms of lophophore united at their extremities; consistence soft, fleshy.

Sub-order 3. *Rhabdopleurea* (marine).

Cœnœcium branched, adherent, membranous, with a solid chitinous rod on its adherent side, to which the polypites are attached by their funiculi. Lophophore completely hippocrepian, with a peculiar shield-like body on its hæmal side. No epistome (?)

ORDER II. GYMNOLÆMATA.

Lophophore orbicular, or nearly so; no epistome.

Sub-order 4. *Paludicellea* (fresh-water).

Polypide completely retractile; evagination of tentacular sheath imperfect; consistence horny or sub-calcareous.

Sub-order 5. *Cheilostomata* (marine).

Polypide completely retractile; evagination perfect; orifice of cell sub-terminal, of less diameter than the cell, and usually closed with a movable lip or shutter, sometimes by a contractile sphincter; cells not tubular; consistence calcareous, horny, or fleshy.

Sub-order 6. *Cyclostomata* (marine).

Cell tubular; orifice terminal, of the same diameter as the cell, without any movable apparatus for its closure; consistence calcareous.

Sub-order 7. *Ctenostomata* (marine).

Orifice of the cell terminal, furnished with a usually setose fringe for its closure; cells distinct, arising from a common tube; consistence horny or carnose.

CHAPTER XLIII

TUNICATA.

CLASS II. TUNICATA (*Ascidioida*).—The members of this class of the *Molluscoida* are defined as follows:—"Alimentary canal suspended in a double-walled sac, but not capable of protrusion and retraction; mouth opening into the bottom of a respiratory sac, whose walls are more or less completely lined by a network of blood-vessels."—(Allman.) Animal simple or composite. An imperfect heart in the form of a simple tube open at both ends.

The Tunicaries are all marine, and are protected by a leathery, elastic integument, which takes the place of a shell. In appearance a solitary Ascidian (fig. 103) may be compared to a double-necked jar with two prominent apertures situated close to one another at the free extremity of the animal, one of these being the mouth, whilst the other serves as an excretory aperture. The covering of an Ascidian is composed of two layers. Of these the outer is called the "external tunic," or "test," and is distinguished by its coriaceous or cartilaginous consistence. It is also remarkable for containing a substance which gives the same chemical reactions as cellulose, and is probably identical with this characteristic vegetable product. The test is lined by a second coat, which is termed the "second tunic," or "mantle," and which is mainly composed of longitudinal and circular muscular fibres. By means of these the animal is endowed with great contractility, and has the power of ejecting water from its branchial aperture with considerable force. The mantle lines the test, but is only slightly and loosely attached to it, especially near the apertures. The mouth is generally surrounded by a circlet of small, non-ciliated, non-retractile tentacles, and opens into a large chamber (fig. 103, 1, *c*), which usually occupies the greater part of the cavity of the mantle, and has its walls perforated by numerous apertures. This is known variously as the "pharynx," the "respiratory sac," or the "branchial sac." (It must be remembered that the aperture here spoken of as the mouth can only be looked upon in this light provided that the respiratory sac is looked upon as the pharynx. By Professor Allman, whose definition is given at the head of this chapter, this view is not accepted, and consequently the internal or *inferior* opening of the respiratory sac is regarded as the true mouth.) Inferiorly the respiratory sac leads by a second aperture into an œsophagus, which opens into a capacious stomach. From the

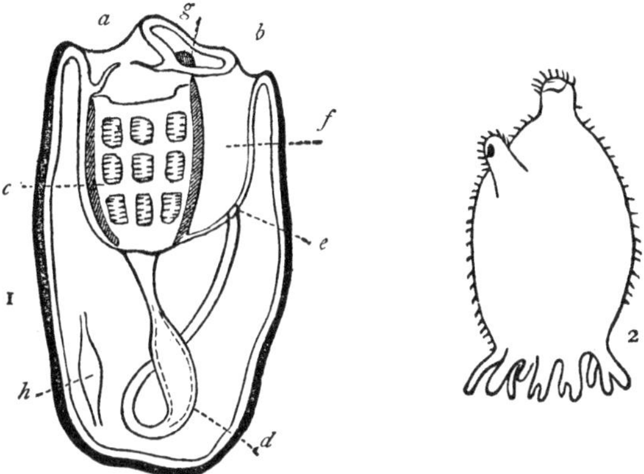

Fig. 103.—Morphology of Tunicata. 1. Diagram of a Tunicary (after Allman): *a* Ora aperture; *b* Atrial aperture; *c* Pharyngeal or branchial sac, with its rows of ciliated apertures, *d* Alimentary canal, with its hæmal flexure; *e* Anus; *f* Atrium; *g* Nervous ganglion. 2. *Cynthia papillosa*, a simple Ascidian (after Woodward).

stomach an intestine is continued, generally with few flexures, to the anal aperture, which does not communicate directly with the exterior, but opens into the bottom of a second chamber, which is called the "cloaca" (fig. 103, 1, *f*). Superiorly the cloaca communicates with the external medium, by means of the second aperture in the test. The first bend of the intestine is such that, if continued, it would bring the anus on the opposite side of the mouth to that on which the nervous ganglion is situated. The intestine, therefore, is said to have a "hæmal flexure;" whereas the flexure in the case of the *Polyzoa* is "neural." The intestine, however, in the *Tunicata* does not preserve this primary hæmal flexure, but is again bent to the neural side of the body, the nervous ganglion coming finally to be situated between the mouth and the rectum. As just stated, the anus is not in direct communication with the exterior, but opens into a large cavity, called the "cloaca," or "atrial chamber," which, in turn, opens externally by the second aperture of the animal. This cloaca is a large sac lined by a membrane which "is reflected like a serous sac on the viscera, and constitutes the 'third tunic,' or 'peritoneum.'" From the cloaca "it is reflected over both sides of the pharynx" (respiratory sac), "extending towards its dorsal part very nearly as far as that structure which has been termed the 'endostyle.' It then passes from the sides of the pharynx to the body-walls, on which the right and left lamellæ become continuous, so as to form the lining of the chamber into which the second aperture leads, or the 'atrial chamber.' Posteriorly, or at the opposite end of the atrial chamber to its aperture, its lining membrane (the 'atrial tunic') is reflected to a greater or less extent over the intestine and circulatory organs. . . . . Where the atrial tunic is reflected over the sides of the pharynx, the two enter into a more or less complete union, and the surfaces of contact become perforated by larger or smaller, more or less numerous, apertures. Thus the cavity of the pharynx acquires a free communication with that of the atrium; and as the margins of the pharyngo-atrial apertures are fringed with cilia working towards the interior of the body, a current is produced, which sets in at the oral aperture and out by the atrial opening, and may be readily observed in a living Ascidian."—(Huxley.)

As regards some points in the above description, Professor Allman does not agree with Huxley, but believes, on the other hand, "that the walls of the atrium simply surround the branchial sac, without being reflected on its sides, and that the branchial sac is therefore properly *within* the cavity of the atrium."

In structure, the pharyngeal or "branchial" sac is composed of a series of longitudinal and transverse bars, which cross each other at right angles, and thus give rise to a series of quadrangular meshes, the margins of which are fringed with vibratile cilia. These bars are hollow, and are really vessels which open on each side into two main longitudinal sinuses, the so-called "branchial" or "thoracic" sinuses—one of which is placed along the hæmal side of the pharynx, whilst the other runs along its neural aspect. The function of the entire perforated pharynx is clearly respiratory.

The *Tunicata* possess a distinct heart, consisting of a simple muscular tube, which is open at both ends, and is not provided with valves. In consequence of this, the circulation in the majority of Tunicaries is periodically reversed, the blood being propelled in one direction for a certain number of contractions, and being then driven for a like period in an opposite direction; "so that the two ends of the heart are alternately arterial and venous."

The nervous system consists of a single ganglion placed on one side of the oral aperture, between it and the anus, in all known *Tunicata*, except in the aberrant form *Appendicularia*.

The only organs of sense are pigment-spots, or ocelli, placed between the oral tentacles, and an auditory capsule, sometimes containing an otolith. These organs, however, do not appear to be constantly present.

With the exception of *Doliolum* and *Appendicularia*, all the *Tunicata* are hermaphrodite. The reproductive organs are situated in the fold of the intestine, and their efferent duct opens into the atrium. The embryo Tunicate is at first generally free, and is mostly shaped like the tadpole of a frog, swimming by means of a long caudal appendage. In one species (*Molgula tubulosa*) the larval form is destitute of a tail, inactive, and amœboid, and it almost immediately attaches itself by means of little outward processes which it develops. Lastly, in several instances the larval caudal appendage has been shown to exhibit a cylindrical rod-like body, which has been paralleled with the *chorda dorsalis* of Vertebrates.

Amongst the Salpians a species of alternation of generations has been observed. A solitary Salpian produces long chains of embryos, which remain organically connected throughout their entire life. Each individual of these associated specimens produces solitary young, which are often very unlike their parents, and these again give rise to the aggregated forms.

The *Tunicata* are often spoken of as exhibiting three main types of structure, which give origin to as many sections, known respectively as the *solitary*, the *social*, and the *compound* forms. In the "solitary" Tunicaries, the individuals, however produced, remain entirely distinct, or, if not so primitively, they become so. In the "social" Ascidians the organism consists of a number of zoöids, produced by gemmation and permanently connected together by a vascular canal, or "stolon," composed of a prolongation of the common tunic, through which the blood circulates. Finally, in the "compound" forms, the zoöids become aggregated into a common mass, their tests being fused together, but there being no internal union. The Botrylli, which are familiar examples of the compound Tunicates, form semi-transparent masses, often of brilliant colours, attached to various submarine objects, and consisting of numerous zoöids arranged in star-shaped groups. They are almost always "very small, soft, irritable, and contractile, changing their form with the slightest movement."—(Stark.)

HOMOLOGIES OF THE TUNICATA.—The general resemblance between a solitary Ascidian and a single polypide of a *Polyzoön* is extremely obvious; each consisting of a double-walled sac. containing a freely suspended alimentary canal, with a distinct mouth and anus, and a nervous ganglion placed between the two. The chief feature in the *Tunicata*, as to the exact nature of which there is much difference of opinion, is the branchial or respiratory sac. By Professor Allman this is believed to be truly homologous with the tentacular crown of the *Polyzoa*, and the oral tentacles of the Tunicaries are believed to be something superadded, and not represented at all in the *Polyzoa*. By Professor Huxley, on the other hand, the branchial sac is looked upon as an enormously developed pharynx, and the oral tentacles are regarded as a rudimentary representative of the tentacular crown of the *Polyzoa*. Probably the most correct view of the homologies of the *Tunicata* is taken by Rolleston, who regards the "branchial sac" as the homologue of the gills of the ordinary Bivalve Molluscs (*Lamellibranchiata*), whilst the oral and atrial apertures are looked upon as corresponding to the respiratory apertures of these same animals.

DIVISIONS OF THE TUNICATA.—By Professor Huxley the following arrangement of the Tunicaries is adopted:—

CLASS TUNICATA.

Order I. *Ascidia Branchialia*.
Branchial sac occupying the whole, or nearly the whole, length of the body; intestine lying on one side of it. (*Ascidiadæ, Botryllus, &c.*)

Order II. *Ascidia Abdominalia*.
Alimentary canal completely behind the branchial sac, which is comparatively small. (*Clavellina, Doliolum, &c.*)

Order III. *Ascidia Larvalia*.
Permanent larval form. (*Appendicularia.*)

The following subdivisions are those adopted by Mr Woodward:—

CLASS TUNICATA.

Fam. I. *Ascidiadæ* (Simple Ascidians).
Animal simple, fixed, solitary, or gregarious; oviparous; sexes united; branchial sac simple; or disposed in (8—18) deep and regular folds.

Fam. II. *Clavellinidæ* (Social Ascidians).
Animal compound, fixed; individuals connected by creeping tubular prolongations of the common tunic through which the blood circulates

(or by a common gelatinous base). Reproduction effected by ova, or by gemmation from the common tube ; the new individuals remaining attached to the parent, or becoming completely free.

Fam. III. *Botryllidæ* (Compound Ascidians).

Animals compound, fixed, their tests fused, forming a common mass in which they are imbedded in one or more groups. Individuals not connected by any internal union ; oviparous and gemmiparous.

Fam. IV. *Pyrosomidæ.*

Animal compound, free and oceanic.

Fam. V. *Salpidæ.*

Animals free and oceanic ; alternately solitary and aggregated.

CHAPTER XLIV.

BRACHIOPODA.

CLASS III.—BRACHIOPODA (*Palliobranchiata*).—The members of this class are defined by the possession of a body protected by a bivalve shell, which is lined by an expansion of the integument, or "mantle." The mouth is furnished with two long cirriferous arms. The nervous system consists of a single ganglion, placed in the re-entering angle between the gullet and the rectum, so that the intestine has a "neural flexure."

The *Brachiopoda* are essentially very similar in structure to the *Polyzoa*, from which they are distinguished by the fact that they are never composite, and by the possession of a bivalve, calcareous, or sub-calcareous shell. They are commonly known as "Lamp-shells," and are all inhabitants of the sea. All the living forms are fixed to some solid object in their adult condition; but there is good reason to believe that many of the fossil forms were unattached and free in their fully-grown condition. From the presence of a bivalve shell, the Brachiopods have often been placed near the true bivalve Mollusca (the *Lamellibranchiata*); but their organisation is very much inferior, and there are also sufficient differences in the shell to justify their separation.

The two valves of the shell of any *Brachiopod* are articulated together by an apparatus of teeth and sockets, or are kept in apposition by muscular action alone. One of the valves is always slightly, sometimes greatly, larger than the other, so that the shell is said to be "inequivalve." As regards the contained animal, the position of the valves is anterior and posterior, so that they are therefore termed respectively the "ventral" and "dorsal" valves. In the ordinary bivalve

Mollusca (*Lamellibranchiata*), on the other hand, the two valves of the shell are usually of the same size (equivalve), and they are situated upon the sides of the animal; so that, instead of being dorsal and ventral, they are now termed "right" and "left" valves. The ventral valve in the shell of the *Brachiopoda* is usually the largest, and usually possesses a prominent curved beak. The beak is sometimes perforated by a "foramen," or terminal aperture, through which there is transmitted a muscular peduncle, whereby the shell is attached to some foreign object. In some cases, however (as in *Lingula*, fig. 104), the peduncle simply passes between the apices of the valves, and there is no foramen; whilst in others (as in *Crania*) the shell is merely attached by the substance of the ventral valve. The dorsal or smaller valve is always free, and is never perforated by a foramen.

Fig. 104.—*Lingula anatina*, showing the muscular peduncle by which the shell is attached.

In intimate structure, the shell of most of the *Brachiopoda* consists "of flattened prisms, of considerable length, arranged parallel to one another with great regularity, and at a very acute angle—usually only about 10° or 12°—with the surfaces of the shell."—(Carpenter.) In most cases, also, the shell is perforated by a series of minute canals, which pass from one surface of the shell to the other, in a more or less vertical direction, usually widening as they approach the external surface. These canals give the shell a "punctated" structure, and in the living animal they contain cæcal tubuli, or prolongations, from the mantle, which are considered by Huxley as analogous to the vascular processes by which in many Ascidians the muscular tunic, or "mantle," is attached to the outer tunic, or "test." In some of the *Brachiopoda* (as in the *Rhynchonellidæ*) the shell is "impunctate," or is devoid of this singular canal system.

The inner surface of the valves of the shell is lined by expansions of the integument which secrete the shell, and are called the "lobes" of the "pallium," or "mantle." The digestive organs and muscles occupy a small space near the beak of the shell, which is partitioned off by a membranous septum, which is perforated by the aperture of the mouth. The remainder of the cavity of the shell is almost filled by two long oral processes, which are termed the "arms," and from which the name of the class has been derived (fig. 105, 1). These organs are lateral prolongations of the margins of the mouth, usually of great length, closely coiled up, and fringed on one side with lateral processes, or "cirri." In many Brachiopods the arms are supported upon a more or less complicated internal calcareous framework or skeleton, which is sometimes called the "carriage-spring apparatus."

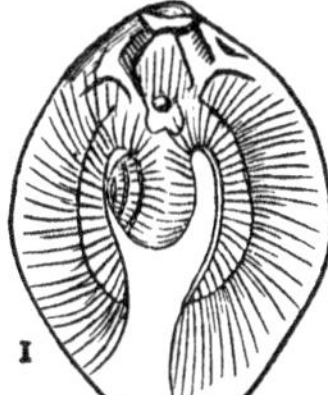

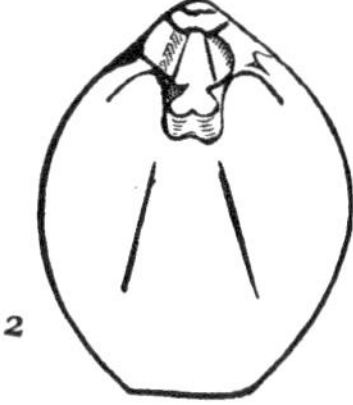

Fig. 105.—Brachiopoda (*Terebratula vitrea*). 1. Showing the ciliated "arms;" 2. Showing the shell with its loop. (After Woodward.)

The mouth conducts by an œsophagus into a distinct stomach, surrounded by a well-developed granular liver. The intestine has a "a neural flexure," and "either ends blindly in the middle line, or else terminates in a distinct anus between the pallial lobes."—(Huxley.)

Within the pallial lobes there is a remarkable system of more or less branched tubes, anastomosing with one another, and ending in cæcal extremities. This, which has been termed by Huxley the "atrial system," communicates with the perivisceral cavity by means of two or four organs which are called "pseudo-hearts," and which were at one time supposed to be true hearts. "Each pseudo-heart is divided into a narrow, elongated, external portion (the so-called 'ventricle'), which communicates, as Dr Hancock has proved, by a small apical aperture, with the pallial cavity; and a broad, funnel-shaped, inner division (the so-called 'auricle') communicating, on the one hand, by a constricted neck, with the so-called 'ventricle;' and, on the other, by a wide, patent mouth, with a chamber which occupies most of the cavity of the body proper, and sends more or less branched diverticula into the pallial lobes." —(Huxley.) This system of the atrial canals has been looked upon as a rudimentary respiratory apparatus; but its function is more probably to act as an excretory organ, and also to convey away the reproductive elements, the organs for which are developed in various parts of its walls. By Woodward the pseudo-hearts are regarded as oviducts, and it is stated that they have been found to contain mature ova, so that there can be little doubt but that this view of their nature is the correct one. By Rolleston the pseudo-hearts are looked upon as corresponding with the so-called "organ of Bojanus" of the *Lamellibranchiata*.

The function of respiration is probably performed, mainly, if not entirely, by the cirriferous oral arms, as it appears chiefly to be by the homologous tentacular crown of the *Polyzoa*. A true vascular system and a distinct heart are present in some, at any rate, of the *Brachiopoda*, but this subject is still involved in considerable obscurity. In *Terebratula* the heart is in the form of a unilocular, pyriform vesicle, placed on the dorsal surface of the stomach.

The nervous system consists of a principal ganglion of no great size, placed in the re-entering angle between the gullet and the rectum. In those Brachiopods in which the valves of the shell are united by a hinge, the nervous system attains a greater development, and consists of a gangliated œsophageal collar.

The sexes are said to be ordinarily distinct, but in some cases they appear to be united in the same individual. The development of the *Brachiopoda* is still shrouded in considerable obscurity, but in some cases the young have been observed to move from place to place, either by protruding their ciliated arms, or by means of spines developed in the ventral lobe of the mantle.

The *Brachiopoda* may be divided into two groups, called respectively the *Articulata* and *Inarticulata*. In the former the valves of the shell are united along a hinge-line, the lobes of the mantle are not completely free, and the intestine ends cæcally. In this group are the recent *Terebratulidæ* and *Rhynchonellidæ*. In the *Inarticulata* the valves of the shell are not united along a hinge-line, the mantle-lobes are completely free, and the intestine terminates in a distinct anus. In this group are the *Craniadæ*, *Discinidæ*, and *Lingulidæ*.

Affinities of the Brachiopoda.—There can be no question as to the close relationships subsisting between the *Brachiopoda* and *Polyzoa*, and until recently most naturalists held that both these groups had strongly-marked affinities with the *Lamellibranchiata*. This view is still held by the generality of naturalists; but recently Mr Edward Morse has brought forward evidence to show that the *Brachiopoda* and *Polyzoa* are most nearly related to the Tubicolar Annelides; and this opinion had been previously advanced as regards the latter group by

Leuckhart. Amongst the more striking facts adduced in support of this view may be mentioned the assertion that the long and worm-like peduncle of *Lingula pyramidata* is normally encased in a sand-tube resembling that of a Tubicolous Annelide. The peduncle of this form is also contractile and hollow, admitting the blood into its interior, and the blood is stated to be red. In the meanwhile, however, the affinities between the *Brachiopoda* and *Polyzoa*, on the one hand, and the *Tunicata* on the other, are too strong to allow of our unhesitating acceptance of this sweeping change.

Classification of the Brachiopoda (after Davidson).

Class Brachiopoda.

Fam. I. *Terebratulidæ.*

Shell minutely punctate ; ventral valve with a prominent beak perforated by a foramen for the emission of a muscular peduncle, whereby the animal is fixed to some solid object. Foramen partially surrounded by a deltidium of one or two pieces. Oral appendages entirely or partially supported by calcified processes, usually in the form of a loop, and always fixed to the dorsal valve.

Genera.—*Terebratula* (with *Terebratulina*, and *Waldheimia*), *Terebratella*, *Stringocephalus*, *&c.*

Fam. II. *Thecididæ.*

Shell fixed to the sea-bottom by the beak of the larger or ventral valve; structure punctated. Oral processes united in the form of a bridge over the visceral cavity; cirrated arms folded upon themselves, and supported by a calcareous loop.

Genus.—*Thecidium.*

Fam. III. *Spiriferidæ.*

Animal free, or rarely attached by a muscular peduncle. Shell punctated or unpunctated. Arms largely developed, and entirely supported by a thin, shelly, spirally rolled lamella.

Genera.—*Spirifer*, *Spiriferina*, *Cyrtia*, *Athyris*, *&c.*

Fam. IV. *Koninckidæ.*

Animal unknown. Shell free; valves unarticulated (?). Oral arms supported by two lamellæ, spirally coiled.

Genus.—*Koninckia.*

Fam. V. *Rhynchonellidæ.*

Animal free, or attached by a muscular peduncle issuing from an aperture situated under the extremity of the beak of the ventral valve. Arms spirally rolled, flexible, and supported only at their origin by a pair of short, curved, shelly processes. Shell-structure fibrous and impunctate.

Genera.—*Rhynchonella*, *Pentamerus*, *Porambonites*, *&c*

Fam. VI. *Strophomenidæ.*

Animal unknown ; some probably free, others attached, during the whole or a portion of their existence, by a muscular peduncle. No calcified supports for the arms. Shell with a straight hinge-line, and a low triangular area in each valve. Shell-structure fibrous and punctated.

Genera.—*Orthis*, *Orthisina*, *Strophomena*, and *Leptæna.*

Fam. VII. *Productidæ.*

Animal unknown. Shell entirely free, or attached to marine bottoms by the substance of the beak; valves either regularly articulated, or kept in place by muscular action. No calcified support for the oral appendages.

Genera.—*Producta, Chonetes, Strophalosia, Aulosteges.*

Fam. VIII. *Craniadæ.*

Animal fixed to submarine objects by the substance of the shell of the ventral valve. Arms fleshy and spirally coiled; no hinge or articulating processes; upper or dorsal valve patelliform (*i.e.* limpet-shaped).

Genus.—*Crania.*

Fam. IX. *Discinidæ.*

Animal attached by means of a muscular peduncle passing through the ventral or lower valve by means of a slit in its hinder portion, or a circular foramen excavated in its substance. Arms fleshy, valves unarticulated.

Genera.—*Discina, Trematis, Siphonotreta, Acrotreta.*

Fam. X. *Lingulidæ.**

Animal fixed by a muscular peduncle passing out between the beaks of the valves; arms fleshy, unsupported by calcified processes. Shell unarticulated, sub-equivalve, texture horny.

Genera.—*Lingula, Obolus.*

CHAPTER XLV.

DISTRIBUTION OF MOLLUSCOIDA.

Distribution of Molluscoida in Space.—The *Polyzoa*, like all the *Molluscoida*, are exclusively aquatic in their habits, but, unlike the remaining two classes, they are not exclusively confined to the sea. The marine *Polyzoa* are of almost universal occurrence in all seas. The fresh-water *Polyzoa*, however, not only differ materially from their marine brethren in structure, but appear to have a much more limited range, being, as far as is yet known, confined to the north temperate zone. Britain can claim the great majority of the described species of fresh-water *Polyzoa*, but this is probably due to the more careful scrutiny to which this country has been subjected.

The *Tunicata* are cosmopolitan in their distribution, and are found in all seas, the Mediterranean appearing to be especially rich in members of this class. Four genera are pelagic in their habits, and several are found in the Arctic regions.

* Another family was formerly constituted for the reception of the singular Devonian fossils known as *Calceolæ*. It has been shown, however, that these are probably operculate Corals.

The *Brachiopoda*, though of very partial occurrence, have a wide range in space, being found both in tropical seas and in the Arctic Ocean. Their bathymetrical range is also very wide, extending from the littoral zone almost to the greatest depths at which animal life has hitherto been detected.

DISTRIBUTION OF MOLLUSCOIDA IN TIME.—The *Polyzoa* have left abundant traces of their past existence in the stratified series, commencing in the Lower Silurian Rocks and extending up to the present day. The *Oldhamia* of the Cambrian Rocks of Ireland, and the *Graptolites*, have been supposed to belong to the *Polyzoa;* but the former is very possibly a plant, and the latter should be referred to the *Hydrozoa.* Of undoubted *Polyzoa,* the marine orders of the *Cheilostomata* and *Cyclostomata* are alone known with certainty to be represented. Several Palæozoic genera—such as *Fenestella* (the Lace-coral), *Ptilodictya*, *Ptilopora*, &c.—are exclusively confined to this epoch, and do not extend into the Secondary Rocks. Amongst the Mesozoic formations, the Chalk is especially rich in *Polyzoa,* over two hundred species having been already described from this horizon alone. In the Tertiary period, the Coralline Crag (Pleiocene) is equally conspicuous for the great number of the members of this class.

The *Tunicata*, from the nature of their bodies, are not known to occur in a fossil condition.

The *Brachiopoda* are found from the Cambrian Rocks up to the present day, and present us with an example of a group which appears to be slowly dying out. Nearly two thousand extinct species have been described, and the class appears to have attained its maximum in the Silurian epoch, which is, for this reason, sometimes called the "Age of Brachiopods." Numerous genera and species are found also in both the Devonian and Carboniferous formations. In the Secondary Rocks *Brachiopoda* are still abundant, though less so than in the Palæozoic period. In the Tertiary epoch a still further diminution takes place, and at the present day we are not acquainted with a hundred living forms. Of the families of *Brachiopoda*, the *Productidæ*, *Strophomenidæ*, and *Spiriferidæ* are the more important extinct types. Of the genera, the most persistent is the genus *Lingula*, which commences in the Cambrian Rocks, and has maintained its place up to the present day, though it appears to be gradually dying out.

According to Woodward:—"The hingeless genera attained their maximum in the Palæozoic age, and only three now survive (*Lingula*, *Discina*, *Crania*)—the representatives of as many distinct families. Of the genera with articulated valves, those provided with spiral arms appeared first, and attained their maximum while the *Terebratulidæ* were still few in number. The subdivision with calcareous spires disappeared with the Liassic period, whereas the genus *Rhynchonella* still exists. Lastly, the typical group, *Terebratulidæ*, attained its maximum in the Chalk period, and is scarcely yet on the decline."

MOLLUSCA PROPER.

CHAPTER XLVI.

LAMELLIBRANCHIATA.

DIVISION II. MOLLUSCA PROPER.—This division includes those members of the sub-kingdom *Mollusca* in which *the nervous system consists of three principal pairs of ganglia; and there is always a well-developed heart, which is never composed of fewer than two chambers.*

The *Mollusca proper* may be roughly divided into two great sections, respectively termed the *Acephala* and the *Encephala* (or *Cephalophora*), characterised by the absence or presence of a distinctly differentiated head. The headless, or Acephalous, Molluscs correspond to the class *Lamellibranchiata;* also distinguished, at first sight, by the possession of a bivalve shell. The Encephalous Molluscs are more highly organised, and are divided into three classes—viz., the *Gasteropoda*, the *Pteropoda*, and the *Cephalopoda*. The shell in these three classes is of very various nature, but they all possess a singular and complicated series of lingual teeth; hence they are grouped together by Professor Huxley under the name of *Odontophora*.

CLASS I. LAMELLIBRANCHIATA, or CONCHIFERA. — The members of this class are characterised by the absence of a distinctly differentiated head, and by having the body more or less completely protected in a bivalve shell. There are two lamellar gills on each side of the body, the intestine has a neural flexure, and there is no odontophore.

The *Lamellibranchiata* are commonly known as the bivalve shell-fish, such as Mussels, Cockles, Oysters, Scallops, &c., and they are all either marine or inhabitants of fresh water.

Though they agree with the *Brachiopoda* in possessing a shell which is composed of two pieces or valves, there are, nevertheless, many points in which the shell of a Lamellibranch is distinguishable from that of a Brachiopod, irrespective of the great difference in the structure of the animal in each. The shell in the *Brachiopoda*, as we have seen, is rarely or never quite equivalve, and always has its two sides equally developed (equilateral); whilst the valves are placed antero-posteriorly as regards the animal, one in front and one behind, so that they are "dorsal" and "ventral." In the *Lamellibranchiata*, on the other hand, the two valves are usually of nearly equal size (equivalve), and are more developed on one side than on the other (inequilateral); whilst their position as regards the animal is always *lateral*, so that they are properly termed "right" and "left" valves, instead of "ventral" and "dorsal."

The following are the chief points to be noticed in connection with the shell of any Lamellibranch: Each valve of the shell may be regarded as essentially a hollow cone, the apex of which is turned more or less to one side; so that more of the shell is situated on one side of the apex than on the other. The apex of the valve is called the "umbo," or "beak," and is always turned towards the mouth of the animal. Consequently, the side of the shell towards which the umbones are turned is the "anterior" side, and it is usually the shortest half of the shell. The longer half of the shell, from which the umbones turn away, is called the "posterior" side, but in some cases this is equal to, or even shorter than, the anterior side. The side of the shell where the beaks are situated, and where the valves are united to one another, is called the "dorsal" side; and the opposite margin, along which the shell opens, is called the "ventral" side, or "base." The *length* of the shell is measured from its anterior to its posterior margin, and its *breadth* from the dorsal margin to the base.

At the dorsal margin the valves are united to one another, for a shorter or longer distance, along a line which is called the "hinge-line." The union is effected in most shells by means of a series of parts which interlock with one another (the "teeth"), but these are sometimes absent, when the shell is said to be "edentulous." Posterior to the umbones, in most bivalves, is another structure passing between the valves, which is called the "ligament," and which is usually composed of two parts, either distinct or combined with one another. These two parts are known as the "external ligament" (or the ligament proper) and the "cartilage," and they constitute the agency whereby the shell is opened, but one or other of them may be absent. The ligament proper is outside the shell, and consists of a band of horny fibres, passing from one valve to the other just *behind* the beaks, in such a manner that it is put upon the stretch when the shell is closed. The cartilage, or internal ligament, is lodged between the hinge-lines of the two valves, generally in one or more "pits," or in special processes of the shell. It consists of elastic fibres placed perpendicularly between the surfaces by which it is contained, so that they are necessarily shortened and compressed when the valves are shut. To open the shell, therefore, it is simply necessary for the animal to relax the muscles which are provided for the closure of the valves, whereupon the elastic force of the ligament and cartilage is sufficient of itself to open the shell.

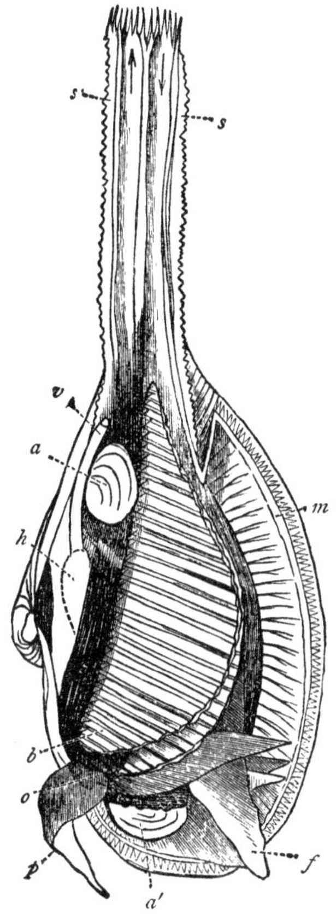

Fig. 106.—Anatomy of a bivalve Mollusc (*Mya arenaria*). The left valve and mantle-lobe and half the siphons are removed. *s s* Respiratory siphons, the arrows indicating the direction of the currents; *a a'* Adductor muscles; *b* Gills; *h* Heart; *o* Mouth, surrounded by (*p*) labial palpi; *f* Foot; *v* Anus; *m* Cut edge of the mantle. (After Woodward.)

The body in the *Lamellibranchiata* is always enclosed in an expansion of the dorsal integument, which constitutes the "mantle," or "pallium," whereby the shell is secreted. The lobes of the mantle are right and left, and not anterior and posterior as are the mantle-lobes of the *Brachiopoda*. Towards its circumference the mantle is more or less completely united to the shell, leaving in its interior, when the soft parts are removed, a more or less distinctly impressed line, which is called the "pallial line," or "impression" (fig. 107).

There is no distinctly differentiated head in any of the *Lamellibranchiata*, and the mouth is simply placed at the anterior extremity of the body. It is furnished with membranous processes or "palpi" (usually four in number), but there is no dental apparatus. The mouth opens into a gullet, which conducts to a distinct stomach. On the right side of the stomach, and opening into it, is, in many cases, a blind sac containing a peculiar transparent glassy body, which is known as the "crystalline stylet," but the functions of which are absolutely unknown. The intestine has its first flexure neural, perforates the wall of the heart, and terminates posteriorly in a distinct anus, which is always placed near the respiratory aperture. The liver is large and well developed, but there are no salivary glands.

There is always a distinct heart, composed either of an auricle and ventricle, or of two auricles and a ventricle. The ventricle propels the blood into the arteries, by which it is distributed through the body. From the arteries it passes into the veins, and is conducted to the gills, where it is aerated, and is finally returned to the auricles.

The respiratory organs in all the *Lamellibranchiata* consist of two lamelliform gills, placed on each side of the body (fig. 106, *b*). In some cases there is only one gill on each side of the body, the external pair of branchiæ being absent. The gills are in the form of membranous plates, composed usually of tubular rods, which support a network of capillary vessels, and are covered with vibrating cilia, whereby a circulation of the water is maintained over their surfaces. In some bivalves the margins of the mantle are united to one another, so that a closed branchial chamber is produced; and in the others the arrangements for the admission of fresh and the expulsion of effete water are equally perfect, though there is no such chamber. In those in which the mantle-lobes are united at their margins, there are two orifices, one of which serves to admit fresh water, whilst the effete water is expelled by the other. The margins of these "inhalant" and "exhalant" apertures are often drawn out and extended into long muscular tubes or "siphons," which may be either free, or may be united to one another along one side (fig. 106, *s s*), and which can usually be partially or entirely retracted within the shell by means of special muscles, called the "retractor-muscles of the siphons." These siphons are more especially characteristic of those Lamellibranchs which spend their existence buried in the sand, protruding their respiratory tubes in order to obtain water, and with it such nutrient particles as the water may contain. The presence or absence of retractile siphons can be readily determined merely by inspection of the dead shell. In those bivalves in which siphons are not present, or if present are not retractile, the "pallial line" in the interior of the shell is unbroken in its curvature, and presents no indentation (*Integro-palliata*). In those, on the other hand, in which retractile siphons exist, the pallial line does not run in an unbroken curve, but is deflected inwards posteriorly, so as to form an indentation or bay, which is termed the "pallial sinus," or "siphonal impression," and is caused by the insertion of the retractor-muscle of the siphon. Those bivalves in which this sinus exists form the section *Sinu-pallialia* (fig. 107, 2).

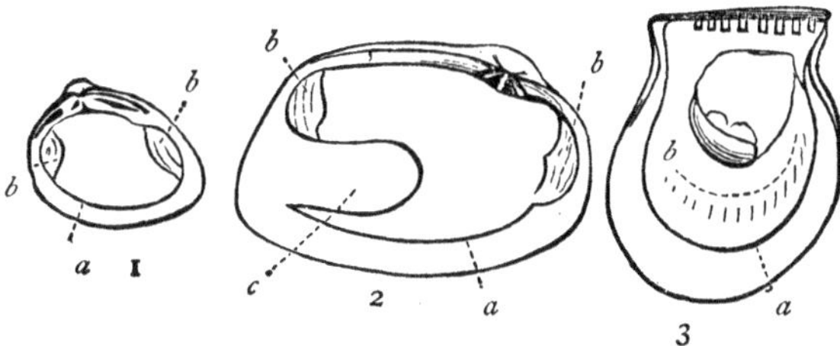

Fig. 107.—Shells of Lamellibranchiata. 1. *Cyclas amnica*, a dimyary shell with an entire pallial line. 2. *Tapes pullastra*, a dimyary shell with an indented pallial line. 3. *Perna ephippium*, a monomyary shell (after Woodward). *a* Pallial line; *b* Muscular impressions left by the adductors; *c* Siphonal impression.

The nervous system of the *Lamellibranchiata* is composed of the three normal ganglia—the cephalic, the pedal, and the parieto-splanchnic or branchial. The so-called "organ of Bojanus" of the bivalves is doubtless mainly concerned in excretion, and in all probability represents the kidney. There is one of these organs on each side of the body, each composed of two sacs separated from those of the opposite side by a venous sinus. Or it may be looked upon as a double organ composed of two bilaterally symmetrical halves. It is situated just below the "pericardium," and communicates with it and also with the mantle-cavity. Though undoubtedly performing the functions of a kidney, the organ of Bojanus is also connected in some cases with reproduction, and it appears to correspond to the "pseudo-hearts" of the *Brachiopoda.*

The majority of the bivalves are diœcious, but in some the sexes are united in the same individual. The young are hatched before they leave the parent, and are, when first liberated, ciliated and free-swimming.

The muscular system of the Lamellibranchs is well developed. Besides the muscular margin of the mantle, and the muscles of the siphons (when these exist), there are also present other muscles, of which the most important are the muscles which close the shell and those which form the "foot" (fig. 106, *f*). The "foot" is present in the majority of bivalves, though it is not such a striking feature as in the *Gasteropoda.*

It is essentially a muscular organ, developed upon the ventral surface of the body, its retractor-muscles usually leaving distinct impressions or scars (the "pedal impressions") in the interior of the shell. In many the foot subserves locomotion, but in the attached bivalves it is rudimentary, and in others (as in the Scallops) locomotion is effected by the alternate opening and closure of the valves. In some—such as the ordinary Mussel—the foot is subsidiary to a special gland, which secretes the tuft of silky threads ("byssus") whereby the shell is attached to foreign objects. This gland secretes a viscous material, which the foot moulds into threads.

The valves of the shell are brought together by one or two muscles, which are called the "adductor muscles"—those bivalves with only one being called *Monomyaria*, whilst those which possess two are termed *Dimyaria*. In most there are two adductor muscles (fig. 106, *a a'*) passing between the inner surfaces of the valves, one being placed anteriorly in front of the mouth, the other posteriorly on the neural side of the intestine. In the monomyary bivalves the posterior adductor is the one which remains, and the anterior adductor is absent. The adductors leave distinct "muscular impressions" in the interior of the shell, so that it is easy to determine whether there has been one only in any given specimen, or whether two were present.

The habits of the *Lamellibranchiata* are very various. Some, such as the Oyster (*Ostrea*), and the Scallop (*Pecten*), habitually lie on one side, the lower valve being the deepest, and the foot being wanting, or rudimentary. Others, such as the Mussel (*Mytilus*) and the *Pinna*, are attached to some foreign object by an apparatus of threads, which is called the "byssus," and is secreted by a special gland. Others are fixed to some solid body by the substance of one of the valves. Many, such as the *Myas*, spend their existence sunk in the sand of the seashore or in the mud of estuaries. Others, as the *Pholades* and *Lithodomi*, bore holes in rock or wood, in which they live. Finally, many are permanently free and locomotive.

The *Lamellibranchiata* are divided into two sections, according as respiratory siphons are absent or present, as follows:—

SECTION A. ASIPHONIDA. — Animal without respiratory siphons; mantle-lobes free; the pallial line simple and not indented (*Integro-pallialia*).

This section comprises the families *Ostreidæ*, *Aviculidæ*, *Mytilidæ*, *Arcadæ*, *Trigoniadæ*, and *Unionidæ*.

SECTION B. SIPHONIDA.—Animal with respiratory siphons; mantle-lobes more or less united.

Two subdivisions are comprised in this section. In the first *the siphons are short, and the pallial line is simple* (*Integro-pallialia*); as is seen in the families *Chamidæ*, *Hippuritidæ*, *Tridacnidæ*, *Cardiadæ*, *Lucinidæ*, *Cycladidæ*, and *Cyprinidæ*.

The second subdivision (*Sinu-pallialia*) is distinguished by the possession of *long respiratory siphons*, and *a sinuated pallial line*, and it comprises the families *Veneridæ*, *Mactridæ*, *Tellinidæ*, *Solenidæ*, *Myacidæ*, *Anatinidæ*, *Gastrochænidæ*, and *Pholadidæ*.

Synopsis of the Families of the Lamellibranchiata.

SECTION A. ASIPHONIDA.

Fam. 1. *Ostreidæ.*—Shell inequivalve, slightly inequilateral, free or attached; hinge usually edentulous. Ligament internal. Lobes of the mantle entirely separated; the foot small and byssiferous, or wanting. A single adductor. Ill. Gen. *Ostrea*, *Pecten*, *Spondylus*, *&c.*

Fam. 2. *Aviculidæ.*—Shell inequivalve, very oblique, attached by a byssus; hinge nearly, or quite, edentulous. Mantle-lobes free; anterior adductor small, leaving its impression within the umbo; posterior adductor large and sub-central. Foot small. Ill. Gen. *Avicula*, *Inoceramus*, *Pinna.*

Fam. 3. *Mytilidæ.*—Shell equivalve, umbones *anterior*, hinge edentulous; anterior muscular impression small, posterior large. Shell attached by a byssus. Mantle-lobes united between the siphonal apertures. Foot cylindrical, grooved, and byssiferous. Ill. Gen. *Mytilus*, *Modiola*, *Dreissena.*

Fam. 4. *Arcadæ.*—Shell equivalve; hinge long, with many comb-like equal teeth. Muscular impressions nearly equal. Mantle-lobes separated; foot large, bent, and deeply grooved. Ill. Gen. *Arca*, *Pectunculus*, *Cucullæa.*

Fam. 5. *Trigoniadæ.*—Shell equivalve trigonal; hinge-teeth few, diverging; umbones directed posteriorly. Mantle open; foot long and bent. Ill. Gen. *Trigonia*, *Axinus.*

Fam. 6. *Unionidæ.*—Shell usually equivalve, with a large external ligament. Anterior hinge-teeth thick and striated; posterior laminar, or wanting. Mantle-lobes united between the siphonal apertures. Foot very large, compressed, byssiferous in the fry, Ill. Gen. *Unio*, *Anodon*, *Mülleria.*

SECTION B. SIPHONIDA.

Subdivision I. *Integro-pallialia.*—Siphons short, pallial line simple.

Fam. 7. *Chamidæ.*—Shell inequivalve, attached; hinge-teeth 2–1 (two in one valve and one in the other). Adductor impressions large. Mantle closed; pedal and siphonal orifices small and nearly equal. Foot very small. Ill. Gen. *Chama*, *Diceras.*

Fam. 8. *Hippuritidæ.*—"Shell inequivalve, unsymmetrical, thick, attached by the *right* umbo; umbones frequently camerated; structure and sculpturing of valves dissimilar; ligament internal; hinge-teeth 1–2; adductor impressions 2, large, those of the left valve on prominent apophyses; pallial line simple, sub-marginal."—(Woodward.) Ill. Gen. *Hippurites*, *Radiolites*, *Caprinella.*

Fam. 9. *Tridacnidæ.*—Shell equivalve; ligament external; muscular impressions blended, sub-central. Animal attached by a byssus, or free. Mantle-lobes extensively united; pedal aperture large; siphonal orifices surrounded by a thickened pallial border. Foot finger-like and byssiferous. Ill. Gen. *Tridacna.*

Fam. 10. *Cardiadæ.*—Shell equivalve, heart-shaped, with radiating ribs; cardinal teeth 2; lateral teeth 1–1, in each valve. Mantle open in front, siphons usually very short; foot large, sickle-shaped. Ill. Gen. *Cardium, Hemicardium, Conocardium.*

Fam. 11. *Lucinidæ.*—Shell orbicular, and free; hinge-teeth 1 or 2; lateral teeth 1–1, or obsolete. Mantle-lobes open below, with one or two siphonal orifices behind; foot elongated, cylindrical, or strap-shaped. Ill. Gen. *Lucina, Diplodonta, Kellia.*

Fam. 12. *Cycladidæ.* — Shell sub-orbicular, closed; hinge with cardinal and lateral teeth; ligament external. Mantle open in front; 1–2 siphons, more or less united. Foot large, tongue-shaped. Ill. Gen. *Cyclas, Cyrena.*

Fam. 13. *Cyprinidæ.*—Shell equivalve, closed; ligament external; cardinal teeth 1–3 in each valve, and usually a posterior tooth. Mantle-lobes united behind by a curtain pierced with two siphonal orifices. Foot thick, and tongue-shaped. Ill. Gen. *Cyprina, Astarte, Isocardia.*

Subdivision II. *Sinu-pallialia.*—Respiratory siphons large; pallial line sinuated.

Fam. 14. *Veneridæ.*—Shell regular, sub-orbicular or oblong; ligament external; hinge with usually 3 diverging teeth in each valve. Animal usually free and locomotive; mantle with a rather large anterior opening; siphons unequal, more or less united. Foot tongue-shaped, compressed, sometimes grooved and byssiferous. Ill. Gen. *Venus, Cytherea, Venerupis.*

Fam. 15. *Mactridæ.*—Shell equivalve, trigonal; hinge with two diverging cardinal teeth, and usually with anterior and posterior lateral teeth. Mantle more or less open in front; siphons united, with fringed orifices; foot compressed. Ill. Gen. *Mactra, Lutraria.*

Fam. 16. *Tellinidæ.*—Shell free, usually equivalve and closed; cardinal teeth 2 at most, laterals 1–1, sometimes wanting. Ligament on the shortest side of the shell, sometimes internal. Mantle widely open in front. Siphons separate, long and slender; foot tongue-shaped, compressed. Ill. Gen. *Tellina, Psammobia, Donax.*

Fam. 17. *Solenidæ.*—Shell elongated, gaping at both ends; ligament external; hinge-teeth usually 2–3. Siphons short and united (in the long-shelled genera), or longer and partly separate (in the genera with shorter shells). Foot very large and powerful. Gills prolonged into the branchial siphon. Ill. Gen. *Solen, Cultellus, Solecurtus.*

Fam. 18. *Myacidæ.*—Shell gaping posteriorly. Mantle almost entirely closed; siphons united, partly or wholly retractile. Foot very small. Ill. Gen. *Mya, Panopæa, Glycimeris.*

Fam. 19. *Anatinidæ.*—Shell often inequivalve, with an external ligament. Mantle-lobes more or less united; siphons long, more or less united. Foot small. Ill. Gen. *Anatina, Pholadomya, Myochama.*

Fam. 20. *Gastrochænidæ.*—Shell equivalve, gaping, with thin edentulous valves, sometimes cemented to a calcareous tube. Mantle-margins thick in front, united, with a small pedal aperture. Siphons very long, united. Foot finger-shaped. Ill. Gen. *Gastrochæna*, *Saxicava*, *Aspergillum*.

Fam. 21. *Pholadidæ.*—Shell gaping at both ends, without hinge or ligament, often with accessory valves. Animal club-shaped or worm-like, with a short truncated foot. Mantle closed in front. Siphons long, united to near their extremities. Ill. Gen. *Pholas*, *Xylophaga*, *Teredo*.

CHAPTER XLVII.

GASTEROPODA.

DIVISION ENCEPHALA, OR CEPHALOPHORA. — The remaining three classes of the Mollusca proper all possess a distinctly differentiated head, and are all provided with a peculiar masticatory apparatus, which is known as the "odontophore." For the first of these reasons they are often grouped together under the name *Encephala;* and for the second reason they are united by Huxley into a single great division, under the name of *Odontophora*. Whichever name be adopted, the three classes in question (viz., the *Gasteropoda*, *Pteropoda*, and *Cephalopoda*) certainly show many points of affinity, and form a very natural division of the *Mollusca*. The *Pteropoda*, as being the lowest class, should properly be treated of first, but it will conduce to a clearer understanding of their characters if the *Gasteropoda* are considered first.

CLASS II. GASTEROPODA.—The members of this class are characterised by being never included in a bivalve shell; locomotion being effected by means of a broad, horizontally flattened, ventral disc—the "foot;" or by a vertically flattened, ventral, fin-like organ. Flexure of intestine hæmal or neural.

This class includes all those Molluscous animals which are commonly known as "univalves," such as the land-snails, sea-snails, whelks, limpets, &c. The shell, however, is sometimes composed of several pieces (multivalve), and in many there is either no shell at all, or nothing that would be generally recognised as such. In none is there a bivalve shell.

In their habits the Gasteropods show many differences, some being sedentary, but the great majority being free and locomotive. In these latter, locomotion may be effected by the successive contractions and expansions of a muscular foot; but some possess the power of swimming freely by means of a modified fin-like foot.

In most of the *Gasteropoda* the body is unsymmetrical, and is coiled up spirally, "the respiratory organs of the left side being usually atrophied."—(Woodward.) The body is enclosed in a "mantle," which is not divided into two lobes as in the *Lamellibranchiata*, but is continuous round the body. Locomotion is effected by means of the "foot," which is usually a broad muscular disc, developed upon the ventral surface of the body, and not exhibiting any distinct division into parts. In the *Heteropoda*, however, and in the Wing-shells (*Strombidæ*), the foot exhibits a division into three portions—an anterior, the "propodium;" a middle, the "mesopodium;" and a posterior lobe, or "metapodium."

In some, again, the upper and lateral surfaces of the foot are expanded into muscular side-lobes, which are called "epipodia." In many cases the metapodium, or posterior portion of the foot, secretes a calcareous, horny, or fibrous plate, which is called the "operculum" (fig. 109, *o*), and which serves to close the orifice of the shell when the animal is retracted within it.

The head in most of the *Gasteropoda* is very distinctly marked out, and is provided with two tentacles and with two eyes, which are often placed upon long stalks. Very often there is an elongated retractile proboscis with ear-sacs, containing otoliths, at its base. The mouth is sometimes furnished with horny jaws, and is always provided with a singular masticatory apparatus, called the "tongue" or "odontophore" (fig. 108). "It consists essentially of a cartilaginous cushion, supporting, as on a pulley, an elastic strap, which bears a long series of transversely-disposed teeth. The ends of the strap are connected with muscles attached to the upper and lower surface of the hinder extremities of the cartilaginous cushions; and these muscles, by their alternate contractions, cause the toothed strap to work backwards and forwards over the end of the pulley formed by its anterior end. The strap consequently acts, after the fashion of a chain-saw, upon any substance to which it is applied, and the resulting wear and tear of its anterior teeth are made good by the incessant development of new teeth in the secreting sac in which the

Fig. 108.—Fragment of the lingual ribbon or odontophore of the common Whelk (*Buccinum undatum*), magnified. (After Woodward.)

hinder end of the strap is lodged."—(Huxley.) The teeth of the odontophore ("lingual teeth") are composed of silica, and are usually arranged in a central ("rachidian") and two lateral ("pleural") rows. The mouth leads by a gullet into a distinct stomach, which is sometimes provided with calcareous plates for the trituration of the food. The intestine is long, and its first flexure is commonly "hæmal," or towards that side of the body on which the heart is situated; though in some the flexure is "neural." Distinct salivary glands are usually present, and the liver is well developed.

A distinct heart is usually present, composed of an auricle and ventricle. In many Gasteropods it has been shown that the blood-vessels form closed tubes, and that the arteries and veins are connected by an intermediate system of capillaries, instead of merely communicating through the interstices and lacunæ between the tissues. It seems also certain that, in general at any rate, there is no direct connection between the blood-vessels and the outer medium, though, in some cases, such a communication seems undoubtedly to exist. Respiration is very variously effected; one great division (*Branchiogasteropoda*) being constructed to breathe air by means of water; whilst in another section (*Pulmogasteropoda*) the respiration is aerial. In the former division respiration may be effected in three ways. Firstly, there may be no specialised respiratory organ, the blood being simply exposed to the water

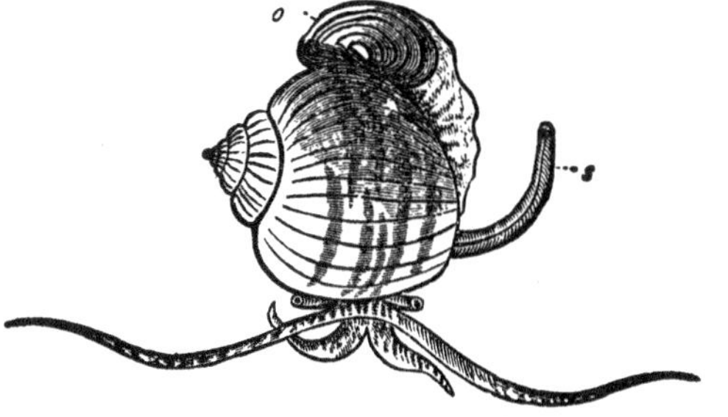

Fig. 109.—*Ampullaria canaliculata*, one of the Apple-shells. *o* Operculum; *s* Respiratory siphon.

in the thin walls of the mantle-cavity (as in some of the *Heteropoda*). Secondly, the respiratory organs may be in the form of outward processes of the integument, exposed in tufts on the back and sides of the animal (as in the *Nudibranchiata*). Thirdly, the respiratory organs are in the form of pectinated or plume-like branchiæ, contained in a more or less complete branchial chamber formed by an inflection of the mantle. In many members of this last section the water obtains access to the gills by means of a tubular prolongation or folding of the mantle, forming a "siphon," the effete water being expelled by another posterior siphon similarly constructed. In the air-breathing Gasteropods, the breathing organ is in the form of a pulmonary chamber, formed by an inflection of the mantle, and having a distinct aperture for the admission of air.

The nervous system in the *Gasteropoda* has its normal composition of three principal pairs of ganglia, the supra-œsophageal or cerebral, the infra-œsophageal or pedal, and the parieto-splanchnic; but there is a tendency to the aggregation of these in the neighbourhood of the head. The organs of sense are the two eyes, and auditory capsules placed at the bases of the tentacles, the latter being tactile organs.

The sexes are mostly distinct, but in some they are united in the same individual. The young, when first hatched, are *always* provided with an embryonic shell, which in the adult may become concealed in a fold of the mantle, or may be entirely lost. In the branchiate Gasteropods the embryo is protected by a small nautiloid shell, within which it can entirely retract itself; and it is enabled to swim freely by means of two ciliated lobes arising from the sides of the head; thus, in many respects, resembling the permanent adult condition of the *Pteropoda*. In the branchiate *Gasteropoda*, however, of fresh waters, the young do not possess these ciliated buccal lobes.

Shell of the Gasteropoda.—The shell of the Gasteropods is composed either of a single piece (univalve), or of a number of plates succeeding one another from before backwards (multivalve). The univalve shell is to be regarded as essentially a cone, the apex of which is more or less oblique. In the simplest form of the shell the conical shape is retained without any alteration, as is seen in the common Limpet (*Patella*). In the great majority of cases, however, the cone is considerably elongated, so as to form a tube, which may retain this shape (as in *Dentalium*), but is usually coiled up into a spiral. The "spiral univalve" (fig. 110) may, in fact, be looked upon as the typical form of the shell in the *Gasteropoda*. In some cases the coils of the shell—termed technically the "whorls" —are hardly in contact with one another (as in *Vermetus*). More commonly the whorls are in contact, and are so amalgamated that the inner side of each convolution is formed by the pre-existing whorl. In some cases the whorls of the shell are coiled round a central axis *in the same plane*, when the shell is said to be "discoidal" (as in the common fresh-water shell *Planorbis*). In most cases, however, the whorls are wound round an axis in an oblique manner, a true spiral being formed, and the shell becoming "turreted," "trochoid," "turbinated," &c. This last form is the one which may be looked upon as most characteristic of the Gasteropods, the shell being composed of a number of whorls passing obliquely round a central axis or "columella," having the embryonic shell or "nucleus" at its apex, and having the mouth or "aperture" of the shell placed at the extremity of the last and largest of the whorls, termed the "body-whorl." The lines or grooves formed by the junction of the whorls are termed the "sutures," and the whorls above the body-whorl constitute the "spire" of the shell. The axis of the shell (columella) round which the whorls are coiled is usually solid, when the shell is said to be "imperforate;" but it is sometimes hollow, when the shell is said to be "perforated," and the aperture of the axis near the mouth of the shell is called the "umbilicus." The margin of the "aperture" of the shell is termed the "peristome," and is composed of an outer and inner lip, of which the former is often expanded or fringed with spines. When these expansions or fringes are periodically formed, the place of the mouth of the shell at different stages of its growth is marked by ridges or rows of spines, which cross the whorls, and are called "varices." In most of the phytophagous Gasteropods (*Holostomata*) the aperture of the shell (fig. 110, *a*) is unbrokenly round or "entire," but in the carnivorous forms (*Siphonostomata*) it is notched, or produced into a canal (fig. 110, *b*). Often there are two of these canals, an anterior and a posterior, but they do not necessarily indicate the nature of the food, as their function is to protect the respiratory siphons. The animal withdraws into its shell by a retractor-muscle, which passes into the foot, or is attached to the operculum; its scar or impression being placed, in the spiral univalves, upon the columella.

In the multivalve Gasteropods, the shell is composed of eight transverse imbricated plates, which succeed one another from before backwards, and are imbedded in the leathery or fibrous border of the mantle, which may be plain, or may be beset with bristles, spines, or scales.

CHAPTER XLVIII.

DIVISIONS OF THE GASTEROPODA.

THE *Gasteropoda* are divided into two primary sections or sub-classes, according as the respiratory organs are adapted for breathing air directly, or dissolved in water: termed respectively the *Pulmonifera* or *Pulmogasteropoda*, and the *Branchifera* or *Branchiogasteropoda*.

SUB-CLASS A. BRANCHIFERA OR BRANCHIOGASTEROPODA.—In this sub-class *respiration is aquatic*, effected by the thin walls of the mantle-cavity, by external branchial tufts, or by pectinated or plume-like gills, contained in a more or less complete branchial chamber. *Flexure of intestine hæmal.*

This sub-class comprises three orders — viz., the *Prosobranchiata*, the *Opisthobranchiata*, and the *Nucleobranchiata* or *Heteropoda*.

ORDER I. PROSOBRANCHIATA.—The members of this order are defined as follows:—"*Abdomen* well developed, and protected by a shell, into which the whole animal can usually retire. *Mantle* forming a vaulted chamber over the back of the head, in which are placed the excretory orifices, and in which the branchiæ are almost always lodged. *Branchiæ* pectinated or plume-like, situated (*proson*) in advance of the heart. *Sexes* distinct."—M.-Edwards. (See Woodward's 'Manual.')

The order *Prosobranchiata* includes all the most characteristic members of the Branchiate Gasteropods, and is divisible into two sections, termed respectively *Siphonostomata* and *Holostomata*, according as the aperture of the shell is notched or produced into a canal, or is simply rounded and "entire."

The *Siphonostomata*, of which the common Whelk (*Buccinum undatum*) may be taken as an example, are all marine, and are mostly carnivorous in their habits. The following families are comprised in this section:—*Strombidæ* (Wing-shells), *Muricidæ*, *Buccinidæ* (Whelks), *Conidæ* (Cones), *Volutidæ*, and *Cypræidæ* (Cowries).

The *Holostomata*, of which the Common Periwinkle (*Littorina littorea*) is a good example, are either spiral or limpet-shaped, in some few instances tubular, or multivalve; the aperture of the shell being in most cases entire. They are mostly plant-eaters, and they may be either marine or inhabitants of fresh water. The following families are included in this section:—*Naticidæ*, *Pyramidellidæ*, *Cerithiadæ*, *Melaniadæ*, *Turri-*

tellidæ, *Littorinidæ* (Periwinkles), *Paludinidæ* (River-snails) *Neritidæ*, *Turbinidæ* (Top-shells), *Haliotidæ* (Ear-shells), *Fissurellidæ* (Key-hole Limpets), *Calyptræidæ* (Bonnet Limpets), *Patellidæ* (Limpets), *Dentalidæ* (Tooth-shells), and *Chitonidæ*.

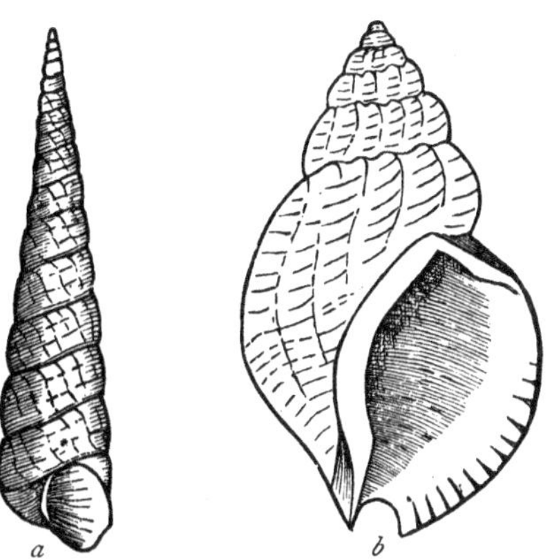

Fig. 110.—Gasteropoda. *a* Holostomatous shell (*Turritella communis*); *b* Siphonostomatous shell (*Buccinum undatum*).

ORDER II. OPISTHOBRANCHIATA.—This order is defined as follows :—

"*Shell* rudimentary, or wanting. *Branchiæ* arborescent or fasciculated, not contained in a special cavity, but more or less completely exposed on the back and sides, towards the rear (*opisthen*) of the body. Sexes united."—M.-Edwards. (See Woodward's 'Manual.')

The *Opisthobranchiata*, or "Sea-slugs," may be divided into two sections, the *Tectibranchiata* and *Nudibranchiata*, according as the branchiæ are protected or are uncovered.

The first section, that of the *Tectibranchiata*, is distinguished by the fact that the animal is usually provided with a shell, both in the larval and adult state, and that the branchiæ are protected by the shell or by the mantle. Under this family are included the families of the *Tornatellidæ*, *Bullidæ* (Bubble shells), *Aplysiadæ* (Sea-hares), *Pleurobranchidæ* and *Phyllidiadæ*.

In the second section, that of the *Nudibranchiata* (fig. 111), the animal is destitute of a shell, except in the embryo condition, and the branchiæ are always placed externally on the back or sides of the body. This section comprises the families *Doridæ* (Sea-lemons), *Tritoniadæ*, *Æolidæ*, *Phylilrhoidæ*, and

Elysiadæ. Specimens of the Sea-slugs and Sea-lemons may at any time be found creeping about on sea-weeds, or attached to the under surface of stones at low water. The head is furnished with tentacles, which appear to be rather connected with the sense of smell than to be used as tactile organs; and behind the tentacles are generally two eyes. The nervous system is extremely well developed, and would lead to the belief that the *Nudibranchs* are amongst the highest of the *Gasteropoda*. Locomotion is effected, as in the true Slugs, by creeping about on the flattened foot.

Fig. 111.—Nudibranchiata. *Doris Johnstoni*, one of the Sea-lemons.

ORDER III. NUCLEOBRANCHIATA OR HETEROPODA.—This order is defined by the following characteristics:—Animal provided with a shell, or not, free-swimming and pelagic; locomotion effected by a fin-like tail, or by a fan-shaped, vertically flattened, ventral fin.

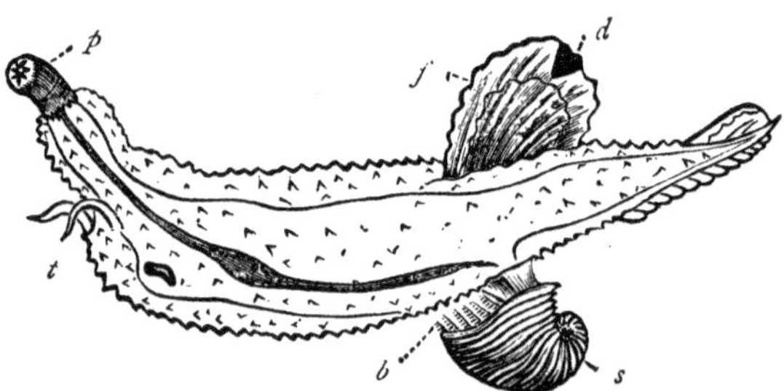

Fig. 112.—Heteropoda. *Carinaria cymbium*. *p* Proboscis; *t* Tentacles; *b* Branchiæ; *s* Shell; *f* Foot; *d* Disc. (After Woodward.)

The *Heteropoda* are pelagic in their habits, and are found swimming at the surface of the sea. They are to be regarded as the most highly organised of all the *Gasteropoda*, at the same time that they are not the most typical members of the class. Some of them can retire completely within their shells, closing them with an operculum; but most have large bodies, and the shell is either small or entirely wanting. They swim by means of a flattened ventral fin, or by an elongated tail, and adhere at pleasure to sea-weed by a small sucker situated on the side of the fin. These organs are merely modifications of the foot of the ordinary Gasteropods; the fin-like tail being the "metapodium" (as shown by its occasionally carrying an operculum), the sucker being the "mesopodium," and the ventral fin being a modified "propodium." The "epipodia" are apparently altogether wanting. Respiration is sometimes carried on by distinct branchiæ, but in many cases these are wanting, and the function is performed simply by the walls of the pallial chamber.

The *Heteropoda* are divided into the two families *Firolidæ* and *Atlantidæ*, the former characterised by having a small shell covering the circulatory and respiratory organs, or by having no shell at all; whilst in the latter there is a well-developed shell, into which the animal can retire, and an operculum is often present.

SUB-CLASS B. PULMONIFERA OR PULMOGASTEROPODA.—In this sub-class of the *Gasteropoda respiration is aerial*, and is carried on by an inflection of the mantle, forming a pulmonary chamber, into which air is admitted by an external aperture. The *flexure of the intestine is neural*, and the sexes are united in the same individual.

Fig. 113.—*Limax Sowerbyi*, one of the Slugs. (After Woodward.)

The *Pulmonifera* include the ordinary land-snails, slugs, pond-snails, &c., and are usually provided with a well-developed shell, though this may be rudimentary (as in the slugs), or even wanting. Though formed to breathe air directly, many of the members of this sub-class are capable of inhabiting fresh water. The common Pond-snails are good examples of these last. The condition of the shell varies greatly. Some, such as the common Land-snails, have a well-developed shell, within which the animal can withdraw itself completely. Others, such as the common Slugs (fig. 113) have a rudimentary shell, which is completely concealed within the mantle. Others are entirely destitute of a shell. They are divided into two sections as follows :—

Section I. *Inoperculata.*—*Animal not provided with an operculum to close the shell.* In this section are included the families

Helicidæ (Land-snails), *Limacidæ* (Slugs), *Oncidiadæ*, *Limnæidæ* (Pond-snails), and *Auriculidæ*.

Section II. *Operculata.—Shell closed by an operculum.* In this section are included the families *Cyclostomidæ* and *Aciculidæ*.

SYNOPSIS OF THE FAMILIES OF THE GASTEROPODA.
(AFTER WOODWARD.)

SECTION A. BRANCHIFERA. Respiration aquatic, by the walls of the mantle-cavity, or by branchiæ.

ORDER I. PROSOBRANCHIATA. The branchiæ situated (*proson*) in advance of the heart.

Division a. Siphonostomata. Margin of the shell-aperture notched or produced into a canal.

Fam. 1. *Strombidæ.* Shell with an expanded lip, deeply notched near the canal. Operculum claw-shaped. Foot narrow, adapted for leaping. Ill. Gen. *Strombus, Pteroceras.*

Fam. 2. *Muricidæ.* Shell with a straight anterior canal, the aperture entire posteriorly. Foot broad. Ill. Gen. *Murex, Triton, Pyrula, Fusus.*

Fam. 3. *Buccinidæ.* Shell notched anteriorly, or with the canal abruptly reflected, producing a kind of varix on the front of the shell. Ill. Gen. *Buccinum, Nassa, Purpura, Cassis, Harpa, Oliva.*

Fam. 4. *Conidæ.* Shell inversely conical, with a long narrow aperture, the outer lip notched at or near the suture. Operculum minute, lamellar. Ill. Gen. *Conus, Pleurotoma.*

Fam. 5. *Volutidæ.* Shell turreted or convolute, the aperture notched in front; the columella obliquely plaited. No operculum. Foot very large; mantle often reflected over the shell. Ill. Gen. *Voluta, Mitra, Marginella.*

Fam. 6. *Cypræidæ.* Shell convolute, enamelled; spire concealed, aperture narrow, channelled at each end. Outer lip thin in the young shell, but thickened and inflected in the adult. Foot broad; mantle forming lobes which meet over the back of the shell. Ill. Gen. *Cypræa, Ovulum.*

Division b. Holostomata. Margin of the shell-aperture "entire," rarely notched or produced into a canal.

Fam. 1. *Naticidæ.* Shell globular, of few whorls, with a small spire, outer lip acute, pillar often callous. Foot very large; mantle-lobes hiding more or less of the shell. Gen. *Natica, Sigaretus.*

Fam. 2. *Pyramidellidæ.* Shell turreted, with a small aperture, sometimes with one or more prominent plaits on the columella. Operculum horny, imbricated. Ill. Gen. *Pyramidella, Chemnitzia, Eulima.*

Fam. 3. *Cerithiadæ.* Shell spiral, turreted; aperture channelled in front, with a less distinct posterior canal. Lip generally expanded in the adult. Operculum horny and spiral. Ill. Gen. *Cerithium, Potamides, Aporrhais.*

Fam. 4. *Melaniadæ.* Shell spiral, turreted; aperture often channelled or notched in front; outer lip acute. Operculum horny and spiral. Ill. Gen. *Melania, Paludomus.*

Fam. 5. *Turritellidæ.* Shell tubular, or spiral, often turreted; upper part partitioned off; aperture simple. Operculum horny, many-whorled. Foot very short. Branchial plume single. Ill. Gen. *Turritella, Vermetus, Scalaria.*

Fam. 6. *Littorinidæ.* Shell spiral, top-shaped, or depressed; aperture rounded and entire; operculum horny and pauci-spiral. Ill. Gen. *Littorina, Solarium, Rissoa, Phorus.*

Fam. 7. *Paludinidæ.* Shell conical or globular; aperture rounded and entire; operculum horny or shelly. Ill. Gen. *Paludina, Ampullaria, Valvata.*

Fam. 8. *Neritidæ.* Shell thick, globular, with a very small spire; aperture semi-lunate, its columellar side expanded; outer lip acute. Operculum shelly, sub-spiral. Ill. Gen. *Nerita, Pileolus, Neritina.*

Fam. 9. *Turbinidæ.* Shell turbinated (top-shaped), or pyramidal, nacreous inside. Operculum horny and multi-spiral, or calcareous and pauci-spiral. Ill. Gen. *Turbo, Trochus, Delphinula, Euomphalus.*

Fam. 10. *Haliotidæ.* Shell spiral, ear-shaped, or trochoid; aperture large, nacreous. Outer lip notched or perforated. No operculum. Mantle-margin with a posterior fold or siphon, occupying the slit or perforation in the shell. Metapodium rudimentary. Ill. Gen. *Haliotis, Scissurella, Pleurotomaria, Murchisonia, Ianthina.*

Fam. 11. *Fissurellidæ.* Shell conical, patelliform, with a notch in the anterior margin, or a perforation at its apex, which is occupied by an anal siphon. Muscular impression horse-shoe-shaped, open in front. Ill. Gen. *Fissurella, Emarginula, Parmophorus.*

Fam. 12. *Calyptræidæ.* Shell patelliform, with a more or less spiral apex; interior simple, or divided by a shelly process to which the adductor muscles are attached. Ill. Gen. *Calyptræa, Pileopsis.*

Fam. 13. *Patellidæ.* Shell conical, with the apex turned forwards; muscular impression horse-shoe-shaped, open in front. Foot as large as the margin of the mantle. Respiratory organ in the form of one or two branchial plumes, lodged in a cervical cavity; or of a series of lamellæ surrounding the animal between the body and the mantle. Ill. Gen. *Patella, Acmæa.*

Fam. 14. *Dentalidæ.* Shell tubular, symmetrical, curved, open at both ends. Aperture circular. Foot pointed, with symmetrical side-lobes. Gen. *Dentalium.**

Fam. 15. *Chitonidæ.* Shell multivalve, composed of eight transverse imbricated plates. Animal with broad creeping foot; branchiæ forming a series of lamellæ between the foot and the mantle, round the posterior part of the body. Ill. Gen. *Chiton, Cryptochiton.*

ORDER II. OPISTHOBRANCHIATA. Branchiæ placed towards the rear (*opisthen*) of the body.

Section a. Tectibranchiata. Branchiæ covered by the shell or mantle; a shell in most. Sexes united.

Fam. 1. *Tornatellidæ.* Shell external, spiral, or convoluted; aperture long and narrow; columella plaited. Ill. Gen. *Tornatella, Cinulia.*

---

* *Dentalium* is placed by Professor Huxley amongst the *Pteropoda*, from its rudimentary head, the neural flexure of the intestine, the nature of the epipodia, and the characters of the larva.

Fam. 2. *Bullidæ.* Shell convoluted, thin; spire small or concealed, lip sharp. Animal more or less investing the shell. Ill. Gen. *Bulla*, *Cylichna*, *Philine*.

Fam. 3. *Aplysiadæ.* Shell absent, or rudimentary and concealed by the mantle. Animal slug-like, with extensive side-lobes, (epipodia), reflected over the back and shell. Ill. Gen. *Aplysia*, *Dolabella*.

Fam. 4. *Pleurobranchidæ.* Shell patelliform, or concealed, rarely wanting. Mantle or shell covering the back of the animal. Ill. Gen. *Pleurobranchus*, *Umbrella*, *Tylodina*.

Fam. 5. *Phyllidiadæ.* Animal shell-less, covered by a mantle. Ill. Gen. *Phyllidia*, *Diphyllidia*.

Section b. Nudibranchiata. Animal destitute of a shell in the adult condition. Branchiæ external, on the back or sides of the body.

Fam. 6. *Doridæ.* Ill. Gen. *Doris*.

Fam. 7. *Tritoniadæ.* Ill. Gen. *Tritonia*, *Scyllæa*.

Fam. 8. *Æolidæ.* Ill. Gen. *Æolis*, *Glaucus*.

Fam. 9. *Phyllirhoidæ.* Gen. *Phyllirhoe*.

Fam. 10. *Elysiadæ.* Ill. Gen. *Elysia*, *Acteonia*.

ORDER III. NUCLEOBRANCHIATA or HETEROPODA. Shell present or absent. Animal free-swimming and pelagic, with a fin-like tail, or a flattened ventral fin.

Fam. 1. *Firolidæ.* Body large; branchiæ exposed on the back, or covered by a small hyaline shell: locomotion by means of a ventral fin and a tail-fin. Ill. Gen. *Carinaria*, *Firola*.

Fam. 2. *Atlantidæ.* Animal furnished with a well-developed shell into which it can retire. Branchiæ contained in a dorsal mantle-cavity. Shell symmetrical, discoidal, sometimes with an operculum. Ill. Gen. *Atlanta*, *Bellerophon*, *Maclurea*.

SECTION B. PULMONIFERA. Respiration aerial, by means of a pulmonary chamber.

DIVISION I. INOPERCULATA. *Shell not provided with an operculum.*

Fam. 1. *Helicidæ.* Shell external, capable of containing the whole animal. Ill. Gen. *Helix*, *Bulimus*, *Clausilia*, *Pupa*.

Fam. 2. *Limacidæ.* Shell rudimentary, usually internal or partly concealed by the mantle. Ill. Gen. *Limax*, *Parmacella*, *Testacella*.

Fam. 3. *Oncidiadæ.* Shell wanting. Animal slug-like. Ill. Gen. *Oncidium*, *Vaginulus*.

Fam. 4. *Limnæidæ.* Shell thin, horn-coloured, well developed. Aperture simple, lip sharp. Ill. Gen. *Limnæa*, *Physa*, *Ancylus*, *Planorbis*.

Fam. 5. *Auriculidæ.* Shell spiral, with a horny epidermis; aperture elongated, denticulated. Ill. Gen. *Auricula*, *Conovulus*.

DIVISION II. OPERCULATA. *Shell with an operculum.*

Fam. 6. *Cyclostomidæ.* Shell spiral, rarely elongated, often depressed. Aperture nearly circular. Operculum spiral. Ill. Gen. *Cyclostoma*, *Cyclophorus*, *Pupina*.

Fam. 7. *Aciculidæ.* Shell elongated, cylindrical; operculum thin, and sub-spiral. Gen. *Acicula*, *Geomelania*.

CHAPTER XLIX.

PTEROPODA.

CLASS III. PTEROPODA.—The *Pteropoda* are defined by being *free and pelagic*, swimming by means of two wing-like appendages (epipodia), developed from each side of the anterior extremity of the body. The flexure of the intestine is neural.

As to the position of the *Pteropoda* in the Molluscan scale, they must be looked upon as inferior in organisation to any of the *Gasteropoda*, of which class they are often regarded as the lowest division. They permanently represent, in fact, the transient larval stage of the sea-snails.

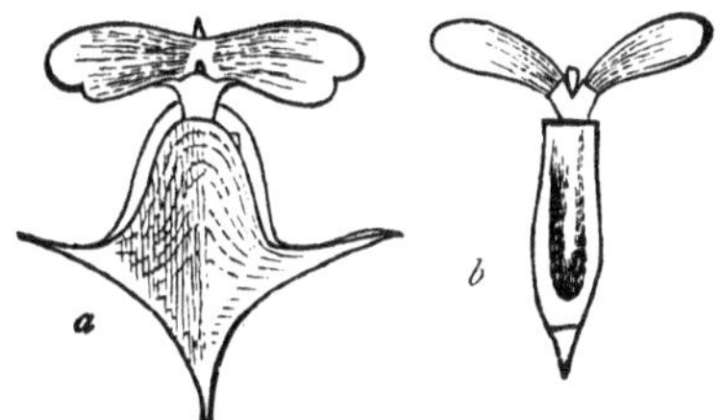

Fig. 114.—Pteropoda. *a Cleodora pyramidata; b Cuvieria columnella.* (After Woodward.)

The Pteropods are all of small size, and are found swimming at the surface of the open ocean, often in enormous numbers. Locomotion is effected by two wing-like fins, developed from the sides of the head, and composed of the greatly developed "epipodia." The true "foot" is rudimentary and rarely distinct, but the "metapodium" is sometimes provided with an operculum. There is usually a symmetrical glassy shell (fig. 114), either consisting of a dorsal and ventral plate united, or forming a spiral, but in some cases the body is naked. The head is rudimentary, and bears the mouth, which is occasionally tentaculate, and which is furnished with an odontophore. There is a muscular stomach and a well-developed liver; and the flexure of the intestine is neural, so that the anus is situated on the ventral surface of the body.

The heart consists of an auricle and ventricle. The respiratory organ is very rudimentary, and consists of a ciliated surface, which is either entirely unprotected, or may be contained in a branchial chamber.

The ganglia of the nervous system "are concentrated into

a mass *below* the œsophagus" (Woodward), and the eyes are rudimentary.

The sexes are united in all the Pteropods, and the young pass through a metamorphosis, having at first a bilobed ciliated veil attached to the sides of the head.

The *Pteropoda* are divided into two orders, termed *Thecosomata* and *Gymnosomata*; the former characterised by possessing an external shell and an indistinct head; the latter by being devoid of a shell, and by having a distinct head, with fins attached to the neck.

The *Pteropoda*, as already said, are found swimming near the surface in the open ocean, and they are found in all seas from the tropics to within the arctic circle, sometimes in such numbers as to discolour the water for many miles. They are nocturnal in their habits, and, minute as they are, they constitute in high latitudes one of the staple articles of diet of the whale. They themselves are, in turn, probably carnivorous, feeding upon small Crustaceans and other diminutive animals. Though all the living forms are small, Geology leads us to believe that there formerly existed comparatively gigantic representatives of this class of the *Mollusca*.

SYNOPSIS OF THE FAMILIES OF THE PTEROPODA. (AFTER WOODWARD.)

ORDER I. THECOSOMATA.

Animal with an external shell; head indistinct; foot and tentacles rudimentary; mouth situated in a cavity formed by the union of the locomotive organs. Respiratory organs contained within a mantle-cavity.

Fam. 1. *Hyaleidæ*.

Shell symmetrical, straight or curved, globular or needle-shaped. Ill. Gen. *Hyalea, Cleodora, Theca, Conularia.*

Fam. 2. *Limacinidæ*.

Shell minute, spiral, sometimes operculate. Ill. Gen. *Limacina, Spirialis.*

ORDER II. GYMNOSOMATA.

Animal naked, without mantle or shell; head distinct; fins attached to the sides of the neck; gill indistinct.

Fam. 3. *Cliidæ*.

Body fusiform, foot distinct, with a central and posterior lobe; head with tentacles. Ill. Gen. *Clio, Pneumodermon.*

CHAPTER L.

CEPHALOPODA.

CLASS IV. CEPHALOPODA.—The members of this class are defined by the possession of eight or more arms placed in a circle round the mouth ; the body is enclosed in a muscular mantle-sac, and there are two or four plume-like gills within the mantle. There is an anterior tubular orifice (the "infundibulum" or "funnel"), through which the effete water of respiration is expelled. The flexure of the intestine is neural.

The *Cephalopoda*, comprising the Cuttle-fishes, Squids, Pearly Nautilus, &c., constitute the most highly organised of the classes of the *Mollusca*. They are all marine and carnivorous, and are possessed of considerable locomotive powers. At the bottom of the sea they can walk about, head downwards, by means of the arms which surround the mouth, and which are usually provided with numerous suckers or "acetabula." They are also enabled to swim, partly by means of lateral expansions of the integument or fins (not always present), and partly by means of the forcible expulsion of water through the tubular "funnel," the reaction of which causes the animal to move in the opposite direction.

The majority of the living Cephalopods are naked, possessing only an internal skeleton, and this often a rudimentary one; but the Argonaut (Paper Nautilus), and the Pearly Nautilus, are protected with an external shell, though the nature of this is extremely different in the two forms.

The integument in the Cuttle-fishes is provided with numerous little sacs, containing pigment-granules of different colours, and termed "chromatophores." By means of these many species can change their colours rapidly, under irritation or excitement.

The body in the *Cephalopoda* is symmetrical, and is enclosed in an integument which may be regarded as a modification of the mantle of the other *Mollusca*. Ordinarily there is a tolerably distinct separation of the body into an anterior cephalic portion (*prosoma*), and a posterior portion, enveloped in the mantle, and containing the viscera (*metasoma*). The head is very distinct, bearing a pair of large globular eyes, and having the mouth in its centre. The mouth is surrounded by a circle of eight, ten, or more, long muscular processes, or "arms" (fig. 115), which are generally provided with rows of suckers. Each sucker, or "acetabulum," consists of a cup-shaped cavity, the muscular fibres of which converge to the centre, where there is a little muscular eminence or papilla. When the sucker is applied to any surface, the contraction of the radiating muscular fibres depresses the papilla so as to produce a vacuum below it, and in this way each sucker acts most efficiently as an adhesive organ. In some forms (*Decapoda*) the base of the papilla, or piston, is surrounded by a horny dentated ring, and in some others (as in *Onychoteuthis*) the papillæ are produced into long claws. In the Octopod Cuttle-fishes there are only eight arms, and these are all nearly alike. In the Decapod Cuttle-fishes there are ten arms, but two of these—called "tentacles"—are much longer than the others, and bear suckers only at their extremities, which are enlarged and club-shaped. In the Pearly Nautilus the arms are numerous, and are devoid of suckers.

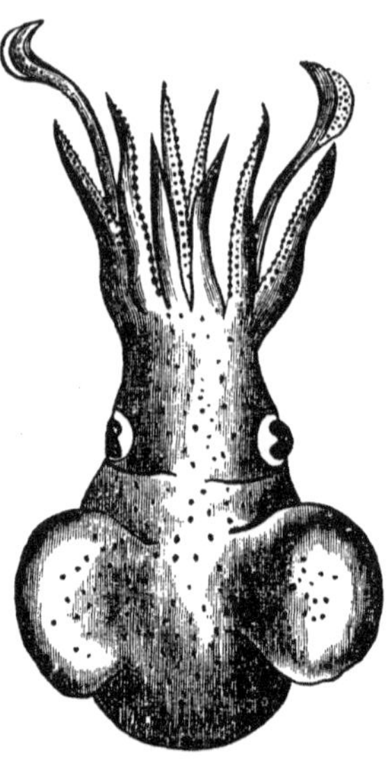

Fig. 115.—Cephalopoda. *Sepiola Atlantica*, one of the Cuttle-fishes (after Woodward).

The arms are really produced by an extension of the margins of the "foot," or of the part corresponding to the foot of the other *Mollusca*. The "antero-lateral parts of each side of the foot extend forwards beyond the head, uniting with it and with one another; so that, at length, the mouth, from having been situated, as usual, above the anterior margin of the foot, comes to be placed in the midst of it. The two epipodia, on the other hand, unite posteriorly above the foot, and where they coalesce, give rise either to a folded muscular expansion, the edges of which are simply in apposition, as in the *Nautilus*; or to an elongated flexible tube, the apex of which projects beyond the margin of the mantle, called the 'funnel,' or 'infundibulum,' as in the dibranchiate *Cephalopoda*."—(Huxley.)

The mouth leads into a buccal cavity, containing two powerful, horny, or partially calcareous mandibles, working vertically like the beak of a bird; together with an "odontophore" or "tongue," the anterior part of which is sentient, whilst the remainder is covered with recurved spines. The buccal cavity conducts by an œsophagus—into which salivary glands usually pour their secretion—to a stomach, from which an intestine is continued, with a neural flexure, to open on the ventral surface of the animal at the base of the funnel. In many cases there is also a special gland, called the "ink-bag," for the secretion of an inky fluid, which the animal discharges into the water, so as to enable it to escape when menaced or pursued. The duct of the ink-bag opens at the base of the funnel; but this apparatus is entirely wanting in the Tetrabranchiate Cephalopods, where, in consequence of the presence of an external shell, this means of defence is not needed.

The respiratory organs are in the form of two or four plume-like gills, placed on the sides of the body in a branchial cavity which opens anteriorly on the under surface of the body. At the base of each gill, in the Cuttle-fishes, is a special contractile cavity, whereby the venous blood returned from the body is driven through the branchiæ. In addition to these accessory organs—the so-called "branchial hearts"—there is a true systemic heart, by which the aerated blood received from the gills is propelled through the body. In the higher Cephalopods a capillary system of vessels intervenes, in most cases at any rate, between the arteries and the veins. The admission of water to the branchiæ is effected by the expansion of the mantle so as to allow the entrance of the outer water into the pallial chamber. The mantle then contracts, and the water is forcibly expelled through the funnel, which is provided with a valve permitting the egress of water, but preventing its ingress. By a repetition of this process, not only is respiration effected, but locomotion is simultaneously subserved; the jets of water expelled from the funnel, by their reaction, driving the animal in the opposite direction.

The nervous system is formed upon essentially the same plan as in the other *Mollusca*, but it is more concentrated, and the supra-œsophageal or cerebral ganglia are protected by a cartilage, which is to be regarded as a rudimentary cranium. This structure, therefore, presents us with the nearest approach which we have yet met with to the Vertebrate type of organisation.

The sexes in all the *Cephalopoda* are in different individuals, and the reproductive process in the Dibranchiate section of the class (Cuttle-fishes) is attended with some very singular phenomena. In this order the ducts of the generative organs open into the pallial chamber, and each individual, besides the essential organs of reproduction (testis or ovary), generally possesses an accessory gland; that of the female secreting a viscid material which unites the eggs together, whilst that of the male coats the spermatozoa, and aggregates them into peculiar worm-like filaments, termed "spermatophores," or the "moving filaments of Needham." The spermatophore is filled with spermatozoa, and possesses the power of expanding when moistened, rupturing, and expelling the contained spermatozoa with considerable force. During the congress of the sexes the male transfers the spermatophores to the pallial chamber of the female, true intromission not being possible. Further, in all the male Cuttle-fishes one of the arms is specially modified to subserve reproduction; being in many cases so altered as to become useless as a locomotive organ. The arm so affected, in the more striking forms, is said to be "hectocotylised," and—like the metamorphosed palpi of the male spiders—it serves to convey the seminal fluid to the female. The mode in which this is effected varies in different species. Thus, in the male *Octopus* (the Poulpe) the third right arm is primitively developed in a cyst, which ultimately ruptures and liberates the metamorphosed arm, which then appears to be of greater size than the corresponding arm on the left side, and to terminate in an oval plate (fig. 116). To this terminal plate the spermatophore is probably transmitted, but the arm itself probably remains permanently attached to the animal. It is asserted, however, that in the form figured below (*Octopus carena*) the hectocotylised arm is detached and deposited in the pallial chamber of the female; being reproduced after each generative act. In *Tremoctopus* the third right arm of the male is "hectocotylised," and is converted into a vermiform body, with two rows of ventral suckers, and an oval appendage or sac behind, which contains spermatozoa. Besides the suckers, the anterior part of the back is fringed with a number of so-called "branchial" filaments.

In the *Argonaut* the male is not more than an inch in length, is devoid of a shell, and has its third left arm hectocotylised. This arm is developed in a cyst, which is ruptured by the movements of the "hectocotylus," which then appears as a small worm-like body, with a filiform appendage in front, with two rows of alternating suckers, and a dorsal sac with numerous "chromatophores." The duct of the testis probably opens into the base of the hectocotylus, which is ultimately detached, and is deposited by the male within the pallial chamber of the female. When first discovered in this position, it was described as a parasitic worm under the name of "Hectocotylus;" subsequently it was described as the entire male, and it is only recently that its true nature has been fully ascertained.

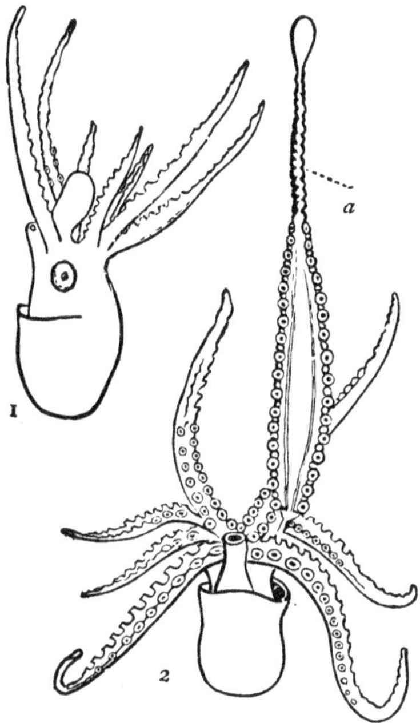

Fig. 116.—1. *Octopus carena* (male), showing cyst in place of the third arm. 2. Ventral side of an individual, more developed, with the hectocotylus (*a*). (After Woodward.)

The *shell* of the *Cephalopoda* is sometimes external, sometimes internal. The internal skeleton is known as the "cuttle-bone," "sepiostaire," or "pen" (*gladius*), and may be either corneous or calcareous. In some cases it is rendered complex by the addition of a chambered portion or "phragmacone," which is to be regarded as a visceral skeleton or "splanchno-skeleton." In *Spirula* the phragmacone is the sole internal skeleton, and is coiled into a spiral, the coils of which lie in one plane, and are near one another, but not in contact. It thus resembles the shell of the Pearly Nautilus, but it is *internal*, and differs, therefore, entirely from the *external* shell of the latter. The only living Cephalopods which are provided with an external shell are the Paper Nautilus (*Argonauta*), and the Pearly Nautilus (*Nautilus pompilius*); but not only is the structure of the animal different in each of these, but the nature of the shell itself is entirely different. The shell of the Argonaut (fig. 117) is involuted, but is not divided into chambers, and it is secreted by the webbed extremities of two of the dorsal arms of the female. The arms are bent backwards, so as to allow the animal to live in the shell, but there is in reality no organic connection between the shell and the body of the animal. In fact, the shell of the Argonaut, being confined to the female, and serving by its empty apex as a receptacle for the ova, may be looked upon as a "nidamental shell," or as it is secreted by a modified portion of the foot, it may more properly be regarded as a "pedal shell." The shell of the Pearly Nautilus (fig. 119), on the other hand, is a true pallial shell, and is secreted by the body of the animal, to which it is organically connected. It is involuted, but it differs from the shell of the Argonaut in being divided into a series of chambers by shelly partitions or septa, which are pierced by a tube or "siphuncle," the animal itself living in the last chamber only of the shell.

The *Cephalopoda* are divided into two extremely distinct and well-marked orders, termed the *Dibranchiata* and the *Tetrabranchiata*. The former is characterised by the possession of two branchiæ only, and comprises the Cuttle-fishes, Squids, and the Paper Nautilus. The latter is distinguished by the presence of four gills, and, though abundantly represented in past time, has no other living representative than the Pearly Nautilus alone.

CHAPTER LI.

DIVISIONS OF THE CEPHALOPODA.

ORDER I. DIBRANCHIATA.—The members of this order of the *Cephalopoda* are characterised as being swimming animals, almost invariably naked, with never more than eight or ten arms, which are always provided with suckers. There are two branchiæ, which are furnished with branchial hearts; an ink-sac is always present; the funnel is a complete tube, and the shell is internal, or, if external, is not chambered.

The Cuttle-fishes are rapacious and active animals, swimming freely by means of the jet of water expelled from the funnel. The arms constitute powerful offensive weapons, being excessively tenacious in their hold, and being sometimes provided with a sharp claw in the centre of each sucker. They are mostly nocturnal or crepuscular animals, and they sometimes attain to a great size. They may be divided into two sections, *Octopoda* and *Decapoda*, according as they have simply eight arms, or eight arms and two additional "tentacles."

SECTION A. OCTOPODA.—The Cephalopods comprised in this section are distinguished by the possession of not more than eight arms, which are provided with sessile suckers. The shell is internal and rudimentary; in one instance only (the Argonaut) external. The body is short and bursiform, and ordinarily without fins.

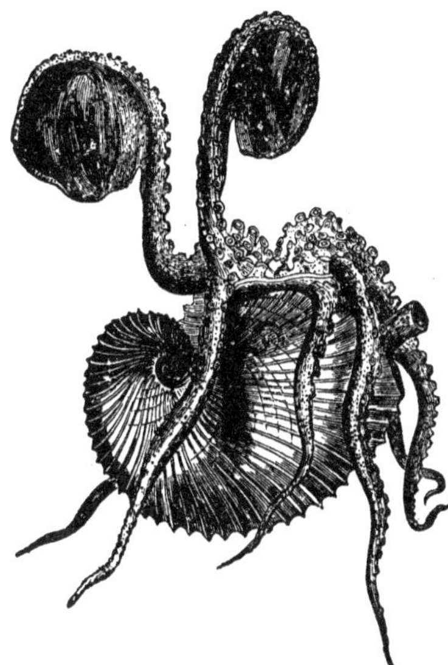

Fig. 117.—*Argonauta argo*, the "Paper Nautilus," female. The animal is represented in its shell, but the webbed dorsal arms are separated from the shell, which they ordinarily embrace.

This section comprises the two families of the *Argonautidæ*, and the *Octopodidæ*. In the former of these there is only the single genus *Argonauta* (the Paper Sailor, or the Paper Nautilus), of which the female and male differ greatly from one another. The female Argonaut (fig. 117) is protected by a thin *single-chambered* shell, in form symmetrical and involuted, which is secreted by the webbed extremities of the dorsal arms, but is not attached in any way to the body of the animal. It sits in its shell with the funnel turned toward the keel, and the webbed arms applied to the shell. The male Argonaut is much smaller than the female (about an inch in length), and is not protected by any shell. The third left arm is developed in a cyst, and ultimately becomes a "hectocotylus," and is deposited by the male in the pallial chamber of the female.

In the *Octopodidæ* (or Poulpes) there are eight arms, all similar to one another, and united at the base by a web. There is an internal rudimentary shell, represented by two short styles encysted in the substance of the mantle.—(Owen.) The body is seldom provided with lateral fins. The third right arm of the male is primarily developed in a cyst, and ultimately becomes "hectocotylised."

SECTION B. DECAPODA.—The Cephalopods of this section have eight arms and two additional "tentacles," which are much longer than the true arms, are retractile, and have expanded club-shaped extremities (fig. 115). The suckers are pedunculated; the body is always provided with lateral fins, and the shell is always internal.

This section comprises the three living families of the *Teuthidæ*, *Sepiadæ*, and the *Spirulidæ*, and the extinct family of the *Belemnitidæ*.

The family of the *Teuthidæ* comprises the Calamaries or Squids, characterised by the possession of an elongated body with lateral fins. The shell is internal and horny, consisting of a median shaft and of two lateral wings; it is termed the "gladius" or "pen," and in old specimens several may be found lodged in the mantle, one behind the other. In the common Calamary (*Loligo*) the fourth left arm of the male is metamorphosed towards its extremity to subserve reproduction.

In the family of the *Sepiadæ* the internal shell is calcareous ("cuttlebone" or "sepiostaire"), and is in the form of a broad plate, having an imperfectly chambered apex. The broad laminated plate is extremely light and spongy, and the chambered apex is called the "mucro." In the living members of the family the body is provided with long lateral fins, sometimes as long and as wide as the body itself.

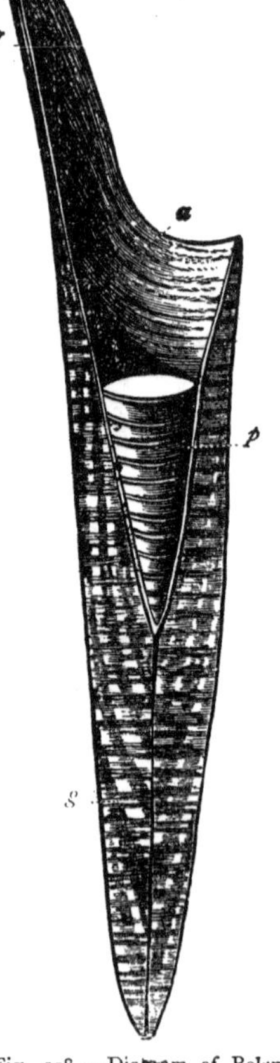

Fig. 118.—Diagram of Belemnite (after Professor Phillips). *r* Horny pen or "pro-ostracum;" *p* Chambered "phragmacone" in its cavity (*a*) or "alveolus;" *g* "Guard."

In the singular family of the *Spirulidæ* the internal skeleton is in the form of a nacreous, discoidal shell, the whorls of which are not in contact with one another, and which is divided into a series of chambers by means of partitions or septa which are pierced by a ventral tube or "siphuncle." The body is provided with minute terminal fins, and the arms have six rows of small suckers. The shell of the *Spirula*—commonly known as the "post-horn"—is similar in structure to the shell of the *Nautilus*, but it is lodged in the posterior part of the body of the animal, and is therefore *internal*, whereas the shell of the latter is *external*. It really corresponds to the "phragmacone" of the Belemnite. Though the shell occurs in enormous numbers in certain localities, a single perfect specimen of the animal is all that has been hitherto obtained.

In the extinct family of the *Belemnitidæ*, our knowledge is chiefly confined to the hard parts. Certain specimens, however, have been discovered which show that the *Belemnite* had essentially the structure of a Cuttle-fish, such as the recent *Sepia*. The body was provided with lateral fins; the arms were eight, furnished with horny hooks, with two "tentacles;" and probably the mouth was provided with horny mandibles. An ink-bag was present. The internal skeleton of a *Belemnite* (fig. 118) consists of a chambered cone—the "phragmacone"—the septa of which are pierced with a marginal tube or "siphuncle." In the last chamber of the phragmacone is contained the ink-bag, often in a well-preserved condition. Anteriorly the phragmacone is continued into a horny lamina or "pen" (the "pro-ostracum" of Huxley), and posteriorly it is lodged in a conical sheath or "alveolus," which is excavated in the substance of a nearly cylindrical, fibrous body, the "guard" (fig. 118, *g*), which projects backwards for a longer or shorter distance, and is the part most usually found in a fossil condition.

ORDER II. TETRABRANCHIATA.—The members of this order of the *Cephalopoda* are characterised by being creeping animals, protected by an external, many-chambered shell, the septa between the chambers of which are perforated by a membranous or calcareous tube, termed the "siphuncle." The arms are numerous, and are devoid of suckers; the branchiæ are four in number, two on each side of the body; the funnel does not form a complete tube; and there is no ink-bag.

Though abundantly represented by many and varied extinct forms, the only living member of the *Tetrabranchiata* is the Pearly Nautilus, which has been long known by its beautiful chambered shell, but the soft parts of which can hardly be said to be known by more than one perfect specimen, which was examined by Professor Owen.

The soft structures in the Pearly Nautilus may be divided into a posterior, soft, membranous mass (*metasoma*), containing the viscera, and an anterior muscular division, comprising the head (*prosoma*); the whole being contained in the outermost, capacious chamber (the body-chamber) of the shell, from which the head can be protruded at will. The shell itself (fig. 119) is involuted and many-chambered, the animal being contained successively in each chamber, and retiring from it as its size becomes sufficiently great to necessitate the acquisition of more room. Each chamber, as the animal retires from it, is walled off by a curved, nacreous septum; the communication between the chambers being still kept up by a membranous

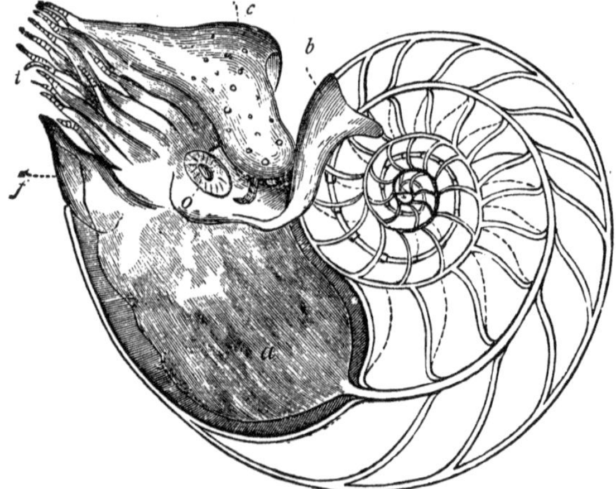

Fig. 119.—Pearly Nautilus (*Nautilus pompilius*). *a* Mantle; *b* Its dorsal fold; *c* Hood; *o* Eye; *t*. Tentacles; *f* Funnel.

tube or siphuncle, which opens at one extremity into the pericardium, and is continued through the entire length of the shell. The position of the siphuncle is in the centre of each septum.

Posteriorly the mantle of the Nautilus is very thin, but it is much thicker in front, and forms a thick fold or collar surrounding the head and its appendages. From the sides of the head spring a great number of muscular prehensile processes or "arms," which are annulated, but are not provided with cups or suckers. In the centre of the head is the mouth, surrounded by a circular fleshy lip, external to which is a series of labial processes. The mouth opens into a buccal cavity, armed with two horny mandibles, partially calcified towards their extremities, and shaped like the beak of a parrot, except that the under mandible is the longest. There is also a "tongue," which is fleshy and sentient in front, but is armed with recurved teeth behind. The gullet opens into a large crop, which in turn conducts to a gizzard, and the intestine terminates at the base of the funnel. On each side of the crop is a well-developed liver.

The heart is contained in a large cavity, divided into several chambers, and termed the "pericardium."—(Owen.) The respiratory organs are in the form of four pyramidal branchiæ, two on each side.

The chief masses of the nervous system are the cerebral and infra-œsophageal ganglia, which are partially protected by a cartilaginous plate, which is to be regarded as a rudimentary cranium, and which sends out processes for the attachment of muscles. The organs of sense are two large eyes, attached by short stalks to the sides of the head, and two hollow plicated subocular processes, believed to be olfactory in their function.

The reproductive organs of the female consist of an ovary, oviduct, and accessory nidamental gland.

There is no ink-bag, and the funnel does not form a complete tube, but consists of two muscular lobes, which are simply in apposition. It is the organ by which swimming is effected, the animal being propelled through the water by means of the reaction produced by the successive jets emitted from the funnel. The function of the chambers of the shell appears to be that of reducing the specific gravity of the animal to near that of the surrounding water, since they are most probably filled with some gas secreted by the animal. The function of the siphuncle is unknown, except in so far as it doubtless serves to maintain the vitality of the shell.

SHELL OF THE TETRABRANCHIATA.—The shells of all the *Tetrabranchiata* agree in the following points:—

1. The shell is external.

2. The shell is divided into a series of chambers by plates or "septa," the edges of which, where they appear on the shell, are termed the "sutures."

3. The outermost chamber of the shell is the largest, and is the one inhabited by the animal.

4. The various chambers of the shell are united by a tube, termed the "siphuncle."

Agreeing in all these fundamental points of structure, two very distinct types of shell may be distinguished as characteristic of the two families *Nautilidæ* and *Ammonitidæ*, into which the order *Tetrabranchiata* is divided.

In the family *Nautilidæ* (fig. 120), the "septa" of the shell are simple, curved, or slightly lobed; the "sutures" are more or less completely plain; and the "siphuncle" is central, sub-central, or internal (*i.e.*, on the *concave* side of the curved shells).

In the family *Ammonitidæ* (fig. 120), on the other hand, the septa are folded and complex; the sutures are angulated, zig-zag, lobed, or foliaceous; and the siphuncle is external (*i.e.*, on the *convex* side of the curved shells).

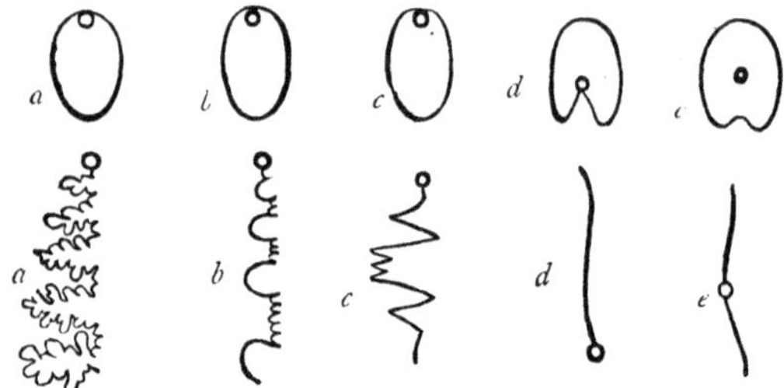

Fig. 120.—Diagram to illustrate the position of the siphuncle and the form of the septa in various Tetrabranchiate Cephalopoda. The upper row of figures represents transverse sections of the shells, the lower row represents the edges of the septa. *a a Ammonite* or *Baculite; b b Ceratite; c c Goniatite; d d Clymenia; e e Nautilus* or *Orthoceras.*

In both these great *types* of shell, a series of representative *forms* exists, resembling each other in the manner in which the shell is folded or coiled, but differing in their fundamental structure. All these different forms may be looked upon as produced by the modification of a greatly elongated cone, the structure of which may be in conformity with the type either of the *Nautilidæ* or of the *Ammonitidæ*. The following table (after Woodward) exhibits the representative forms in the two families:—

	Nautilidæ.	*Ammonitidæ.*
Shell straight	Orthoceras . .	Baculites.
„ bent on itself . .	Ascoceras . .	Ptychoceras.
„ curved	Cyrtoceras . .	Toxoceras.
„ spiral	Trochoceras .	Turrilites.
„ discoidal	Gyroceras . .	Crioceras.
„ discoidal and produced	Lituites . . .	Ancyloceras.
„ involute	Nautilus . .	Ammonites.

After the *Nautilus* itself, the most important form of the *Nautilidæ* is the *Orthoceras* (fig. 121). In structure this was doubtless essentially identical with the Nautilus, but the shell, instead of being coiled into a spiral lying in one plane, was extended in a straight, or nearly straight, line. Orthoceratites of more than six feet in length have been discovered, but in all, the body-chamber, in which the animal was lodged, appears to have been comparatively small. The siphuncle is usually very complex in structure, and was calcareous throughout its entire length.

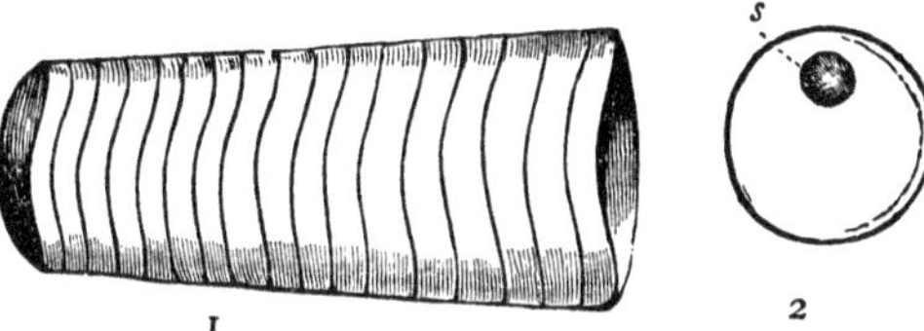

Fig. 121.—*Orthoceras explorator*, Billings. 1. Side view of a fragment, showing the septa. 2. Transverse section of the same, showing (*s*) the siphuncle.

The structure of the shell in the *Ammonitidæ* is exactly that of the Pearly Nautilus, consisting of an outer porcellanous and an inner nacreous layer. The body-chamber was rather elongated than laterally expanded or dilated. The simplest form of the *Ammonitidæ* is the *Baculite*, in which the shell is straight, like that of an *Orthoceras*, whilst the septa have the characters of those of an *Ammonite*, and the siphuncle is external. In the *Turrilite* the structure of the shell is the same, but it is coiled into a turreted spiral. In the *Ammonite* itself, the shell is discoidal and involuted, corresponding (in *form*) to the shell of the *Nautilus;* the body-chamber was of comparatively large size, and had its aperture closed, in some species at any rate, by an operculum. The shell sometimes attained a gigantic size, and several hundred species of the genus have been described. In *Crioceras* the shell was a flat spiral, like that of the Ammonites, but the whorls are not in contact. In *Toxoceras* the shell is shaped like a bow. In *Ancyloceras* the shell is at first discoidal, with separate whorls, then produced into a straight line, and finally bent forwards into a hook.

Synopsis of the Families of the Cephalopoda.

Class Cephalopoda.

Order I. Dibranchiata.

Animal with two branchiæ; not more than eight or ten arms, provided with suckers; an ink-bag; shell commonly internal and rudimentary; rarely external, but not chambered.

Section A. Octopoda.

Arms eight, suckers sessile.

Fam. 1. *Argonautidæ.*
Female provided with a calcareous, external, monothalamous shell, secreted by the webbed extremities of the dorsal arms. Gen. *Argonauta.*

Fam. 2. *Octopodidæ.*
Shell internal, rudimentary, uncalcified. No pallial fins in most. Ill. Gen. *Octopus, Tremoctopus, Eledone, Pinnoctopus.*

SECTION B. DECAPODA.
Arms eight, with two clavate "tentacles;" suckers pedunculated.

Fam. 3. *Teuthidæ.*
Shell an internal horny "pen" or "gladius." Fins mostly terminal. Ill. Gen. *Loligo, Onychoteuthis, Ommastrephes.*

Fam. 4. *Belemnitidæ.*
Shell internal, composed of a conical chambered portion ("phragmacone") with a marginal siphuncle, sometimes produced into a horny plate or "pen," and lodged in a cylindrical fibrous "guard." Ill. Gen. *Belemnites, Belemnitella, Belemniteuthis.*

Fam. 5. *Sepiadæ.*
Shell calcareous, consisting of a broad, laminar plate, terminating posteriorly in an imperfectly chambered apex ("phragmacone"). Ill. Gen. *Sepia, Beloptera, Spirulirostra.*

Fam. 6. *Spirulidæ.*
Shell internal, nacreous, chambered, discoidal; the whorls separate; a ventral siphuncle. Gen. *Spirula.*

ORDER II. TETRABRANCHIATA.
Animal with four gills; arms more than ten, without suckers; no ink-bag; shell external, chambered, and siphuncled.

Fam. 1. *Nautilidæ.*
Sutures of the shell simple; the siphuncle central, sub-central, or near the concavity of the curved shells, simple.

Sub-family Nautilidæ proper.
Body-chamber capacious; aperture simple; siphuncle central or internal. Ill. Gen. *Nautilus, Lituites, Trochoceras.*

Sub-family Orthoceratidæ.
Shell straight, curved, or discoidal; body-chamber small; aperture contracted; siphuncle complicated. Ill. Gen. *Orthoceras, Phragmoceras, Cyrtoceras.*

Fam. 2. *Ammonitidæ.*
Shell discoidal, curved, spiral, or straight; body-chamber elongated; aperture guarded by processes, or closed by an operculum; sutures angulated, lobed, or foliaceous; siphuncle external or dorsal (on the convex side of the curved shells). Ill. Gen. *Ammonites, Ceratites, Baculites, Turrilites, Scaphites, Ancyloceras.*

CHAPTER LII.

DISTRIBUTION OF THE MOLLUSCA PROPER IN TIME.

REMAINS of the *Mollusca proper* are found in greater or less abundance in almost all the stratified rocks from the commencement of the Silurian period up to the present day. Speaking generally, the Tetrabranchiate *Cephalopoda* are the chief representatives of the *Mollusca* in the Palæozoic rocks, the *Lamellibranchiata* and the Dibranchiate *Cephalopoda* in the Mesozoic rocks, and the *Gasteropoda* in the Kainozoic period; but all the primary classes are represented even in the Lower Silurian rocks. The folllowing are the more noticeable facts relating to the distribution of the various classes in past time.

Lamellibranchiata.—The Lamellibranchs are known to have existed in the Lower Silurian period, and have steadily increased up to the present day, when the class appears to have attained its maximum, both as regards numbers and as regards variety of type. The recent bivalves are also superior in organisation to those which have preceded them. Upon the whole the Asiphonate bivalves are more characteristically Palæozoic, whilst those in which the mantle-lobes are united, and there are respiratory siphons, are chiefly found in the Secondary and Tertiary epochs. One very singular and aberrant family—viz., the *Hippuritidæ*—is exclusively confined to the Secondary rocks, and is, indeed, not known to occur beyond the limits of the Cretaceous formation. The *Veneridæ*, which are perhaps the most highly organised of the families of the *Lamellibranchiata*, appear for the first time in the Oolitic rocks, and, increasing in the Tertiary period, have culminated in the Recent period.

Gasteropoda.—The *Gasteropoda* are represented in past time from the Lower Silurian rocks up to the present day. Of the *Branchifera* the *Holostomata* are more abundant in the Palæozoic period, the *Siphonostomata* abounding more in the Secondary and Tertiary rocks, but not attaining their maximum till the present day. The place of the carnivorous *Siphonostomata* in the Palæozoic seas appears to have been filled by the Tetrabranchiate Cephalopods. The branchiate Gasteropods of fresh water are chiefly represented as fossils by the genera *Paludina*, *Valvata*, and *Ampullaria*.

The *Heteropoda* are likewise of very ancient origin, having commenced their existence in the lowest Silurian deposits.

The genera *Bellerophon*, *Porcellia*, *Cyrtolites*, and *Maclurea*, are almost exclusively Palæozoic; *Bellerophina* is found in the Gault (Secondary), and *Carinaria* has been detected in the Tertiaries.

The Pulmonate *Gasteropoda*, as was to be anticipated, are not found abundantly as fossils, occurring chiefly in lacustrine and estuarine deposits, in which the genera *Limnæa*, *Physa*, *Ancylus*, &c., are amongst those most commonly represented. These, however, are entirely Mesozoic and Kainozoic. In the Palæozoic period the sole known representatives of the *Pulmonifera* are the *Pupa vetusta* and *Zonites priscus* of the Carboniferous rocks.

Pteropoda.—The Pteropods are not largely represented in fossiliferous deposits, but they have a wide range in time, extending from the Lower Silurian rocks up to the present day. The *Theca* and *Conularia* of the Palæozoic period, if truly Pteropods, are of comparatively gigantic size, and extend from the Lower Silurian to the Carboniferous period. The Silurian fossil, *Tentaculites*, is asserted by M. Barrande to be a Pteropod, but it is usually looked upon as a tubicolous Annelide. The recent genus *Hyalea* is represented in the Tertiary period (Miocene).

Cephalopoda.—The Cephalopods are largely represented in all the primary groups of stratified rocks from the Lower Silurian up to the present day. Of the two orders of *Cephalopoda*, the *Tetrabranchiata* is the oldest, attaining its maximum in the Palæozoic period, decreasing in the Mesozoic and Kainozoic epochs, and being represented at the present day by the single form *Nautilus pompilius*. Of the sections of this order, the *Nautilidæ* proper and the *Orthoceratidæ* are pre-eminently Palæozoic, and the *Ammonitidæ* are not only pre-eminently but are almost exclusively Secondary. Of the abundance of the two former families in the Silurian seas some idea may be obtained when it is mentioned that over a thousand species have been described by M. Barrande from the Silurian basin of Bohemia alone. The *Nautilidæ* proper have gradually decreased in numbers from the Palæozoic through the Secondary and Tertiary periods to the present day. The *Orthoceratidæ* died out much sooner, being exclusively Palæozoic, with the exception of the genera *Orthoceras* itself and *Cyrtoceras*, which survived into the commencement of the Secondary period, finally dying out in the Trias.

The second family of the *Tetrabranchiata*—viz., the *Ammonitidæ*—is almost exclusively Secondary, being very largely represented by numerous species of the genera *Ammonites*,

Ceratites, *Baculites*, *Turrilites*, &c. The only Palæozoic genera are *Goniatites* and *Bactrites*, of which the former is found from the Upper Silurian to the Trias, whilst the latter is a Devo-

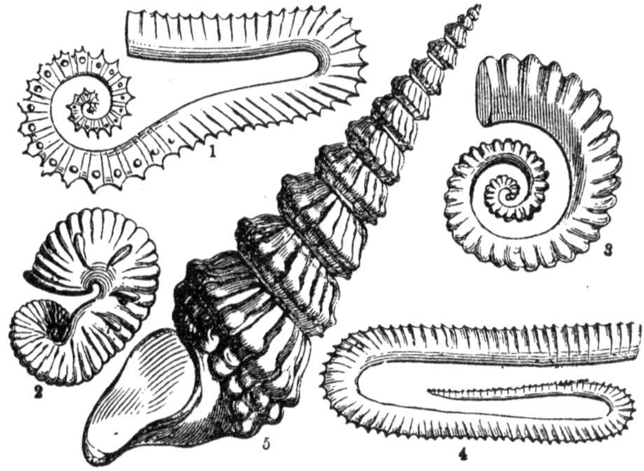

Fig. 122.—Shells of Secondary Cephalopods. 1 *Ancyloceras Matheronianus*; 2 *Scaphites æqualis*; 3 *Crioceras Duvalii*; 4 *Hamites attenuatus*; 5 *Turrilites catenatus*.

nian form. The genus *Ceratites* is characteristically Triassic, but it is said to occur in the Devonian rocks. All the remaining genera are exclusively Secondary, the genera *Baculites*, *Turrilites*, *Hamites*, and *Ptychoceras* being confined to the Cretaceous period.

Of the Dibranchiate Cephalopoda the record is less perfect, as they have few structures which are capable of preservation. They attain their maximum, as fossils, shortly after their first appearance in the Secondary rocks, where they are represented by the large and important family of the *Belemnitidæ*. Some of the *Teuthidæ* and *Sepiadæ* are found both in the Secondary and in the Tertiary rocks, and two species of Argonaut have been discovered in the later Tertiaries. No example of a Dibranchiate Cephalopod is known from the Palæozoic deposits, and the order attains its maximum at the present day.

TABULAR VIEW OF THE CHIEF SUBDIVISIONS OF THE INVERTEBRATA.

SUB-KINGDOM I.—PROTOZOA.

CLASS I. GREGARINIDÆ.

CLASS II. RHIZOPODA.

Order 1. Monera.
2. Amœbea.
3. Foraminifera.
4. Radiolaria.
5. Spongida.

CLASS III. INFUSORIA.

Order 1. Suctoria.
2. Ciliata.
3. Flagellata.

SUB-KINGDOM II.—CŒLENTERATA.

CLASS I. HYDROZOA.

Sub-class A. *Hydroida.*

Order 1. Hydrida.
2. Corynida.
3. Sertularida.
4. Campanularida.

Sub-class B. *Siphonophora.*

Order 5. Calycophoridæ.
6. Physophoridæ.

Sub-class C. *Discophora.*

Order 7. Medusidæ.

Sub-class D. *Lucernarida.*

Order 8. Lucernariadæ.
9. Pelagidæ.
10. Rhizostomidæ.

Sub-class E. *Graptolitidæ.*

CLASS II. ACTINOZOA.

Order 1. Zoantharia.

Sub-order *a.* Z. Malacodermata.
b. Z. Sclerobasica.
c. Z. Sclerodermata.

Order 2. Alcyonaria.

Fam. *a.* Alcyonidæ.
b. Tubiporidæ.
c. Pennatulidæ.
d. Gorgonidæ.

Order 3. Rugosa.

Order 4. Ctenophora.

Sub-order *a.* Stenostomata.
b. Eurystomata.

SUB-KINGDOM III.—ANNULOIDA.

CLASS I. ECHINODERMATA.
Order 1. Cystoidea.
2. Blastoidea.
3. Crinoidea.
4. Echinoidea.
5. Asteroidea.
6. Ophiuroidea.
7. Holothuroidea.

CLASS II. SCOLECIDA.
Division A. *Platyelmia.*
Order 1. Tæniada.
2. Trematoda.
3. Turbellaria.
Sub-order *a.* Planarida.
b. Nemertida.
Division B. *Nematelmia.*
Order 4. Acanthocephala.
5. Gordiacea.
6. Nematoda.
Division C. *Rotifera.*
Order. Rotifera.

SUB-KINGDOM IV.—ANNULOSA.

DIVISION A. ANARTHROPODA.
CLASS I. GEPHYREA.
CLASS II. ANNELIDA.
Order 1. Hirudinea } Abranchiata.
2. Oligochæta }
3. Tubicola } Branchiata.
4. Errantia }
CLASS III. CHÆTOGNATHA.

DIVISION B. ARTHROPODA or ARTICULATA.
CLASS I. CRUSTACEA.
Sub-class I. *Epizoa.*
Order 1. Ichthyophthira.
2. Rhizocephala.
Sub-class II. *Cirripedia.*
Order 3. Thoracica. { Balanidæ. Verrucidæ. Lepadidæ.
4. Abdominalia.
5. Apoda.

Sub-class III. *Entomostraca.*

Order 6. Ostracoda } Legion Lophyropoda.
7. Copepoda }
8. Cladocera }
9. Phyllopoda } Legion Branchiopoda.
10. Trilobita }
11. Merostomata.
Sub-order *a.* Xiphosura.
b. Eurypterida.

Sub-class IV. *Malacostraca.*

Order 12. Læmodipoda } Division A. Edriophthalmata.
13. Isopoda }
14. Amphipoda }
15. Stomapoda } Division B. Podophthalmata.
16. Decapoda }
Tribe *a.* Macrura.
b. Anomura.
c. Brachyura.

CLASS II. ARACHNIDA.

Division A. *Trachearia.*
Order 1. Podosomata.
2. Acarina.
3. Adelarthrosomata.

Division B. *Pulmonaria.*
Order 4. Pedipalpi.
5. Araneida.

CLASS III. MYRIAPODA.

Order 1. Chilopoda.
2. Chilognatha.
3. Pauropoda.

CLASS IV. INSECTA.

Sub-class I. *Ametabola.*
Order 1. Anoplura.
2. Mallophaga.
3. Thysanura.

Sub-class II. *Hemimetabola.*
Order 4. Hemiptera.
5. Orthoptera.
6. Neuroptera.

Sub-class III. *Holometabola.*
Order 7. Aphaniptera.
8. Diptera.

9. Lepidoptera.
10. Hymenoptera.
11. Strepsiptera.
12. Coleoptera.

SUB-KINGDOM V.—MOLLUSCA.

DIVISION A. MOLLUSCOIDA.

Class I. *Polyzoa.*
Order 1. Phylactolæmata.
2. Gymnolæmata.

Class II. *Tunicata.*
Order 1. Ascidia branchialia.
2. Ascidia abdominalia.
3. Ascidia larvalia.

Class III. *Brachiopoda.*
Order 1. Articulata.
2. Inarticulata.

DIVISION B. MOLLUSCA PROPER.

Class IV. *Lamellibranchiata.*
Section *a.* Asiphonida.
b. Siphonida.

Class V. *Gasteropoda.*
Sub-class I. *Branchifera.*
Order 1. Prosobranchiata.
Section *a.* Siphonostomata.
b. Holostomata.
Order 2. Opisthobranchiata.
Section *a.* Tectibranchiata.
b. Nudibranchiata.
Order 3. Nucleobranchiata (Heteropoda)
Fam. *a.* Firolidæ.
b. Atlantidæ.
Sub-class II. *Pulmonifera.*
Section *a.* Inoperculata.
b. Operculata.

Class VI. *Pteropoda.*
Order 1. Thecosomata.
2. Gymnosomata.

Class VII. *Cephalopoda.*
Order 1. Tetrabranchiata.
2. Dibranchiata.

PART II.

VERTEBRATE ANIMALS

VERTEBRATE ANIMALS.

CHAPTER LIII.

General Characters and Divisions of the Vertebrata.

The five sub-kingdoms which we have previously considered—viz., the *Protozoa*, *Cœlenterata*, *Annuloida*, *Annulosa*, and *Mollusca*—were grouped together by the French naturalist Lamarck to form one great division, which he termed *Invertebrata*, the remaining members of the animal kingdom constituting the division *Vertebrata*. The division *Vertebrata*, though including only a single sub-kingdom, is so compact and well-marked a division, and its distinctive characters are so numerous and so important, that this mode of looking at the animal kingdom is, at any rate, a very convenient one.

The sub-kingdom *Vertebrata* may be shortly defined as comprising *animals in which the body is composed of a number of definite segments, arranged along a longitudinal axis; the nervous system is in its main masses dorsal, and the neural and hæmal regions of the body are always completely shut off from one another by a partition; the limbs are never more than four in number, and are always turned away from the neural aspect of the body; mostly there is the bony axis known as the "spine" or "vertebral column," and in all the structure known as the "notochord" is present—in the embryo, at any rate.* These characters distinguish the *Vertebrata*, as a whole, from the *Invertebrata;* but it is necessary to define these broad differences more minutely, and to consider others which are of little less importance.

One of the most obvious, as it is one of the most fundamental, of the distinctive characters of *Vertebrates*, is to be found in the shutting off of the main masses of the nervous system from the general cavity of the body. In all *Invertebrate* animals, without exception, the body (fig. 123, A) may be regarded as a *single* tube, enclosing all the viscera; and consequently, in this case, the nervous system is contained within the general cavity of the body, and is not in any way shut off from the alimentary canal. The transverse section, however, of a *Vertebrate* animal exhibits *two* tubes (fig. 123, B), one of which contains the great masses of the nervous system —that is, the cerebro-spinal axis, or brain and spinal cord —whilst the other contains the alimentary canal and the chief circulatory organs, together with certain portions of the nervous system, known as the "ganglionic" or "sympathetic" system. Leaving the cerebro-spinal centres out of sight for a moment, we see that the larger or visceral tube of a Vertebrate animal contains the digestive canal, the hæmal system, and a

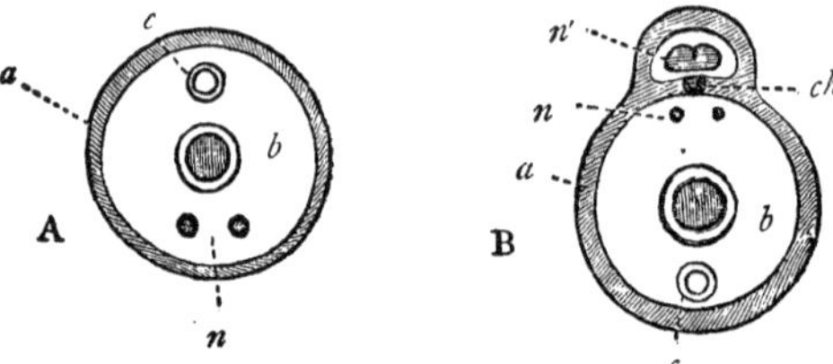

Fig. 123.—A, Transverse section of the body of one of the higher *Invertebrata*; *a* Body-wall; *b* Alimentary canal; *c* Hæmal system; *n* Nervous system. B, Transverse section of the body of a *Vertebrate* animal: *a* Body-wall; *b* Alimentary canal; *c* Hæmal system; *n* Sympathetic system of nerves; *n'* Cerebro-spinal system of nerves; *ch* Notochord.

gangliated nervous system. Now this is exactly what is contained in the visceral cavity of any of the higher Invertebrate animals; and it follows from this, as pointed out by Von Baer, that it is the sympathetic nervous system of Vertebrates which is truly comparable to, and homologous with, the nervous system of Invertebrates. The cerebro-spinal nervous centres of the *Vertebrata* are to be regarded as something superadded, and not represented at all amongst the *Invertebrata*.

The tube containing the cerebro-spinal centres is formed as follows:—At an early period in the development of the embryo of any Vertebrate animal, the portion of the ovum in which development is going on—the "germinal area"—becomes elevated into two parallel ridges, one on each side of the middle line, enclosing between them a long groove, which is known as the "primitive groove" (fig. 124, A, B). The ridges which bound the primitive groove are known as the "laminæ dorsales;" and they become more and more raised up, till they ultimately meet in the middle line, and unite to form a tube, within which the cerebro-spinal nervous centres are developed. It follows from its mode of formation that the inner wall of the tube formed by the primitive groove, which remains as the septum between the cerebro-spinal canal and the body-cavity, is nothing more than a portion of the primitive wall of the body of the embryo. And there appears to be little doubt, as believed by Remak and Huxley, that the cerebro-spinal nervous centres are "the result of a modification of that serous layer of the germ, which is continuous elsewhere with the epidermis" (Huxley).

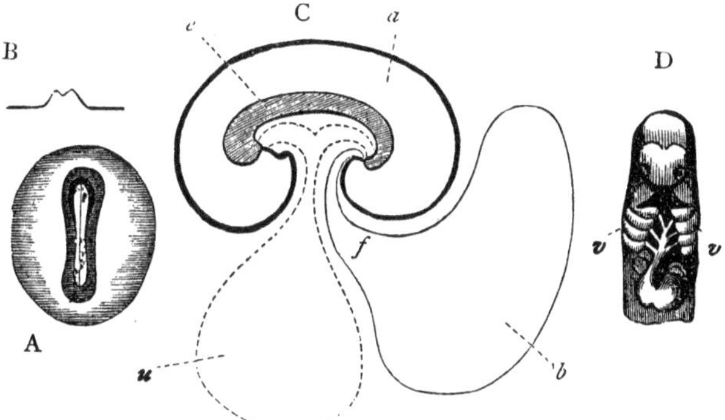

Fig. 124.—Embryology of Vertebrata. A, Portion of the germinal area of the ovum of a Bitch, showing the primitive groove (after Bischoff). B, Profile view of the same. C, Diagram representing the amnion and allantois: *e* Embryo; *a* Amnion; *u* Umbilical vesicle; *b* Allantois; *f* Pedicle of the allantois, afterwards the urinary bladder. D, Head of an embryo, showing the visceral arches (*v v*).

Another remarkable peculiarity as regards the nervous system is found in the fact that in no Vertebrate animal does the alimentary canal pierce the main masses of the nervous system, but turns away to open on the opposite side of the body. In most Invertebrates, on the other hand, in which there is a well-developed nervous system, this is perforated by the gullet, so that an œsophageal nerve-collar is formed, and some of the nervous centres become præ-œsophageal, whilst others are post-œsophageal.

Furthermore, the floor of the "primitive groove" in the embryo of all Vertebrates has developed in it at an early period the structure known as the "notochord" or "chorda dorsalis" (fig. 123, B, *ch*). This structure, doubtfully present in any Invertebrate, is a semi-gelatinous or cartilaginous collection of cells, forming a rod-like axis, which tapers at both ends, and extends along the floor of the cerebro-spinal canal, supporting the cerebro-spinal nervous centres. In some Vertebrates, such as the Lancelet (*Amphioxus*), the notochord is persistent throughout life. In the majority of cases, however, the notochord is replaced before maturity by the structure known as the "vertebral column" or "backbone," from which the sub-kingdom *Vertebrata* originally derived its name. This is not the place for an anatomical description of the spinal column, and it is sufficient to state here that it is essentially composed of a series of cartilaginous, or more or less completely ossified, segments or *vertebræ*, arranged so as to form a longitudinal axis, which protects the great masses of the nervous system. It is to be remembered, however, that all Vertebrate animals do not possess a vertebral column. They all possess a *notochord;* but this may be persistent, and in many cases the development of the spinal column is extremely imperfect.

Another embryonic structure which is characteristic of all Vertebrates, is found in the so-called "visceral arches" and "clefts" (fig. 124, D). The "visceral arches" are a series of parallel ridges running transversely to the axis of the body, situated at the sides of, and posterior to, the mouth. As development proceeds, the intervals between these ridges become grooved by depressions which gradually deepen, until they become converted into a series of openings or "clefts," whereby a free communication is established between the upper part of the alimentary canal (pharynx) and the external medium.

The limbs of Vertebrate animals are always articulated to the body, and they are always turned away from the neural aspect of the body. They may be altogether wanting, or they may be partially undeveloped; but there are never more than two pairs, and they always have an internal skeleton for the attachment of the muscles of the limb.

A specialised blood-vascular or "hæmal" system is present in all the *Vertebrata*, and in all except one—the *Amphioxus*—there is a contractile cavity or *heart*, which never consists of less than two chambers provided with valvular apertures. In all the *Vertebrata* the heart is essentially a *respiratory* heart—that is to say, it is concerned with driving the impure or venous blood to the breathing-organs; and in its simplest form (fishes) it is nothing more than this. In the higher Vertebrates, however, there is superadded to this a pair of cavities which are concerned in driving the pure or arterial blood to the body. In the case of the Mammals, these two circulations are often spoken of as the "lesser" or "pulmonary" circulation, and the "greater" or "systemic" circulation.

In all Vertebrates there is that peculiar modification of the venous system which is known as the "hepatic portal system." That is to say, a portion of the blood which is sent to the alimentary canal, instead of returning to the heart by the ordinary veins, is carried to the liver by a special vessel—the *vena portæ*—which ramifies through this organ after the manner of an artery.

In all Vertebrates, also, is found the peculiar system of vessels known as the "lacteal system." This is to be regarded as an appendage of the venous system of blood-vessels, and consists of a series of vessels which take up the products of digestion from the alimentary canal, elaborate them, and finally empty their contents into the veins.

Lastly, the masticatory organs of Vertebrates are modified portions of the walls of the head, and never "hard productions of the alimentary mucous membrane or modified limbs" (Huxley), as they are amongst the Invertebrata.

The above are the leading characters of the *Vertebrata* as a whole; but before going on to consider the primary divisions of the sub-kingdom, it may be as well to give a very brief and general description of the anatomy of the higher and more typical Vertebrates, commencing with their bony framework or skeleton.

The *skeleton* of the *Vertebrata* may be regarded as consisting essentially of the bones which go to form the head and trunk on the one hand (sometimes called the "axial" skeleton), and of those which form the supports for the limbs ("appendicular" skeleton) on the other hand. The bones of the head and trunk may be looked upon as essentially composed of a series of bony rings or segments, arranged longitudinally, one behind the other. Anteriorly these segments are much expanded, and likewise much modified, to form the bony case which encloses the brain, and which is termed the *cranium* or skull. Behind the head the segments enclose a much smaller cavity, which is called the "neural" or spinal canal, as it encloses the spinal cord; and they are arranged one behind the other, forming the vertebral column. The segments which form the vertebral column are called "vertebræ," and they have the following general structure:—Each vertebra (fig. 125, A) consists of a central piece, which is the fundamental and essential element of the vertebra, and is known as the "body" or "centrum" (*c*). From the upper or posterior surface of the centrum spring two bony arches (*n n*), which are called the "neural arches" or "neurapophyses" because they form with the body a canal—the "neural canal"—which encloses the spinal cord. From the point where the neural arches meet behind, there is usually developed a longer or shorter spine, which is termed the "spinous process" or "neural spine" (*s*). From the neural arches there are also developed in the typical vertebra two processes (*a a*), which are known as the "articular" processes, or "zygapophyses." The vertebræ are united to one another partly by these, but to a greater extent by the bodies or "centra." From the sides of the vertebral body, at the point of junction with the neural arches, there proceed two lateral processes (*d d*), which are known as the "transverse processes." (In the typical

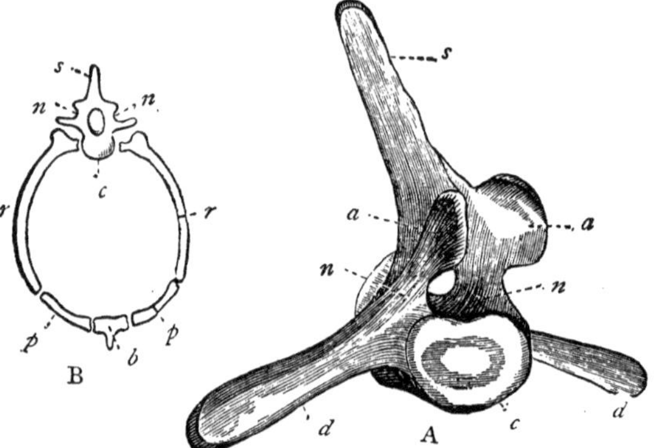

Fig. 125.—A, Lumbar vertebra of a Whale: *c* Body or centrum; *n n* Neural arches; *s* Neural spine; *a a* Articular processes; *d d* Transverse processes. B, Diagram of a thoracic vertebra: *c* Centrum; *n n* Neural arches enclosing the neural canal; *s* Neural spine; *r r* Ribs, assisting in the formation of the hæmal arch; *p p* Costal cartilages; *b* Sternum, with hæmal spine. (After Owen.)

vertebra the transverse processes consist each of two pieces, an anterior piece or "parapophysis," and a posterior piece or "diapophysis." These elements form the *vertebra* of the human anatomist, but the "vertebra" of the transcendental anatomist is completed by a second arch which is placed beneath the body of the vertebra, and which is called the "hæmal" arch, as it includes and protects the main organs of the circulation. This second arch is often only recognisable with great difficulty, as its parts are generally much modified, but a good example may be obtained in the human chest, or in the caudal vertebra of a bony fish.

The hæmal arch in the case of the human thorax (fig. 125, B) is formed by the ribs (*r r*) and the costal cartilages (*p p*).

and is completed in front by the breast-bone or sternum (*b*), which in some cases—but not in man—develops a spine (the hæmal spine) which corresponds to the neural spine on the opposite aspect of the vertebra.

It follows from the above, that the typical vertebra consists of a central piece or body from which two arches are given off, one of which protects the great masses of the nervous system, and is therefore said to be "neural;" whilst the other protects the main organs of the circulation, and is therefore said to be "hæmal." The correspondence of the typical bony

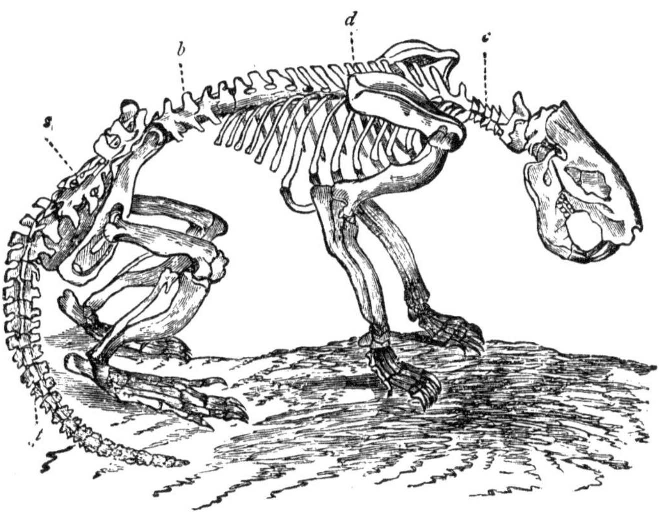

Fig. 126.—Skeleton of the Beaver (*Castor fiber*), showing the different regions of the vertebral column. *c* Cervical region; *d* Dorsal region; *b* Lumbar region; *s* Sacrum; *t* Caudal region.

segment or vertebra with the doubly tubular structure of the body in all Vertebrates is thus too obvious to require to be specially pointed out.

As a general rule, the vertebral column is divisible into a number of distinct regions, of which the following are recognisable in man and in the higher *Vertebrata*:—1. A series of vertebræ which compose the neck, and constitute the "cervical region" of the spine (fig. 126, *c*). 2. A number of vertebræ which usually carry well-developed ribs, and form the "dorsal region" (*d*). 3. A series of vertebræ which form the region of the loins, or "lumbar region" (*b*). 4. A greater or less number of vertebræ which constitute the "sacral region," and are usually amalgamated or "anchylosed" together to form a single bone, the "sacrum" (*s*). 5. The spinal column is completed by a variable number of vertebræ which constitute the "caudal" region, or tail (*t*).

As regards the *skull* of the *Vertebrata*, it has been thought advisable not to enter into any general details here, partly because the subject is one which can only be properly discussed in a work specially devoted to Human or Comparative Anatomy, and partly because there is still much diversity of opinion as to the exact composition of the skull. There is, however, a very general concurrence of opinion that the skull is composed of a number of separate segments, and this is a point which it is important to remember. By Owen, and by many other competent authorities, these cranial segments are looked upon as being nothing more than so many *vertebræ*, the neural canals of which are greatly expanded to enclose the brain, whilst the hæmal arches are very greatly modified to serve different purposes. This view is not accepted by Huxley, but the general fact that the skull is composed of separate segments appears to be universally admitted. The only portion of the bony framework of the head which it is absolutely essential to understand, is the lower jaw or "mandible." The lower jaw is sometimes wanting, but when present, it consists in all *Vertebrata* of two halves or "rami," which are united to one another in front, and articulate separately with the skull behind. In many cases, each half, or "ramus," of the lower jaw consists of several pieces united to one another by sutures; but in the *Mammalia* each ramus consists of no more than a single piece. The two rami are very variously connected with one another, being sometimes only joined by ligaments and muscles, sometimes united by cartilage or by bony suture, and sometimes fused or anchylosed with one another, so as to leave no evidence of their true composition. The mode by which each ramus of the lower jaw articulates with the skull also varies. In the *Mammalia* the lower jaw articulates with a cavity formed on what is known to human anatomists as the temporal bone; but in Birds and Reptiles the lower jaw articulates with the skull, not directly, but by the intervention of a special bone, known as the "quadrate bone" or "*os quadratum*.

As regards the *limbs* of Vertebrates, whilst many differences exist, which will be afterwards noticed, there is a general agreement in the parts of which they are composed. As a rule, each pair of limbs is joined to the trunk by means of a series of bones which also correspond to one another in general structure. The fore-limbs, often called the "pectoral" limbs, are united with the trunk by means of a bony arch, which is called the "pectoral" or "scapular" arch; whilst the hind-limbs are similarly connected with the trunk by means of the "pelvic arch." In giving a general description of the parts which compose the limbs and their supporting arches, it will be best to take the case of a Mammal, and the departures from this type will then be readily recognised.

The pectoral or scapular arch consists usually of three bones, the "scapula" or shoulder-blade, the "coracoid," and the "clavicle" or collar-bone; but in the great majority of the Mammals, the coracoid is anchylosed with the scapula, of which it forms a mere process. The scapula or shoulder-blade (fig. 127, *s*) is usually placed outside the ribs, and it forms, either alone or in conjunction with the other bones of the shoulder-girdle, the cavity with which the upper arm is articulated. The coracoid, though rarely existing as a distinct bone in the Mammals, plays a very important part in other Vertebrates, as we shall see hereafter. The clavicles are often wanting, or rudimentary, and they are the least essential elements of the scapular arch. The fore-limb proper consists, firstly, of a single bone which forms the upper arm, and which is known as the *humerus* (*h*). This articulates above with the shoulder-girdle, and is followed below by the fore-arm, which consists of two bones, called the *radius* and *ulna*. Of these the *radius* is chiefly concerned with carrying the hand. The radius and ulna are followed by the bones of the wrist, which are usually composed of several bones, and constitute what is called the *carpus* (*d*). These support the bones of the root of the hand, which vary in number, but are always more or less cylindrical in shape. They constitute what is called the *metacarpus*. The bones of the metacarpus carry the digits,

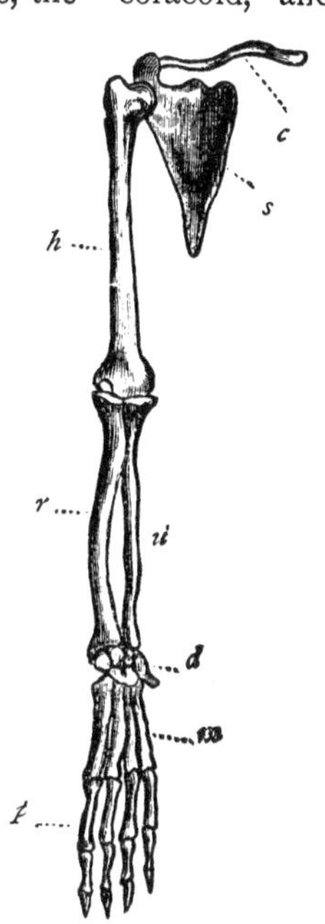

Fig. 127. — Pectoral limb (arm) of Chimpanzee. (After Owen.) *c* Clavicle; *s* Scapula or shoulder-blade; *h* Humerus; *r* Radius; *u* Ulna; *d* Bones of the wrist, or carpus; *m* Metacarpus; *p* Phalanges of the fingers.

which also vary in number, but are composed each of from two to three cylindrical bones, which are known as the *phalanges* (*p*).

Homologous parts are, as a rule, readily recognisable in the hind-limb. The pelvic arch, by which the hind-limb is united with the trunk, consists of three pieces—the *ilium*, *ischium*, and *pubes*—which are usually anchylosed together, and form conjointly what is known as the *innominate bone* (fig. 128, *i*). In most Mammals, the two innominate bones unite in front by a ligamentous or cartilaginous union, and they constitute, with the sacrum, what is known as the *pelvis*. The hind-limb proper consists of the following parts:—1. The thigh-bone or *femur*, corresponding with the humerus in the fore-limb. 2. The bones of the shank, corresponding with the radius and ulna of the fore-limb, and known as the *tibia* and *fibula*. Of these, the tibia is mainly or altogether concerned in carrying the foot, and it is thus shown to correspond to the radius, whilst the fibula corresponds to the ulna. 3. The small bones of the ankle, known as the *tarsus*, and varying in number in different cases. 4. A variable number of cylindrical bones (normally five), which are called the *metatarsus*, and which correspond to the metacarpus. 5. Lastly, the metatarsus carries the digits, which consist of from two to three small bones or *phalanges*, as in the fore-limb.

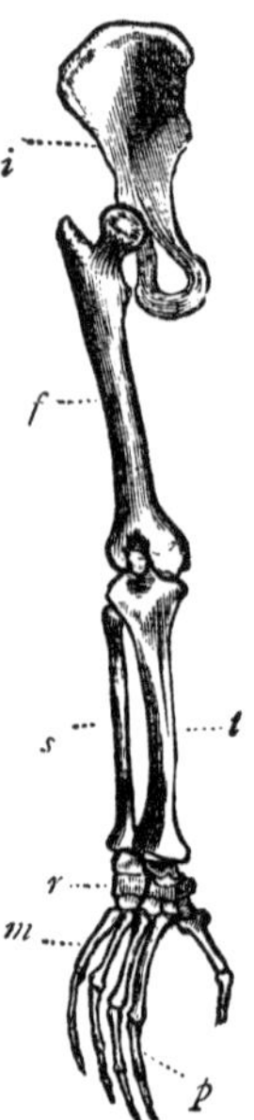

Fig. 128.—Pelvic limb (hind-limb of Chimpanzee) (after Owen). *i* Innominate bone; *f* Femur, or thigh-bone; *t* Tibia; *s* Fibula; *r* Tarsus; *m* Metatarsus; *p* Phalanges of the toes.

The *digestive system* of Vertebrates will be spoken of at greater length hereafter; but a brief sketch may be given here of the general phenomena of digestion. All Vertebrate animals are provided with a mouth for the reception of food, and in the great majority of cases the mouth is furnished with *teeth*, which are used sometimes merely to hold the prey, but more commonly to cut and bruise the food, and thus render it capable of digestion. The food is also generally subjected in the mouth to the action of "salivary" glands, the secretion of which serves not only to moisten the food, and thus mechanically assist deglutition, but also to render soluble the starchy elements of the food. The food is next swallowed, or, in other words, is transferred from the mouth to the stomach, this being effected by a complicated arrangement of muscles, whereby the food is forced down the gullet (*œsophagus*) to the proper digestive cavity or stomach. In the stomach (fig. 129, *s*) the food is subjected to two sets of actions; it is mechanically triturated and ground down by the constant contractions of the muscular walls of the stomach; and it is subjected to the chemical action of a special fluid secreted by the stomach, and called the "gastric juice." This fluid has the power of reducing albuminoid substances to a soluble form, and by its action the food is ultimately reduced to a thick acid fluid, called the "chyme." Leaving the stomach by its lower aperture (the *pylorus*), the chyme passes into the intestine, the first portion of which is divided into several sections, but is collectively known as the "small intestine." Here the chyme is subjected to the action of three other digestive fluids; the *bile*, secreted by a special organ, the liver; the *pancreatic juice*, secreted by another gland, the pancreas; and the *intestinal juice*, secreted by certain glands situated in the mucous membrane of the intestine itself. The result of the whole process is that the "chyme" is ultimately converted into a white, alkaline, milky fluid, which is called "chyle." The indigestible portions of the food pass from the small intestine into a tube of larger dimensions, called the "large intestine." Such portions of the food as are still soluble, and capable of being employed in nutrition, are here taken up into the blood, the useless remainder being ultimately expelled by an anal aperture. The last portion of the large intestine is usually less convoluted than the rest, and is called the "rectum."

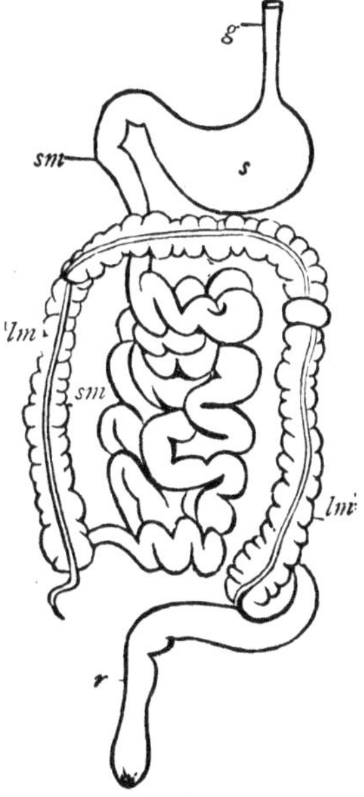

Fig. 129.—Diagram of the digestive system of a Mammal. *g* Gullet; *s* Stomach; *sm* Small intestine; *lm* Large intestine; *r* Rectum, terminating in the aperture of the anus.

The fluid and originally soluble portions of the food, and the chyle which is formed in the process of digestion, are taken into the blood, the losses of which they serve to repair. Part of the nutritive materials of the food is taken up directly by the blood-vessels, and is conveyed by the "vena portæ" to the liver, whence it ultimately reaches the great veins which go to the heart. The greater part, however, of the liquefied food, constituting the chyle, is taken up, not by the blood-vessels, but by a special set of tubes, which form a network in the walls of the intestine, and are known as the "lacteals." In these vessels, and in certain glands which are developed upon them, the chyle undergoes still further elaboration, and is made more similar in composition to the blood itself. All the lacteal vessels ultimately unite into one or more large vessels which open into one of the veins, so that all the chyle is thus finally added to the mass of the circulating blood.

The *blood*, then, or nutrient fluid from which the tissues are built up, is formed in this way out of the materials which are taken into the alimentary canal as food. In all the Vertebrata, with the single exception of the Lancelet (*Amphioxus*), the blood is of a red colour when viewed in mass. This is due to

Fig. 130.—Blood-corpuscles of Vertebrata. *a* Red blood-discs of man; *b* Blood-discs of Goose; *c* Crocodile; *d* Frog; *e* Skate.

the presence in it of an incredible number of microscopical bodies, which are known as the "blood-corpuscles," the fluid in which these float being itself colourless (fig. 130).

In all the *Vertebrata* the blood is distributed through the body by means of a system of closed tubes, which constitute the "blood-vessels;" and in all except the Lancelet, the means of propulsion are derived from a contractile muscular cavity or "heart," furnished with valvular apertures. In the most complete form of circulation, as seen in Birds and Mammals, the heart is essentially a double organ, composed of two halves, each of which consists of two cavities, an auricle and a ventricle. The right side of the heart is wholly concerned with the "lesser" or pulmonary circulation, whilst the left side is concerned with driving the blood to all parts of the body (systemic circulation). The modifications of the circulatory process will be noticed in speaking of the different classes of Vertebrates, but a brief sketch may be given here of the circulation in its most complete form, as in a Mammal. In such a case, the venous or impure blood, which has circulated through the body and has parted with its oxygen, is returned by the great veins to the right auricle. From the right auricle (fig. 131, *a*) the blood passes by a valvular aperture into the right ventricle (*v*), whence it is driven through the pulmonary artery to the lungs. The right side of the heart is therefore wholly respiratory in its function. Having been submitted to the action of the lungs, and having given off carbonic acid and taken up oxygen, the blood now becomes arterial, and is returned by the pulmonary veins to the left auricle (*a'*). From the left auricle the aerated blood passes through a valvular aperture into the left ventricle (*v'*), whence it is propelled to all parts of the body by means of a great systemic vessel, the "aorta." The left side of the heart is therefore wholly occupied in carrying out the "greater" or systemic circulation.

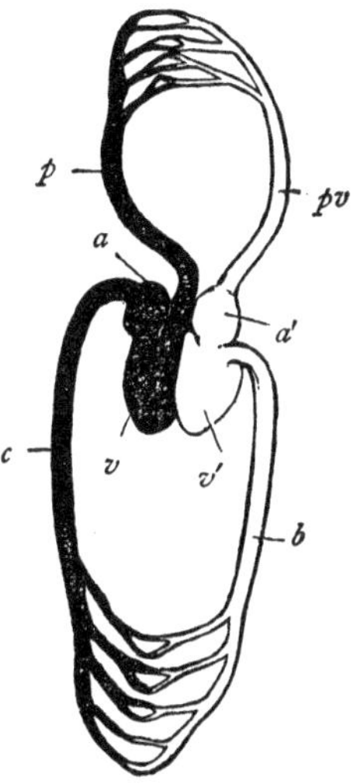

Fig. 131.—Diagram of the circulation of a Mammal. The venous system is marked black; the arterial system is left white. *a* Right auricle; *v* Right ventricle; *p* Pulmonary artery, carrying venous blood to the lungs; *pv* Pulmonary veins carrying arterial blood from the lungs; *a'* Left auricle; *v'* Left ventricle; *b* Aorta, carrying arterial blood to the body; *c* Vena cava carrying venous blood to the heart.

The purification of the blood is carried out in all Vertebrates by means of distinct respiratory organs, assisted to a greater or less extent by the skin. In the Fishes, and in the Amphibians to some extent, the process of respiration is carried on by means of *branchiæ* or gills—that is, by organs adapted for breathing air dissolved in water. These are, therefore, often spoken of as "Branchiate" Vertebrates; but the Amphibians always develop true lungs in the later stages of their existence. In the Reptiles, Birds, and Mammals, branchiæ are never developed, and the respiration is always carried on by means of true lungs—that is, by organs adapted for breathing air directly. These are therefore often spoken of as the "Abranchiate" Vertebrates.

The waste substances of the body—of which the most important are water, carbonic acid, and urea—are got rid of by the skin, lungs, and kidneys. Under ordinary circumstances, the lungs are mainly occupied with the excretion of carbonic acid and watery vapour. The skin chiefly gets rid of superfluous moisture, but can also in many animals excrete carbonic acid as well. The kidneys are present in almost all Vertebrate animals, and their function is mainly to excrete water, and the nitrogenous substance known as urea. In the majority of cases the fluid excreted by the kidneys is conveyed to the exterior by means of two tubes known as the ureters, which empty themselves into a common receptacle, the urinary bladder. In some cases, however, the ureters open into the termination of the alimentary canal (rectum).

The nervous system of Vertebrate animals usually exhibits a well-marked division into two parts—the cerebro-spinal system, and the sympathetic system. The cerebro-spinal system of nerves constitutes the great mass of the nervous system of Vertebrates, and usually exhibits a well-marked separation into spinal cord (*myelon*) and brain (*encephalon*). The proportion borne by the brain to the spinal cord differs much in different cases; and in the Lancelet a brain can hardly be said to be present at all. As already said, the brain and spinal cord are always completely shut off from the visceral cavity, and they are placed upon the dorsal surface of the body. The nerves given off from the cerebro-spinal axis are symmetrically disposed on the two sides of the body, and they are mainly concerned with the functions of "animal" life—that is to say, with sensation and locomotion. The sympathetic system of nerves is unsymmetrically disposed to a greater or less extent, and presides mainly over the functions of "organic" or "vegetative" life, being mainly concerned with regulating the functions of digestion and respiration, and the circulation of the blood. In its most fully developed form it consists of a double gangliated cord placed in the visceral cavity on the under surface of the spine, and of a series of nervous ganglia, united by nervous cords, and scattered mainly over the great viscera of the thorax and abdomen.

The organs of the senses are well developed in the *Vertebrata*, and those appropriated to the senses of *sight*, *hearing*, *smell*, and *taste* are protected within bony cavities of the head. The perfection of the senses differs much in different cases, but they are probably never wholly wanting in any Vertebrate animal. There are cases in which vision must be of the most rudimentary character; but even in these cases it is probable that there is a perception of light, even if there is no power of distinguishing objects. The only cases in which it would appear that vision is really altogether absent, are those of animals placed under the wholly abnormal condition of spending their existence in darkness (such as the *Proteus anguinus* of the caves of

Illyria). Smell, hearing, and taste are probably rarely, if ever, altogether absent in Vertebrates; though in many cases their organs are very rudimentary. Touch, or "tactile sensibility," is usually possessed to a greater or less degree by the entire surface of the body; but the sense of touch is generally localised in certain particular parts, such as the appendages of the mouth, the lips, the tongue, or the digits.

In all *Vertebrata* without exception reproduction is carried on by means of the sexes, and in all the sexes are in different individuals. No Vertebrate animal possesses the power of reproducing itself by fission or gemmation; and in no case are composite organisms or colonies produced. Most of the Vertebrates are *oviparous*, that is to say, the *ova* are expelled from the body of the parent either before or very shortly after impregnation. In other cases, the eggs are retained within the body of the parent until the young are hatched, and in these cases the animals are said to be *ovo-viviparous*. In other cases, again, not only is the egg hatched within the parent, but the embryo is retained within the body of the mother until its development has been carried out to a greater or less extent; and these animals are said to be *viviparous*.

DIVISIONS OF THE VERTEBRATA.—The sub-kingdom *Vertebrata* is divided into the five great classes of the Fishes (*Pisces*), Amphibians (*Amphibia*), Reptiles (*Reptilia*), Birds (*Aves*), and Mammals (*Mammalia*). So far there is perfect unanimity; but when it is inquired into what larger sections the Vertebrata may be divided, there is much difference of opinion. Here, the divisions proposed by Professor Huxley will be adopted, but it is necessary that those employed by other writers should be mentioned and explained.

One of the commonest methods of classifying the *Vertebrata* is to divide them into the two primary sections of the *Branchiata* and *Abranchiata*. Of these, the Branchiate section includes the fishes and Amphibians, and is characterised by the fact that the animal is always provided at some period of its life with branchiæ or gills. The Abranchiate section includes the Reptiles, Birds, and Mammals, and is characterised by the fact that the animal is never provided at any time of its life with gills. Additional characters of the Branchiate Vertebrates are, that the embryo is not furnished with the structures known as the *amnion* and *allantois*. Hence the Branchiate Vertebrates are often spoken of as the *Anamniota* and as the *Anallantoidea*. In the Abranchiate Vertebrates, on the other hand, the embryo is always provided with an amnion and allantois, and hence this section is spoken of as the *Amniota* or as the *Allantoidea.**

By Professor Owen the *Vertebrata* are divided into the two primary sections of the *Hæmatocrya* and the *Hæmatotherma*, the characters of the blood being taken as the distinctive character. The *Hæmatocrya* or Cold-blooded Vertebrates comprise the Fishes, Amphibia, and Reptiles, and are characterised by their cold blood and imperfect circulation. The *Hæmatotherma* or Warm-blooded Vertebrates comprise the Birds and the Mammals, and are characterised by their hot blood, four-chambered heart, and complete separation of the pulmonary and systemic circulations. The chief objection to this division lies in the separation which is effected between the Reptiles and the Birds, two classes which are certainly very nearly allied to one another.

By Professor Huxley the *Vertebrata* are divided into the following three primary sections:—

I. ICHTHYOPSIDA.—This section comprises the Fishes and the Amphibians, and is characterised by the presence at some period of life of gills or branchiæ, the absence of an amnion, the absence or rudimentary condition of the allantois, and the possession of nucleated red blood-corpuscles.

II. SAUROPSIDA.—This section comprises the Birds and the Reptiles, and is characterised by the constant absence of gills, the possession of an amnion and allantois, the articulation of the skull with the vertebral column by a single occipital condyle; the composition of each ramus of the lower jaw of several pieces, and the articulation of the lower jaw with the skull by the intervention of an "os quadratum;" and, lastly, the possession of nucleated red blood-corpuscles.

III. MAMMALIA.—This section includes the single class of the Mammals, and agrees with the preceding in never possessing gills, and in having an amnion and allantois. The *Mam-*

* The *amnion* (fig. 124) is a membranous sac, containing a fluid—the liquor amnii—and completely enveloping the embryo. It constitutes one of the so-called "fœtal membranes," and is thrown off at birth. The *allantois* (fig. 124, C) is an embryonic structure, which is developed out of the middle or "vascular" layer of the germinal membrane. It appears at first as a solid, pear-shaped, cellular mass, arising from the under part of the body of the embryo. In the process of development, the allantois increases largely in size, and becomes converted into a vesicle which envelops the embryo in part or wholly. It is abundantly supplied with blood, and is the organ whereby the blood of the fœtus is aerated. The part of the allantois which is external to the body of the embryo is cast off at birth; but the portion which is within the body is retained, and is converted into the urinary bladder.

malia, however, differ from the *Sauropsida* in the fact that the skull articulates with the vertebral column by two occipital condyles; each ramus of the lower jaw is simple, composed of a single piece, and the lower jaw is united with the temporal (squamosal) element of the skull, and is not articulated to a quadrate bone. There are special glands—the mammary glands—for the nourishment of the young for a longer or shorter period after birth, and the red blood-corpuscles are non-nucleated.

DIVISION I. ICHTHYOPSIDA.

CHAPTER LIV.

CLASS I.—PISCES.

THE first class of the *Vertebrata* is that of the Fishes (*Pisces*), which may be broadly defined as including *Vertebrate animals which are provided with gills throughout the whole of life; the heart, when present, consists (with one exception) of a single auricle and a single ventricle; the blood is cold; the limbs, when present, are in the form of fins, or expansions of the integument; and there is neither an amnion nor allantois in the embryo, unless the latter is represented by the urinary bladder.*

In form, Fishes are adapted for rapid locomotion in water, the shape of the body being such as to give rise to the least possible friction in swimming. To this end also, as well as for purposes of defence, the body is usually enveloped with a coating of scales developed in the inferior or dermal layer of the skin. The more important modifications in the form of these dermal scales are as follows: I. *Cycloid* scales (fig. 132, *a*), consisting of thin, flexible, horny scales, circular or elliptical in shape, and having a more or less completely smooth outline. These are the scales which are characteristic of most of the ordinary bony fishes. II. *Ctenoid* scales (fig. 132, *b*), also consisting of thin horny plates, but having their posterior margins fringed with spines, or cut into comb-like projections. III. *Ganoid* scales, composed of an inferior layer composed of bone, covered by a superficial layer of hard polished enamel (the so-called

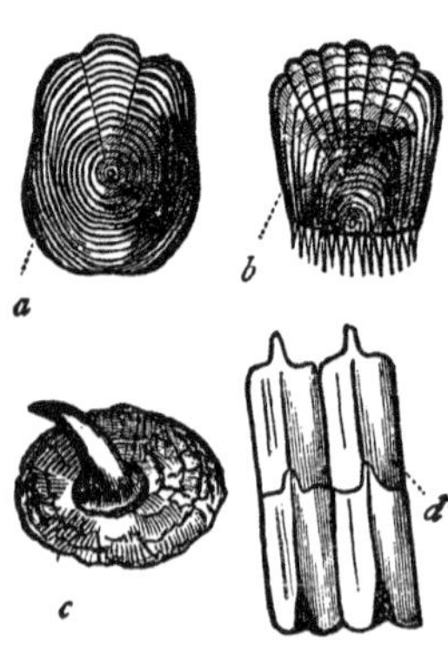

Fig. 132.—Scales of different fishes. *a* Cycloid scale (Pike); *b* Ctenoid scale (Perch); *c* Placoid scale (Thornback); *d* Ganoid scales (*Palæoniscus*).

"ganoine"). These scales (fig. 132, *d*) are usually much larger and thicker than the ordinary scales, and though they are often articulated to one another by special processes, they only rarely overlap. IV. *Placoid* scales, consisting of detached bony grains, tubercles, or plates, of which the latter are not uncommonly armed with spines (fig. 132, *c*).

In most fishes there is also to be observed a line of peculiar scales, forming what is called the "lateral line." Each of the scales in this line is perforated by a tube leading down to a longitudinal canal which runs along the side of the body, and is connected with cavities in the head. The function of this singular system has been ordinarily believed to be that of secreting the mucus with which the surface of the body is covered; but it seems to be more probably sensory in function, and to be connected with the sense of touch.

As regards their true osseous system or endoskeleton, Fishes vary very widely. In the Lancelet there can hardly be said to be any skeleton, the spinal cord being simply supported by the gelatinous notochord, which remains throughout life. In others the skeleton remains permanently cartilaginous; in others it is partially cartilaginous and partially ossified; and, lastly, in most modern fishes it is entirely ossified, or converted into bone. Taking a bony fish (fig. 133) as in this respect a typical example of the class, the following are the chief points in the osteology of a fish which require notice:—

The *vertebral column* in a bony fish consists of vertebræ, which are hollow at both ends, or biconcave, and are technically said to be "amphicœlous." The cup-like margins of the vertebral bodies are united by ligaments, and the cavities formed between contiguous vertebræ are filled with the gelatinous remains of the notochord. This elastic gelatinous substance acts as a kind of ball-and-socket joint between the bodies of the vertebræ, thus giving the whole spine the extreme mobility which is requisite for animals living in a watery medium. The ossification of the vertebræ is often much more imperfect than the above, but in no case except that of the Bony Pike (*Lepidosteus*) is ossification carried to a greater extent than this. In this fish, however, the vertebral column is composed of "opisthocœlous" vertebræ—that is, of vertebræ the bodies of which are concave behind and convex in front. The entire spinal column is divisible into not more than two distinct regions, an *abdominal* and a *caudal region*. The abdominal vertebræ possess a superior or neural arch (through which passes the spinal cord), a superior spinous process (neural spine), and two transverse processes to which the ribs are usually attached.

The caudal vertebræ (fig. 133) have no marked transverse processes; but, in addition to the neural arches and spines, they give off an inferior or *hæmal* arch below the body of the vertebra, and the hæmal arches carry inferior spinous processes (hæmal spines).

The *ribs* of a bony fish are attached to the transverse processes, or to the bodies of the abdominal vertebræ, in the form of slender curved bones which articulate with no more than one vertebra each, and that only at a single point. Unlike the ribs of the higher Vertebrates, the ribs do not enclose a thoracic cavity, but are simply embedded in the muscles which bound the abdomen. Usually each rib gives off a spine-like bone,

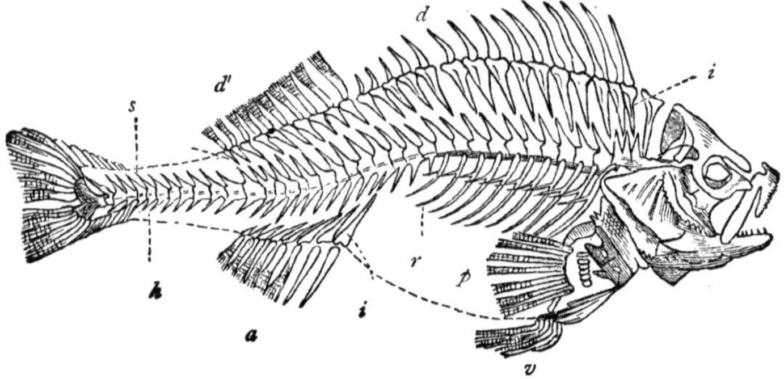

Fig. 133.—Skeleton of the common Perch (*Perca fluviatilis*). *p* One of the pectoral fins; *v* One of the ventral fins; *a* Anal fin, supported upon interspinous bones (*i*); *c* Caudal fin; *d* First dorsal fin; *d'* Second dorsal fin, both supported upon interspinous bones; *i i* Interspinous bones; *r* Ribs; *s* Spinous processes of vertebræ; *h* Hæmal processes of vertebræ.

which is directed backwards amongst the muscles. Inferiorly the extremities of the ribs are free, or are rarely united to dermal ossifications in the middle line of the abdomen; but there is never any breast-bone or *sternum* properly so called.

The only remaining bones connected with the skeleton of the trunk are the so-called *interspinous bones* (fig. 133, *i i*). These form a series of dagger-shaped bones plunged in the middle line of the body between the great lateral muscles which make up the greater part of the body of a fish. The internal ends or points of the interspinous bones are attached by ligament to the spinous processes of the vertebræ; whilst to their outer ends are articulated the "rays" of the so called "median" fins, which will be hereafter described. As a rule, there is only one interspinous bone to each spinous process, but in the Flat-fishes (Sole, Turbot, &c.) there are two.

Beside the fins which represent the limbs (pectoral and ventral fins), fishes possess other fins placed in the middle line of the body, and all of these alike are supported by bony spines or "rays," which are of two kinds, termed respectively "spinous rays" and "soft rays." The "spinous rays" are simple bony spines, apparently composed of a single piece each, but really consisting of two halves firmly united along the middle line. The "soft rays" are composed of several slender spines proceeding from a common base, and all divided transversely into numerous short pieces. The soft rays occur in many fishes in different fins, but they are invariably found in the caudal fin or tail (fig. 133, *c*). The rays of the median fins, whatever

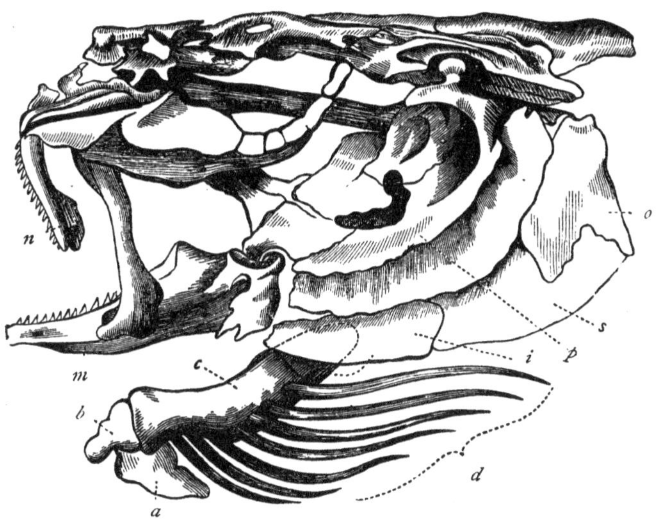

Fig. 134.—Skull of Cod (*Morrhua vulgaris*)—Cuvier. *a* Urohyal; *b* Basihyal; *c* Ceratohyal; *d* Branchiostegal rays; *p* præ-operculum; *o* Operculum proper; *s* Sub-operculum; *i* Inter-operculum; *m* Mandible; *n* Inter-maxillary bone.

their character may be, always articulate by a hinge-joint with the heads of the interspinous bones.

The *skull* of the bony fishes is an extremely complicated structure, and it is impossible to enter into its composition here. The only portions of the skull which require special mention are the bones which form the gill-cover or operculum, and the hyoid bone with its appendages. For reasons connected with the respiratory process in fishes, as will be afterwards seen, there generally exists between the head and the scapular arch a great cavity or gap on each side, within which are contained the branchiæ. The cavity thus formed opens externally on each side of the neck by a single vertical fissure or "gill-slit," closed by a broad flap, called the "gill-cover" or "operculum," and by a membrane termed the "branchi-ostegal membrane."

The gill-cover (fig. 134, *p*, *o*, *s*, *i*) is composed of a chain of broad flat bones, termed the opercular bones. Of these, the innermost articulates with the skull (tympano-mandibular arch), and is called the "præ-operculum;" the next is a large bone called the "operculum" proper; and the remaining two bones called respectively the "sub-operculum" and "inter-operculum," form, with the operculum proper, the edge of the gill-cover. These various bones are united together by membrane, and they form collectively a kind of movable door, by means of which the branchial chamber can be alternately opened and shut. Besides the gill-cover, however, the branchial chamber is closed by a membrane called the "branchiostegal membrane," which is attached to the os hyoides. The membrane is supported and spread out by a number of slender curved spines, which are attached to the lateral branches of the hyoid bone, act very much as the ribs of an umbrella, and are known as the "branchiostegal rays" (fig. 134, *d*).

The hyoid arch of fishes is attached to the temporal bones of the skull by means of two slender styliform bones, which correspond to the styloid processes of man, and are called the "stylohyal" bones (fig. 135, *f*). The rest of the hyoid arch is composed of a central portion and two lateral branches. Each branch is composed of the following parts:—1. A triangular bone attached above to the stylohyal, and termed the "epihyal bone" (fig. 135, *e*); 2. A much longer bone, known as the "ceratohyal" (*d*). The central portion of the hyoid arch is made up of two small polyhedral bones—the "basihyals" (*b*). From the basihyal there extends *forwards* in many fishes a slender bone, which supports the tongue, and is termed the "glossohyal" or "lingual" bone (*a*). There is also another compressed bone, which extends *backwards* from the basihyals, and which is known as the "urohyal bone" (*c*). This last-mentioned bone is of importance, as it often extends backwards to the point of union of the coracoid bones, and thus forms the isthmus which separates the two branchial apertures.

From the outer margins of the epihyal and ceratohyal bones on each side arise the slender curved "branchiostegal rays," which have been previously mentioned. There are usually seven of these on each side. Above the urohyal, and attached in front of the body of the os hyoides, is a chain of bones, placed one behind the other, and termed by Owen the "basi-branchial bones." Springing from these are four bony arches —the "branchial arches"—which proceed upwards to be connected superiorly by ligament with the under surface of the skull. The branchial arches—as will be subsequently described—carry the branchiæ, and each is composed of two

Fig. 135.—Os hyoides, branchiostegal rays, and scapular arch of the Perch (after Cuvier). *ss* Supra-scapula; *s* Scapula, *co* Coracoid; *cl* Supposed representative of the clavicle; *a* Glossohyal bone; *b* Basihyal; *c* Urohyal; *d* Ceratohyal; *e* Epihyal; *f* Stylohyal; *br* Branchial arches; *t* Branchiostegal rays.

main pieces, termed respectively the "cerato-branchial" and "epi-branchial" bones. The second and third arches are connected with the skull by the intervention of two small bones, often called the "superior pharyngeal bones," but termed by Owen the "pharyngo-branchial" bones.

The *limbs* of fishes depart considerably from the typical form exhibited in the higher Vertebrates. One or both pairs of limbs may be wanting, but when present the limbs are always in the form of *fins*—that is, of expansions of the integument strengthened by bony or cartilaginous fin-rays. The anterior limbs are known as the *pectoral* fins, and the posterior as the *ventral* fins; and they are at once distinguished from the so-called "median" fins by being always symmetrically disposed in pairs. Hence they are often spoken of as the *paired* fins. The scapular arch (figs. 135, 136) supporting the pectoral

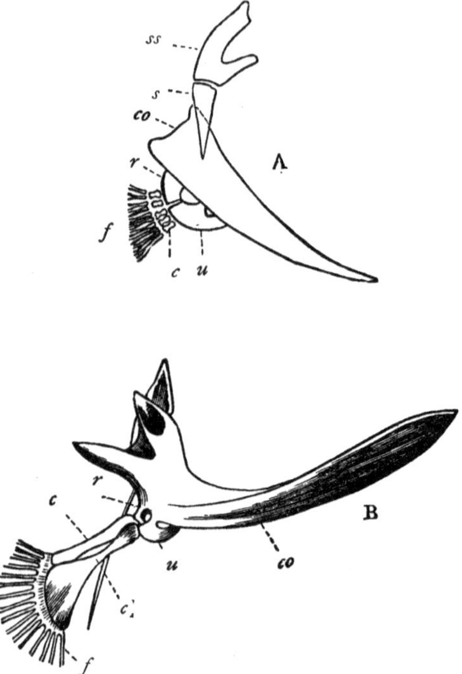

Fig. 136.—Pectoral limbs of Fishes (after Owen). A, Cod (*Morrhua vulgaris*); B, Angler (*Lophius*). *ss* Supra-scapula; *s* Scapula; *co* Coracoid; *r* Radius; *u* Ulna; *cc* Carpal bones; *f* Fin-rays, representing the metacarpus and phalanges of the fingers.

limbs is usually joined to the skull (occipital bone), and consists of the following pieces on each side:—1. The supra-scapula (*ss*); 2. The scapula (*s*), articulating with the former; and, 3. The coracoid (*co*), attached above with the scapula, and united below, by ligament or suture, with the coracoid of the opposite side, thus completing the pectoral arch. Lastly, there is often another bone, sometimes single, but oftener of two pieces, attached to the upper end of the coracoid, and this is believed to represent the collar-bone or clavicle.*

The fore-limb possesses in a modified form most of the bones which are present in the higher *Vertebrata*. The *humerus*, or bone of the upper arm, is usually wanting, or it is altogether rudimentary. A radius and ulna (fig. 136, *r*, *u*) are usually present, and are followed by a variable number of bones, which represent the carpus, and some of which sometimes articulate directly with the coracoid. The carpus is followed by the "rays" of the fin proper, these representing the metacarpal bones and phalanges. The pectoral fins vary much in size and in other characters. In the Flying Gurnard (*Dactylopterus*), and the true Flying Fish (*Exocœtus*), the pectorals are enormously developed, and enable the fish to take extensive leaps out of the water.

The hind-limbs or "ventral fins" are wanting in many fishes, and they are less developed and less fixed in position than are the pectoral fins. In the ventral fins no representatives of the tarsus, tibia and fibula, or femur, are ever developed. The rays of the ventral fins—representing the metatarsus and the phalanges of the toes—unite directly with a pelvic arch, which is composed of two sub-triangular bones, united in the middle line and believed to represent the *ischia*. The imperfect pelvic arch, thus constituted, is never united to the vertebral column in any fish. In those fishes in which the ventral fins are "abdominal" in position (*i.e.*, placed near the hinder end of the body) the pelvic arch is suspended freely amongst the muscles. In those in which the ventral fins are "thoracic" or "jugular" (*i.e.*, placed beneath the pectoral fins, or on the sides of the neck), the pelvic arch is attached to the coracoid bones of the scapular arch, and is therefore wholly removed from its proper vertebra.

In addition to the pectoral and ventral fins—the homologues of the limbs—which may be wanting, fishes are furnished with certain other expansions of the integument, which are "median" in position, and must on no account be confounded with the true "paired" fins. These median fins are variable in number, and in some cases there is but a single fringe running round the posterior extremity of the body. In all cases, however, the median fins are "azygous"—that is to say, they occupy the middle line of the body, and are not sym-

* These are the views entertained by Owen as to the composition and nature of the pectoral arch of fishes, but they are dissented from by Mr Parker, one of the greatest living authorities on this subject.

metrically disposed in pairs. Most commonly, the median fins consist of one or two expansions of the dorsal integument, called the "dorsal fins" (fig. 137, *d d'*); one or two on the ventral surface near the anus—the "anal fins" (fig. 137, *a*); and

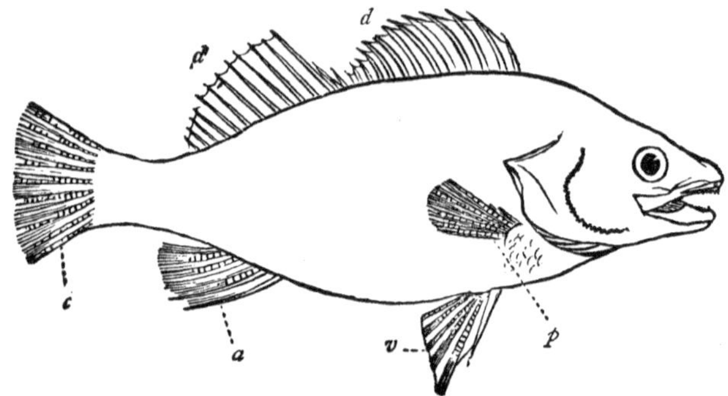

Fig. 137.—Outline of a fish (*Perca granulata*), showing the paired and unpaired fins. *p* One of the pectoral fins; *v* One of the ventral fins; *d* First dorsal fin; *d'* Second dorsal fin; *a* Anal fin; *c* Caudal fin.

a broad fin at the extremity of the vertebral column, called the "caudal fin" or tail (*c*). In all cases, the rays which support the median fins are articulated with the so-called interspinous bones, which have been previously described. Though called "median," from their position in the middle line of the body, and from their being unpaired, the median fins of Fishes, as shown by Goodsir and Humphrey, are truly to be regarded as formed by the coalescence of two lateral elements in the mesial plane of the body.

Fig. 138.—Tails of different fishes. *a* Homocercal tail (Sword-fish); *b* Heterocercal tail (Sturgeon).

The caudal fin or tail of fishes is always set vertically at the extremity of the spine, so as to work from side to side, and it is the chief organ of progression in the fishes. In its vertical position, and in the possession of fin-rays, it differs altogether from the horizontal integumentary expansion which constitutes the tail of the Whales, Dolphins, and *Sirenia* (Dugong and Manatee). In the form of the tail fishes exhibit two very distinct types of structure, termed respectively the "homocercal" and "heterocercal" type of tail (fig. 138). The homocercal tail is the one which most commonly occurs in our modern fishes, and it is characterised by the fact that the two lobes of the tail are equal, and the vertebral column, instead of being prolonged into the upper lobe of the tail, stops short at its base. In the heterocercal tail, on the other hand, the vertebral column is prolonged into the upper lobe of the tail, so that the tail becomes unequally lobed, its greater portion being placed below the spine. Even where the vertebral column is not prolonged into the upper lobe, the tail may nevertheless become heterocercal, in consequence of a great development of the hæmal spines as compared with the neural spines of the vertebræ.

The process of *respiration* in all fishes is essentially aquatic, and is carried on by means of branchial plates or tufts developed upon the posterior visceral arches, which are persistent, and do not disappear at the close of embryonic life, as they do in other Vertebrates. In the Lancelet alone, respiration is effected partly by branchial filaments placed round the commencement of the pharynx, and partly by the pharynx itself, which is greatly enlarged, and has its walls perforated by a series of transverse ciliated fissures. The arrangement and structure of the branchiæ differ a good deal in the different orders of Fishes, and these modifications will be noticed subsequently. In the meanwhile it will be sufficient to give a brief description of the branchial apparatus in one of the bony fishes. In such a fish, the branchiæ are connected with the hyoid arch, and are situated in two special chambers, situated one on each side of the neck. The branchiæ are carried upon the outer convex sides of what have been already described as the "branchial arches;" that is to say, upon a series of bony arches which are connected with the hyoid arch inferiorly, and are united above with the base of the skull. The internal concave sides of the branchial arches are usually furnished with a series of processes, constituting a kind of fringe, the function of which is to prevent foreign substances finding their way amongst the branchiæ, and thus interfering with the proper action of the respiratory organs. The branchiæ, themselves, usually have the form of a double series of cartilaginous leaflets or laminæ. The branchial laminæ are flat, elongated, and pointed in shape, and they are covered with a highly vascular mucous membrane, in which the branchial capillaries ramify. The blood circulates through the branchial laminæ, and is here subjected to the action of ærated water, whereby it is oxygenated. The water is constantly taken in at the mouth by a movement analogous to swallowing, and it gains admission to the branchial chambers by means of a series of clefts or slits, the "branchial fissures," which are situated on both sides of the pharynx. Having passed over the gills, the deoxygenated water makes its escape posteriorly by an aperture called the "gill-slit" or "opercular aperture," one of which is situated on each side of the neck. As we have seen before, the gill-slit is closed in front by a chain of flat bones, collectively constituting the "gill-cover" or "operculum;" and the gill-covers are finally completed by a variable number of bony spines—the "branchiostegal rays"—which articulate with the hyoid arch, and support a membrane—the "branchiostegal membrane."

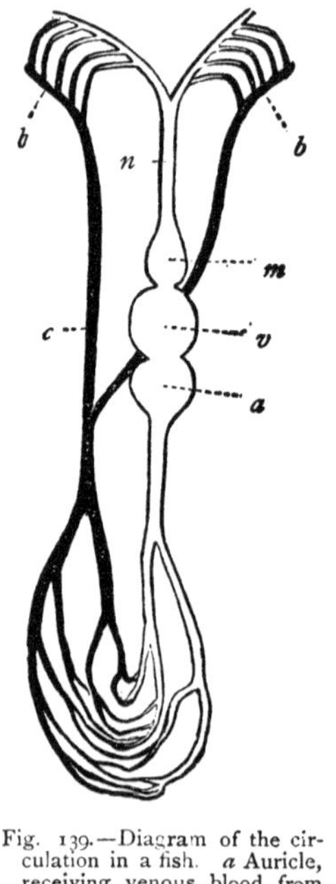

Fig. 139.—Diagram of the circulation in a fish. *a* Auricle, receiving venous blood from the body; *v* Ventricle; *m* Bulbus arteriosus, at the base of the branchial artery; *n* Branchial artery, carrying the venous blood to the gills (*b b*); *c* Aorta, carrying the arterialised blood to all parts of the body.

The *heart* of Fishes is, properly speaking, a branchial or respiratory heart. It consists of two cavities, an auricle and a ventricle (fig. 139, *a*, *v*), and the course of the circulation is as follows:—The venous blood derived from the liver and from the body generally is poured by the vena cava into the auricle (*a*), and from this it is propelled into the ventricle (*v*). From the ventricle arises a single aortic arch (the right), and the base of this is usually dilated into a cavity or sinus, called the "bulbus arteriosus" (*m*). The arterial bulb is sometimes covered with a special coat of striated muscular fibres, and is provided with several transverse rows of valves. In these cases, the bulbus acts as a kind of continuation of the ventricle, being capable of rhythmical contractions. The blood is driven by the ventricle through the branchial artery (*n*) to the gills, through which it is distributed by means of the branchial vessels, the number of which varies (there are *three* on each side in a few fishes, *four* in most of the bony fishes, *five* in the Skates and Sharks, and *six* or *seven* in the Lampreys). The aerated blood which has passed through the gills is not returned to the heart, but is driven from the branchiæ through all parts of the body; the propulsive force necessary for this being derived chiefly from the heart, assisted by the contractions of the voluntary muscles. In some fishes (as in the Eel) the return of the blood to the heart is assisted by a rhythmically contractile dilatation of the caudal vein. The essential peculiarity, then, of the circulation of fishes depends upon this—that the arterialised blood returned from the gills is propelled through the systemic vessels of the body, without being sent back to the heart.

The Lancelet (*Amphioxus*), alone of all Fishes, has no special heart, and the circulation is effected by contractile dilatations developed upon several of the blood-vessels. In the Mud-fish (*Lepidosiren*) the heart consists of *two* auricles and a single ventricle. The blood-corpuscles of Fishes are nucleated (fig. 130, *e*), and the blood is red in all except the *Amphioxus*.

As regards the *digestive system* of Fishes there is not much of peculiar importance. The mouth is usually furnished with a complicated series of teeth, which, in the Bony Fishes, are not only developed upon the jaws proper, but are also situated upon other bones which enter into the composition of the buccal cavity (such as the palate, the pterygoids, vomer, branchial arches, the glossohyal bone, &c.) The œsophagus is usually short and capacious, and generally opens into a large and well-marked stomach. The pyloric aperture of the stomach is usually furnished with a valve, and behind it there is usually a number (from one to sixty) of blind appendages, termed the "pyloric cæca." These are believed to represent the pancreas, but there may be a recognisable pancreas either alone or in addition to the pyloric cæca. The intestinal canal is a longer or shorter, more or less convoluted tube, the absorbing surface of which, in certain fishes, is largely increased by a spiral reduplicature of the mucous membrane, which winds like a screw in close turns from the pylorus to the anus. The liver is usually large, soft, and oily, and a gall-bladder is almost universally present; but in the *Amphioxus* the liver is doubtfully represented by a hollow sac-like organ.

The kidneys of fishes are usually of great size, and form two elongated organs, which are situated beneath the spine, and extend along the whole length of the abdominal cavity. The ureters often dilate, and form a species of bladder, the doubtful representative of the allantois.

Whilst the respiration of all fishes is truly aquatic, most of them are, nevertheless, furnished with an organ which is doubtless the homologue of the lungs of the air-breathing Vertebrates. This—the "air" or "swim bladder"—is a sac containing gas, situated beneath the alimentary tube, and often communicating with the gullet by a duct. In the great majority of fishes the functions of the air-bladder are certainly hydrostatic—that is to say, it serves to maintain the necessary accordance between the specific gravity of the fish and that of the surrounding water. In the singular Mud-fishes, however, it acts as a respiratory organ, and is therefore not only the homologue, but also the analogue, of the lungs of the higher Vertebrates. In most fishes the air-bladder is an elongated sac with a single cavity, but in many cases it is variously subdivided by septa. In the Mud-fish the air-bladder is composed of two sacs, completely separate from one another, and divided into a number of cellular compartments. The duct leading in many fishes from the air-bladder (*ductus pneumaticus*) opens into the œsophagus, and is the homologue of the wind-pipe (*trachea*). The air contained in the swim-bladder is composed mainly of nitrogen in most fresh-water fishes, but in the sea-fishes it is mainly made up of oxygen. The fishes which live habitually at the bottom of the sea, such as the Flat-fishes, possess no swim-bladder, and it is much reduced in size in those which live principally at the surface.

The *nervous system* of fishes is of an inferior type of organisation, the brain being of small size, and consisting mainly of ganglia devoted to the special senses. As regards the special senses, there is one peculiarity which deserves special notice, and this is the conformation of the nasal sacs. The cavity of the nose is usually double, and is lined by an olfactory membrane, folded so as to form numerous plicæ. Anteriorly, the water is admitted into the nasal sacs by a single or double nostril, usually by two apertures; but posteriorly the nasal sacs are closed, and do not communicate with the pharynx by any aperture. The only exceptions to this statement are to be found in the Myxinoids and in the Lepidosiren. The essential portion of the organ of hearing (*labyrinth*) is present in almost all fishes, but in none is there any direct communication between the ear and the external medium. In some cases, however, there is a communication between the ear and the swim-bladder, thus foreshadowing the Eustachian tube in man.

As regards their *reproductive system*, fishes are, for the most part, truly *oviparous*, the ovaries being familiarly known as the "roe." The testes of the male are commonly called the "soft roe" or "milt." The products of the reproductive organs are often set free into the peritoneal cavity, ultimately finding their way to the external medium, either by means of an abdominal pore (or pores), or by being taken up by the open mouths of the "Fallopian tubes." In other cases the generative products are directly conveyed to the exterior by the proper ducts of the reproductive organs.

DIVISIONS OF FISHES.

CHAPTER LV.

PHARYNGOBRANCHII AND MARSIPOBRANCHII.

THE class *Pisces* has been very variously subdivided by different writers; but the classification here adopted is the one proposed by Professor Huxley, who divides the class into the following six orders, in the subdivisions of which Professor Owen has been followed:—*

ORDER I. PHARYNGOBRANCHII (= *Cirrostomi*, Owen; and *Leptocardia*, Müller).—This order includes but a single fish, the anomalous *Amphioxus lanceolatus*, or Lancelet (fig. 140), the organisation of which differs in almost all important points from that of all the other members of the class. The order is defined by the following characters, which, as will be seen, are mostly negative:—No skull is present, nor lower jaw (mandible), nor limbs. The notochord is persistent; and there are no vertebral centra nor arches. No distinct brain nor auditory organs are present. In place of a distinct heart, pulsating dilatations are developed upon several of the great blood-vessels. The blood is pale. The mouth is in the form of a longitudinal fissure, surrounded by filaments or cirri. The walls of the pharynx are perforated by numerous clefts or fissures, the sides of which are ciliated, the whole exercising a respiratory function.

The Lancelet is a singular little fish which is found burrowing in sandbanks, in various seas, but especially in the Mediterranean. The body is lanceolate in shape, and is provided with a narrow membranous border, of the nature of a median fin, which runs along the whole of the dorsal and part of the

* Cuvier divided the class *Pisces* into the great orders of the *Chondropterygii* (or Cartilaginous Fishes), the *Acanthopterygii* (or Fishes with spinous rays in the paired fins), and *Malacopterygii* (or Fishes with soft rays in the paired fins). Agassiz divides Fishes, from the character of the scales, into the four orders, *Cycloidei*, *Ctenoidei*, *Ganoidei*, and *Placoidei*. Müller divides the Fishes into the five orders *Leptocardia* (Lancelet), *Cyclostomata* (Lampreys and Hag-fishes), *Teleostei* (Bony Fishes), *Ganoidei* (Ganoid Fishes), and *Selachia* (Sharks and Rays).

ventral surface, and expands at the tail to form a lancet-shaped caudal fin. No true paired fins, representing the anterior and posterior limbs, are present. The mouth is a longitudinal fissure, situated at the front of the head, and destitute of jaws. It is surrounded by a cartilaginous ring, composed of many pieces, which give off prolongations, so as to form a number of cartilaginous filaments or "cirri" on each side of the mouth. (Hence the name of *Cirrostomi*, proposed by Professor Owen for the order.) The throat is provided on each side with vascular lamellæ, which are believed by Owen to perform the function of free branchial filaments. The mouth leads into a dilated chamber, which is believed to represent the pharynx, and is termed the "pharyngeal" or "branchial sac." It is an elongated chamber, the walls of which are strengthened by numerous cartilaginous filaments, between which is a series

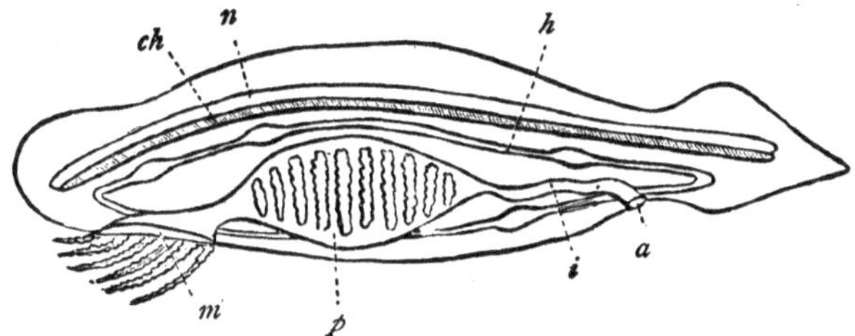

Fig. 140.—Diagram of the Lancelet (*Amphioxus*). *m* Mouth, surrounded by cartilaginous cirri; *p* Greatly-dilated pharynx, perforated by ciliated clefts; *i* Intestine, terminating in anus (*a*); *h* Hæmal system, with pulsating dilatations; *ch* Notochord; *n* Spinal cord.

of transverse slits or clefts, the whole covered by a richly ciliated mucous membrane. This branchial dilatation has given rise to the name *Branchiostoma*, often applied to the Lancelet. Posteriorly the branchial sac opens into an alimentary canal, to which is appended a long and capacious sac or cæcum, which is believed to represent the liver. The intestinal tube terminates posteriorly by a distinct anus. Respiration is effected by the admission of water taken in by the mouth into the branchial sac, having previously passed over the free branchial filaments before mentioned. The water passes through the slits in the branchial sac, and thus gains access to the abdominal cavity, from which it escapes by means of an aperture with contractile margins situated a little in front of the anus, and called the "abdominal pore." There is no distinct heart, and the circulation is entirely effected by means of several rhythmically contractile dilatations which are developed upon several of the great blood-vessels. The blood itself is colourless. No kidneys have as yet been discovered, and there is no lymphatic system. There is no skeleton properly so called. In place of the vertebral column, and constituting the whole endoskeleton, is the semi-gelatinous cellular notochord, enclosed in a fibrous sheath, and giving off fibrous arches above and below. The notochord is, further, peculiar in this, that it is prolonged quite to the anterior end of the body, whereas in all other Vertebrates it stops short at the pituitary fossa. There is no cranium, and the spinal cord does not expand anteriorly to form a distinct cerebral mass. The brain, however, may be said to be represented, since the anterior portion of the nervous axis gives off nerves to a pair of rudimentary eyes, and another branch to a ciliated pit, believed to represent an olfactory organ. The generative organs (ovaria and testes) are not furnished with any efferent ducts (oviduct or vas deferens). The generative products, therefore, must be admitted into the abdominal cavity, and gain the external medium by the "abdominal pore."

ORDER II. MARSIPOBRANCHII (= *Cyclostoma*, Owen; and *Cyclostomata*, Müller). — This order includes the Lampreys (*Petromyzonidæ*) and the Hag-fishes (*Myxinidæ*), and is defined by the following characters:—The body is cylindrical, worm-like, and destitute of limbs. The skull is cartilaginous, without cranial bones, and having no lower jaw (mandible). The notochord is persistent, and there are either no vertebral centra, or but the most rudimentary traces of them. The heart consists of one auricle and one ventricle, but the branchial artery is not furnished with a bulbus arteriosus. The gills are sac-like, and are not ciliated.

The type of piscine organisation displayed in the *Marsipobranchii* is of a very low grade, as indicated chiefly by the persistent notochord without vertebral centra, the absence of any traces of limbs, the absence of a mandible, and the structure of the gills.

Both the Lampreys (fig. 141, A) and the Hag-fishes are vermiform, eel-like fishes, which agree in possessing no paired fins, to represent the limbs, but in having a median fin running round the hinder extremity of the body. The skeleton remains throughout life in a cartilaginous condition, the chorda dorsalis is persistent, and the only traces of bodies of vertebræ are found in hardly perceptible rings of osseous matter developed in the sheath of the notochord. The neural arches of the vertebræ, enclosing the spinal cord, are only represented by cartilaginous prolongations. The mouth in the Hag-fish

(*Myxine*) is of a very remarkable character, and enables it to lead a very peculiar mode of life. It is usually found, namely, embedded in the interior of some other large fish, into which

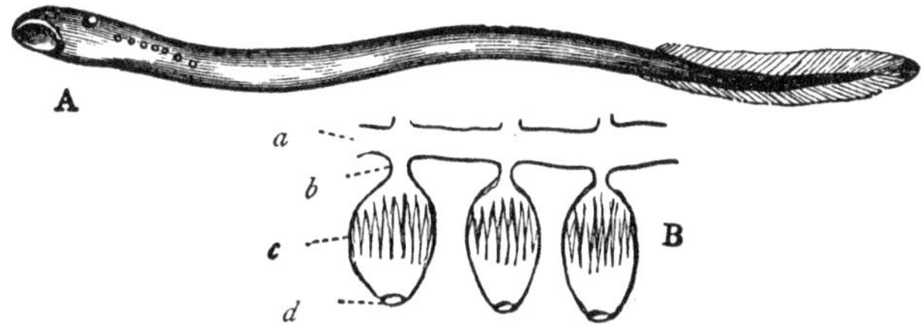

Fig. 141.—A, Lamprey (*Petromyzon*), showing the sucking-mouth and the apertures of the gill-sacs. B, Diagram to illustrate the structure of the gills in the Lamprey: *a* Pharynx; *b* Tube leading from the pharynx into one of the gill-sacs; *c* One of the gill-sacs, showing the lining membrane thrown into folds; *d* External opening of the gill-sac. (In reality the gill-sacs do not open directly into the pharynx, but into a common respiratory tube, which is omitted for the sake of clearness.

it has succeeded in penetrating by means of its singular dental apparatus. The mouth is sucker-like, destitute of jaws, but provided with tactile filaments or cirri. In the centre of the palate is fixed a single, large, recurved fang, which is firmly attached to the under surface of the cranium. The sides of this fang are strongly serrated, and it is by means of this that the Hag-fish bores its way into its victim, having previously attached itself by its sucker-like mouth. In the Lampreys the mouth has also the form of a circular cup or sucker, and is also destitute of jaws; but in addition to the palatine fang of the *Myxine*, the margins of the lips bear a number of horny processes, which are not really true teeth, but are hard structures developed in the labial mucous membrane. The tongue, also, is armed with serrated teeth, and acts as a kind of piston; so that the Lampreys are in this manner enabled to attach themselves firmly to solid objects.

A very remarkable peculiarity in the Hag-fishes, and one very necessary to remember, is found in the structure of the nasal sacs. In all fishes, namely, except these and the Mud-fishes (*Lepidosiren*), the nasal sacs are closed behind, and do not open posteriorly into the throat. In the Myxinoids, however, such a communication exists, and the nasal sac—for there is only one—is placed in communication with the cavity of the mouth by means of a canal which perforates the palate. In front the nasal cavity communicates with the external medium by a second tube, which opens on the top of the head by a single aperture, which is often called the "spiracle," and which is in reality an unpaired nostril. In the Lampreys, on the other hand, the single nasal sac has the same structure as in the typical fishes—that is to say, it is closed behind, and does not communicate in any way with the cavity of the mouth.

Another very remarkable point in the Hag-fishes and Lampreys is to be found in the structure of the gills, from which the name of the order is derived. In the Lampreys, in place of the single gill-slit, covered by a gill-cover, as seen in the ordinary bony fishes, the side of the neck, when viewed externally, exhibits six or seven round holes placed far back in a line on each side (fig. 141, A). In the Hag fishes the external apertures of the gills are reduced to one on each side, placed below the head; but the internal structure of the gills is the same in both cases. In both the Lampreys and the Hag-fishes, namely, the gills are in the form of sacs or pouches (fig. 141, B), the mucous membrane of which is thrown into folds or plaits like the leaves of a book, over which the branchial vessels ramify. Internally the sacs communicate with the cavity of the pharnyx, either directly or by the intervention of a common respiratory tube. It follows from this, that the gill-pouches on the two sides, with their included fixed branchial laminæ, communicate freely with one another through the pharynx. The object of this arrangement appears to be mainly that of obviating the necessity of admitting water to the gills through the mouth, as is the case with the ordinary bony fishes. These fishes are in the habit of fixing themselves to foreign objects by means of the suctorial mouth; and when in this position, it is, of course, impossible that they can obtain the necessary water of respiration through the mouth. As the branchial pouches, however, on the two sides of the neck, communicate freely with one another through the pharynx, water can readily pass in and out. This, in the Lampreys, is further assisted by a kind of elastic cartilaginous framework upon which the respiratory apparatus is supported, and which acts somewhat like the ribs of the higher *Vertebrata*. Water can also be admitted to the pharynx, and thence to the branchial sacs, by means of a tube which leads from the pharynx to an aperture placed on the top of the head.

The Lampreys are, some of them, inhabitants of rivers; but the great Sea-lamprey (*Petromyzon marinus*) only quits the salt water in order to spawn. The mouth in the *Petromyzonidæ* is a circular cartilaginous ring, formed by the amalgamation of the palatine and mandibular arches, and carrying numerous teeth and small tubercles. The tongue is armed with a double series of small teeth, and acts like a piston, enabling the animal to attach itself to stones and rocks. There is no air-bladder. The body is cylindrical, compressed towards the tail, and destitute of scales. The skeleton consists of a series of cartilaginous rings without ribs.

In the *Myxinidæ* the mouth is circular, membranous, with eight cirrhi. The palate carries a single fang, and the tongue is armed with a double row of small teeth on each side. There may be seven branchial apertures on each side (*Heptatrema*), or the branchial pouches open into a common tube on each side, and each of these terminates in a distinct aperture situated under the heart on the lower surface of the body (*Gastrobranchus*). The Hags pour out so much mucus through the lateral line that they can surround themselves with jelly; hence the name of the common species (*Myxine glutinosa*). The Glutinous Hag is a native of the North and British seas, and is chiefly found in the interior of the Cod and Haddock (often five or six individuals in one fish).

CHAPTER LVI.

TELEOSTEI.

ORDER III. TELEOSTEI.—This order includes the great majority of fishes in which there is a well-ossified endoskeleton, and it corresponds very nearly with Cuvier's division of the "osseous" fishes. The *Teleostei* are defined as follows:—The skeleton is usually well ossified; the cranium is provided with cranial bones; and a mandible is present; whilst the vertebral column almost always consists of more or less completely ossified vertebræ. The pectoral arch has a clavicle; and the two pairs of limbs, when present, are in the form of fins supported by rays. The gills are *free*, pectinated or tufted in shape; a bony gill-cover and branchiostegal rays being always developed. The branchial artery has its base developed into a bulbus arteriosus; but this is never rhythmically contractile, and is separated from the ventricle by no more than a single row of valves.

The order *Teleostei* comprises almost all the common fishes; and it will be unnecessary to dilate upon their structure, as they were taken as the types of the class in giving a general description of the Fishes. It may be as well, however, to recapitulate very briefly some of the leading characters of the order.

I. The *skeleton*, instead of remaining throughout life more or less completely cartilaginous, is now always more or less thoroughly ossified. The notochord is not persistent, and the vertebral column, though sometimes cartilaginous, consists of a number of vertebræ. The bodies of the vertebræ are what is called "amphicœlous"—that is to say, they are concave at both ends. It follows from this, that between each pair of vertebræ there is formed a doubly-conical cavity, and this is filled with the cartilaginous or semi-gelatinous remains of the notochord. By this means an extraordinary amount of flexibility is given to the entire vertebral column. In no fish except the Bony Pike (which belongs to the order *Ganoidei*) is the ossification of the vertebral centra carried further than this. The skull is of an extremely complicated nature, being composed of a number of distinct cranial bones; and a mandible or lower jaw is invariably present.

II. The anterior and posterior pairs of limbs are usually, but not always, present, and when developed they are always in the form of fins. The fins may be supported by "spinous" or "soft" rays, of which the former are simple undivided spines of bone, whilst the latter are divided transversely into a number of short transverse pieces, and also are broken up into a number of longitudinal rays proceeding from a common root. (The Fishes with soft rays in their paired fins are termed "*Malacopterygii*"—those with spinous rays, "*Acanthopterygii.*")

III. Besides the paired fins, representing the limbs, there is a variable number of unpaired or azygous integumentary expansions, which are known as the "median fins." When fully developed (fig. 137), they consist of one or two fins on the back—the "dorsal" fins; one or two on the ventral surface—the "anal" fins; and one clothing the posterior extremity of the body—the "caudal" fin. The caudal fin is set vertically, and not horizontally, as in the Whales and Dolphins, and in all the bony fishes its form is "homocercal"—that is, it consists of two equal lobes, and the vertebral column is not prolonged into the superior lobe. In all the median fins the fin-rays are supported upon a series of dagger-shaped bones, which are plunged in the flesh of the middle line of the body, and are attached to the spinous processes of the vertebræ. These are the so-called "interspinous" bones.

IV. The *heart* consists of two chambers, an auricle and a ventricle, and the branchial artery is furnished with a bulbus arteriosus. The arterial bulb, however, is not furnished with a special coat of striated muscular fibres, is not rhythmically contractile, and is separated from the ventricle by no more than a single row of valves (fig. 142, A).

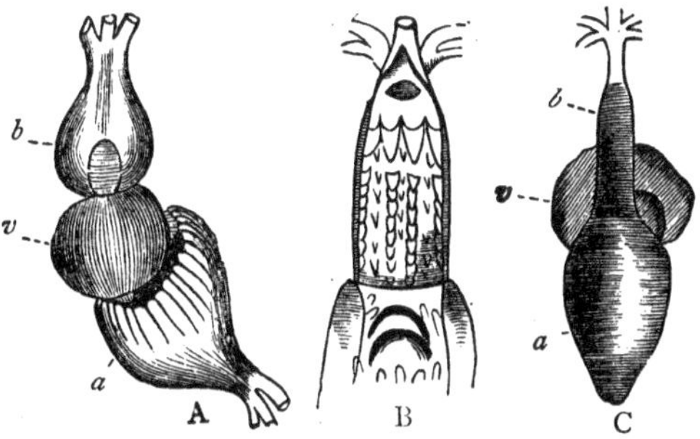

Fig. 142.—A, Heart of the Angler (*Lophius piscatorius*). B, Arterial bulb of Bony Pike (*Lepidosteus*) cut open. C, Heart of the same, viewed externally; *a* Auricle; *v* Ventricle; *b* Arterial bulb.

V. The *respiratory organs* consist of free, pectinated, or tufted branchiæ, situated in two branchial chambers, each of which communicates internally with the pharynx by a series of clefts, and opens externally on the side of the neck by a single aperture (or "gill-slit"), which is protected in front by a bony gill-cover, and is also closed by a "branchiostegal membrane," supported upon "branchiostegal rays." The branchiæ are attached to a series of bony branchial arches, which are connected inferiorly with the hyoid bone and superiorly with the skull; and the water required in respiration is taken in at the mouth by a process analogous to swallowing.

VI. The nasal sacs never communicate posteriorly with the cavity of the pharynx.

The subdivisions of the osseous fishes are so numerous, and they contain so many families, that it will be sufficient to run over the more important sub-orders, and to mention the more familiar examples of each.

SUB-ORDER A. MALACOPTERI, Owen (= *Physostomata*, Müller).—This sub-order is defined by usually possessing a complete set of fins, supported by rays, all of which are "*soft*" or many-jointed, with the occasional exception of the first rays in the dorsal and pectoral fins. A swim-bladder is always present, and always communicates with the œsophagus by means of a duct, which is the homologue of the windpipe. The skin is rarely naked, and is mostly furnished with cycloid scales; but in some cases ganoid plates are present.

This sub-order is one of great importance, as comprising

many well-known and useful fishes. It is divided into two groups, according as ventral fins are present or not. In the first group—*Apoda*—there are no ventral fins; and the most familiar examples are the common Eels of our own country. The Eels (*Murænidæ*) have an elongated, almost cylindrical, body, with the scales deeply sunk in the skin, and scarcely apparent. A swim-bladder is generally present, and the operculum is small and mostly enveloped in the skin. More remarkable, however, than the ordinary Eels is the *Gymnotus electricus*, or great Electric Eel, which inhabits the marshy waters of those wonderful South American plains, the so-called "Llanos." This extraordinary fish (fig. 143) is from five to

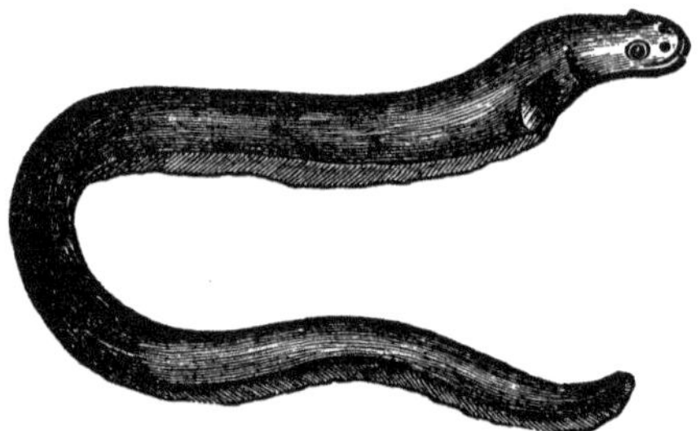

Fig. 143.—Electric Eel (*Gymnotus electricus*).

six feet in length, and the discharge of its electrical organs is sufficiently powerful to kill even large animals. The following striking account is given by Humboldt of the manner in which the *Gymnoti* are captured by the Indians:—"A number of horses and mules are driven into a swamp which is closely surrounded by Indians, until the unusual disturbance excites the daring fish to venture an attack. Serpent-like, they are seen swimming along the surface of the water, striving cunningly to glide under the bellies of the horses. By the force of their invisible blows numbers of the poor animals are suddenly prostrated; others, snorting and panting, their manes erect, their eyes wildly flashing terror, rush madly from the raging storm; but the Indians, armed with long bamboo staves, drive them back into the midst of the pool.

"By degrees the fury of this unequal contest begins to slacken. Like clouds which have discharged their electricity, the wearied eels disperse. They require long rest and nourishing food, to repair the galvanic force which they have so lavishly expended. Their shocks gradually become weaker and weaker. Terrified by the noise of the trampling horses, they timidly approach the banks of the morass, where they are wounded by harpoons, and drawn on shore by non-conducting pieces of dry wood.

"Such is the remarkable contest between horses and fish. That which constitutes the invisible but living weapon of these inhabitants of the waters—that which, awakened by the contact of moist and dissimilar particles, circulates through all the organs of animals and plants—that which, flashing amid the roar of thunder, illuminates the wide canopy of heaven—which binds iron to iron, and directs the silent recurring course of the magnetic needle—all, like the refracted lays of light, flow from one common source, and all blend together into one eternal all-pervading power."

The second group of the *Malacopteri* is that of the *Abdominalia*, in which there are ventral fins, and these are abdominal in position. Space will not permit of more here than merely mentioning that in this section are contained amongst others the well-known and important groups of the *Clupeidæ* (Herring Tribe), the Pikes (*Esocidæ*), the Carps, Barbels, Roach, Chub, Minnow, &c. (*Cyprinidæ*), and the *Salmonidæ*, comprising the various species of Salmon and Trout. Also belonging to this group are the Sheat-fishes (*Siluridæ*), which are chiefly noticeable because they are amongst the small number of living fishes possessed of structures of the same nature as the fossil spines known as "ichthyodorulites." The structure in question consists of the first ray of the pectoral fins, which is largely developed, and constitutes a formidable spine, which the animal can erect and depress at pleasure. Unlike the old "ichthyodorulites," however, the spines of the *Siluridæ* have their bases modified for articulation with another bone, and they are not simply hollow and implanted in the flesh. The "Siluroids" are also remarkable for their resemblance to certain of the extinct ganoid fishes (*e.g.*, *Pterichthys*, *Coccosteus*, &c.), caused by the fact that the head is protected with an exoskeleton of dermal bones. The largest European species is the *Silurus glanis* of the Swiss lakes, and of various European rivers. Another remarkable member of this family is the *Malapterurus* of the Nile and west coast of Africa, which is endowed with electrical powers.

SUB-ORDER B. ANACANTHINI.—This sub-order is distinguished by the fact that the fins are entirely supported by "soft" rays, and never possess "spiny" rays; whilst the ventral fins are either wanting, or, if present, are placed under the throat, beneath or in advance of the pectorals, and supported by the pectoral arch. The swim-bladder may be wanting, but when present it does not communicate with the œsophagus by a duct.

As in the preceding order, the *Anacanthini* are divided into two groups, distinguished by the presence or absence of the ventral fins. In the first of these groups (*Apoda*) are only a few fishes, of which one of the most familiar examples is the little Sand-eel (*Ammodytes lancea*), which occurs on all our coasts. In the second group (*Sub-brachiata*) in which ventral fins exist, are the two important families of the *Gadidæ* and *Pleuronectidæ*. The *Gadidæ* or Cod family, comprising the Haddock, Whiting, Ling, and Cod itself, is of great value to man, most of its members being largely consumed as food. In the *Pleuronectidæ* or Flat-fishes are comprised the Sole, Plaice, Turbot, Halibut, Brill, and others, in all of which there is a very curious modification in the form of the body. The body, namely, in all the Flat-fishes (fig. 144) is very much compressed

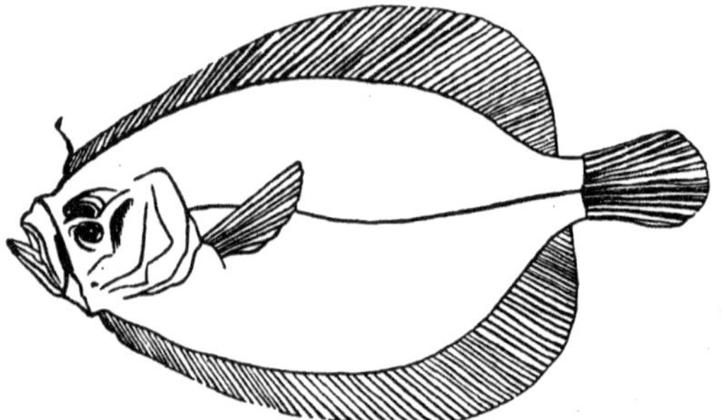

Fig. 144.—Pleuronectidæ. *Rhombus punctatus.* Natural size (after Gosse).

from side to side, and is bordered by long dorsal and anal fins. The bones of the head are twisted in such a manner that the two eyes are both brought to one side of the body, which is sometimes the right side, sometimes the left. The fish usually keeps this side uppermost, and is dark-coloured on this aspect; whilst the opposite side, on which it rests, is white. From this habit of the Flat-fishes of resting upon one flat surface, the sides are often looked upon as the dorsal and ventral surfaces of the body. This, however, is erroneous, as they are shown by the position of the paired fins to be truly the *lateral* surfaces of the body. The mouth has its two sides unequal, the pectorals are rarely of the same size, the ventrals look like a continuation of the anal fin, and the branchiostegal rays are six in number.

SUB-ORDER C. ACANTHOPTERI.—This sub-order is characterised by the fact that one or more of the first rays in the fins are in the form of true, unjointed, inflexible, "spiny" rays. The exoskeleton consists, as a rule, of ctenoid scales. The ventral fins are generally beneath or in advance of the pectorals, and the duct of the swim-bladder is invariably obliterated.

This sub-order comprises two families :—

a. The *Pharyngognathi*, in which the inferior pharyngeal bones are anchylosed so as to form a single bone, which is usually armed with teeth. The family is not of much importance, the only familiar fishes belonging to it being the "Wrasses" (*Cyclolabridæ*).

b. The *Acanthopteri veri*, characterised by having always spiny rays in the first dorsal fin, and usually in the first rays of the other fins, whilst the inferior pharyngeal bones are never anchylosed into a single mass. This family includes many subordinate groups, and may be regarded as, on the whole, the most typical division of the Teleostean fishes. It will not be necessary, however, to do more than mention as amongst the more important fishes contained in it, the Perch family (*Percidæ*), the Mullets (*Mugilidæ*), the Mackerel family (*Scomberidæ*), the Gurnards (*Sclerogenidæ*), the Gobies (*Gobiidæ*), the Blennies (*Blenniidæ*), and the Anglers (*Lophiidæ*). The *Percidæ* form by far the most important member of this group, and are distinguished by having ctenoid scales, the operculum and præ-operculum variously armed with spines, teeth on the vomer and palate as well as on the jaws, and the branchiostegal rays from five to seven in number.

SUB-ORDER D. PLECTOGNATHI.—This sub-order is characterised by the fact that the maxillary and premaxillary bones

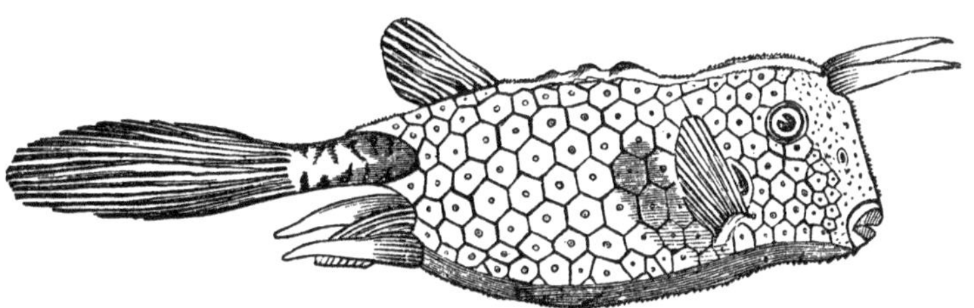

Fig. 145.—Ostraciontidæ. Horned Trunk-fish (*Ostracion cornutus*).

are immovably connected on each side of the jaw. The endoskeleton is only partially ossified, and the vertebral column often remains permanently cartilaginous. The exoskeleton is in the form of ganoid plates, scales, or spines. The ventral fins are generally wanting, and the air-bladder is destitute of a duct.

The most remarkable fishes of this section are the Trunk-fishes (*Ostraciontidæ*, fig. 145), in which the body is entirely enclosed, with the exception of the tail, in an immovable case, composed of large ganoid plates, firmly united to one another at their edges.

Besides the Trunk-fishes, this section also includes the File-fishes (*Balistidæ*) and the Globe-fishes (*Gymnodontidæ*).

SUB-ORDER E. LOPHOBRANCHII.—This is a small and unimportant group, mainly characterised by the peculiar structure of the gills, which are arranged in little tufts upon the branchial arches, instead of the comb-like plates of the typical bony fishes. The endoskeleton is only partially converted into bone, and the exoskeleton, by way of compensation, consists of ganoid plates. The swim-bladder is destitute of an air-duct.

The singular Sea-horses (*Hippocampidæ*), now kept in most of our large aquaria, belong to this sub-order, but the only point about them which requires notice is the curious fact that the males in this family are provided with a sort of marsupial pouch, into which the eggs are placed by the female, and to which the young, when hatched, can retire if threatened by any danger. This singular cavity is only found in the males, and is situated at the base of the tail. More familiar than the Sea-horses are the Pipe-fishes (*Syngnathidæ*), of which one species occurs commonly on our shores.

CHAPTER LVII.

GANOIDEI.

ORDER IV. GANOIDEI.—The fourth order of fishes is the large and important one of the *Ganoid* fishes, represented, it is true, by few living forms, but having an enormous development in past geological epochs. For this reason the study of the Ganoid fishes is one which claims considerable attention.

The order *Ganoidei* may be defined by the following characters:—The endoskeleton is only partially ossified, the vertebral column mostly remaining cartilaginous throughout life, especially amongst the extinct forms of the Palæozoic period, in which the notochord is persistent. The skull is furnished with distinct cranial bones, and the lower jaw is present. The exoskeleton is in the form of ganoid scales, plates, or spines. There are usually two pairs of limbs, in the form of fins, each supported by fin-rays. The first rays of the fins are mostly in the form of strong spines. The pectoral arch has a clavicle, and the posterior limbs (ventral fins) are placed close to the anus. The caudal fin is mostly unsymmetrical or "heterocercal." The swim-bladder is always present, is often cellular, and is provided with an air-duct. The intestine is often furnished with a spiral valve. The gills and opercular apparatus are essentially the same as in the Bony fishes. The heart has one auricle and a ventricle, and the base of the branchial artery is dilated into a bulbus arteriosus, which is rhythmically contractile, is furnished with a distinct coat of striated muscular fibres, and is provided with several transverse rows of valves.

Of these characters, the ones which it is most important to remember are the following:—

I. The *endoskeleton* is rarely thoroughly ossified, but varies a good deal as to the extent to which ossification is carried. In some forms, including most of the older members of the order, the chorda dorsalis is persistent, no vertebral centra are developed, and the skull is cartilaginous, and is protected by ganoid plates. Even in these forms, however, the peripheral elements of the vertebræ are ossified. In others, the bodies of the vertebræ are marked out by osseous or semi-cartilaginous rings, enclosing the primitive matter of the notochord. In others, the vertebræ are like those of the Bony fishes—that is to say, deeply biconcave or "amphicœlous." In one Ganoid, however—the Bony Pike (*Lepidosteus*)—the vertebral column consists of a series of "opisthocœlous" vertebræ,—that is to say, vertebræ which are convex in front and concave behind. This is the highest point of development reached in the spinal column of any fish, and its structure is more Reptilian than Piscine.

II. The *exoskeleton* consists in all Ganoid fishes of scales, plates, or spines, which are said to possess *ganoid* characters. The peculiarities of these scales are that they are composed of two distinct layers—an inferior layer of bone and a superficial covering of a kind of enamel, somewhat similar to the enamel of the teeth, called "ganoine." In form the ganoid scales most generally exhibit themselves as rhomboidal plates, placed edge to edge, without overlapping, in oblique rows, the plates of each row being often articulated to those of the next by distinct processes (fig. 132, *d*). In other cases the ganoid structures are simply in the form of detached plates, tubercles, or spines; and in some cases their *shape* is even undistinguishable from the horny scales of the typical Teleostean fishes. In all cases, however, whatever their form may be, they have the distinctive ganoid structure, being composed of an inferior layer of true bone and a superior layer of enamel. It is to be remembered, however, that these *ganoid* plates and scales are not confined to the fishes of the order *Ganoidei*, but that they occur in two sub-orders of the Bony fishes—namely, the *Plectognathi* and *Lophobranchii*—and in some others of the *Teleostei* as well.

III. As to the *fins*, both pectorals and ventrals are usually present, and the ventrals are always placed far back in the neighbourhood of the anus, and are never situated in the immediate vicinity of the pectorals. In some living and many extinct forms the fin-rays of the paired fins are arranged so as to form a fringe round a central lobe (fig. 146). This structure characterises a division of Ganoids called by Huxley, for this reason, *Crossopterygidæ*, or "fringe-finned." The form of the caudal fin varies, the Ganoids being in this respect intermediate between the Bony fishes, in which the tail is "homocercal," and the Sharks and Rays, in which there is a "heterocercal" caudal fin. In the majority of Ganoids, then, the tail is unsymmetrical or "heterocercal," but it is sometimes equi-lobed or "homocercal."

IV. As to the structure of the *respiratory organs*, the Ganoid fishes agree essentially with the Bony fishes. They all possess *free* pectinated gills attached to branchial arches, and enclosed in a branchial chamber, which is protected by an operculum, and is closed by a branchiostegal membrane, usually supported by branchiostegal rays. Besides the ordinary branchiæ there is frequently an additional gill, called the "opercular branchia," attached to the interior of each operculum, and below this a false gill or "pseudo-branchia," which receives arterialised blood only.

V. There is always a swim-bladder, which is often divided by partitions into several cells, and is always connected with the gullet by an air-duct, as in the Malacopterous division of the Teleostean fishes.

VI. As to the structure of the *heart*, the Ganoids differ from the Bony fishes, and agree with the Sharks and Rays in having a rhythmically contractile bulbus arteriosus, which is furnished with a special coat of striated muscular fibres, and is separated from the ventricle by several rows of valves (fig. 142, B, C). This is a decided advance in structure, as in this way the arterial bulb is enabled to act as a continuation of the ventricle.

VII. The intestine is often furnished with a spiral reduplication of its mucous membrane, forming a spiral valve, such as we shall afterwards see in the Sharks and Rays.

The order of the Ganoid fishes is divided by Owen into the two divisions of the *Lepidoganoidei* and the *Placoganoidei*. The best-known living fishes belonging to the Lepidoganoids are the Bony Pike and the *Polypterus*. The Bony Pike (*Lepidosteus*) inhabits the rivers and lakes of North America, and attains a length of several feet. The body is entirely clothed with an armour of ganoid scales, arranged in obliquely transverse rows. The vertebral column is exceedingly well ossified, and is reptilian in its characters, the bodies of the vertebræ being "opisthocœlous." The jaws form a long narrow snout, armed with a double series of teeth; and the tail is heterocercal.

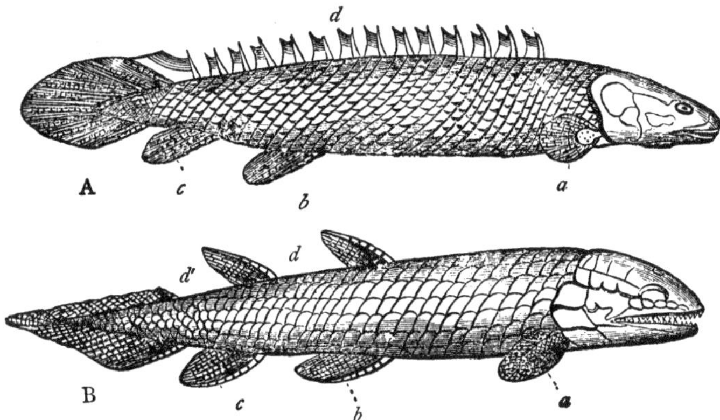

Fig. 146.—Ganoid Fishes. A, *Polypterus*; B, *Osteolepis* (extinct). *a* One of the pectoral fins, showing the fin-rays arranged round a central lobe; *b* One of the ventral fins; *c* Anal fin; *d* Dorsal fin; *d'* Second dorsal fin.

The *Polypteri*, of which several species are known, inhabit the Nile, Senegal, and other African rivers, and are remarkable for the peculiar structure of the dorsal fin (fig. 146, A), which is broken up into a number of separate portions, each composed of a single spine in front, with a soft fin attached to it behind. Two species of *Polypterus* have recently been stated to possess *external* branchiæ when young, losing them when fully grown. This observation, if confirmed, will bring the Ganoids into a nearer relationship with the Mud-fishes (*Lepidosiren*). Another group of Lepidoganoids is formed by the Trout-like *Amiæ* of the fresh waters of the United States.

The section *Placoganoidei* includes the largest and best known of all the living Ganoid fishes—namely, the Sturgeons—and it also contains some highly singular fossil forms. The sub-order is defined by the fact that the skeleton is always imperfectly ossified, and often retains the notochord, whilst the head and more or less of the body are protected by large ganoid plates, which in many cases are united together at their edges by sutures. The tail is heterocercal.

The family *Sturionidæ* comprises the various species of Sturgeon, which are found in the North, Black, and Caspian seas, whence they ascend the great rivers for the purpose of spawning. Other allied forms are peculiar to the North American continent (*e.g.*, the Paddle-fish, *Spatularia*). The vertebral column in the Sturgeon remains permanently in an embryonic condition. The notochord is persistent, and the vertebral centra are wanting, but the neural arches of the vertebræ reach the condition of cartilage. The mouth is destitute of teeth, and the head is covered with an armour of large ganoid plates joined together at their edges by suture. Rows of detached ganoid plates also occur on the body. The various species of Sturgeon attain a great size, one—the Beluga—often measuring twelve or fifteen feet in length. They are commercially of considerable importance, the swimming-bladder yielding most of the *isinglass* of commerce, whilst the roe is largely employed as a delicacy under the name of *caviare.*

Two or three fossil forms belonging to the *Sturionidæ* are all that are at present known; and by far the greater number of extinct Placoganoids belong to the family *Ostracostei*, established by Owen, and characterised by the fact that the head, and generally the anterior part of the trunk as well, was encased in a strong armour composed of numerous large ganoid plates, immovably joined to one another. The posterior extremity of the body was more or less completely unprotected, and, whilst the notochord was persistent, the peripheral elements of the vertebræ—namely, the neural and hæmal spines—were ossified. The following are the more remarkable forms belonging to this section:—

a. Pterichthys.—This is one of the most singular of fossil fishes, and was first discovered in the Old Red Sandstone by the late Hugh Miller. The whole of the head and the anterior part of the trunk were defended by a buckler of large ganoid plates, those covering the trunk forming a back-plate and a breast-plate, articulated together at the sides.

The rest of the body was covered with small ganoid scales (fig. 148). A small dorsal fin, a pair of ventrals, a pair of pectorals, and a heterocercal tail-fin were present. The form of the pectoral fins is the peculiar characteristic of the *Pterichthys*. These were in the form of two long curved spines, something like wings, covered by finely-tuberculated ganoid plates. From their form, they cannot have been of much use in swimming; but they probably, as suggested by Owen, enabled the fish to shuffle along the sandy bottom of the sea, if left dry at low water.

b. Pteraspis.—In most respects this genus was not unlike *Pterichthys*, but it did not possess the peculiar pectoral fins of the latter. One species of Pteraspis has been found in the Upper Silurian Rocks (Ludlow), and is as yet one of the earliest known indications of the appearance of the great sub-kingdom *Vertebrata* upon the globe.

c. Cephalaspis (fig. 147). This, again, is not unlike *Pterichthys* in many respects. The cephalic buckler, however, has

Fig. 147.—*Cephalaspis Lyellii.*

its posterior angles produced backwards, so as to give it the shape of a "saddler's knife," whilst the pectoral limbs have not the form of spines.

d. Coccosteus (fig. 148).—This is another characteristic

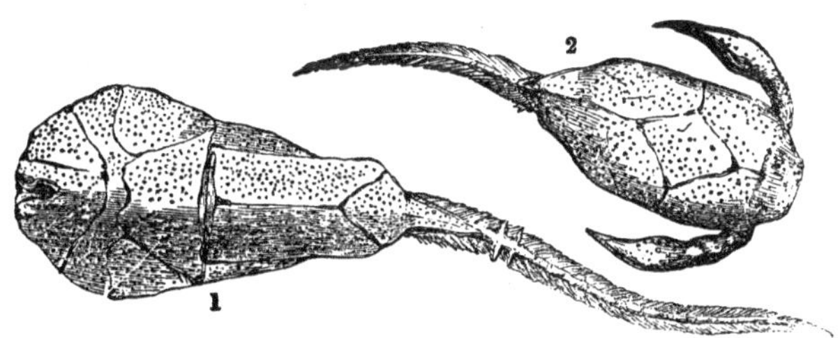

Fig. 148.—1. *Coccosteus decipiens*; 2. *Pterichthys Milleri.*

genus of the Old Red Sandstone. In this genus, as in the preceding, there is a cephalic buckler, the plates of which are covered with small hemispherical tubercles. The notochord was persistent, but the neural and hæmal spines, and the rays of the dorsal and ventral fins, are well ossified. A large heterocercal tail-fin was doubtless present as well.

CHAPTER LVIII.

ELASMOBRANCHII AND DIPNOI.

ORDER V. ELASMOBRANCHII (= *Selachia*, Müller; *Placoidei*, Agassiz; *Holocephali* and *Plagiostomi*, Owen).—This order includes the Sharks, Rays, and Chimæræ, and corresponds with the greater and most typical portion of the *Chondropterygidæ* or Cartilaginous fishes of Cuvier. The order is distinguished by the following characters:—The skull and lower jaw are well developed, but there are no cranial bones, and the skull consists of a single cartilaginous box, without any indication of sutures. The vertebral column is sometimes composed of distinct vertebræ, sometimes cartilaginous or sub-notochordal. The exoskeleton is in the form of placoid granules, tubercles, or spines. There are two pairs of fins, representing the limbs, and supported by cartilaginous fin-rays; and the ventral fins are placed far back near the anus. The pectoral arch has no clavicle. The heart consists of a single auricle and ventricle, and the bulbus arteriosus is rhythmically contractile, is provided with a special coat of striated muscular fibres, and is furnished with several transverse rows of valves. The gills are pouch-like.

In most of the above characters it will be seen at once that the *Elasmobranchii* agree with the Ganoid fishes, especially as regards the structure of the heart. The following points of difference, however, require more special notice:—

I. The *exoskeleton* is what is called by Agassiz "placoid." It consists, namely, of no continuous covering of scales or ganoid plates, but of more or less numerous detached grains, tubercles, or spines, composed of bony matter, and scattered here and there in the integument. In the case of the Rays, these placoid ossifications often take a very singular shape, consisting (fig. 132, *c*) of an osseous or cartilaginous disc, from the upper surface of which springs a sharp recurved spine, composed of dentine.

II. The *gills* are fixed and pouch-like, and differ very materially from those of the Bony and Ganoid fishes. In the case of the Sharks and Rays, the structure of the gills is as follows: —The branchial arches are fixed, and the branchial laminæ are not only attached by their bases to the branchial arches, but are also fixed by the whole of one margin to a series of partitions, which divide the branchial chamber into a number of distinct pouches (fig. 149). Each partition, therefore, carries a series of branchial laminæ attached to each side like the leaves of a book. By means of these septa a series of branchial sacs or pouches are formed, each of which opens internally into the pharynx by a separate slit, and communicates externally with the water by a separate aperture placed on the side of the neck (fig. 149, B). The arrangement of the gills being such, there is, of course, no gill-cover, and no branchiostegal membrane or rays. In one section of the order, however—viz., the *Holocephali*—though the *internal* structure of the gills is the same as the above, there is only a single branchial aperture or gill-slit *externally*, and this is protected by a rudimentary operculum and branchiostegal rays.

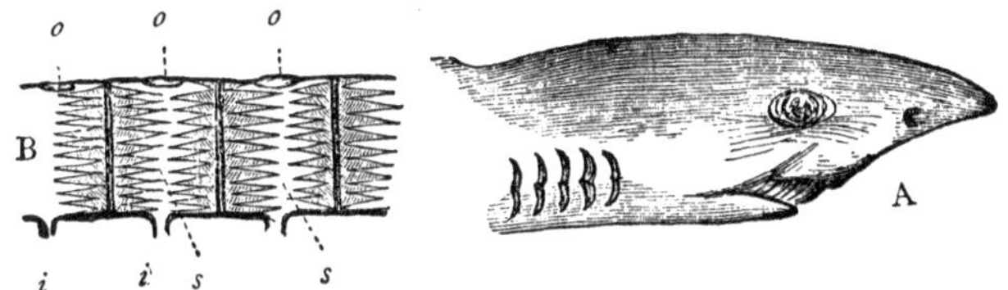

Fig. 149 —A, Head of Piked Dog-fish (*Spinax*), showing the transverse mouth on the under surface of the head, and the apertures of the gill-pouches. B, Diagram of the structure of the gill-pouches: *o o* External apertures; *i i* Apertures leading into the pharynx; *s s* Gill-sacs, containing the fixed gills.

III. Another character in the *Elasmobranchii*, shared, however, by many of the Ganoids, is the structure of the intestinal canal. The intestine is extremely short; but, to compensate for this, there is a peculiar folding of the mucous membrane, constituting what is known as the "spiral valve." The mucous membrane, namely, from the pylorus to the anal aperture, is folded into a spiral reduplication, which winds in close coils round the intestine, like the turns of a screw. By this means the absorptive surface of the intestine is enormously increased, and its shortness is thus compensated for.

The order *Elasmobranchii* is divided into two sub-orders—the *Holocephali*, characterised by the mouth being terminal in position, and there being only a single gill-slit; and the *Plagiostomi*, in which the mouth is transverse, and placed on the under surface of the head (fig. 149, A), and there are several branchial apertures on each side of the neck.

SUB-ORDER A. HOLOCEPHALI.—This sub-order includes certain curious fishes, of which the only living forms are the *Chimæridæ*. The notochord is persistent; but the neural arches and transverse processes are cartilaginous. The jaws are bony, and are covered by broad plates representing the teeth. The exoskeleton consists of placoid granules. The first ray of the anterior dorsal fin is in the form of a powerful defensive spine, like the "ichthyodorulites" of many fossil fishes. The ventral fins are abdominal, and the tail is heterocercal. There is only a single external gill-aperture, covered with a gill-cover and branchiostegal membrane; but only a small portion of the borders of the branchial laminæ is free. The mouth is placed at the extremity of the head.

The best-known living representative of the sub-order is the *Chimæra monstrosa* (fig. 150, B), commonly known as the

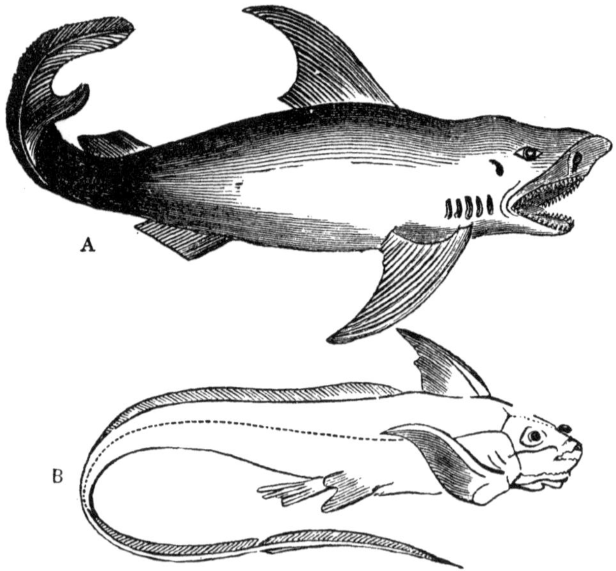

Fig. 150.—Plagiostomi and Holocephali. A, White Shark (*Carcharias*); B, *Chimæra monstrosa*. (After Gosse.)

"king of the Herrings." In *Chimæra* there is only one apparent gill-slit, but the gills really adhere to the integument by a large portion of their borders, and there are consequently five holes communicating with the gill-slit. A rudimentary operculum is present, covered by the skin. In the closely-allied *Callorhynchus* from the South Seas, there is a large fleshy appendage at the end of the snout. In the Secondary and Tertiary Rocks, are found several fossil forms, constituting the genera *Edaphodus*, *Elasmodus*, and *Ischiodus*.

SUB-ORDER B. PLAGIOSTOMI.—This sub-order is of considerably greater importance, as it includes the well-known Sharks and Rays. The vertebral centra are usually more or less ossified, and even when quite cartilaginous, the centra are marked out by distinct rings. The skull is in the form of a cartilaginous capsule, without distinct cranial bones. The mouth is transverse, and is placed on the under surface of the head (fig. 149, A). The exoskeleton consists of placoid granules, tubercles, or spines. The branchial sacs open externally by as many distinct apertures as there are sacs, and there is no operculum. A pair of tubes proceed from the pharynx to open on the upper surface of the head by two apertures, which are termed "spiracles." By means of these water can be admitted to the pharynx, and thence to the gills.

By Professor Owen the Plagiostomi are divided into three sections, termed respectively the *Cestraphori*, the *Selachii*, and the *Batides*.

a. Cestraphori.—In this division there is a strong spine in front of each dorsal fin, and the back teeth are obtuse. The only living representative of this group is the Port Jackson Shark (*Cestracion Philippi*), characterised by its pavement of plate-like crushing teeth, adapted for comminuting small Molluscs and Crustaceans. It is exclusively an inhabitant of the Australian seas, and is remarkable for its close resemblance to a large group of extinct forms, of which the best known are the genera *Hybodus* and *Acrodus* from the Secondary Rocks.

b. Selachii.—This group comprises the formidable Sharks and Dog-fishes, and is characterised by the lateral position of the branchiæ on the side of the neck, and by the fact that the pectoral fins have their ordinary form and position. The Dog-fishes are of common occurrence in British seas, but are of little value. Their egg-cases are frequently cast up on our shores, and are familiarly known as "Mermaid's purses." The true Sharks are not infrequently found in various European seas, but they are mostly inhabitants of warmer waters. One of the largest is the "White Shark" (*Carcharias vulgaris*), which attains a length of over thirty feet (fig. 150, A). The body in the Sharks (*Squalidæ*) is not rhomboidal, but is elongated; the nostrils are placed on the under side of the snout, and the teeth are arranged in several rows, of which the outermost alone is employed, the inner ones serving to replace the former when worn out.

c. Batides.—This group includes the Rays and Skates, and is distinguished by the fact that the branchial apertures are placed on the under surface of the body, forming two rows of openings a little behind the mouth. In the typical members of the group, the body is flattened out so as to form a kind of rhomboidal disc (fig. 151), the greater part of which is made up of the enormously-developed pectoral fins. Upon the upper surface of the disc are the eyes and spiracles; upon the lower surface are the nostrils, mouth, and branchial apertures. The flattened bodies of the Rays, however, must be carefully distinguished from those of the Flat-fishes (*Pleuronectidæ*). In the former, the flat surfaces of the body are truly the dorsal and ventral surfaces. In the latter, as before remarked, the body is flattened, not from above downwards, but from side to side, and the head is so twisted that both eyes are brought to one side of the body. The tail in the Rays is long and slender, usually armed with spines, and generally with two or three fins (the homologues of the dorsal fins). The mouth is paved with flat teeth of a more or less rhomboidal shape.

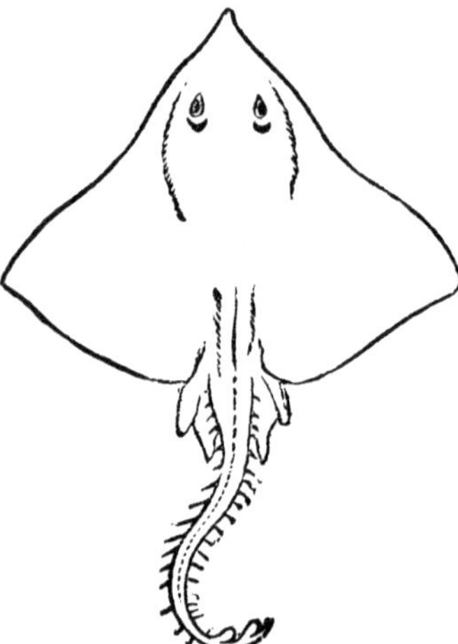

Fig. 151.—Batides. *Raia marginata*, one of the Skates. Reduced one-sixth. (After Gosse.)

The typical members of the *Batides* are the Skates and Rays, of which the common Thornback (*Raia clavata*) may be taken as a familiar example. More remarkable than the common Rays is the Electric Ray or *Torpedo*, which has the power of discharging electrical shocks, if irritated. The identity of the force produced in this way with the electricity of the machine has been demonstrated by many careful experiments. The Torpedo owes its remarkable powers to two special organs —the "electrical organs," which consist of two masses placed on each side of the head, and consisting each of numerous vertical gelatinous columns, separated by membranous septa, and richly furnished with nerves from the eighth pair; the whole arrangement presenting a singular resemblance to the cells of a voltaic battery. There is no doubt, however, but that the force which is expended in the production of the electricity is only nerve-force. For every equivalent of electricity which is generated, the fish loses an equivalent of nervous energy; and for this reason, the production of the electric force is strictly limited by the amount of nerve-force possessed by the animal.

Other well-known members of the family are the Sting-rays (*Trygon*), the Eagle-rays (*Myliobatis*), the Horned Rays (*Cephaloptera*), and the Beaked Rays (*Rhinobatis*).

In the Saw-fish (*Pristis antiquorum*) the body has not the typical flattened form of the Rays, and the snout is elongated so as to form a long sword-like organ, the sides of which are furnished with strong tooth-like spines. This constitutes a powerful weapon, with which the Saw-fish attacks the largest marine animals.

Before leaving the *Elasmobranchii*, a few words may be said as to their position in the class of fishes. From the cartilaginous nature of the endoskeleton, and the similarity between the form of their gills and those of the Lampreys and Myxinoids, the *Elasmobranchii* were long placed low down in the scale of fishes, to which also the permanently heterocercal tail conduced. When we come, however, to take into consideration the sum of all their characters, there can be little hesitation in placing the order nearly at the summit of the entire class. The nervous system, and especially the cerebrál mass, is very much more highly developed proportionately than is the case with any other division of the fishes. The organs of sense are, comparatively speaking, of a very high grade of organisation, the auditory organs being more than ordinarily elaborate, the eyes being sometimes furnished with a third eyelid (*membrana nictitans*), and the nasal sacs having a very complex structure. The structure of the heart agrees with that of the Ganoids, and is a decided advance upon the heart of the more typical bony fishes. Finally, the embryo, before its exclusion from the egg, is furnished with external filamentous branchiæ, this being a decided approximation to the *Amphibia*.

ORDER VI. DIPNOI (= *Protopteri*, Owen).—This order is a very small one, and includes only the singular Mud-fishes (*Lepidosiren*); but it is nevertheless of great importance as exhibiting a distinct transition between the fishes and the *Amphibia*. So many, in fact, and so striking, are the points of resemblance between the two, that until recently the *Lepidosiren* (fig. 152) was always made to constitute the lowest class of the *Amphibia*. The highest authorities, however, now concur in placing it amongst the fishes, of which it constitutes the highest order. The order *Dipnoi* is defined by the following characters :—The body is fish-like in shape. There is a skull with distinct cranial bones and a lower jaw, but the notochord is persistent, and there are no vertebral centra, nor an occipital condyle. The exoskeleton consists of small, horny, overlapping scales, having the "cycloid" character. The pectoral and ventral limbs are both present, but have the form of awl-shaped, filiform, many-jointed organs, of which the former only have a membranous fringe inferiorly. The ventral limbs are attached close to the anus, and the pectoral arch has a clavicle; but the scapular arch is attached to the occiput. The hinder

Fig. 152.—Dipnoi. *Lepidosiren annectens.*

extremity of the body is fringed by a vertical median fin. The heart has two auricles and one ventricle. The respiratory organs are twofold, consisting, on the one hand, of free filamentous gills contained in a branchial chamber, which opens externally by a single vertical gill-slit; and, on the other hand, of true lungs in the form of a double cellular air-bladder, communicating with the œsophagus by means of an air-duct or trachea. The branchiæ are supported upon branchial arches, but these are not connected with the hyoid bone; and in some cases, at any rate, rudimentary *external* branchiæ exist as well. The nasal sacs open posteriorly into the throat.

If these characters are examined a little more minutely, it is easy to point to those in which the *Lepidosiren* approaches the Fishes, and to those in which it resembles the Amphibians. It resembles the Fishes in the shape of the body, and in the possession of a covering of horny overlapping scales of the true cycloid character; whilst the limbs are more like those of fishes than of reptiles. The fin, also, which clothes the posterior extremity of the body, is of a decided fish-like character. The most marked piscine feature, however, is the presence of free branchiæ, attached to branchial arches, and placed in a branchial cavity, which opens internally into the pharynx by a number of slits, and communicates externally with the outer world by means of a single vertical gill-slit.

On the other hand, the *Lepidosiren* approximates to the Amphibians in the following important points:—The heart consists of *three* cavities, two auricles and a single ventricle. True lungs are present, with a trachea and glottis, returning their blood to the heart by a distinct pulmonary vein, and in every respect discharging the functions of the lungs of the higher Vertebrates. It is true that the lungs of the *Lepidosiren* are merely a modification of the swim-bladder of the other fishes, but the significance of the change of function is not affected by this. Lastly, sometimes, at any rate, there are rudimentary external branchiæ placed on the side of the neck. This feature, as will be seen shortly, is characteristic of all the Amphibians, either permanently or in their immature state.

Upon the whole, then, whilst for the purposes of systematic classification the *Lepidosiren* must be placed amongst the Fishes, it is not to be forgotten that many of its characters are those of a higher class, and that it may justly be looked upon as a connecting link, or transitional form, between the two great divisions of the Fishes and the Amphibians.

As regards their distribution and mode of life, two species at least of *Lepidosiren* are known—the *L. paradoxa* from the Amazon, and the *L. annectens* from the Gambia. They both inhabit the waters of marshy tracts, and appear to be able in the dry season to bury themselves in the mud, forming a kind of chamber, in which they remain dormant till the return of the rains. Recently there has been discovered in the rivers of Queensland (Australia) a fish which has been described under the name of *Ceratodus* (?) *Fosteri*, and which would appear to be very closely related to the *Lepidosiren*. This singular fish is from three to six feet long, and has the body covered with large cycloid scales. The skeleton is notochordal, all the bones remaining permanently cartilaginous. There is a well-developed operculum, but—as in *Lepidosiren*—no branchiostegal rays. The tail is homocercal, and the pectoral and ventral fins are supported by a median, many-jointed, cartilaginous rod, with numerous lateral branches on each side. The heart consists of a single auricle and ventricle, with a "Ganoid" bulbus arteriosus. There are five branchial arches, of the Teleostean type, but cartilaginous. The swim-bladder is single, composed of two symmetrical halves, cellular in structure, with a pneumatic duct and glottis, as in *Lepidosiren*. The intestine has a spiral valve, and there are no pyloric cæca.

Upon the whole, Dr Günther concludes that the *Dipnoi* are to be regarded as a simple sub-order of Ganoids, and that the entire order *Ganoidei* may be united with the *Elasmobranchii* into a single order, called *Palæichthyes*, characterised by having a "heart with a contractile bulbus arteriosus, intestine with a spiral valve, and optic nerves non-decussating."

CHAPTER LIX.

DISTRIBUTION OF FISHES IN TIME.

THE geological history of fishes presents some points of peculiar interest. Of all the classes of the great sub-kingdom *Vertebrata*, the fishes are the lowest in point of organisation. It might therefore have been reasonably expected that they would present us with the first indications of vertebrate life upon the globe; and such is indeed the case. After passing through the enormous group of deposits known as the Laurentian, Huronian, Cambrian, and Lower Silurian formations—representing an immense lapse of time, during which, so far as we yet know, no vertebrate animal had been created, we find in the Upper Silurian rocks the first traces of fish. The earliest of these, in Britain, is found in the base of the Ludlow rocks (Lower Ludlow Shale), and belongs to the placoganoid genus *Pteraspis*. Also in the Ludlow rocks, but at the summit of their upper division, are found fin-spines and shagreen, probably belonging to Cestraciont fishes—that is to say, to fishes of as high a grade of organisation as the *Elasmobranchii*. So abundant are the remains of fishes in the next great geological epoch—namely, the Devonian or Old Red Sandstone—that this period has frequently been designated the "Age of Fishes." Most of the fishes of the Old Red Sandstone belong to the order *Ganoidei*. In the Carboniferous and Permian rocks which close the Palæozoic period, most of the fishes are still Ganoid, but the former contain the remains of many Plagiostomous fishes. At the close of the Palæozoic and the commencement of the Mesozoic epoch, the Ganoid fishes begin to lose that predominant position which they before occupied, though they continue to be represented through the whole of the Mesozoic and Kainozoic periods up to the present day. The Ganoids, therefore, are an instance of a family which has endured through the greater part of geological time, but which early attained its maximum, and has been slowly dying out ever since. Towards the close of the Mesozoic period (in the Cretaceous period) the great family of the Teleostean or Bony fishes is for the first time known certainly to have made its appearance. The families of the *Marsipobranchii*, *Pharyngobranchii*, and *Dipnoi*, have not left, so far as is known, any traces of their existence in past time. Judging from analogy, however, it is highly probable that the two former of these must have had a vast antiquity, and it is not impossible that the so-called "Conodonts" from the Lower Silurian rocks of Russia may yet be shown to be the horny teeth of fishes allied to the Lampreys. At present, however, the weight of evidence is in favour of looking upon these problematical little bodies as probably referable to some of the *Invertebrata*.

Leaving these unrepresented orders out of consideration, the following are the chief facts as to the geological distribution of the other great groups:—

I. *Ganoidei.*—As far as is yet known with certainty, the oldest representatives of the fishes belong to this order. The order is represented, namely, in the Upper Silurian rocks by the remains of at least four genera. In the Devonian rocks, or Old Red Sandstone, the Ganoids attain their maximum both in point of numbers and development. The Placoganoid division of the order is represented by the singular genera *Pterichthys* (fig. 148), *Cephalaspis* (fig. 147), *Pteraspis*, and *Coccosteus* (fig. 148). The Lepidoganoid division of the order is now also abundantly represented for the first time, the genera *Dipterus*, *Osteolepis* (fig. 146), *Glyptolepis*, *Holoptychius*, *Diplacanthus*, and many others belonging to this section. As regards the further distribution of the Placoganoids, the section of the *Ostracostei*, characterised by the great development of the cephalic buckler, appears to have died out at the close of the Devonian period. The other section, however—namely, that of the *Sturionidæ*—is represented in the Liassic period (*Mesozoic*) by the genus *Chondrosteus*, and in the Eocene (*Kainozoic*) by a true Sturgeon, the *Acipenser toliapicus.*

The Lepidoganoids continue from the period of the Old Red in great profusion, and they are represented by very many genera in the Carboniferous and Permian rocks. In the earlier portion of the Mesozoic period—*i.e.*, in the Lias and Trias—they are still represented, but all the forms are as yet heterocercal. In the Oolitic rocks, for the first time, Lepidoganoids with homocercal tails appear, and they continue to be represented up to the present day.

II. *Elasmobranchii.*—Like the *Ganoidei*, the great order of the Sharks and Rays is one of vast antiquity. At the top of the Upper Ludlow rocks, or at the close of the Upper Silurian epoch, there have been discovered the remains of undoubted Plagiostomous fishes, most nearly allied to the existing Port Jackson Shark (*Cestracion Philippi*). These remains consist chiefly of defensive spines, which formed the first rays in the dorsal fins, and upon these the genus *Onchus* has been founded. Besides these there have been found portions of skin or "shagreen," with little placoid tubercles, like the skin of a living shark. These have been referred to the genus *Sphagodus.* They are the earliest-known remains of Plagiostomous fishes, and with the exception of the few remains from the Lower Ludlow rocks, they are the earliest known remains of fishes in the stratified series. The discovery of these remains, at that time the earliest known traces of Vertebrate life, is due to the genius of Sir Roderick Murchison, the author of 'Siluria.'

Most of the fossil *Elasmobranchii* belong to the division *Cestraphori* of Owen, so called because they are provided with the large fin-spines, which are known to geologists as "ichthyo-

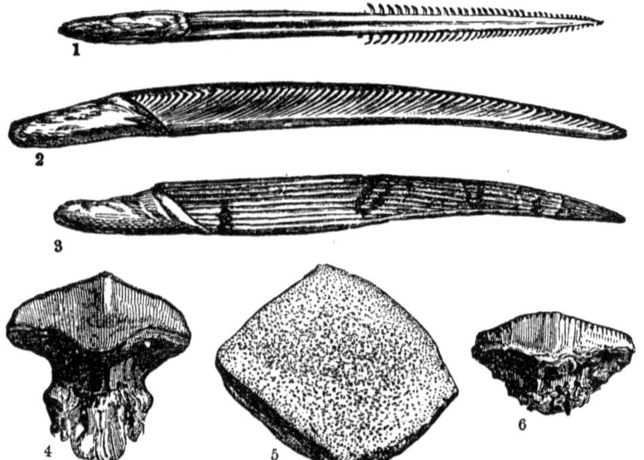

Fig. 153.—1. Fin-spine of *Pleuracanthus* (one of the rays); 2. *Gyracanthus*; 3. *Ctenacanthus*; 4. Tooth of *Petalodus*; 5. *Psammodus*; 6. *Ctenoptychius.* All from the Carboniferous Rocks.

dorulites." The two families of this division—the Cestracionts and Hybodonts—are largely represented in past time, the former chiefly in the Palæozoic period, the latter chiefly in the

Mesozoic rocks. Subjoined is an illustration of the "ichthyodorulites" and teeth of some of the Palæozoic *Cestraphori.*

The true Sharks are represented in the later Mesozoic deposits (*e.g.*, by teeth of *Notidanus* in the Oolites); but they are chiefly Tertiary. The teeth of *Odontaspis*, *Galeocerdo*, and *Carcharodon*, are good examples from the Eocene of the Isle of Sheppey. The true Rays are older than the true Sharks, the Carboniferous fossil, *Pleuracanthus*, being probably the spine of a Ray (fig. 153). Numerous remains of Rays, chiefly in the form of the pavement-like teeth, are known, both from the Secondary and Tertiary rocks. The last division of the *Elasmobranchii*—viz., that of the *Holocephali*, is poorly represented in past time by the Mesozoic and Kainozoic *Ischiodus*, *Elasmodus*, *Ganodus*, and *Edaphodus*.

III. The *Bony* or *Teleostean* Fishes do not make their appearance sooner than the Cretaceous period—that is, towards the close of the Mesozoic epoch. From this time on, however, Bony fishes with cycloid or ctenoid scales are the chief representatives of the whole class, and the order appears to have attained its maximum in our present seas.

DIVISION I. ICHTHYOPSIDA.

CHAPTER LX.

CLASS II.—AMPHIBIA.

THE class *Amphibia* comprises the Frogs and Toads, the Salamandroids, the *Cæciliæ*, and the extinct *Labyrinthodonts*, and may be briefly defined as follows:—As is the case with the Fishes, the embryo is not furnished with an amnion, and the urinary bladder is the only representative of the allantois. As in Fishes, also, *branchiæ or filaments adapted for breathing air dissolved in water are always developed upon the visceral arches for a longer or shorter time.* On the other hand, the Amphibians differ from the Fishes in the fact that *true lungs are always present in the adult; the limbs are never converted into fins; and when median fins are present, as is sometimes the case, these are never furnished with fin-rays.* The limbs, when present, exhibit in their skeleton the same parts as do the limbs of the higher Vertebrates. *The skull always articulates with the vertebral column by means of two occipital condyles. The heart consists of two auricles and a single ventricle. The nasal sacs communicate posteriorly with the pharynx; and the rectum, ureters, and ducts of the reproductive organs open into a common chamber or "cloaca."*

The great and distinguishing character of the *Amphibia* is the fact that they undergo a *metamorphosis* after their exclusion from the egg. They commence life as water-breathing larvæ, provided with gills or branchiæ; but in their adult state they invariably possess lungs—the branchiæ in the higher forms disappearing when the lungs are developed—but being in other cases permanently retained throughout life.

In the earliest embryonic condition the branchiæ are *external*, placed on the side of the neck, and not situated in an internal chamber as in Fishes. In some cases the external branchiæ only are present, and they are, in any case, the gills which are retained in those forms in which the branchiæ are permanent (*Perennibranchiata*). In the tailed Amphibians

(*Urodela*), and in the Frogs and Toads (*Anoura*), two sets of gills are developed—an external set, which is very soon lost, and an internal set, which is retained for a longer or shorter

Fig. 154.—*Anoura*. *Hyla leucotænia*, one of the Tree-frogs (after Günther).

period. As maturity is approached, true lungs adapted for breathing air are developed. The development, however, of the lungs varies with the completeness with which aerial respiration has to be accomplished; being highest in those forms which lose their gills when grown up (*Caducibranchiata*), and lowest in those in which the branchiæ are retained throughout life (*Perennibranchiata*).

In accordance with the change from an aquatic or branchial to a more or less completely aerial or pulmonary mode of respiration, considerable changes are effected in the course and distribution of the blood-vessels. In the larval condition, when the respiration is entirely effected by means of the gills, the circulation is carried on very much as it is in Fishes. The heart is composed of a single auricle and ventricle, and the blood is propelled through a bulbus arteriosus and branchial artery to the gills. The aerated blood is then collected in the branchial veins, and instead of being returned to the heart, is forthwith propelled to all parts of the body, the descending aorta being formed out of the branchial veins. At this stage, therefore, the heart is a branchial one, and the single contraction of the heart is sufficient to drive the blood through both the branchial and systemic circulations, just as we saw was permanently the case with all the Fishes except the *Lepi-*

dosiren. The pulmonary arteries are at first very small, and take their origin from the last pair of branchial arteries. When the lungs, however, are developed, and the respiration commences to be aerial, the pulmonary arteries increase proportionately in size, and more and more blood is gradually diverted from the gills and carried to the lungs, so that the branchiæ suffer a proportionate diminution in size. In those Amphibians in which branchiæ are permanently retained (*Perennibranchiata*), this state of affairs remains throughout life—that is to say, a portion of the venous blood is sent by the pulmonary artery to the lungs, and a portion goes to the gills. In those Amphibians, however, in which the adult breathes by lungs alone (*Caducibranchiata*), further changes ensue. In these the pulmonary arteries increase so much in size that they ultimately divert all the blood from the branchiæ, and these organs, having fulfilled their temporary function, become atrophied and disappear. The vessels which return the aerated blood from the lungs (the pulmonary veins) increase in size proportionately with their increased work, and ultimately come to open into a second auricle formed at their point of union. The heart, therefore, of the *Amphibia* in their adult state consists of *two* auricles and a common ventricle. The right auricle receives the venous blood from the body, and the left receives the arterial blood from their lungs, and both empty their contents into the single ventricle. As in Reptiles, therefore, the ventricular cavity of the heart in adult Amphibians contains a mixed fluid, partly venous and partly arterial, and from this both the body and the lungs are supplied with blood.

As regards the digestive system of the *Amphibia* there is little to say, except that the rectum opens, as it does in Reptiles, into a common chamber or "cloaca," into which are also discharged the secretions of the kidneys and generative organs. A liver, gall-bladder, spleen, and pancreas are always present. Singular pulsating cavities, belonging to the lymphatic system, and known as "lymph-hearts," are also present in the higher Amphibians.

CHAPTER LXI.

ORDERS OF AMPHIBIA.

THE *Amphibia* are usually divided by modern writers into four orders, the old order *Lepidota*, comprising the *Lepidosiren*, being now placed at the head of the Fishes, under the name of *Dipnoi.* Whilst there is a general agreement as to the number and characters of the Amphibian orders, the names employed to designate them are very various, and it really matters little which are adopted.

ORDER I. OPHIOMORPHA, Owen (= *Gymnophiona*, Huxley; *Apoda* of older writers; *Ophidobatrachia*).—This is a small order, including only certain snake-like, vermiform animals, which are found in various tropical countries, burrowing in marshy ground, something like gigantic earthworms. They form the family *Cæciliadæ* (so called by Linnæus from their supposed blindness), and are characterised by their snake-like form, and by having the anus placed almost at the extremity of the body. The skin is quite soft, but differs from that of the typical Amphibians in mostly having small horny scales embedded in it. Another fish-like character is that the vertebræ are amphicœlous or biconcave, and the cavities formed by their apposition are filled with the cartilaginous or gelatinous remains of the notochord. The body is cylindrical and worm-like, and is completely destitute of limbs. The skin is glandular, naked, and viscous, thrown into numerous folds, and containing nu-

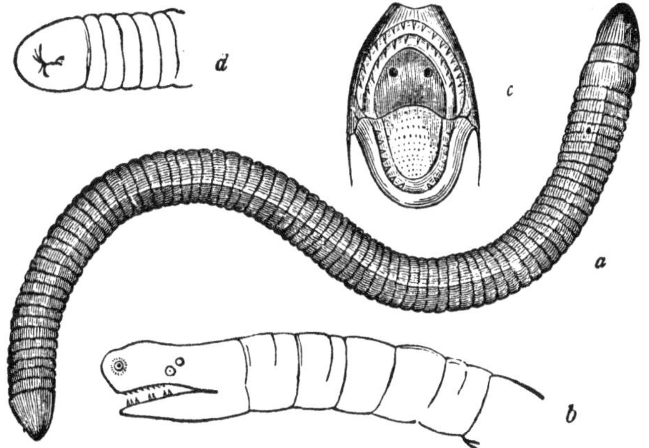

Fig. 155.—Ophiomorpha. *a Siphonops annulatus*, one of the Cæcilians, much reduced *b* Head; *c* Mouth, showing the tongue, teeth, and internal openings of the nostrils; *d* Tail and cloacal aperture. (After Dumeril and Bibron.)

merous delicate, rounded, horny scales, which are dermal in their character, and are wanting in *Siphonops annulatus.* The mandibular rami are short, and are united in front by a symphysis. The teeth are long, sharp, and generally recurved;

and a row of palatine teeth forms a concentric series with the maxillary teeth. The tongue is fleshy, fixed to the concavity of the lower jaw, and not protrusible (fig. 155). The ribs are numerous, but there is no sternum. The adult possesses lungs, one of which is smaller than the other, and the nose opens behind into the mouth. The eyes are rudimentary, nearly concealed beneath the skin, or altogether wanting.

The position of the *Cæciliæ* was long doubtful; but their Amphibian character was ultimately proved by the discovery that whilst the adult breathes by lungs, the young possess internal branchiæ, communicating with the external world by a branchial aperture on each side of the neck. Only a few species of *Cæcilia* are known, and they are all inhabitants of hot climates, such as South America, Java, Ceylon, and the Guinea coast. They sometimes attain a length of several feet.

ORDER II. URODELA (= *Ichthyomorpha*, Owen; *Saurobatrachia*).—This order is commonly spoken of collectively as that of the "Tailed" Amphibians, from the fact that the larval tail is always retained in the adult. The *Urodela* are characterised by having the skin naked, and destitute of any exoskeleton. The body is elongated posteriorly to form a compressed or cylindrical tail, which is permanently retained throughout life. The dorsal vertebræ are biconcave (*amphicœlous*), or concave behind and convex in front (*opisthocœlous*), and they have short ribs attached to the transverse processes. The bones of the fore-arm (*radius* and *ulna*) on the one hand, and those of the shank (*tibia* and *fibula*) on the other, are not anchylosed to form single bones.

In one section of the order—formerly called *Amphipneusta*—the gills are retained throughout life, and the animal is therefore "perennibranchiate." In this section are the *Proteus*, *Siren*, and *Menobranchus*. In the remaining members of the order the gills disappear at maturity, and the animal is therefore "caducibranchiate." In this section are the land and water salamanders. One form, however—the Axolotl of Mexico—appears to be sometimes caducibranchiate, though generally perennibranchiate. The genera *Amphiuma* and *Menopoma*, also, exhibit a partially intermediate state of parts; for though they lose their branchiæ when adult, they nevertheless retain the branchial apertures behind the head.

Of the perennibranchiate *Urodela*, one of the best known is the singular *Proteus* (*Hypochthon*) *anguinus* (fig. 156), which is only found inhabiting pools in certain caves in Illyria and Dalmatia. It is of a pale flesh-colour, or nearly white, with three pairs of scarlet branchiæ on each side of the neck. It attains a length of about a foot, and has two pairs of weak limbs, of which the anterior have three toes, and the posterior only two. From its habitat, the power of vision must be quite unnecessary, and, as a matter of fact, the eyes are altogether rudimentary, and are covered by the skin. Several varieties of *Proteus* are known, and the one figured above has been described as a distinct species (*P. xanthostictus*). The blood-corpuscles in *Proteus* are oval in shape, and are larger than those of any other Vertebrate animal.

Fig. 156.—Head and fore-part of the body of *Proteus anguinus*, showing the external branchiæ and tridactylous fore-limb.

Of the *Sirenidæ*, the most familiar are the Sirens and the Axolotls. The Siren, or Mud-eel, is found abundantly in the rice-swamps of South Carolina, and attains a length of three feet. The branchiæ are persistent, and the hinder pair of legs wholly wanting. Two other species are known, but they are likewise confined to North America.

The Mexican Axolotl (*Siredon pisciforme*, fig. 157) is a native of the Mexican lakes, and attains a length of about a foot or fourteen inches. It possesses both pairs of limbs, the anterior pair having four toes and the hinder pair five toes. As ordinarily known in its native country, the Axolotl is certainly perennibranchiate, and they breed in this condition freely. There is no doubt, however, that individual specimens may lose their gills, without thereby suffering any apparent change, except it be one of colour. The Axolotl, therefore, is in the singular position of being sometimes "caducibranchiate," whilst it is ordinarily "perennibranchiate." * Nearly allied

* Professor Marsh of New Haven has recently shown that the *Siredon lichenoides* of the western States of America, when kept in confinement, loses its gills, and dorsal and caudal fins, whilst it changes much in colour, and undergoes various minor modifications in structure. Its habits, also, become less aquatic, and it becomes apparently absolutely identical with *Amblystoma mavortium*, a Salamandroid. This discovery has thrown considerable doubt upon the value of the distinction between perennibranchiate and caducibranchiate Amphibians, and has rendered it probable that all the species of *Siredon* are merely larval Salamanders, as long ago suspected by Cuvier. At the same time, the Axolotls certainly breed freely whilst in possession of their branchiæ, and there is as yet no proof that they lose their gills whilst in a state of nature.

to the Axolotl is the *Menobranchus* of North America, in which the branchiæ are persistent. *Amphiuma* and *Menopoma*, as already remarked, differ from the forms just mentioned in losing the gills when adult, but in retaining the external branchial apertures on the side of the neck. The former is exclusively North American, whilst the latter is represented by different but nearly-related species in both North America and Java. Both possess the normal two pairs of limbs. The Javanese species (*Menopoma maxima*) has the gill-slit closed in the adult. It is about three feet long, and is the nearest living relative of the extinct *Andrias*.

Fig. 157.—The Axolotl (*Siredon pisciforme*)—after Tegetmeier. The ordinary form, with persistent branchiæ.

In the second section of the *Urodela*, comprising those forms in which the gills are caducous, and both pairs of limbs are always present, are the Water-salamanders or Tritons, and the Land-salamanders. The Tritons are the only examples of the aquatic Salamanders which occur in Britain, and every one, probably, is acquainted with the common Newt.

The Water-salamanders or Newts (fig. 159) are distinguished from the terrestrial forms by being furnished with a compressed fish-like tail, and by being strictly oviparous. The larvæ are tadpole-like, with external branchiæ, which they retain till about the third month. The adult is destitute of gills, and breathes by lungs alone, but the larval tail is retained throughout the life of the animal. The tongue is small, free, and pointed behind, and there are two rows of palatine teeth. The fore-feet are four-toed, the hind-feet five-toed; and the males have a crest on the back and tail.

The development of the Newts is so like that of the Frogs that it is unnecessary to dilate further upon it here; but there are these two points of difference to be noticed:—1*stly*, That the embryonic tail is not cast off in the adult; and, 2*dly*, That the fore-limbs are developed sooner than the hind-limbs—the reverse of this being the case amongst the *Anoura*.

Fig. 158.—Great Water-newt (*Triton cristatus*)—after Bell.

The Land-salamanders form the genus *Salamandra*, and are distinguished from their aquatic brethren by having a cylindrical instead of a compressed tail, and by bringing forth their young alive, or by being ovo-viviparous, in which case the larvæ have sometimes shed their external branchiæ prior to birth. The head is thick, the tongue broad, and the palatine teeth in two long series. The skin is warty, with many glands secreting a watery fluid. The best-known species is the *S. maculosa* of Southern Europe. Another species (*S. alpina*) lives upon lofty mountains. The chief thing to remember about the Land-salamanders, and, indeed, about all the *Urodela*, is their complete distinctness from the true Lizards (*Lacertilia*). They are often completely lizard-like in form when adult, but they always possess gills in the earlier stages of their existence, and this distinguishes them from all the Lacertilians.

ORDER III. ANOURA (= *Batrachia*, Huxley; *Theriomorpha*, Owen; *Chelonobatrachia*, &c.)—This order includes the Frogs and Toads, and is perhaps best designated by the name of *Anoura*, or "Tail-less" Amphibians. The name *Batrachia*, employed by Huxley, is inexpedient, partly because it is used by Owen to designate the entire class *Amphibia*, and partly because, in common language, it is usual to understand by a "Batrachian" any of the higher Amphibians—such, for instance, as a Labyrinthodont.

The *Anoura*, or Tail-less Amphibians, are characterised by the following points:—The adult is destitute of both gills and tail, both of which structures exist in the larva, whilst the two pairs of limbs are always present. The skin is soft, and there are rarely any traces of an exoskeleton. The dorsal vertebræ are "procœlous," or concave in front, and are furnished with long transverse processes, which take the place of ribs, which are only present in a rudimentary form. The radius and ulna in the fore-limb, and the tibia and fibula in the hind-limb, are anchylosed to form single bones (fig. 159). The mouth is sometimes edentulous, but the upper jaw has usually small teeth, and the lower jaw sometimes. The hind-limbs usually have the digits webbed for swimming, and are generally much larger and longer than the fore-limbs. The vertebral column is short (of ten vertebræ in the Frogs, but only eight in *Pipa*). The tongue is soft and fleshy, not supported by an os hyoides, but fixed to the symphysis of the lower jaw in front. *Pipa* has a sort of valve over the tympanum; *Hyla* and *Rana* have the tympanum shown externally; and *Bufo* has the tympanum concealed.

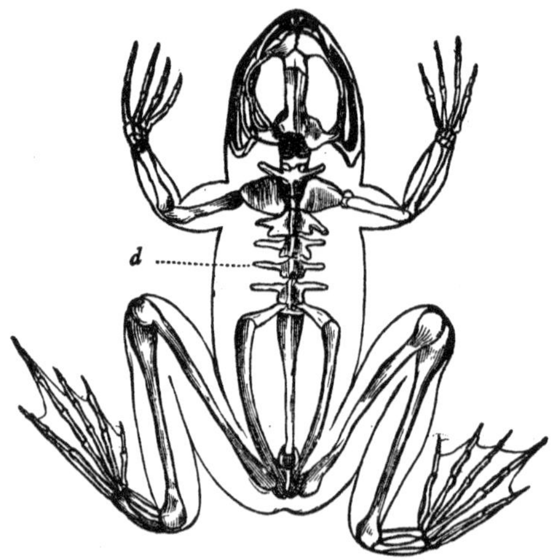

Fig. 159.—Skeleton of the common Frog (*Rana temporaria*). *d* Dorsal Vertebræ, with long transverse processes.

In the adult *Anoura*, respiration is purely aerial, and is carried on by means of lungs, which are, comparatively speaking, well developed. As there are no movable ribs by which the thoracic cavity can be expanded, the process of respiration is somewhat peculiar. The animal first closes its mouth, and fills the whole buccal cavity with air taken in through the nostrils. The posterior nares are then closed, and by the contraction of the muscles of the cheeks and pharynx the inspired air is forcibly driven into the windpipe through the open glottis. The process, in fact, is one of swallowing; and it is possible to suffocate a frog simply by holding its mouth open, and thereby preventing the performance of the above-mentioned actions. There can be no doubt, also, that the skin in these animals plays a very important part in the aeration of the blood, and that the frogs especially can carry on their respiration cutaneously, without the assistance of the lungs, for a very lengthened period. This undoubted fact, however, should not lead to any credence being given to the often-repeated stories of the occurrence of frogs and toads in cavities in solid rock, no authenticated instance of such a phenomenon being as yet known to science.

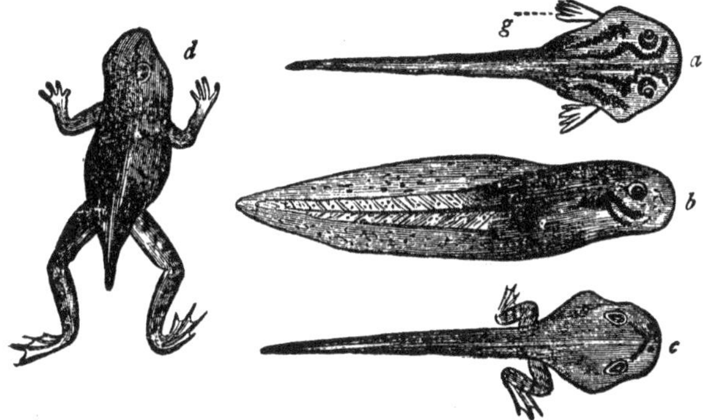

Fig. 160.—Development of the common Frog (*Rana temporaria*). *a* Tadpole, viewed from above, showing the external branchiæ (*g*); *b* Side view of a somewhat older specimen, showing the fish-like tail; *c* Older specimen, in which the hind-legs have appeared; *d* Specimen in which all the limbs are present, but the tail has not been wholly absorbed. (After Bell.)

The young or larvæ of the Frogs and Toads are familiarly known as "Tadpoles." The ova of the Frog are deposited in masses in water, and the young form, upon exclusion from the egg, presents itself as a "tailed" Amphibian, completely fish-like in form, with a broad rounded head, a sac-like abdomen, and a compressed swimming-tail (fig. 160, *a*). There are at first two sets of gills, one external and the other internal. The external branchiæ (fig. 160, *a*) have the form of filaments attached to the side of the neck, and they disappear very shortly after birth. The internal branchiæ are attached to cartilaginous arches, which are connected with the hyoid bone, and they are contained in a gill-cavity, protected by a flap of integument, which differs from the gill-cover of fishes in never developing any opercular bones or branchiostegal rays. Within the branchial chamber thus formed the fore-limbs are budded forth, but the hind-limbs are the first to appear, instead of the fore-limbs, as is the case with the *Urodela*. Even after the first appearance of the limbs, the tail is still retained as an instrument of progression; but as the limbs become fully developed, the tail is gradually absorbed (fig. 160, *d*), until in the adult it has wholly disappeared.

The development of the Frog is thus a good illustration of the general zoological law that the transient embryonic stages of the higher members of any division of the animal kingdom are often represented by the permanent condition of the lower members of the same division. Thus the transitory condition of the young Frog in its earliest stage, when the branchiæ are external, is permanently represented by the adult perennibranchiate *Urodela*, such as the Proteus or the Siren. The final stage, again, when the gills have disappeared and the limbs have been developed, but the tail has not been wholly absorbed, is represented by the caducibranchiate *Urodela*, such as the common Newt.

The order *Anoura* comprises a considerable number of forms, but may be divided into the three sections of the *Pipidæ*, *Bufonidæ*, and *Ranidæ*. In the *Pipidæ*, or Surinam Toads, there are rarely teeth, and the mouth is destitute of a tongue. A singular and hideous species (*Pipa Americana*) is the best known, and it inhabits Brazil and Surinam. In this curious Amphibian the eggs are placed by the male on the back of the female, in the integument of which, in cell-like cavities, the eggs are hatched and the young developed. In the aberrant form *Dactylethra* the upper jaw is furnished with small teeth, and the three inner toes of the hind-feet are furnished with nails, as is the case with no other Amphibian, except *Salamandra unguiculata* amongst the *Urodela*. This curious form is found at the Cape of Good Hope and in Mozambique.

In the Toads, or *Bufonidæ*, a tongue is present, but the jaws are not armed with teeth. The tongue agrees with that of the Frogs in being fixed to the *front* of the mouth, whilst it is *free behind*, so that it can be protruded for some distance from the mouth. The hind-limbs are not disproportionately developed, whilst the toes are only imperfectly webbed, and the toes of the fore-limb are free. The skin is warty and glandular. The common Toad (*Bufo vulgaris*) is an excellent example of this family. The Natter-jack Toad is the only other British species, but about fifty other forms are known, of which many are American.

In the *Ranidæ* the tongue has the same form as in the Toads, but the upper jaw always carries teeth. The hind-limbs are much larger than the fore-limbs, and are fitted for leaping, whilst the toes are webbed. The toes of the fore-limbs are free. The common Frog (*Rana temporaria*) is a good example of the typical *Ranidæ*. Larger than the common Frog is the Eatable Frog (*Rana esculenta*) of Europe, and larger again than this is the Bull-frog (*Rana pipiens*) of North America. The Tree-frogs (*Hyla*) are adapted for a wholly different mode of life, having the toes of all the feet furnished with terminal suckers, by the help of which they climb with ease. They are mostly found in warm countries, especially in America, but one species (*Hyla arborea*) is European (fig. 154).

ORDER IV. LABYRINTHODONTIA.—The members of this, the last order of the Amphibia, are entirely extinct. They were Batrachians, probably most nearly allied to the *Urodela*, but all of large size, and some of gigantic dimensions, the skull of one species (*Labyrinthodon Jagaeri*) being upwards of three feet in length and two feet in breadth. The Labyrinthodonts were first known to science simply by their footprints, which were found in certain sandstones of the age of the Trias. These footprints consisted of a series of alternate pairs of hand-shaped impressions, the hinder print of each pair being much larger than the one in front (fig. 161). So like were these impressions to the shape of the human hand that the unknown animal which produced them was at once christened *Cheirotherium*, or "Hand-beast." Further discoveries, however, soon showed that the footprints of *Cheirotherium* had been produced by different species of Batrachians, to which the name of Labyrinthodonts was applied, in consequence of the complex microscopic structure of the teeth.

Fig. 161.—Footprints of a Labyrinthodont (*Cheirotherium*).

The Labyrinthodonts were "salamandriform, with relatively weak limbs and a long tail."—(Huxley.) The vertebral centra and arches were ossified, and the bodies of the dorsal vertebræ are biconcave (amphicœlous). "In the thoracic region three superficially-sculptured exoskeletal plates, one median and two lateral, occupy the place of the interclavicle and clavicles. Between these and the pelvis is a peculiar armour, formed of rows of oval dermal plates, which lie on each side of the middle line of the abdomen, and are directed obliquely forwards and inwards to meet in that line."—(Huxley.)

The head was defended by an external covering or helmet of hard and polished osseous plates, sculptured on their external surface, and often exhibiting peculiar, smooth, symmetrical grooves—the so-called "mucous canals." The skull was articulated to the vertebral column by two occipital condyles. The teeth are rendered complex by numerous foldings of their parietes, giving rise to the "labyrinthine" pattern, from which the name of the order is derived.

The Labyrinthodonts are known to occur from the Carboniferous to the Triassic or Liassic period inclusively; but they are most characteristically and distinctively Triassic.

Distribution of Amphibia in Time.—From a geological point of view, by far the most important of the *Amphibia* are the *Labyrinthodontia*, the distribution of which has just been spoken of. The living orders of *Amphibia* are of much more modern date, being, as far as known, wholly Tertiary and Post-tertiary. The *Anoura* are represented by both Toads and Frogs in Miocene times, and they have survived to the present day. The "Tailed" Amphibians are best known to geologists by a singular fossil, which was described by its original discoverer as human, under the name of *Homo diluvii testis.* The fossil in question is of Miocene age, and it is now known to belong to a Salamander, nearly allied to the giant-salamander of Java (*Menopoma*). It is termed the *Andrias Scheuchzeri.*

DIVISION II. SAUROPSIDA.

CHAPTER LXII.

CLASS III.—REPTILIA.

THE second great division of the Vertebrate sub-kingdom, according to Huxley, is that of the *Sauropsida*, comprising the true Reptiles and the Birds. It is, no doubt, at first sight an almost incredible thing that there should be any near bond of relationship between the Birds and the Reptiles, no two classes of animals being more unlike one another in habits and external appearance. It is, nevertheless, the fact that the Birds are more nearly related to the Reptiles than to any other class of the *Vertebrata*, and it will shortly be seen that many affinities and even transitional forms are known to exist between these great sections. The Reptiles and Birds, then, may be naturally included in a single primary section of Vertebrates, which may be called *Sauropsida* after Huxley, and which is defined by the possession of the following characters:—At no period of existence are branchiæ, or water-breathing respiratory organs, developed upon the visceral arches; the embryo is furnished with a well-developed amnion and allantois; the red corpuscles of the blood are nucleated (fig. 130, *b*, *c*); the skull articulates with the vertebral column by means of a single articulating surface or condyle; and each half or "ramus" of the lower jaw is composed of several pieces, and articulates with the skull, not directly, but by the intervention of a peculiar bone, called the "quadrate bone," or "os quadratum" (fig. 162).

These being the common characters of Reptiles and Birds by which they are collectively distinguished from other Vertebrates, it remains to inquire what are the characters by which they are distinguished from one another. The following, then, are the characters which separate the Reptiles from the Birds:—The blood in Reptiles is cold—that is to say, slightly warmer than the external medium—owing mainly to the fact that the pulmonary and systemic circulations are always directly connected together, either within the heart or in its immediate neighbourhood, so that the body is supplied with a mixture of venous and arterial blood, in place of pure arterial blood alone. The terminations of the bronchi at the surface of the lung are closed, and do not communicate with air-sacs, placed in different parts of the body. When the epidermis develops horny structures, these are in the form of horny plates or scales, and never in the form of feathers. The fore-limbs are formed for various purposes, including in some cases even flight, but they are never constructed upon the type of the "wings" of Birds. Lastly, with one or two doubtful exceptions, whilst the ankle-joint is placed between the distal and proximal portions of the tarsus, the tarsal and metatarsal bones of the hind-limb are never anchylosed into a single bone.

These are the leading characters by which Reptiles are distinguished from Birds, but we must not forget the other distinctive peculiarities in which Reptiles agree with Birds, and differ from other Vertebrates—namely, the presence of an amnion and allantois in the embryo, the absence of branchiæ at all times of life, the possession of only one occipital condyle, and the articulation of the complex lower jaw with the skull by means of a quadrate bone.

It is now necessary to consider these characteristics of the *Reptilia* a little more minutely. The class includes the Tortoises and Turtles, the Snakes, the Lizards, the Crocodiles, and a number of extinct forms; and with the exception of the Tortoises and Turtles they are mostly of an elongated cylindrical shape, provided posteriorly with a long tail. The limbs may be altogether absent, as in the Snakes, or quite rudimentary, as in some of the Lizards; but, as a general rule, both pairs of limbs are present, sometimes in the form of ambulatory legs, sometimes as swimming-paddles, and in some extinct forms modified to subserve an aerial life. The endoskeleton is always well ossified, and is never cartilaginous or semi-cartilaginous, as in many fishes, and some Amphibians. The skull articulates with the atlas by a single condyle. The lower jaw is complex, each half or ramus being composed of from four to six pieces, united to one another by sutures (fig. 162). In the Tortoises, however, these are anchylosed into a single piece, and the two rami are also anchylosed. In most Reptiles, however, the two rami of the lower jaw are only loosely united—in the Snakes by ligaments and muscles only, in the Lizards by fibro-cartilage, and in the *Crocodilia* by a regular suture. In all, the lower jaw articulates with the skull by a quadrate bone (fig. 162, *a*); and as this often projects backwards, the opening of the mouth is often very extensive, and may even extend beyond the base of the skull. Teeth are usually present, but are not sunk in separate sockets or alveoli, except in

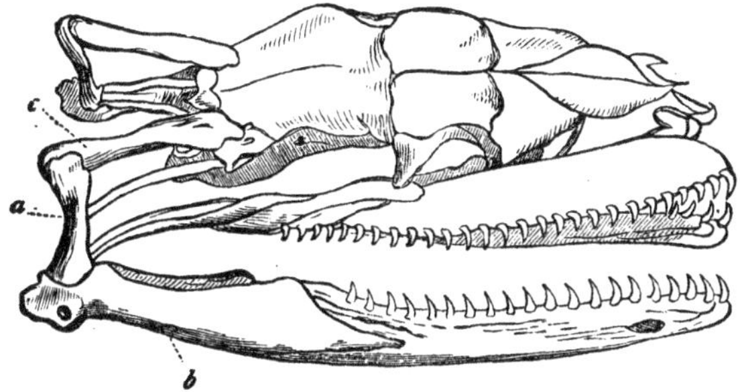

Fig. 162.—Skull of a Serpent (*Python*). *b* Articular portion of the lower jaw; *a* Quadrate bone; *c* Squamosal portion of the temporal bone.

the Crocodiles. In the Tortoises and Turtles alone there are no teeth, and the jaws are simply sheathed in horn, constituting a kind of beak like that of a bird.

Ribs are always present and always well developed, but they differ much in form. It is not correct, however, to regard the presence of ribs as separating the true Reptiles from the *Amphibia*, as is sometimes stated. Some of the most Lizard-like of the Amphibians, such as the Siren, possess short but well-developed ribs, and rudiments of ribs are traceable in other orders; whilst in the *Cæciliæ* they are large and well developed.

As regards the exoskeleton, all Reptiles have horny epidermic scales, and they are divided into two great sections—called respectively *Squamata* and *Loricata*—according as the integumentary skeleton consists simply of these scales, or there are osseous plates developed in the derma as well. In the Tortoises, the epidermic plates unite with the bony exoskeleton and with the true endoskeleton to form the case or box in which the body of these animals is enclosed.

The digestive system of the *Reptilia* possesses few characters of any special importance, except that the rectum opens, as in *Amphibia*, into a common cavity or "cloaca," which not only receives the fæces, but also serves for the discharge of the products of the urinary and generative organs.

The *heart* in the Reptiles consists of two completely separate auricles, and a ventricular cavity, which is divided into two by an incomplete partition. In the *Crocodilia* alone is the septum between the ventricles a perfect one, and even in these, as in all other Reptiles, the heart consists *functionally* of no more than three chambers. The ordinary course of the circulation, where the ventricular septum is imperfect, is as follows:—The impure venous blood returned from the body is, of course, poured by the venæ cavæ into the right auricle (fig. 163, *a*), and thence into the ventricle. The pure arterialised and aerated blood that has passed through the lungs, is, equally of course, poured into the left auricle (*a'*), and thence propelled into the ventricle (*v*). As the ventricular cavity is single, and not divided by a complete partition, it follows of necessity that there is a mixture in the ventricle, resulting in the production of a mixed fluid, consisting partly of venous and partly of arterial blood. This mixed fluid, then, occupies the common ventricular cavity, and by this it is driven both to the lungs (through the pulmonary artery), and to the body (through the systemic aorta). Consequently, in Reptiles, both the lungs and the various tissues and organs of the body are supplied with a mixture of arterial and venous blood, and not with unmixed blood — the lungs with purely venous, and the body with purely arterial blood—as is the case with the higher *Vertebrata*. In the *Crocodilia*, as before said, the partition between the ventricles is a complete one, and consequently this mixture of the arterial and venous blood cannot take place within the heart itself. In these Reptiles, however, a direct communication exists between the pulmonary artery and aorta (the right and left aortæ) by the so-called "foramen Panizzæ," close to the point where these vessels spring respectively from the right and left ventricle. In these Reptiles, therefore, the same mixture of arterial blood with

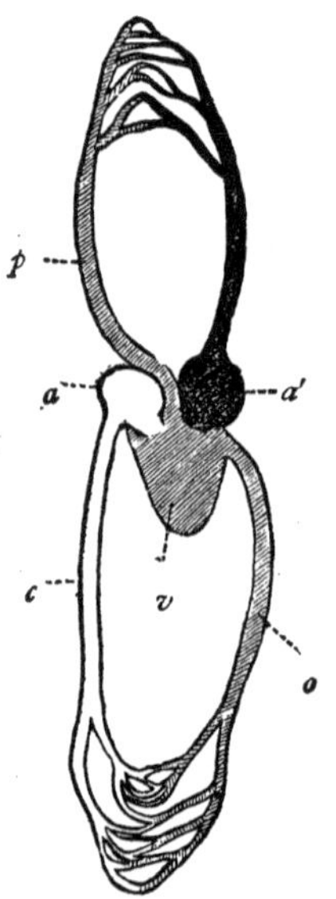

Fig. 163.—Diagram of the circulation in Reptiles. (The venous system is left light, the arterial system is black, and the vessels containing mixed blood are cross-shaded.) *a* Right auricle, receiving venous blood from the body; *a'* Left auricle, receiving arterial blood from the lungs; *v* Arterio-venous ventricle, containing mixed blood, which is driven by (*p*) the pulmonary artery to the lungs, and by (*o*) the aorta to the body.

venous takes place as in the lower *Reptilia*, though probably not to so complete an extent. It is this peculiarity of the circulation in all Reptiles which conditions their low temperature, slow respiration, and generally sluggish vital actions.

The *lungs* in all Reptiles, except the Crocodiles, are less completely cellular than in the Birds and Mammals, and they often attain a very great size. In no Reptile is the cavity of the thorax shut off from that of the abdomen by a complete muscular partition or "diaphragm;" though traces of this structure are found in the Crocodiles. The lungs, therefore, often extend along the whole length of the thoracico-abdominal cavity. In no case are the lungs connected with air-receptacles situated in different parts of the body; and not uncommonly there is only a single active lung, the other being rudimentary or completely atrophied (*Ophidia*).

Lastly, all reptiles are essentially oviparous, but in some cases the eggs are retained within the body till the young are ready to be excluded, and the animals are then ovo-viviparous. The egg-shell is usually parchment-like, but sometimes contains more or less calcareous matter.

CHAPTER LXIII.

DIVISIONS OF REPTILES.

CHELONIA AND OPHIDIA.

THE class *Reptilia* is divided into the following nine orders, of which the first four are represented by living forms, whilst the remaining five are extinct:—

1. *Chelonia* (Tortoises and Turtles). } Recent.
2. *Ophidia* (Snakes). } Recent.
3. *Lacertilia* (Lizards). } Recent.
4. *Crocodilia* (Crocodiles and Alligators). } Recent.
5. *Ichthyopterygia.* } Extinct.
6. *Sauropterygia.* } Extinct.
7. *Anomodontia.* } Extinct.
8. *Pterosauria.* } Extinct.
9. *Deinosauria.* } Extinct.

ORDER I. CHELONIA.—The first order of living Reptiles is that of the *Chelonia*, comprising the Tortoises and Turtles, and distinguished by the following characters :—There is an osseous exoskeleton which is combined with the endoskeleton to form a kind of bony case or box in which the body of the animal is enclosed, and which is covered by a leathery skin, or, more usually, by horny epidermic plates. The dorsal vertebræ, with the exception of the first, are immovably connected together, and are devoid of transverse processes. The ribs are greatly expanded (fig. 164, *r*), and are united to one another by sutures, so that the walls of the thoracic cavity are immovable. All the bones of the skull except the lower jaw and the hyoid bone are immovably united together. There are no teeth, and the jaws are encased in horn so as to form a kind of beak. The tongue is thick and fleshy. The heart is three-chambered, the ventricular septum being imperfect. There is a large urinary bladder, and the anal aperture is longitudinal or circular. The lungs are voluminous, and respiration is by swallowing air, as in the Frogs. All will pass prolonged periods without food, and will live and move, even for months, after the removal of the entire brain.—(Redi.)

Of these characters of the *Chelonia*, the most important and distinctive are the nature of the jaws, and the structure of the exoskeleton and skeleton. As regards the first of these points, the lower jaw in the adult appears to consist of a single piece, its complex character being masked by anchylosis. The separate pieces which really compose each ramus of the jaw are immovably anchylosed together, and the two rami are also united in front by a true bony union. There are also no teeth, and the edges of the jaws are simply sheathed in horn, constituting a sharp beak. In the *Chelydidæ* and *Trionycidæ*, however, the horny jaws are covered with soft skin, constituting a kind of lips. As regards the second of these points, the bony case in which the body of a Chelonian is enclosed consists essentially of two pieces, a superior or dorsal piece, generally convex, called the "carapace," and an inferior or ventral piece, generally flat or concave, called the "plastron." The carapace and plastron are firmly united along their edges, but are so excavated in front and behind as to leave apertures for the head, tail, and fore and hind limbs. The limbs and tail can almost always be withdrawn at will under the shelter of the thoracico-abdominal case formed in this way by the carapace and plastron, and the head is also generally retractile.

The carapace or dorsal shield is composed of the following elements :—

1. *The spinous processes of the dorsal vertebræ*, which are much flattened out laterally and form a series of broad plates.

2. The *ribs*, which are also much flattened and expanded, and constitute what are known as the "costal plates" (fig. 164, *r*). They are generally eight in number on each side, and are commonly united throughout the whole of their lateral margins by sutures. In some cases, however, they leave marginal apertures towards their extremities, and these openings are simply covered by a leathery skin or by horny plates. 3. The margin of the carapace is completed by a series of bony plates, which are called the "marginal plates." These are variously

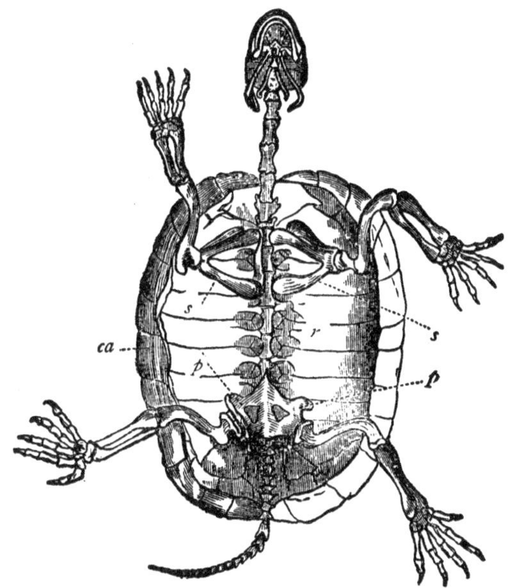

Fig. 164.—Skeleton of Tortoise (*Emys Europæa*), the plastron being removed. *ca* Carapace; *r* Ribs, greatly expanded, and united by their edges; *s* Scapular arch, placed within the carapace, and carrying the fore-limbs; *p* Pelvic arch, also placed within the carapace, and carrying the hind-limbs.

regarded as being dermal bones belonging to the exoskeleton, or as being endoskeletal, and as representing the ossified cartilages of the ribs (in this last case the marginal plates would correspond with what are known as the "sternal ribs" of Birds).

The "plastron" or ventral shield is composed of a number of bony plates (nine in number), the nature of which is doubtful. By some, the plastron is still regarded as a greatly-developed breast-bone or sternum. By others, again, the *Chelonia* are regarded as being wholly without a sternum, and the bones of the plastron are looked upon as exclusively integumentary ossifications. Both the carapace and the plastron are covered by a leathery skin, or more generally by a series of horny plates (fig. 165), which roughly correspond with the bony plates below, and which constitute in some species the "tortoise-shell" of commerce. These *epidermic* plates, however, must on no account be confounded with the true bony box in which the animal is enclosed, and which is produced partly by the true endoskeleton and partly by *dermal* integumentary ossifications.

The other points of importance as regards the endoskeleton are these:—

Firstly, The dorsal vertebræ are immovably joined together, and have no transverse processes, the heads of the ribs uniting directly with the bodies of the vertebræ.

Secondly, The scapular and pelvic arches, supporting the fore and hind limbs respectively (fig. 164, *s* and *p*), are placed *within* the carapace, so that the scapular arch is thus inside the ribs, instead of being outside, as it normally is. The scapular arch consists of the shoulder-blade or scapula, and two other bones, of which one corresponds with the acromion process of human anatomy, and the other to the coracoid process, or to the "coracoid bone" of the Birds. The clavicles, as is also the case with the *Crocodilia*, are absent.

The order *Chelonia* is conveniently divided into three sections, according as the limbs are natatory, amphibious, or terrestrial. In the first of these, the limbs are converted into most efficient swimming-paddles, all the toes being united by a common covering of integument. In this section are the well-known Turtles (*Cheloniidæ*), all of which swim with great ease and power, but are comparatively helpless upon the land (fig. 165). The legs are of unequal length, and the carapace is much depressed and flattened. The best-known species are the "edible" or Green Turtle (*Chelonia mydas*), the Loggerhead Turtle (*Chelonia caouanna*), the Hawk's-bill Turtle (*C. imbricata*), and the Leathery Turtle (*Sphargis coriacea*). The Green Turtle is largely imported into this country as a delicacy, and occurs abundantly in various parts of the Atlantic and Indian Oceans. The Hawk's-bill Turtle is of even greater commercial importance, as the horny epidermic plates of the carapace constitute the "tortoise-shell" so largely used for ornamental purposes. The Leathery Turtle is remarkable in having the carapace covered with a leathery skin in place of the horny plates which are found in other species.

In the second section of the *Chelonia*, in which the limbs are adapted for an amphibious life, are the Mud-turtles or Soft Tortoises (*Trionycidæ*), and the Terrapins (*Emydidæ*). In the *Trionycidæ* the development of the carapace is imperfect, the ribs being expanded and united to one another only near their bases, and leaving apertures near their extremities. The entire carapace is covered by a smooth leathery skin, and the horny jaws are furnished with fleshy lips. All the *Trionycidæ* inhabit fresh water and are carnivorous in their habits. Good examples are found in the Soft-shelled Turtle (*Trionyx ferox*),

Fig. 165.—Hawk's-bill Turtle (*Chelonia imbricata*)—after Bell.

and the large and fierce Snapping Turtle (*Chelydra serpentina*) of the United States; but other species are found in Egypt and in the East Indies. The Terrapins (*Emys*) have a horny beak, and have the shield covered with epidermic plates. They are inhabitants of fresh water, and are most of them natives of America.

The third section of the *Chelonia* comprises only the Land Tortoises (*Testudinidæ*), in which the limbs are adapted for terrestrial progression, and the feet are furnished with short nails. The carapace is strongly convex, and is covered by horny epidermic plates; the head, limbs, and tail can be completely retracted within the carapace. Though capable of swimming, the Tortoises are really terrestrial animals, and are strictly vegetable-feeders. The most familiar species is the *Testudo Græca*, which is indigenous in Spain, Italy, and Greece. A much larger species is the Indian Tortoise (*Testudo Indica*), which attains a length of over three feet.

DISTRIBUTION OF CHELONIA IN TIME.—The earliest-known traces of Chelonians occur in the Permian rocks, in the lower portion, that is, of the New Red Sandstone of older geologists. These traces, however, are not wholly satisfactory, since they consist solely of the footprints of the animal upon the ripple-marked surfaces of the sandstone. Of this nature is the *Chelichnus Duncani*, described by Sir William Jardine in his classical work on the "Ichnology" of Annandale in Dumfriesshire. The earliest unequivocal remains of Chelonians are in the Oolitic rocks (the *Chelonia planiceps* of the Portland Stone). Fossil *Cheloniidæ*, *Emydidæ*, and *Trionycidæ* occur, also, from the Upper Oolites to the present day, the Eocene period being peculiarly rich in their remains. In the Tertiary deposits of India (Sivalik Hills) there occurs a gigantic fossil Tortoise—the *Colossochelys Atlas*—which is believed to have been eighteen to twenty feet in length, and to have possibly survived to within the human period.

ORDER II. OPHIDIA.—The second order of Reptiles is that of the *Ophidia*, comprising the Snakes and Serpents, and distinguished by the following characters:—

The body is always more or less elongated, cylindrical, and worm-like, and whilst possessing a covering of horny scales, is always unprovided with a bony exoskeleton. The dorsal vertebræ are concave in front (procœlous), with rudimentary transverse processes. There is never any sternum, nor pectoral arch, nor fore-limbs, nor sacrum, and as a rule there are no traces of hind-limbs. Rudimentary hind-limbs, however, are occasionally present (*e.g.*, in *Python* and *Tortrix*). There are always numerous ribs. The two halves or rami of the lower jaw are composed of several pieces, and the rami are united anteriorly by ligaments and muscles only, and not by cartilage or suture. The lower jaw further articulates with the skull by means of a quadrate bone (fig. 162, *a*), which is always more or less movable, and is in turn united with the squamous portion of the temporal bone ("mastoid bone"), which is also movable, and is not firmly united with the skull. The superior maxillæ are united with the præmaxillæ by ligaments and muscles only, and the palatine arches are movable and armed with pointed recurved teeth. Hooked conical teeth are always present, but they are never lodged in distinct sockets or alveoli. Functionally, they are capable of performing nothing more than merely holding the prey fast, and the Snakes are provided with no genuine masticatory apparatus. The heart has three chambers, two auricles and a ventricle, the latter imperfectly divided into two cavities by an incomplete septum. The lungs and other paired organs are mostly not bilaterally symmetrical, one of each pair being either rudimentary or absent. There is no urinary bladder, and the cloacal aperture is transverse.

Of these characters of the snakes, the most obvious and striking are to be found in the nature of the organs of locomotion. The front limbs, with the scapular arch and sternum, are invariably altogether absent; and the hind-limbs, if not wholly wanting, are never represented by more than an imperfectly-developed series of bones concealed within the muscles on each side of the anal aperture, and never exhibiting any outward evidence of their existence beyond the occasional presence of short horny claws or spurs ("calcaria"). In the entire absence, then, or rudimentary condition of the limbs, the Snakes progress by means of the ribs. These bones are always extremely numerous (sometimes amounting to more than three hundred pairs), and in the absence of a sternum, they are, of course, extremely movable. Their free extremities, in fact, are simply terminated by tapering cartilages, which are attached by muscular connections to the abdominal scales or "scuta" of the integument. By means of this arrangement the Serpents are enabled to progress rapidly, walking, so to speak, upon the ends of their ribs; their movements being much facilitated by the extreme mobility of the whole vertebral column, conditioned by the cup-and-ball articulation of the bodies of the vertebræ with one another.

The body in the Snakes is covered with numerous scales, developed apparently in the lower layer of the epidermis, and covered by a thin, translucent, superficial pellicle, which is periodically cast off and renewed. On the head and along the abdomen these scales are larger than over the rest of the body, and they constitute what are known as the "scuta" or shields.

The only other points in the anatomy of the *Ophidia* which demand special attention are the structure of the tongue, teeth, and eye.

The tongue in the Snakes is probably an organ more of touch than of taste. It consists of two muscular cylinders, united towards their bases, but free towards their extremities. The bifid organ, thus constituted, can be protruded and retracted at will, being in constant vibration when protruded, and being in great part concealed by a sheath when retracted.

As regards the eye of Serpents (fig. 166, A), the chief

peculiarity lies in the manner in which it is protected externally. There are no eyelids, and hence the stony unwinking

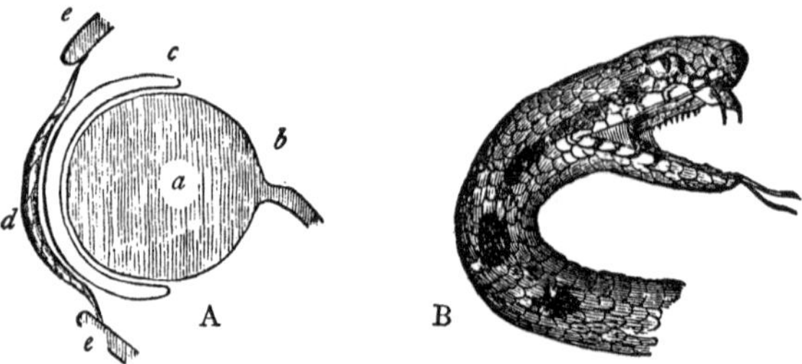

Fig. 166.—A, Diagram of the eye of a Serpent (after Cloquet): *a* Ball of the eye covered by a conjunctival sac, into which the lachrymal secretion is discharged; *b* Optic nerve; *d* Antocular membrane, formed by the epidermis; *e e* Ring of scales surrounding the eye. B, Head of the common Viper (*Pelias berus*)—after Bell—showing the bifid tongue, and the poison fangs in the upper jaw.

stare of all snakes. In place of eyelids, the eye is surrounded by a circle of scales (*e e*), to the circumference of which is attached a layer of transparent epidermis, which covers the whole eye (*d*), and is termed the antocular membrane. This is covered internally by a thin layer of the conjunctiva, which is reflected forwards from the conjunctiva covering the ball of the eye itself. In this way a cavity or chamber is formed between the two layers of conjunctiva, and the lachrymal secretion by which the eye is moistened is received into this. The outer epidermic layer (antocular membrane) covering the ball of the eye in front, is periodically shed with the rest of the epidermis, the animal being rendered thereby blind for a few days. The pupil of the eye is round in most Snakes, but forms a vertical slit in the venomous Serpents and in the Boas.

As regards the dental and maxillary apparatus of the Serpents, the following points require notice:—*Firstly*, in consequence of the articulation of the lower jaw with a movable quadrate bone, which is often directed backwards, in consequence of the quadrate bone being connected with a movable squamosal bone, and in consequence of the rami of the jaw being united in front by ligaments and muscles only, the mouth in the Snakes is capable of opening to an enormous width, and the most astonishing feats in the way of swallowing can be performed. *Secondly*, this structure of the jaws accords exactly with the structure of the teeth, both concurring to render the Snakes wholly incapable of anything like mastication, and at the same time capable of swallowing immense morsels entire. The teeth, namely, are simply fitted for seizing and holding the prey, but not in any way for dividing or chewing it. In the non-venomous and most typical Snakes, the jaws and palatine bones carry continuous rows of solid conical teeth, so that there are four rows above and two below; and the superior maxillæ are very long and are not movable. *Thirdly*, in the Viperine Snakes the ordinary teeth are wanting upon the superior maxillæ, whilst these bones are themselves very much shortened, and are capable of being raised and depressed at will. In place of the ordinary teeth, each maxilla carries a "poison-fang," in the form of a long, conical, curved fang, which is concealed in a fold of the gum when not in use, and has numerous germs or reserve-fangs behind it (fig. 166, B). Each tooth is perforated by a tube, opening by a distinct aperture at the apex of the tooth, and conveying the duct of the so-called poison-gland. (In reality the poison-duct of the fang is formed by an inflection of the tooth upon itself, and not by its actual perforation.) This is a gland, probably produced by a modification of one of the buccal salivary glands, situated behind and under the eye on each side, and secreting the fluid which renders the bite of these snakes dangerous or fatal. When the animal strikes its prey, the poison-fangs are erected, and the poison is forced through the tube which perforates each, partly by the contractions of the muscular walls of the gland, and partly by the muscles of the jaws. In most poisonous snakes the superior maxillæ carry no other teeth except the poison-fangs and their rudimentary successors, but in some cases there are a few teeth behind the fangs; whilst the palatine teeth are always present, as in the harmless species. In some other venomous Snakes, again (*e.g.*, *Naja* and the *Hydrophidæ*), the jaws and teeth agree in most characters with those of the non-venomous Snakes, but the first maxillary teeth are larger than the others, and form canaliculated fangs. Lastly, in a few forms the terminal maxillary teeth are deeply canaliculated, but are not connected with the duct of any poison-gland.

Fourthly, in all the Serpents the teeth are anchylosed to the jaw, and are never sunk into distinct sockets or alveoli.

A good classification of the *Ophidia* is still a desideratum, and probably, in the meanwhile, the one proposed by Dr Gray is the best. This eminent naturalist divides the Snakes into the two sub-orders of the *Viperina* and *Colubrina*, the former having only two perforated poison-fangs on the superior maxillæ, whilst these bones in the latter carry solid teeth, either with or without additional canaliculated fangs.

The sub-order *Viperina* comprises the common Vipers (*Viperidæ*), and the Rattlesnakes (*Crotalidæ*), the former being mostly confined to the Old World, whilst the latter are wholly American. The common Viper (*Pelias berus*), occurs abundantly in England and Scotland, and is capable of inflicting

Fig. 167.—The *Naja Haje*, a venomous Colubrine Snake.

a severe and even dangerous bite, though it is doubtful if fatal effects ever follow except in the case of children or subjects previously debilitated. The Rattlesnakes are exclusively natives of America, and they are highly poisonous. The extremity of the tail in the true Rattlesnake (*Crotalus horridus*) is furnished with a series of horny epidermic cells of an undulated pyramidal shape, articulated one within the other, constituting an appendage which is known as the "rattle." Before striking its prey, the Rattlesnake throws itself into a coil, and shakes its rattle, as it does also when alarmed. The head of the Viperine Snakes (figs. 166, 168) is broad, somewhat triangular in shape, broadest at its middle, and showing a very distinct line of demarcation between the head and neck. The head, also, is usually covered with small scales, rarely interspersed with larger plates or "scuta" (fig. 168). Other well-known members of this group are the Death Adder (*Acanthophis tortor*) of Australia, the Horned Viper (*Cerastes*) of Africa, and the Puff Adder (*Clotho arietans*) of the Cape of Good Hope The *Colubrina* include for the most part harmless snakes, but together with these are some of the most deadly of all the venomous snakes. In accordance with this they are often divided into the three sections of the *Innocua*, *Suspecta*, and *Venenosa*. In the first of these sections (*Innocua*), the superior maxillæ are provided with solid teeth only, and there are no fangs. In this section are the common Ringed Snake of Britain and the Boas and Pythons of warm climates. The common Ringed Snake (*Coluber natrix*) of Britain is a perfectly harmless animal which is commonly found in damp situations, and which lives mainly upon frogs. Closely allied to this is the Black Snake (*Bascanion constrictor*), which attains a length of from three to five feet, but is perfectly harmless, so far as man is concerned. The *Boidæ* or Boas and Pythons are the largest of all living snakes, attaining a length of certainly over twenty feet. Their bite is perfectly harmless, but they are nevertheless highly dangerous and destructive animals, owing to their great size and enormous muscular power. They seize their prey and coil themselves round it in numerous folds, by tightening which they gradually reduce their victim to the condition of a shapeless bolus, fit to be swallowed. In this way a good-sized Python or Boa will certainly dispose of an animal as large as a sheep or goat, and it is asserted that even human beings may be devoured in this way by large individuals of the family. The Boas and Pythons occur in both the Old and New World, the Pythons, however, all belonging to the Old World; and they are amongst the most formidable of all living Ophidians. They possess rudimentary hind-limbs terminating in horny anal spurs, which co-operate with the prehensile tail in enabling the animal to suspend itself from trees. In all, also, the dental apparatus is extremely powerful, giving a firm hold for the constriction of the prey.

In the section *Suspecta*, in which there are canaliculated fangs placed far back on the superior maxillæ, with smaller solid teeth in front of them, are certain unimportant snakes, partly aquatic and partly terrestrial in their habits, and all belonging to the Old World.

In the group *Venenosa*, in which there are canaliculated fangs placed in front of the superior maxillæ with smaller solid teeth behind them, are some of the most deadly of all living serpents. One of the best known of these is the Hooded Snake, or *Cobra di Capello* (*Naja tripudians*), which is commonly found in Hindostan, and is the snake usually carried about by the Indian snake-charmers. It varies from two to six feet in length, and the neck can be extensively dilated, covering the head like a hood. A nearly-allied species is the *Naia Haje* (fig. 167) of Egypt. Also in this section are the venomous Water-snakes (*Hydrophidæ*), which have a compressed tail, and are adapted for an aquatic life. They mostly frequent the mouths of rivers in droves, and they swim with great grace and rapidity.

A very good general character by which the Colubrine snakes may be distinguished from the Viperine snakes, is in the shape and armature of the head. In the *Viperina*, as before said, the head (figs. 166, 168) is triangular, broadest behind, and separated from the neck by a more or less marked diminution in the diameter of this latter part. The scales, too, which cover the head are of small size. In the Colubrine snakes, on the other hand, the head is not markedly triangular, and gradually tapers off into the neck, whilst the upper surface of the

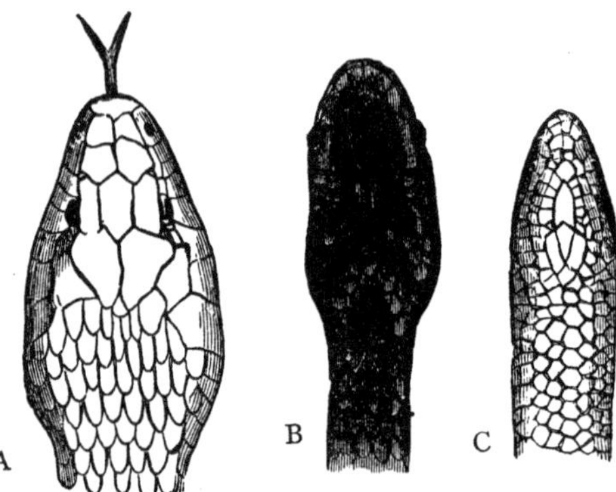

Fig. 168.—A, Head of Colubrine Snake (*Coluber natrix*); B, Head of Viperine Sna (*Pelias berus*); C, Head of Blind-worm (*Anguis fragilis*), one of the serpentiform Lizards. (After Bell.)

head is usually covered with large shield-like plates or "scuta" (fig. 168, A).

Distribution of Ophidia in Time.—The *Ophidia* are not known to occur in any Palæozoic or Mesozoic deposit. The earliest-known traces of any serpent are in the Lower Kainozoic Rocks, the oldest being the *Palæophis toliapicus* of the London Clay of Sheppey. The nearly-allied *Palæophis typhæus* of the Eocene beds of Bracklesham appears to have been a Boa-constrictor-like snake of about twenty feet in length. Other species of *Palæophis* have been described from the Tertiary

Rocks of the United States, and the genus *Dinophis* has been formed for the reception of another gigantic constricting Serpent from the same formation. In some of the later deposits have been found the poison-fangs of a venomous snake. Upon the whole, however, the snakes must be looked upon as a comparatively modern group, and not as one of any great geological antiquity.

CHAPTER LXIV.

LACERTILIA AND CROCODILIA.

ORDER III. LACERTILIA.—The third order of Reptiles is that of the *Lacertilia*, comprising all those animals which are commonly known as Lizards, together with some serpentiform animals such as the Blind-worms. The *Lacertilia* are distinguished by the following characters :—

As a general rule, there are two pairs of well-developed limbs, but there may be only one pair, or all the limbs may be absent. A scapular arch is always present, whatever the condition of the limbs may be. An exoskeleton, in the form of horny scales like those of the Snakes, is almost always present. The vertebræ of the dorsal region are procœlous or concave in front, rarely amphicœlous or concave at both ends. There is a single transverse process at each side, and the heads of the ribs are simple and undivided. There is either no sacrum, or the sacral vertebræ do not exceed two in number. The teeth are not lodged in distinct sockets. The eyes are generally furnished with movable eyelids, and are always so in the completely snake-like forms. The heart consists of two auricles and a ventricle, the latter partially divided by an incomplete partition. There is a urinary bladder, and the aperture of the cloaca is transverse.

As a general rule, the animals included under this order have four well-developed legs (fig. 169), and would therefore be popularly called "Lizards." In some (*Chirotes*) there are no hind-feet; in some (*Bipes*) the fore-limbs are wanting; and others (*Anguis*, *Pseudopus*, and *Amphisboena*) are entirely destitute of limbs, thus coming closely to resemble the true Snakes or Ophidians in external appearance. These serpentiform Lizards, however, can be distinguished from the true Snakes, amongst other characters, by the structure of the jaws. In the Snakes, as before said, the two rami of the lower jaw are loosely united in front by ligaments and muscles, and are attached behind to a movable quadrate bone, which is in turn connected with a movable squamosal, this giving an enormous width of gape to these animals. In the Lizards, however, even in those

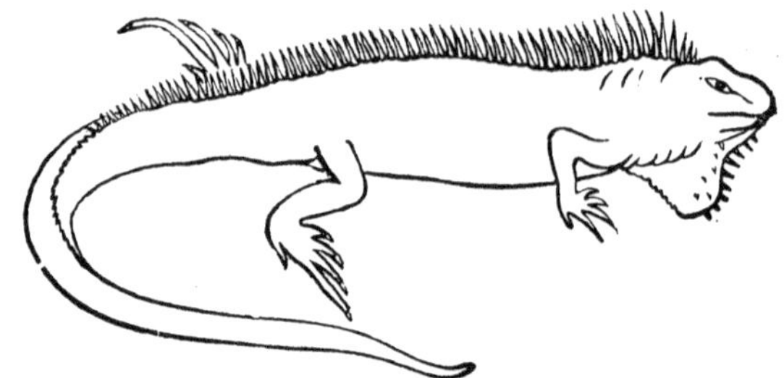

Fig. 169.—Iguana.

most like the Snakes, the halves of the lower jaw are firmly united to one another in front; and though the quadrate bone is usually more or less movable, the jaws can in no case be separated to anything like the extent that characterises the *Ophidia*.

Another good and still more obvious character is to be found in the structure of the protective coverings of the eye. In the Snakes, eyelids are wanting, and the eye is simply covered by a layer of epidermis, constituting the so-called "antocular membrane." In almost all the Lizards, on the other hand, including all the completely snake-like forms, there are movable eyelids, and in few cases is there any structure comparable to the antocular membrane of the true Snakes. Lastly, the typical Lizards all possess a sternum or breast-bone, but this is wanting in some of the snake-like forms, so that it cannot be appealed to as a character by which the *Lacertilia* can be separated from the *Ophidia*.

The whole order of the *Lacertilia* is very often united with the next group of the Crocodilia, under the name of *Sauria*. The term "Saurian," however, is an exceedingly convenient one to designate all the Reptiles which approach the typical Lizards in external configuration, whatever their exact nature may be; and from this point of view it is often very useful as applied to many fossil forms, the structure of which is only imperfectly known. It is therefore perhaps best to employ this term merely in a loose general sense.

The *Lacertilia* are often divided into the two great groups of the *Fissilinguia* and *Brevilinguia*, according as the tongue is bifid and protrusible like that of the Ophidians, or is thick and fleshy, and only protrusible when the mouth is open. These distinctions, however, are not of any very great value, and no good general arrangement of the order has hitherto been proposed. Here, therefore, it will be sufficient to treat very shortly of the more important families of the Lacertilians.

The first family of any importance is that of the *Chalcidæ* or Chalcidian Lizards, comprising a number of snake-like animals which have long occupied a debatable position. In their serpentiform cylindrical form these animals closely resemble the true *Ophidia*, and this likeness is still further increased by the absence or rudimentary condition of the limbs. The scapular arch and sternum, however, are present in a rudimentary form, and one or both pairs of limbs may be present. Another character separating the *Chalcidæ* from the true Snakes is the structure of the lower jaw, the rami of which are united in front by a symphysis so as greatly to restrict the gape. The Chalcidian Lizards are entirely covered with similar scales arranged in rings or whorls; the trunk passes into the tail without any definite line of demarcation, and there is generally a lateral longitudinal fold or groove. In *Chalcides* the body is long and snake-like, but all the limbs are present, though these are small, and have often but a single well-developed toe each. It is represented in both the East Indies and South America. In *Chirotes*, of Mexico, only the fore-limbs are present, and in the African genus *Bipes* only the hind-limbs are present. In the *Amphisbœnæ* of South America the tail is very short, and the vent is placed nearly at the end of the body, whilst there are no limbs. In the Glass Snake (*Ophisaurus*) of the United States there are also no limbs.

The next great family is that of the *Scincidæ*, including a number of small Lacertilians, some of which are completely snake-like, whilst others possess two limbs, and others again have the normal two pairs of limbs in a well-developed condition. All possess movable eyelids, and in all the conformation of the lower jaw is Lacertilian, and not Ophidian. All the Scincoidean Lizards have the body covered by similar scales overlapping one another like the scales of fishes, whilst the head is protected by larger plates. The tongue is free, fleshy, and slightly notched. Of the snake-like forms of this group, none is more familiarly known than the Blind-worm or Slow-worm (*Anguis fragilis*, fig. 170), which is found over almost the whole of Europe, in western Asia, and northern Africa, and which is one of the most abundant of the British Reptiles. The Blind-worm possesses no external appearance of limbs, though the scapular and pelvic arches are present in a rudimentary condition. Its appearance is completely serpentiform, and it is vulgarly regarded as a dangerous and venomous animal, but quite erroneously, as it is even unable to pierce the human skin. It is a perfectly harmless animal, living upon worms, insects, and snails, and hyber-

Fig. 170.—The Blind-worm (*Anguis fragilis*)—after Bell.

nating during the winter. It derives its specific name of *fragilis* from the fact that when alarmed it stiffens its muscles to such an extent that the tail can be readily broken off, as if it were brittle.

Numerous other small Lizards are referable to the *Scincidæ*, but it is only necessary to mention the Skinks themselves (*Scincus*), in which both pairs of limbs are present in a well-developed state. The Skinks are found in almost all the warmer parts of the Old World, and closely-allied forms (such as the West Indian "Galliwasp") are found in the New World. The common Skink (fig. 171) is a native of Arabia and Africa. It attains a length of eight or nine inches, and was formerly used in various diseases as a remedy.

The next family is that of the *Lacertidæ*, comprising the typical Lizards, in which there are always four well-developed limbs, each terminated by five free toes of unequal lengths. The body is covered with scales, which assume the form of shields or "scuta" over the abdomen and on the head. The tail is rounded. The tongue is slender, bifid, and protrusible. The only truly British Lizards are the Sand-lizard (*Lacerta*

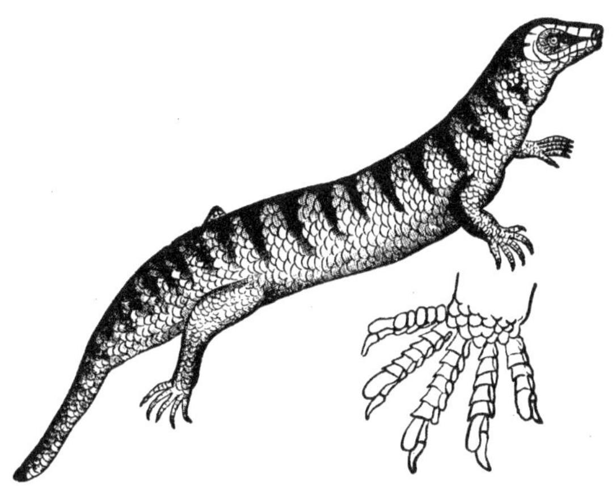

Fig. 171.—The common Skink (*Scincus officinalis*).

agilis), and the Viviparous Lizard (*Zootoca vivipara*); and the commonest form upon the Continent is the graceful little Green Lizard (*Lacerta viridis*), which also occurs in Jersey. The Lizards of the Old World are represented in America by the *Ameivæ*, some of which attain a length of several feet.

Very closely allied to the true Lizards are the *Varanidæ* or Monitors, which indeed are chiefly separated by the comparatively trivial fact that the abdomen and head are covered with ordinary scales, and not with large "scuta." The tongue is protrusible and fleshy, like that of the Snakes. The teeth are lodged in a common alveolar groove, which has no internal border; and there are no palatal teeth. The tail has a double row of carinated scales, and is cylindrical in the terrestrial forms, and compressed in those whose habits are aquatic. The Monitors are exclusively found in the Old World, and are the largest of all the recent *Lacertilia*; the *Varanus Niloticus* of Egypt attaining a length of six feet, and the *Varanus bivittatus* of Java attaining to as much as eight feet. The Safe-guards (*Salvator*) are the Monitory Lizards of the New World, and are also of large size.

The *Geckotidæ* form a large family of Lizards, comprising a great number of species, occurring in almost all parts of the world. The tongue is wide, flat, scarcely notched at its free extremity, and hardly at all protrusible. The eyes are large, with extremely short lids, the pupil mostly linear, but sometimes circular. The teeth are numerous, small, compressed, and implanted on the inner edge of the jaw. The nails are mostly hooked and retractile, and the toes are furnished below with imbricated plates. The animal is capable of running on the smoothest surfaces, or suspending itself back-downwards. They feed on insects, and are found in abundance in the warmer parts of both the Old and New Worlds.

The *Iguanidæ* constitute another large family of Lizards, also belonging partly to the Old and partly to the New World. The tongue is thick, fleshy, notched at its extremity only, and not protrusible. Mostly there is a dorsal crest, and a goitre or throat-pouch. The body is covered with imbricated scales. Only one species of the family is European, but the group is represented by numerous species in N. and S. America, Asia, Africa, and Australia. They are often divided into "ground-iguanas," in which the body is flat and depressed, and "tree-iguanas," in which the body is compressed. The members of the genus *Iguana* itself (fig. 169) are confined to the New World, and are distinguished by having the throat furnished

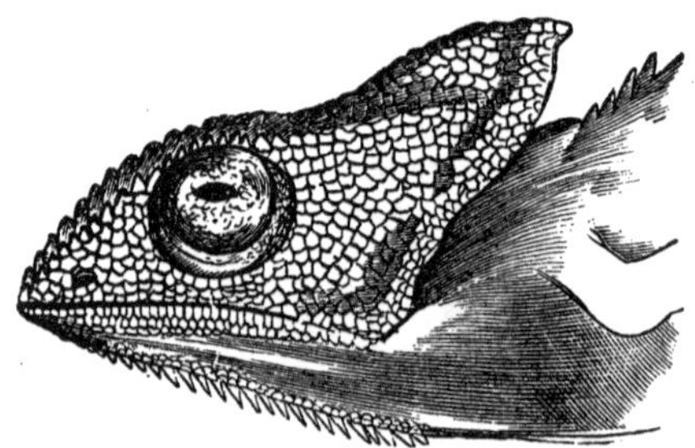

Fig. 172.—Head of a Chameleon (*C. Petersii*)—after Gray.

with a pendulous dewlap or fold of skin, the edge of which is toothed. The back and tail, too, are furnished with an erect crest of pointed scales. The Iguana attains a length of from four to five feet, and though not of a very inviting appearance, is highly esteemed as food. The Basilisks (*Basiliscus*) have the top of the head furnished with a membranous sac, which can be distended with air at will. The *Agamidæ* agree with the Iguanas in most respects, but have two rows of teeth on the hinder margin of the palate, and the tail is covered with imbricated scales. Good examples are the Tapayaxin (*Agama orbicularis*) of South America, and the hideous *Moloch horridus* of Australia. Here also belongs the curious little Frill Lizard (*Chlamydosaurus*) of Australia, which has the neck furnished on each side with a membranous plaited frill, which can be erected at will. More remarkable than the true Iguanas is the little Flying Dragon (*Draco volans*) of the East Indies and Indian Archipelago. In this singular little Lizard there is a broad membranous expansion on each side, formed by a fold of the integument, supported upon the anterior false ribs, which run straight out from the spinal column. By means of these lateral expansions of the skin, the *Draco volans* can take long flying leaps from tree to tree, and can pursue the insects on which it feeds ; but the lateral membranes simply act as parachutes, and there is no power of true flight, properly so called.

The last family of the living Lizards which requires notice is that of the *Chamæleontidæ*, containing the familiar little *Chamæleo Africanus*, which occurs abundantly in the north of Africa and in Egypt, and is so well known for its power of changing its colour under irritation or excitement. In this genus the eye (fig. 172) is of large size, and is covered by a single circular lid, formed by a coalescence of the two lids, and perforated centrally by a small aperture, by which the rays of light reach the pupil. The Chameleon is naturally a sluggish animal, but it catches its food, consisting of insects, by darting out its long, fleshy, and glutinous tongue—an operation which it effects with the most extraordinary rapidity.

The tail in the Chameleons is round and prehensile, the body compressed, and the skin like shagreen. The toes are adapted for the arboreal life and scansorial habits of the animal, being so arranged as to form two equal and opposable sets. The lungs are excessively voluminous. The Chameleons are exceedingly sluggish and slow in their movements, and are confined to the warmer parts of the Old World.

Distribution of Lacertilia in Time.—The geological range of the true *Lacertilia* is not by any means very great, nor, with a single exception, are their remains of much importance. The earliest traces of Lizards in the stratified series are found in the fresh-water strata of the Purbeck beds at the summit of the Jurassic series. Several small Lizards occur here, and have been described under the names of *Nathetes*, *Macellodon*, *Saurillus*, and *Echinodon*. The most remarkable fossil Lizard, however, is the *Mosasaurus* of the Chalk. This gigantic reptile occurs at the very summit of the Cretaceous series, in what is known as the Maestricht Chalk. The skull is no less than five feet long; and as the tail and limbs were formed for swimming, there can be little doubt but that *Mosasaurus*—like the living *Amblyrhynchus*—was aquatic in its habits, and frequented the sea-shore. Several other Reptiles, either belonging to the genus *Mosasaurus*, or nearly allied to it, have been described from the Cretaceous and Tertiary rocks of North America. Recent researches on these Mosasauroid Reptiles by Professor Marsh of New Haven have shown that they possessed fin-like paddles like those of the Ichthyosaur and Plesiosaur.

ORDER IV. CROCODILIA.—The last and highest order of the living *Reptilia* is that of the *Crocodilia*, including the living Crocodiles, Alligators, and Gavials, and characterised by the following peculiarities:—

The body is covered with an outer epidermic exoskeleton composed of horny scales, and an inner dermal exoskeleton consisting of squared bony plates or scutes, which may be confined to the dorsal surface alone, or may exist on the ventral surface as well, and which are disposed on the back of the neck into groups of different form and number in different species. The bones of the skull and face are firmly united together, and the two halves or rami of the lower jaw are united in front by a suture. There is a single row of teeth, which are implanted in distinct sockets, and hollowed at the base for the germs of the new teeth, by which they are successively pushed out and replaced during the life of the animal. The centra of the dorsal vertebræ in all living *Crocodilia* are procœlous, or concave in front, but in the extinct forms they may be either amphicœlous (concave at both ends) or opisthocœlous (concave behind). The vertebral ends of the anterior trunk-ribs are bifurcate. There are two sacral vertebræ. The cervical vertebræ have small ribs (hence the difficulty experienced by the animal in turning quickly); and there are generally false abdominal ribs produced by the ossification of the tendinous intersections of the *recti* muscles. There are no clavicles. The heart consists of four completely distinct and separate cavities, two auricles, and two ventricles, the ventricular septum—as in no other Reptiles—being complete. The right and left aortæ, however—or, in other words, the pulmonary artery and systemic aorta—are connected together close to their origin by a small aperture (*foramen Panizzæ*), so that the two sides of the heart communicate with one another. The aperture of the cloaca is longitudinal, and not transverse, as in the Lizards. All the four limbs are present, the anterior ones being pentadactylous, the posterior tetradactylous. All are oviparous.

The chief points by which the Crocodiles are distinguished from their near allies, the Lacertilians, are the possession of a partial bony dermal exoskeleton in addition to the ordinary epidermic covering of scales, the lodgment of the teeth in

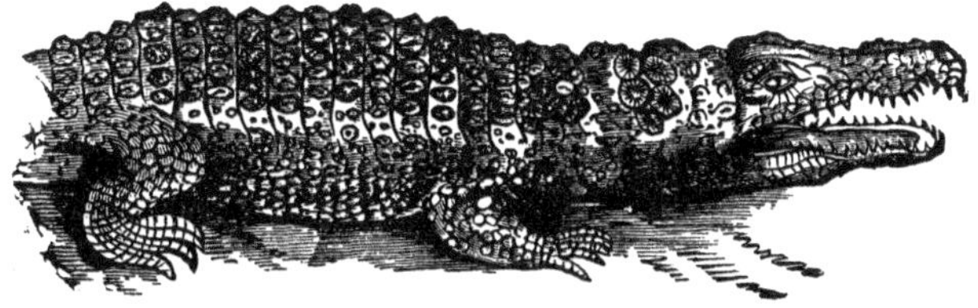

Fig. 173.—Crocodilia. Head and fore-part of the body of the common Crocodile (*Crocodilus vulgaris*).

distinct sockets, and the fact that the mixture of venous and arterial blood, which is so characteristic of Reptiles, takes place, not in the heart itself, but in its immediate neighbourhood, by a communication between the pulmonary artery and aorta directly after their origin.

The only other points about the Crocodiles which require special notice are that the eyes are protected by movable eyelids; the ear is covered by a movable ear-lid; the nasal cavities open in front by a single nostril, and are shut off from the cavity of the mouth, but open far back into the cavity of the pharynx; and lastly, the tongue is large and fleshy, and is immovably attached to the bottom of the mouth. (Hence the belief of the ancients that the Crocodile had no tongue.) The tail is long and compressed, with two rows of keeled plates, which unite about its middle to form a single crest, which is continued to its extremity. The feet are palmate or semipalmate, and only the three inner toes on each foot possess claws. The eyes possess three distinct lids, and there are two glands under the throat secreting a musky substance.

The *Crocodilia* abound in the fresh waters of hot countries, and are the largest of all living Reptiles, not uncommonly attaining a length of twenty feet or upwards. They are divided by Owen into three sub-orders, according to the shape of the dorsal vertebræ, termed the *Procœlia*, *Amphicœlia*, and *Opisthocœlia*.

Sub-order 1. *Procœlia.*—In this sub-order are all the living members of the *Crocodilia* distinguished by having the bodies of the dorsal vertebræ concave in front (procœlous). Three distinct types may be distinguished amongst the living *Croco-*

dilia. The Gavial is distinguished by its elongated snout, at the extremity of which the nostril is placed, and by the fact that the teeth are pretty nearly equal in size and similar in

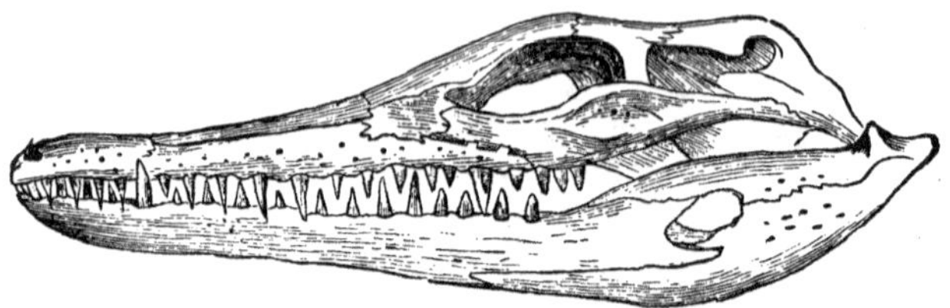

Fig. 174.—Skull of young *Crocodilus biporcatus* (after Van der Höven).

form in the two jaws. In the true Crocodiles (fig. 174) the fourth tooth in the lower jaw is larger than the others, and forms a canine tooth, which is received into a notch excavated in the side of the alveolar border of the upper jaw, so that it is visible externally when the mouth is closed. In the Caimans or Alligators the same tooth in the lower jaw forms a canine, but it is received into a pit in the palatal surface of the upper jaw, where it is entirely concealed when the mouth is shut. The Crocodiles have the hind-legs bordered by a toothed fringe, and the toes completely united by membrane. They are essentially natives of fresh water, but sometimes frequent the mouths of rivers. They occur chiefly in Asia and Africa, but species are found in some of the West Indian Islands. The Alligators have the hind-legs simply rounded, and the feet not completely webbed. They are essentially aquatic, and are voracious animals, living upon fish or Mammals. The best-known species are the Alligator of the Southern United States (*A. Mississippiensis*), the Caiman (*A. palpebrosus*) of Surinam and Guiana, and the "Jacare" or Spectacled Alligator (*A. sclerops*) of Brazil. The Gavial (*Gavialis Gangeticus*) is a native of India, and is the smallest of the living Crocodiles, attaining a length of about ten feet.

True procœlian Crocodiles occur for the first time in the Greensand (Cretaceous series) of North America. In Europe, however, the earliest remains of procœlian Crocodiles are from the Lower Tertiary rocks (Eocene). It is a curious fact that in the Eocene rocks of the south-west of England, there occur fossil remains of all the three living types of the *Crocodilia*—namely, the Gavials, true Crocodiles, and Alligators; though at the present day these forms are all geographically restricted in their range, and are never associated together.

Sub-order 2. *Amphicœlia.*—The Amphicœlian Crocodiles, with biconcave vertebræ, are entirely extinct. They have but a limited geological range, extending only from the Lias to the Chalk inclusive, and being therefore strictly Mesozoic.* The biconcave vertebræ show a decided approach to the structure of the backbone in fishes; and as the rocks in which they occur are marine, there can be little doubt but that these Crocodiles were, in the majority of cases at any rate, marine. The most important genera belonging to this order are *Teleosaurus*, *Steneosaurus*, *Dakosaurus*, *Makrospondylus*, and *Suchosaurus*, the last being from the fresh-water deposits of the Wealden (Cretaceous).

Sub-order 3. *Opisthocœlia.*—This sub-order, like the last, is entirely extinct, and is exclusively Mesozoic, all the known examples occurring in the Liassic, Oolitic, and Cretaceous rocks. The most important genera are *Streptospondylus* and *Cetiosaurus*. The *Cetiosaurus longus* of the Upper Oolites (Portland Stone) must have been the largest of all known *Crocodilia*, the vertebræ of the tail measuring as much as seven inches in length, and more than seven inches across.

CHAPTER LXV.

EXTINCT ORDERS OF REPTILES.

It remains now to consider briefly the leading characters of five wholly extinct orders of Reptiles, the peculiarities of which are very extraordinary, and are such as are exhibited by no living forms.

ORDER V. ICHTHYOPTERYGIA, Owen (= *Ichthyosauria*, Huxley).—The gigantic Saurians forming this order were distinguished by the following characters:—

The body was fish-like, without any distinct neck, and probably covered with a smooth or wrinkled skin, no horny or bony exoskeleton having been ever discovered. The vertebræ were numerous, deeply biconcave or amphicœlous, and having the neural arches united to the centra by a distinct suture. The anterior trunk-ribs possess bifurcate heads. There is no sacrum, and no sternal ribs or sternum, but clavicles were pre-

* If the so-called "Thecodont" Reptiles, such as *Thecodontosaurus* and *Belodon*, belong to this sub-order, then the Amphicœlian *Crocodilia* date from the age of the Triassic rocks.

sent as well as an interclavicle (episternum); and false ribs were developed in the walls of the abdomen. The skull had enormous orbits separated by a septum, and an elongated snout. The eyeball was protected by a ring of bony plates in the sclerotic. The teeth were not lodged in distinct sockets, but in a common alveolar groove. The fore and hind limbs were converted into swimming-paddles, the ordinary number of digits (five) remaining recognisable, but the phalanges being greatly increased in number, and marginal ossicles being added as well. A vertical caudal fin was in all probability present.

The order *Ichthyopterygia* includes only the gigantic and fish-like *Ichthyosauri* (fig. 175), all exclusively Mesozoic, and abounding in the Lias, Oolites, and Chalk, but especially characteristic of the Lias. If, however, the *Eosaurus Acadiensis* (Marsh) of the Coal-measures of Nova Scotia be rightly referred to this order, then the *Ichthyopterygia* date from the Carboniferous period. There is no doubt whatever but that the *Ichthyosauri* were essentially marine animals, and they have been often included with the next order (*Sauropterygia*) in a common group, under the name of *Enaliosauria* or Sea lizards.

In the biconcave vertebræ and probable presence of a vertical tail-fin, the *Ichthyosaurus* approaches the true fishes. There is, however, no doubt as to the fact that the animal was strictly an air-breather, and its reptilian characters cannot be questioned, at the same time that the conformation of the limbs is decidedly Cetacean in many respects. Much has been gathered from various sources as to the habits of the *Ichthyosaurus*, and its history is one of great interest. From the researches of Buckland, Conybeare, and Owen, the following facts appear to be pretty well established:—That the *Ichthyosauri* kept chiefly to open waters may be inferred from their

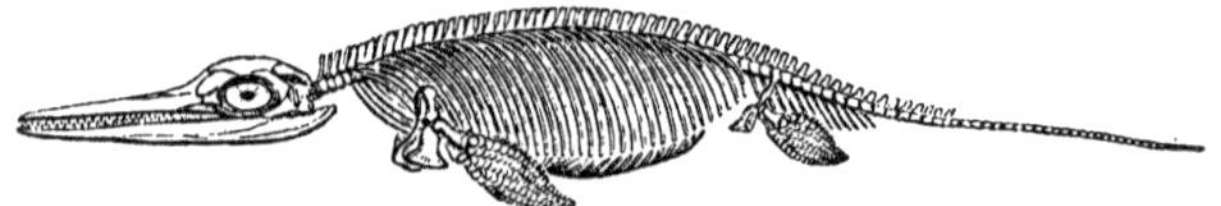

Fig. 175.—*Ichthyosaurus communis.*

strong and well-developed swimming-apparatus. That they occasionally had recourse to the shore, and crawled upon the beach, may be safely inferred from the presence of a strong and well-developed bony arch, supporting the fore-limbs, and closely resembling in structure the scapular arch of the *Orni-*

thorhynchus or Duck-mole of Australia. That they lived in stormy seas, or were in the habit of diving to considerable depths, is shown by the presence of a ring of bony plates in the sclerotic, protecting the eye from injury or pressure. That they possessed extraordinary powers of vision, especially in the dusk, is certain from the size of the pupil, and from the enormous width of the orbits. That they were carnivorous and predatory in the highest degree is shown by the wide mouth, the long jaws, and the numerous, powerful, and pointed teeth. This is proved, also, by an examination of their petrified droppings, which are known to geologists as "coprolites," and which contain numerous fragments of the scales and bones of the Ganoid fishes which inhabited the same seas.

ORDER VI. SAUROPTERYGIA, Owen (=*Plesiosauria*, Huxley).—This order of extinct reptiles, of which the well-known *Plesiosaurus* may be taken as the type, is characterised by the following peculiarities:—

The body, as far as is known, was naked, and not furnished with any horny or bony exoskeleton. The bodies of the vertebræ were either flat or only slightly cupped at each end, and the neural arches were anchylosed with the centra, and did not remain distinct during life. The transverse processes of the vertebræ were long, and the anterior trunk-ribs had simple, not bifurcate, heads. No sternum or sternal ribs are known to have existed, but there were false abdominal ribs. The neck in most was greatly elongated, and composed of numerous vertebræ. The sacrum was composed of two vertebræ. The orbits were of large size, and there was a long snout, as in the *Ichthyosauri*, but there was no circle of bony plates in the sclerotic. The limbs agree with those of the *Ichthyosauri* in being in the form of swimming-paddles (fig. 176), but differ in not possessing any supernumerary marginal ossicles. A pectoral arch, formed of two clavicles and an interclavicle (episternum) appears to have been sometimes, if not always, present. The teeth were simple, and were inserted into distinct sockets, and not lodged in a common groove.

The most familiar and typical member of the *Sauropterygia* is the *Plesiosaurus* (fig. 176), a gigantic marine reptile, chiefly characteristic of the Lias and Oolites. As regards the habits of the *Plesiosaurus*, Dr Conybeare arrives at the following conclusions:—"That it was aquatic is evident from the form of its paddles; that it was marine is almost equally so from the remains with which it is universally associated; that it may have occasionally visited the shore, the resemblance of its extremities to those of the Turtles may lead us to conjecture; its movements, however, must have been very awkward on land; and its long neck must have impeded its progress through the water, presenting a striking contrast to the organisation which so admirably fits the *Ichthyosaurus* to cut through the waves."

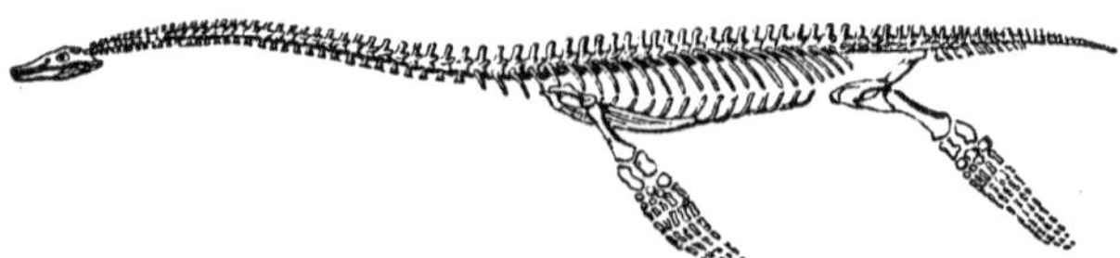

Fig. 176.—*Plesiosaurus dolichodeirus.*

As its respiratory organs were such that it must of necessity have required to obtain air frequently, we may conclude "that it swam upon or near the surface, arching back its long neck like a swan, and occasionally darting it down at the fish which happened to float within its reach. It may, perhaps, have lurked in shoal water along the coast, concealed amongst the sea-weed; and raising its nostrils to a level with the surface from a considerable depth, may have found a secure retreat from the assaults of powerful enemies; while the length and flexibility of its neck may have compensated for the want of strength in its jaws, and its incapacity for swift motion through the water."

The geological range of the *Plesiosaurus* is from the Lias to the Chalk inclusive, and specimens have been found indicating a length of from eighteen to twenty feet.

Of the other genera of the *Sauropterygia*, *Simosaurus* and *Nothosaurus* are from the Trias, and are chiefly characteristic of its middle division, the Muschelkalk. *Placodus* is another genus, also from the Muschelkalk, and is characterised by the extraordinary form of the teeth, which resembled those of many fishes in forming broad crushing plates, constituting a kind of pavement.

ORDER VII. ANOMODONTIA, Owen (= *Dicynodontia*, Huxley).—The leading characters of this order are to be found in the structure of the jaws, which appear to have been sheathed in horn, so as to constitute a kind of beak, very like that of the Chelonians. In the genera *Rhynchosaurus* and *Oudenodon*, both jaws seem to have been altogether destitute of teeth; but in *Dicynodon* there were two long tusks, growing from persistent pulps, placed one on each side in the upper jaw. The pectoral and pelvic arches were very strong, and the limbs were well developed and fitted for walking, and not for swimming.

Dicynodon and *Oudenodon* are known only from strata of

supposed Triassic age in South Africa and India, but *Rhynchosaurus* occurs in the Trias of Europe.

ORDER VIII. PTEROSAURIA.—This order includes a group of extraordinary flying Reptiles, all belonging to the Mesozoic epoch, and exhibiting in many respects a very extraordinary combination of characters. The most familiar members of the order are the so-called "Pterodactyles," and the following are the characters of the order:—

No exoskeleton is known to have existed. The dorsal vertebræ are procœlous, and the anterior trunk-ribs are double-headed. There is a broad sternum with a median ridge or keel, and ossified sternal ribs. The jaws were always armed

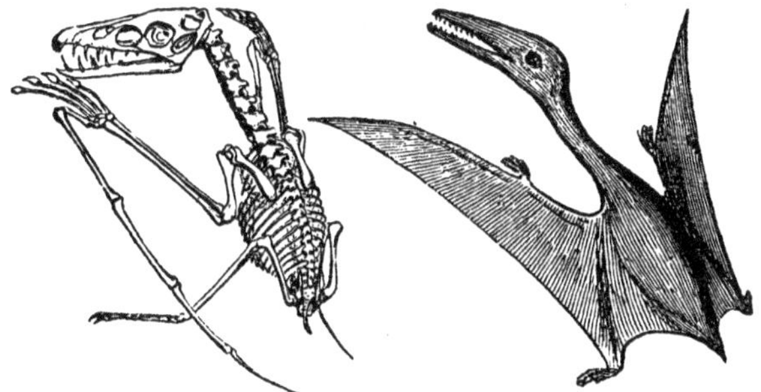

Fig. 177.—*Pterodactylus brevirostris.* Skeleton and restoration.

with teeth, and these were implanted in distinct sockets. In some forms (*Ramphorhynchus*) there appear to have been no teeth in the anterior portion of the jaws, and these parts seem to have been sheathed in horn, so as to constitute a kind of beak. A ring of bony plates occurs in the sclerotic coat of the eye. The pectoral arch consists of a scapula and distinct coracoid bone, articulating with the sternum as in Birds, but no clavicles have hitherto been discovered. The fore-limb (fig. 177) consists of a humerus, ulna and radius, carpus, and hand of four fingers, of which the inner three are short and unguiculate, whilst the outermost is clawless and is enormously elongated. Between this immensely-lengthened finger, the side of the body, and the comparatively small hind-limb, there must have been supported an expanded flying-membrane or "patagium," which the animal must have been able to employ as a wing, much as the Bats of the present day. Lastly, most of the bones were "pneumatic"—that is to say, were hollow and filled with air.

By the presence of teeth in distinct sockets, and, as will be seen hereafter, especially in the structure of the limbs, the Pterodactyles differed from all known Birds, and there can be little question as to their being genuine Reptiles. The only Reptile, however, now existing, which possesses any power of sustaining itself in the air, is the little *Draco volans*, but this can only take extended leaps from tree to tree, and cannot be said to have any power of flight properly so called. That the Pterodactyles, on the other hand, possessed the power of genuine flight, is shown by the presence of a median keel upon the sternum, proving the existence of unusually-developed pectoral muscles; by the articulation of the coracoid bones with the top of the sternum, providing a fixed point or fulcrum for the action of the pectoral muscles; and, lastly, by the existence of air-cavities in the bones, giving the animal the necessary degree of lightness. The apparatus, however, of flight was not a "wing," as in Birds, but a flying-membrane, very similar in its mode of action to the patagium of the Mammalian order of the Bats. The patagium of the Bats, however, differs from that of the Pterodactyles in being supported by the greatly-elongated fingers, whereas in the latter it is only the outermost finger which is thus lengthened out. The difficulty as to the position of the *Pterosauria* is evaded by Mr Seeley by placing them in a distinct class, which he terms *Ornithosauria*, and which he regards as most nearly related to, but coequal with, the class *Aves*.

The *Pterosauria* are exclusively Mesozoic, being found from the Lower Lias to the Middle Chalk inclusive, the Lithographic Slate of Solenhofen (Upper Oolite) being particularly rich in their remains. Most of them appear to have attained no very great size, but the remains of a species from the Cretaceous rocks have been considered to indicate an animal with more than twenty feet expanse of wing, counting from tip to tip.

In the genus *Pterodactylus* proper, the jaws are provided with teeth to their extremities, all the teeth being long and slender.

In *Dimorphodon*, the anterior teeth are large and pointed, the posterior teeth small and lancet-shaped.

In *Ramphorhynchus*, the anterior portion of both jaws is edentulous, and may have formed a horny beak, but teeth are present in the hinder portion of the jaws.

ORDER IX.—DINOSAURIA.—The last order of extinct Reptiles is that of the *Dinosauria*, comprising a group of very remarkable Reptiles, which show many points of decided affinity to the Birds on the one hand, and to the so-called Pachydermatous

Mammals on the other. Most of the *Dinosauria* were of gigantic size, and the order is defined by the following characters:—

The skin was sometimes naked, sometimes furnished with a well-developed exoskeleton, consisting of bony shields, much resembling those of the Crocodiles. A few of the anterior vertebræ were opisthocœlous, the remainder having flat or slightly biconcave bodies. The anterior trunk-ribs were double-headed. The teeth were confined to the jaws and implanted in distinct sockets. There were always two pairs of limbs, and these were strong, furnished with claws, and adapted for terrestrial progression. In some cases the fore-limbs were very small in proportion to the size of the hind-limbs. No clavicles have been discovered.

The most familiar examples of the *Dinosauria* are *Megalosaurus* and *Iguanodon*.

Megalosaurus is a gigantic Oolitic Reptile, which occurs also in the Cretaceous series (Weald Clay). Its length has been estimated at between forty and fifty feet, the femur and tibia each measuring about three feet in length. As the head of the femur is set on nearly at right angles with the shaft, whilst all the long bones contain large medullary cavities, there can be no doubt but that *Megalosaurus* was terrestrial in its habits. That is was carnivorous and destructive in the highest degree is shown by the powerful, pointed, and trenchant teeth.

The *Iguanodon* is mainly, if not exclusively, Cretaceous, being especially characteristic of the great delta-deposit of the Wealden. The length of the *Iguanodon* has been estimated as being probably from fifty to sixty feet, and from the close resemblance of its teeth to those of the living Iguanas, there is little doubt that it was herbivorous and not carnivorous. The femur of a large *Iguanodon* measures from four to five feet in length, with a circumference of twenty-two inches in its smallest part. From the disproportionately small size of the fore-limbs, and from the occurrence of *pairs* of gigantic three-toed footsteps in the same beds, it has been concluded, with much probability, that *Iguanodon*, in spite of its enormous bulk, must have walked temporarily or permanently upon its hind-legs, thus coming to present a most marked and striking affinity to the Birds.

The most remarkable, however, of the *Dinosauria*, is the little *Compsognathus longipes* from the Lithographic Slate of Solenhofen, referred to this order by Professor Huxley. This Reptile is not remarkable for its size, which does not seem to have been much more than two feet, but for the remarkable affinities which it exhibits to the true Birds. The head of *Compsognathus* was furnished with *toothed* jaws, and supported upon a long and slender neck. The fore-limbs were very short, but the hind-limbs were long and like those of Birds. The *proximal* portion of the tarsus resembled that of Birds in being anchylosed to the lower end of the tibia; but the *distal* portion of the tarsus—unlike that of Birds—was free, and was not anchylosed with the metatarsus. Huxley concludes that "it is impossible to look at the conformation of this strange Reptile, and to doubt that it hopped or walked in an erect or semi-erect position, after the manner of a bird, to which its long neck, slight head, and small anterior limbs must have given it an extraordinary resemblance."

DIVISION II. SAUROPSIDA.

CHAPTER LXVI.

CLASS IV.—AVES.

THE fourth class of the *Vertebrata* is that of *Aves*, or Birds. The Birds may be shortly defined as being "oviparous Vertebrates with warm blood, a double circulation, and a covering of feathers" (Owen). More minutely, however, the Birds are defined by the possession of the following characters :—

The embryo possesses an amnion and allantois, and branchiæ or gills are never developed at any time of life upon the visceral arches. The skull articulates with the vertebral column by a single occipital condyle. Each half or ramus of the lower jaw consists of a number of pieces, which are separate from one another in the embryo; and the jaw is united with the skull, not directly, but by the intervention of a quadrate bone (as in the Reptiles). The fore-limb in no existing birds possesses more than three fingers or digits, and the metacarpal bones are anchylosed together. In all living birds the fore-limbs are useless as regards prehension, and in most they are organs of flight. The hind-limbs in all birds have the ankle-joint placed in the middle of the tarsus, the proximal portion of the tarsus coalescing with the tibia, and the distal portion of the tarsus being anchylosed with the metatarsus to constitute a single bone known as the "tarso-metatarsus."

The heart consists of four chambers, two auricles, and two ventricles; and not only are the right and left sides of the heart completely separated from one another, but there is no communication between the pulmonary and systemic circulations, as there is in Reptiles. There is only one aortic arch, the right. The blood is hot, having an average temperature of as much as 103° to 104°. The blood-corpuscles are oval and nucleated.

The respiratory organs are in the form of spongy cellular lungs, which are not freely suspended in pleural sacs; and the bronchi open on their surface into a number of air-sacs, placed in different parts of the body.

All birds are oviparous, none bringing forth their young alive, or being even ovo-viviparous. All birds are, lastly, provided with an epidermic covering, so modified as to constitute what are known as *feathers*.

Professor Huxley's account of the method in which feathers are produced is so remarkably clear, that no apology is necessary for quoting it in its entirety. Feathers "are evolved within sacs from the surface of conical papillæ of the dermis. The external surface of the dermal papilla, whence a feather is to be developed, is provided upon its dorsal surface with a median groove, which becomes shallower towards the apex of the papilla. From this median groove lateral furrows proceed at an open angle, and passing round upon the under surface of the papilla, become shallower, until, in the middle line, opposite the dorsal median groove, they become obsolete. Minor grooves run at right angles to the lateral furrows. Hence the surface of the papilla has the character of a kind of mould, and if it were repeatedly dipped in such a substance as a solution of gelatine, and withdrawn to cool until its whole surface was covered with an even coat of that substance, it is clear that the gelatinous coat would be thickest at the basal or anterior end of the median groove, at the median ends of the lateral furrows, and at those ends of the minor grooves which open into them; whilst it would be very thin at the apices of the median and lateral grooves, and between the ends of the minor grooves. If, therefore, the hollow cone of gelatine, removed from its mould, were stretched from within, or if its thinnest parts became weak by drying, it would tend to give way, along the inferior median line, opposite the rod-like cast of the median groove, and between the ends of the casts of the lateral furrows, as well as between each of the minor grooves, and the hollow cone would expand into a flat feather-like structure, with a median shaft, as a 'vane' formed of 'barbs' and 'barbules.' In point of fact, in the development of a feather, such a cast of the dermal papilla is formed, though not in gelatine, but in the horny epidermic layer developed upon the mould, and, as this is thrust outwards, it opens out in the manner just described. After a certain period of growth the papilla of the feather ceases to be grooved, and a continuous horny cylinder is formed, which constitutes the 'quill.'"

A typical feather (fig. 178) consists of the following parts: —1. The "quill" or "barrel" (*a*), which forms the basal portion of the feather, by which it is inserted in the skin on its own dermal papilla. It is the latest-formed portion of the feather, and consists of a hollow horny cylinder. 2. The "shaft" (*b*), which is simply a continuation of the quill, and which forms the central axis of the feather. The inferior surface of the shaft always exhibits a strong longitudinal groove, and it is composed of a horny external sheath, containing a white spongy substance, very like the pith of a plant. 3. The shaft carries the lateral expansions or "webs" of the feather, collectively constituting the "vane." Each web is composed of a number of small branches, which form an open angle with the shaft, and which are known as the "barbs" (*c*). The margins of each barb are, in turn, furnished with a series of still smaller branches, which are known as the "barbules." As a general rule, the extremities of the barbules are hooked, so that those springing from the one side of each barb interlock with those springing from the opposite side of the next barb. In this way the barbs are kept in apposition with one another over a greater or less portion of the entire web. More or less of the barbs in the lower portion of the feather are, however, disunited, and not connected by their barbules; and these constitute what is known as the "down." In the Ostriches, Emeus, and some others, all the barbs of the feathers are disconnected, giving to the plumage of these birds its peculiarly soft character. At the point where the shaft joins the quill there is very generally found a small feather, known as the "accessory plume," or "plumule." This is usually much the same in structure as the main feather, but considerably smaller. It

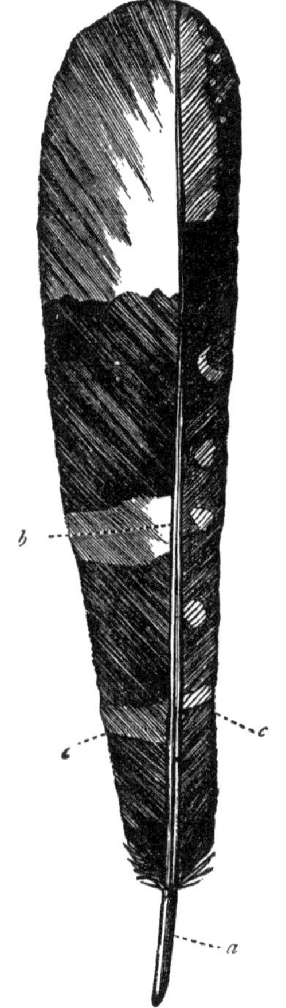

Fig. 178.—Quill-feather (*Stenopsis*). *a* Quill or barrel; *b* Shaft; *c c* Webs, composed of the barbs, and together forming the "vane."

may, however, be as large as the original feather, or it may be reduced to nothing more than a tuft of down.

The feathers vary in different parts of the bird, and are generally divided into those which cover the body—" clothing feathers," and those which occur in the wings and tail—" quill-feathers." As regards the great quill-feathers of the wings, the longest are those which arise from the bones of the hand, and they are called the " primaries." Those which arise from the distal end of the fore-arm (radius and ulna) are termed the " secondaries," and those which are attached to the proximal end of the fore-arm are the " tertiaries." The feathers which lie over the humerus and scapula are the " scapulars." The rudimentary "thumb" also carries some quills, which form what is known as the "alula," or "bastard-wing." The smaller feathers, which cover the bases of the quill-feathers above and below, are the "wing-coverts" — "greater," "lesser," and "under." The great quill-feathers of the tail (" rectrices") form a kind of fan, of great use in steering the bird in flight; and their bases are covered by a series of feathers which constitute the " tail-coverts."

The entire *skeleton* of the Birds is singularly compact, and at the same time singularly light. The compactness is due to the presence of an unusual amount of phosphate of lime; and the lightness, to the absence in many of the bones of the ordinary marrow, and its replacement by air.

As regards the *vertebral column*, birds exhibit some very interesting peculiarities. The cervical region of the spine is unusually long and flexible, since the fore-limbs are useless as organs of prehension—and all acts of prehension must be exercised either by the beak or by the hind-feet, or by both acting in conjunction. In all birds alike, the neck is sufficiently long and flexible to allow of the application of the beak to an oil-gland placed at the base of the tail, this act being necessary for the due performance of the operation of " preening"—that is, of lubricating and cleaning the plumage. The number of vertebræ in the neck varies from nine to twenty-four, and their structure is alway such as to allow of considerable freedom of motion one upon the other. The dorsal vertebræ vary from six to ten in number, and of these the anterior four or five are generally anchylosed with one another, so as to give a base of resistance to the wings. In the Cursorial Birds, however (such as the Ostrich and Emeu), and in some others (such as the Penguin), in which the power of flight is wanting, the dorsal vertebræ are all more or less freely movable one upon another. There are no lumbar vertebræ, but all the vertebræ between the last dorsal and the first caudal (varying from nine to twenty) are anchylosed together to form a bone which is ordinarily known as the "sacrum." To this, in turn, the iliac bones are anchylosed along their whole length, giving perfect immobility to this region of the spine and to the pelvis.

The coccygeal or caudal vertebræ vary in number from eight to ten, and are movable upon one another. The most noticeable feature about this part of the spinal column is what is known as the "ploughshare-bone." This is the last joint of the tail, and is a long, slender, ploughshare-shaped bone, destitute of lateral processes, and without any medullary canal (fig. 182, B). In reality it consists of two or more of the caudal vertebræ, completely anchylosed, and fused into a single mass. It is usually set on to the extremity of the spine at an angle more or less nearly perpendicular to the axis of the body; and it affords a firm basis for the support of the great quill-feathers of the tail ("rectrices"). It also supports the coccygeal oil-glands, and can be raised at pleasure, so as to meet the bill, when the operation of preening is in progress. In the Cursorial Birds, which do not fly, the terminal joint of the tail is not ploughshare-shaped. In the extraordinary Mesozoic bird, the *Archæopteryx macrura*, there is no ploughshare-bone, and the tail consists of twenty separate vertebræ, all distinct from one another, and each carrying a pair of quill-feathers, one on each side (fig. 201). As the vertebræ of the ploughshare-bone are distinct from one another in the embryos of existing birds, the tail of the *Archæopteryx* is to be regarded as a case of the permanent retention in the adult of an embryonic character. In the increased number of caudal vertebræ, however, and in some other characters, the tail of the *Archæopteryx* makes a decided approach to the true Reptiles.

The various bones which compose the *skull* of Birds are amalgamated in the adult so as to form a single piece, and the sutures even are obliterated, the lower jaw alone remaining movable. The occipital bone carries a single occipital condyle only, and this is hemispherical or nearly globular in shape. The "beak" (fig. 179), which forms such a conspicuous feature in all birds, consists of an upper and lower half, or a "superior" and "inferior mandible." The upper mandible is composed almost entirely of the greatly elongated intermaxillary bones, flanked by the comparatively small superior maxillæ. The inferior mandible is primitively composed of twelve pieces, six on each side; but in the adult these are all indistinguishably amalgamated with one another, and the lower jaw forms a single piece. As in the Reptiles, the lower jaw articulates with the skull, not directly, but through the intervention of a distinct bone—the quadrate bone or tympanic bone—which always re-

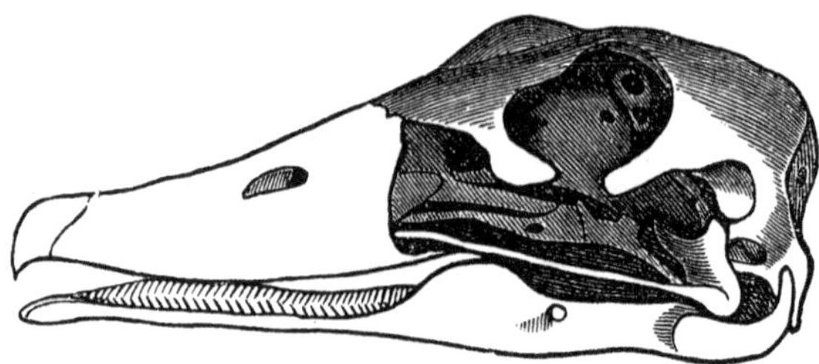

Fig. 179.—Skull of Spur-winged Goose (*Plectropterus Gambensis*).

mains permanently movable, and is never anchylosed with the skull. In no bird are teeth ever developed in either jaw, but both mandibles are encased in horn, forming the beak, and the margins of the bill are sometimes serrated. The quadrate bone (os carré of the French) is in contact at one end with an elongated bone (jugal bone), the other end of which comes against the palate. By the depression of the lower jaw, assisted by proper muscles, the quadrate bone is brought forward, and the jugal bone is thus made to elevate the palate and upper jaw. With one or two exceptions, the upper mandible is thus movable in all birds.

The thoracic cavity is bounded by the dorsal vertebræ, which are usually, as before said, anchylosed to one another to a greater or less extent. Laterally, the thorax is bounded by the ribs, which vary in number from six to ten pairs. In most birds each rib carries a peculiar process—the "uncinate process"—which arises from its posterior margin, is directed upwards and backwards, and passes over the rib next in succession behind, where it is bound down by ligament. The first and last dorsal ribs carry no uncinate processes, and in some cases the processes continue throughout life as separate pieces (fig. 180, B). Anteriorly, the ribs articulate with a series of straight bones, which are called the "sternal ribs," but which in reality are to be looked upon as the ossified "costal cartilages." These sternal ribs (fig. 180, B) are in turn movably articulated to the sternum in front, and "they are the centres upon which the respiratory movements hinge" (Owen). In front the thoracic cavity is completed by an enormously-expanded sternum or breast-bone, which in some birds of great powers of flight extends over the abdominal cavity as well, in some cases even reaching the pelvis. The sternum of all birds which fly, is characterised by the presence of a greatly-developed median ridge or keel (fig. 180, A), to which are attached the great pectoral muscles which move the wings. As a general rule, the size of this sternal crest allows a very tolerable estimate to be formed of the flying powers of the bird to which it may have belonged; and in the Ostriches and other birds which do not fly, there is no sternal keel. At its anterior angles the sternum exhibits two pits for the attachment of the coracoid bones.

The scapular or pectoral arch consists of the shoulder-blade or scapula, the collar-bone or clavicle, and the coracoid bone, on each side. The scapula, as a rule (fig. 180, A, *s s*), is a

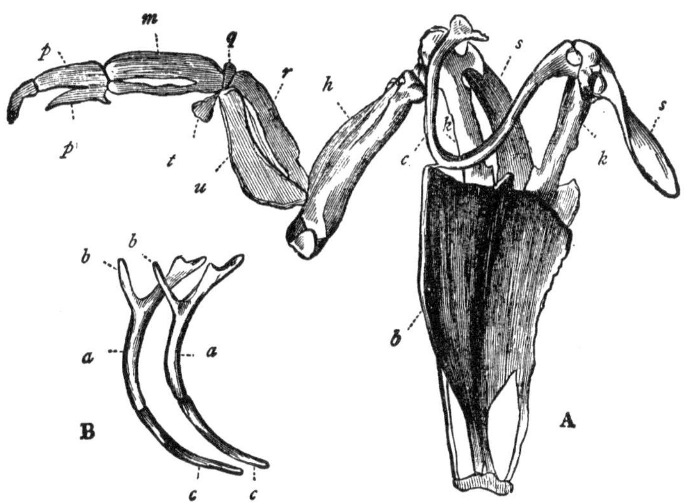

Fig. 180.—A, Breast-bone, shoulder-girdle, and fore-limb of Penguin (after Owen): *b* Sternum, with the sternal keel; *s s* Scapulæ; *k k* Coracoid bones; *c* Furculum or merry-thought, composed of the united clavicles; *h* Humerus; *u* Ulna; *r* Radius; *t* Thumb; *m* Metacarpus; *p* Phalanges of the fingers. B, Ribs of the Golden Eagle: *a a* Ribs giving off (*b b*) uncinate processes; *c c* Sternal ribs.

simple elongated bone, not flattened out into a broad plate, and carrying no transverse ridge, or spinous process. Only a portion of the glenoid cavity for the articulation with the head of the humerus is formed by the scapula, the remainder being formed by the coracoid. The coracoid bones (fig. 180, A, *k k*) correspond with the coracoid processes of man; but in birds they are distinct bones, and are not anchylosed with the scapula. The coracoid bone on each side is always the strongest of the bones forming the scapular arch. Superiorly it articulates with the clavicle and scapula, and forms part of the glenoid cavity for the humerus. Inferiorly each coracoid bone articulates with the upper angle of the sternum. The position of the coracoids is more or less nearly vertical, so that they form fixed points for the action of the wings in their downward stroke. The clavicles (fig 180, A, *c*) are rarely rudimentary or absent, and are in some few cases separate bones. In the great majority, however, of birds, the clavicles are anchylosed together at their anterior extremities, so as to form a single bone, somewhat V-shaped, popularly known as the "merry-thought," and technically called the "furculum" ("fourchette" of the French). The outer extremities of the furculum articulate with the scapula and coracoid; and the anchylosed angle is commonly united by ligament to the top of the sternum. The function of the clavicular or furcular arch is "to oppose the forces which tend to press the humeri inwards towards the mesial plane, during the downward stroke of the wing" (Owen). Consequently the clavicles are stronger, and their angle of union is more open, in proportion to the powers of flight possessed by each bird. The furculum is rudimentary in the Ostrich, Emeu, and Cassowary.

We have next to consider the structure of the bones which compose the fore-limb or "wing" of the bird; and as this organ is the one which chiefly conditions the peculiar life of the bird, it is in it that we find some of the most characteristic points of structure in the whole skeleton. Though considerably modified to suit its function as an organ of aerial progression, the wing of the bird is readily seen to be homologous with the arm of a man or the fore-limb of a Mammal (fig. 180, A, and fig. 181). The upper arm (*brachium*) is supported by a single bone, the humerus, which is short and strong, and articulates above with the articular cavity formed partly by the scapula and partly by the coracoid (fig. 181, *h*). The humerus is succeeded distally by the fore-arm (*antibrachium*) constituted by the normal two bones, the radius and ulna (fig. 181, *r*, *u*), of which the radius is much the smaller and more slender, and the ulna much the larger and stronger. The ulna and radius are followed inferiorly by the bones of the wrist or carpus; but these are reduced in number to *two* small bones, "so wedged in between the antibrachium and metacarpus as to limit the motions of the hand to those of abduction and adduction necessary for the folding up and expansion of the wing; the hand is thus fixed in a state of pronation; all power of flexion, extension, or of rotation, is removed from the wrist-joint, so that the wing strikes firmly, and with the full force of the contraction of the depressor muscles, upon the resisting air" (Owen). One other bone

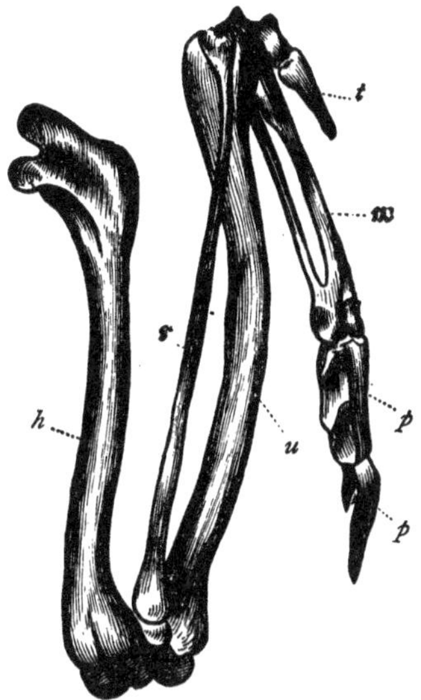

Fig. 181.—Fore-limb of the Jer-falcon. *h* Humerus; *r* Radius; *u* Ulna; *t* "Thumb;" *m* Metacarpals, anchylosed at their extremities; *p p* Phalanges of fingers.

of the normal carpus (namely, the "os magnum") is present, but this is anchylosed with one of the metacarpals. There are thus really *three* carpal bones, though only two appear to be present. The carpus is followed by the metacarpus, the condition of which agrees with that of the carpal bones. The two outermost of the normal five metacarpals are absent, and the remaining three are anchylosed—together with the os magnum—so as to form a single bone (fig. 181, *m*). This bone, however, appears externally as if formed of *two* metacarpals united to one another at their extremities, but free in their median portion. The metacarpal bone which corresponds to the radius is always the larger of the two (as being really composed of two metacarpals), and it carries the digit which has the greatest number of phalanges. This digit corresponds with the "index" finger, and it is composed of two, or sometimes three, phalanges (fig. 181, *p*). At the proximal end of this metacarpal, at its outer side, there is generally attached a single phalanx, constituting the so-called "thumb" (fig. 181, *t*), which carries the "bastard-wing." The digit which is attached to the ulnar metacarpal corresponds to the "ring finger," and never consists of more than a single phalanx (fig. 181).

As regards the structure of the posterior extremity or hind-limb, the pieces which compose the innominate bones (namely, the ilium, ischium, and pubes) are always anchylosed to one another; and the two innominate bones are also always anchylosed, by the medium of the greatly-elongated ilia, to the sacral region of the spine. In no living bird, however, with the single exception of the Ostrich, are the innominate bones united in the middle line in front by a symphysis pubis. The stability of the pelvic arch, necessary in animals which support the weight of the body on the hind-limbs alone, is amply secured in all ordinary cases by the anchylosis of the ilia with the sacrum.

As in the higher Vertebrates, the lower limb (fig. 182, A) consists of a femur, a tibia and fibula, a tarsus, metatarsus, and phalanges; but some of these parts are considerably obscured by anchylosis. The femur or thigh-bone (fig. 182, A, *f*) is generally very short, comparatively speaking. The chief bone of the leg is the tibia (*t*), to which a thin and tapering fibula (*r*) is anchylosed. The upper end of the fibula, however, articulates with the external condyle of the femur. The ankle-joint is placed, as in Reptiles, between the proximal and distal portions of the tarsus. The proximal portion of the tarsus is undistinguishably amalgamated with the lower end of the tibia. The distal portion of the tarsus is anchylosed with the whole of the metatarsus to constitute the most characteristic bone in the leg of the Bird—the "tarso-metatarsus" (*m*). In most of the long-legged birds, such as the waders, the disproportionate length of the leg is given by an extraordinary elongation of the tarso-metatarsus.

The tarso-metatarsus is followed inferiorly by the digits of the foot. In most birds the foot consists of three toes directed forwards and one backwards—four toes in all. In no wild bird are there *more* than four toes, but often there are only three, and in the Ostrich the number is reduced to two. In all birds which have three anterior and one posterior toe, it is the posterior thumb or *hallux* (that is to say, the innermost digit of the hind-limb) which is directed backwards; and it invariably consists of *two* phalanges only. The most internal of the three toes which are directed forwards, consists of *three* phalanges; the next has *four* phalanges; and the outermost or "little" toe is made up of *five* phalanges (fig. 182, A). This

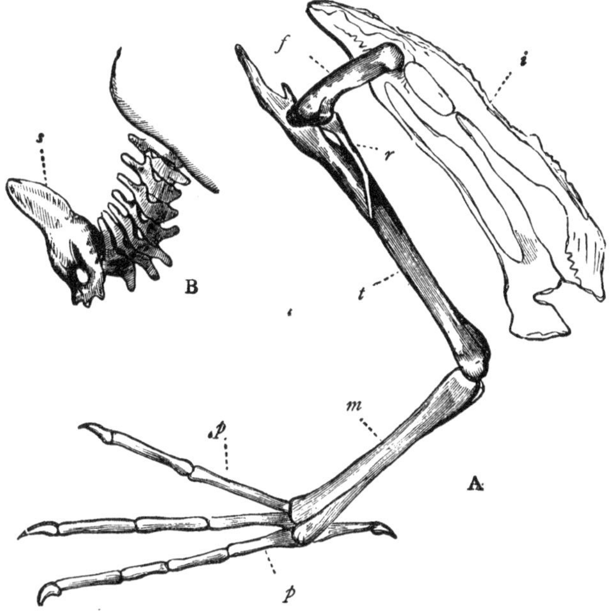

Fig. 182.—A, Hind-limb of the Loon (*Colymbus glacialis*)—after Owen: *i* Innominate bone; *f* Thigh-bone or femur; *t* Tibia, with the proximal portion of the tarsus anchylosed to its lower end; *r* Fibula; *m* Tarso-metatarsus, consisting of the distal portion of the tarsus anchylosed with the metatarsus; *p p* Phalanges of the toes; B, Tail of the Golden Eagle; *s* Ploughshare-bone, carrying the great tail-feathers.

increase in an arithmetical ratio of the phalanges of the toes, in proceeding from the inner to the outer side of the foot, obtains in almost all birds, and enables us readily to detect which digit is suppressed, when the normal four are not all present. Variations of different kinds exist, however, in the number and disposition of the toes. In many birds—such as the Parrots—the outermost toe is turned backwards, so that there are two toes in front and two behind. In others, again, the outer toe is normally directed forwards, but can be turned backwards at the will of the animal. In the Swifts, on the other hand, all four toes are present, but they are all turned forwards. In many cases—especially amongst the Natatorial birds—the hallux is wholly wanting, or is rudimentary. In the Emeu, Cassowary, Bustards, and other genera, the hallux is invariably absent, and the foot is three-toed. In the Ostrich both the hallux and the next toe ("index") are wanting, and the foot consists simply of two toes, these being the outer toe and the one next to it. The toes are mechanically flexed during the sleep of the Bird, in virtue of an arrangement by which, whilst *all* the flexor tendons pass behind the *heel*, *one* of them runs in front of the *knee*. As the muscles, therefore, are relaxed during sleep, and the weight of the body tends to flex the knee, the tendon of this flexor is thereby put on the stretch, and the toes are again bent involuntarily.

The *digestive system* of birds comprises the beak, tongue, gullet, stomach, intestines, and cloaca. Teeth are invariably wanting in birds, and the jaws are encased in horn, constituting the bill. The form of the bill varies enormously in different birds, and it is employed for holding and tearing the prey, for prehensile purposes, for climbing, and in some birds as an organ of touch. In these last-mentioned cases the bill is more or less soft, and is supplied with filaments of the fifth nerve. In many birds, too, in which the bill is not soft, the base of the upper mandible is surrounded by a circle of naked skin, constituting what is called the "cere," and this, no doubt, serves also as a tactile organ.

The tongue of birds can hardly be looked upon as an organ of taste, since it is generally cased in horn like the mandibles. It is, in fact, principally employed as an organ of prehension; but in some cases—as in the Parrots—it is soft and fleshy, and then, doubtless, is to some extent connected with the sense of taste. It is essentially composed of a prolongation of the hyoid bone (the glosso-hyal), which is sheathed in horn, and is variously serrated or fringed.

Salivary glands are invariably present, but they are rarely of large size, and they have often a very simple structure.

In accordance with the structure of the neck, the gullet in birds is usually of great length, and it is generally very dilatable. In the carnivorous, or Raptorial, and in the granivorous birds, the gullet (fig. 183, *o*), is dilated into a pouch, which is situated at the lower part of the neck, just in front of the merry-thought. This is what is known as the "crop" or "ingluvies" (*c*), and it may be either a mere dilatation of the tube of the gullet, or it may be a single or double pouch. The food is detained in the crop for a longer or shorter time, according to its nature, before it is subjected to the action of the proper digestive organs. The œsophagus, after leaving the crop, shortly opens into a second cavity, which is known as the "proventriculus" or "ventriculus succenturiatus" (*p*). This is the true digestive cavity, and its mucous membrane is richly supplied with gastric

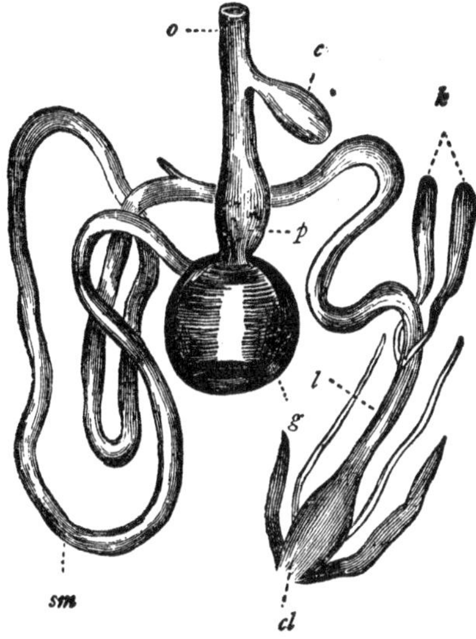

Fig. 183.—Digestive System of the common Fowl (after Owen). *o* Gullet; *c* Crop; *p* Proventriculus; *g* Gizzard; *sm* Small intestine; *k* Intestinal cæca; *l* Large intestine; *cl* Cloaca.

follicles which secrete the gastric juice. The proventriculus, however, corresponds, not with the whole stomach of the Mammals, but only with its cardiac portion; and it opens into a second, muscular cavity, which corresponds to the pyloric division of the Mammalian stomach. The gizzard (*g*) is situated below the liver, and forms in all birds an elongated sac, having two apertures above, of which one conducts into the duodenum or commencement of the small intestine, whilst the other communicates with the proventriculus. The two chief forms of gizzard are exhibited respectively by the Raptorial birds, which feed on easily-digested animal food, and the *Rasores* and some of the *Natatores*, which feed on hardly-digested grains. In the birds of Rapine the gizzard scarcely deserves the name, being, as a rule, nothing more than a wide membranous cavity with thin walls. In the granivorous birds, whose hard food requires crushing, the gizzard is enormously developed; its lining coat is formed of a thick, horny epithelium, and its walls are extremely thick and muscular. This constitutes a grinding apparatus, like the stones of a mill; whilst the "crop" or œsophageal dilatation may be compared to the "hopper" of a mill, since it supplies to the gizzard "small successive quantities of food as it is wanted" (Owen). Supplementing the action of the muscular walls of the gizzard, and acting in the place of teeth, are the small stones or pebbles, which, as is so well known, so many of the granivorous birds are in the habit of swallowing with their food, or at other times. In fact there can be no doubt but that the gravel and pebbles swallowed by these birds are absolutely essential to existence, since the gizzard, without this assistance, is unable properly to triturate the food.

The intestinal canal extends from the gizzard to the cloaca, and is, comparatively speaking, short. The secretions of the liver and pancreas are poured into the small intestine, as in Mammals. The commencement of the large intestine is almost always furnished with two long "cæca" or blind tubes, the length of which varies a good deal in different birds (fig. 183, *k*). They are sometimes wanting; and their exact function is uncertain; though they are most probably connected partly with digestion and partly with excretion. The large intestine is always very short—seldom more than a tenth part of the length of the body—and it terminates in the "cloaca" (fig. 183, *cl*). This is a cavity which in all birds receives the termination of the rectum, the ducts of the generative organs and the ureters; and serves, therefore, for the expulsion of the fæces, the generative products, and the urinary secretion.

Respiration is effected in Birds more completely and actively than in any other class of the *Vertebrata*, and as the result of this, their average temperature is also higher. This extensive development of the respiratory process is conditioned by the fact that, in addition to true lungs, air is admitted into a greater or less number of the bones, and into a number of cavities—the so-called air-receptacles—which are distributed through various parts of the body. By this extensive penetration of air into various parts of the body, the aeration of the blood is effected not only in the lungs, but also over a greater or less extent of the systemic circulation as well; and hence in Birds this process attains its highest perfection. The cavities of the thorax and abdomen are not separated from one another by a complete partition, the diaphragm being only present in a rudimentary form. The lungs are two in number, of a bright-red colour, and spongy texture. They are confined to the back of the thorax, extending along each side of the spine, from the second dorsal vertebra to the kidney. They differ from the lungs of the Mammals in not being freely suspended in a pleural membrane. The pleura, on the other hand, is reflected only over the anterior surface of the lungs. The bronchi, or primary divisions of the windpipe (fig. 184), diminish in size as they pass through the lung, by giving off branches, which, in turn, give off the true air-vesicles of the lung. When the bronchial tubes reach the surface of the lung, they open, by a series of distinct apertures, into a series of "air-sacs." These are a series of membranous sacs formed by the continuation of the lining membrane of the bronchi, and supported by reflections of the serous membrane of the thoracico-abdominal cavity. In those aquatic birds which, like the Penguin, do not enjoy the power of flight, the air-cells are restricted to the abdomen; but in most birds they are continued along the sides of the neck and limbs. In some cases—as the Pelican and Gannet—air-receptacles are situated beneath almost the whole of the integument. The air-cells not only greatly reduce the specific gravity of birds, and thus fit them for an aerial life, but also assist in the mechanical work of respiration, and must also greatly promote the aeration of the blood.

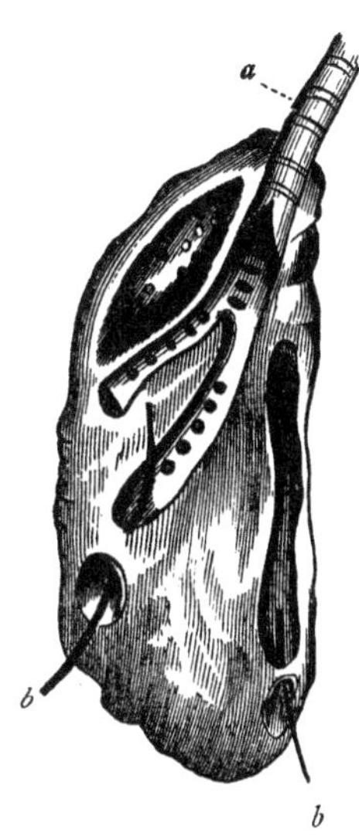

Fig. 184.—Lung of Goose (after Owen). *a* Main bronchus dividing into secondary branches as it enters the lung, these giving off smaller branches, the openings of which are seen on the back of the bronchial tubes; *b b* Bristles passed from the bronchi through the apertures on the surface of the lung by which the bronchi communicate with the air-receptacles.

In connection with the air-receptacles, and as an extension of them, is a series of cavities occupying the interior of a greater or less number of the bones, and also containing air. In young birds these air-cavities do not exist, and the bones are filled with marrow as in the Mammals. The extent also to which the bones are "pneumatic" varies greatly in different birds. In the Penguin—which does not fly—all the bones contain marrow, and there are no air-cavities. In the large

Running birds (*Cursores*), such as the Ostrich, the bones of the leg, pelvis, spine, ribs, skull, and sternum are pneumatic; but the bones of the wings, with the exception of the scapular arch, are without air-cavities, and permanently retain their marrow. All birds which fly, with the singular exception of the Woodcock, have air admitted to the humerus. In the Pelican and Gannet, all the bones of the skeleton, except the phalanges of the toes, are penetrated by air; and in the Hornbill even these are pneumatic. The functions discharged by the air-cavities of the bones appear to be much the same as those of the air-receptacles—namely, that of diminishing the specific gravity of the body and subserving the aeration of the blood.

The *heart* in all Birds consists of four chambers, two auricles and two ventricles. The right auricle and ventricle, constituting the right side of the heart, are wholly concerned with the pulmonary circulation; the left auricle and ventricle, forming the left side of the heart, are altogether occupied with the systemic circulation; and no communication normally exists in adult life between the two sides of the heart. In all essential details, both as regards the structure of the heart itself and the course taken by the circulating fluid, Birds agree with Mammals. The venous blood—namely, that which has circulated through the body—is returned by the venæ cavæ to the right auricle, whence it is poured into the right ventricle. The right ventricle propels it through the pulmonary artery to the lungs, where it is aerated, and becomes arterial. It is then sent back by the pulmonary veins to the left auricle, whence it is driven into the left ventricle. Finally, the left ventricle propels the aerated blood to all parts of the body through the great systemic aorta.

The chief difference between Birds and Reptiles as regards the course of the circulation is, that in the Birds the two sides of the heart are completely separated from one another, the blood sent to the lungs being exclusively venous, whereas that which is sent to the body is exclusively arterial. In Reptiles, on the other hand, the pulmonary and systemic circulations are connected together either in, or in the immediate neighbourhood of, the heart; so that mixed venous and arterial blood is propelled both through the lungs and through every part of the body.

In accordance with their extended respiration and high muscular activity, the complete separation of the greater and lesser circulations, and the perfect structure of the heart, Birds maintain a higher average temperature than is the case with any other class of the *Vertebrata.* This result is also to a considerable extent conditioned by the non-conducting nature of the combined down and feathers which form the integumentary covering of Birds.

The *urinary organs* of Birds consist of two elongated kidneys and two ureters, but there is no urinary bladder. The ureters open into the cloaca, or into a small urogenital sac which communicates with the cloaca.

As regards the *reproductive organs*, the males have two testes placed above the upper extremities of the kidneys, and their efferent ducts (*vasa deferentia*) open into the cloaca alongside of the ureters. A male organ (*penis*) may or may not be present, but there is no perfect urethra. The female bird, as a general rule, is provided with only one ovary and oviduct—that of the left side—the corresponding organs of the right side being rudimentary or absent. The oviduct is very long and tortuous, and the egg, during its passage through it, receives the albuminous covering which serves for the nutrition of the embryo, and which is known as the "white" of the egg. The lower portion of the oviduct is dilated, and the egg receives here the calcareous covering which constitutes the "shell." Finally, the oviduct debouches into the cloaca, into which the egg, when ready, is expelled. The further development of the chick is secured by the process of "incubation" or brooding, for which birds are peculiarly adapted, in consequence of the high temperature of their bodies.

The development of the ovum belongs to physiology, and does not concern us here. It is sufficient to notice the means by which in many cases the chick is ultimately enabled to escape from the egg. When development has reached a stage at which external life is possible, it is of course necessary for the chick to be liberated from the egg, the shell of which is often extremely hard and resistant. To this end, in very many instances, the young bird is provided with a little calcareous knob on the point of the upper mandible, and by means of this it chips out an aperture through the shell. Having effected its purpose, this temporary appendage then disappears, without leaving a trace behind.

The state of the young upon exclusion from the egg is very different in different cases, and in accordance with this, Birds have been divided into the two sections of the *Autophagi* or *Aves præcoces*, and the *Heterophagi* or *Aves altrices.* In the *Autophagi* the young bird is able to run about and help itself from the moment of liberation from the egg. In the *Heterophagi* the young are born in a blind and naked state, unable to feed themselves, or even to maintain unassisted the necessary vital heat. In these birds, therefore, the young require to be brooded over and fed by the parents for a longer or shorter period after exclusion from the egg.

As regards their *nervous system*, the brain of Birds is relatively larger, especially as regards the size of the cerebrum proper, than the brain of Reptiles, but its chief mass consists of the *corpora striata*. The cerebellum, though always present, consists simply of the central lobe (the "vermiform process"), and is not provided with the lateral lobes which occur in the Mammals, or they are only present in a rudimentary form. The corpus callosum is absent, and the surface of the cerebral hemispheres is devoid of convolutions.

As regards the *organs of the senses*, the eyes are always well developed, and in no bird are they ever rudimentary or absent. The chief peculiarity of the eye is that the cornea forms a segment of a much smaller sphere than does the eyeball proper, so that the anterior part of the eye is obtusely conical, whilst the posterior portion is spheroidal. Another peculiarity is that the form of the eye is maintained by a ring of from thirteen to twenty bony plates, which are placed in the anterior portion of the sclerotic coat. Eyelashes are almost universally absent; but in addition to the ordinary upper and lower eyelids, Birds possess a third membranous eyelid—the "membrana nictitans"—which is sometimes pearly-white, sometimes more or less transparent.* This third eyelid is placed on the inner side of the eye, and possesses a special muscular apparatus, by which it can be drawn over the anterior surface of the eye like a curtain, moderating the intensity of the light. As to the organ of hearing, most birds possess no external ear or concha, by which sounds can be collected and transmitted to the internal ear. In some birds, however, as in the Ostrich and Bustard, the external meatus auditorius is surrounded by a circle of feathers, which can be raised and depressed at will. The Nocturnal Birds, also, especially Owls, have the external meatus auditorius protected by a musculo-membranous valve, which foreshadows the cartilaginous concha of the majority of Mammals. The external nostrils in Birds are usually placed on the sides of the upper mandible, near its base, in the form

* The membrana nictitans is simply a fold of the conjunctiva on the inner side of the eye. It occurs in some Fishes (*e.g.*, Sharks), in some Reptiles and Amphibians, in Birds, in Monotremes and Marsupials, and in some of the higher Mammals. In Man, however, in Monkeys, and in most of the higher Mammals, it is rudimentary, and constitutes the so-called "plica semilunaris."

of simple perforations, which sometimes communicate from side to side by the deficiency of the septum narium. In the singular *Apteryx* of New Zealand, the nostrils are placed at the extreme end or tip of the elongated upper mandible. Sometimes the nostrils are defended by bristles, and sometimes by a scale (*Rasores*). Taste must be absent, or almost absent, in the great majority of birds, the tongue being nothing more than a horny sheath surrounding a process of the hyoid bone, and serving for deglutition or to seize the prey. In the Parrots, however, the tongue is thick and fleshy, and some perception of taste may be present. Touch or tactile sensibility, too, as already remarked, is very poorly developed in Birds. The body is entirely, or almost entirely, covered with feathers; the anterior limbs are converted into wings, and rendered thereby useless as organs of touch; and the posterior limbs are covered with horny scales or feathers. The bill certainly officiates as an organ of touch, but it cannot possess any acute sensibility, as in most birds it is encased in a rigid horny sheath. In some birds, however, such as the common Duck, the texture of the bill is moderately soft, and it is richly supplied with filaments of the fifth nerve; so that in these cases the bill doubtless constitutes a tolerably efficient tactile organ. The "cere," too, or the fleshy scale found at the base of the bill in some birds, is in all probability also used as a tactile organ.

The last anatomical peculiarity of Birds which requires notice is the peculiar apparatus known as the "inferior larynx," by which the song of the singing birds is conditioned. "The air-passages of birds commence by a simple *superior larynx*, from which a long *trachea* extends to the anterior aperture of the thorax, where it divides into the two *bronchi*, one for each lung. At the place of its division there exists in most birds a complicated mechanism of bones and cartilages, moved by appropriate muscles, and constituting the true organ of voice; this part is termed the *inferior larynx*."—(Owen.) The structure of the vocal apparatus is extremely complicated, and there is no necessity for entering upon it here. It is to be remembered, however, that those modifications of the voice which constitute the song of birds, are produced in a special and complex cavity placed at the point where the trachea divides into the two bronchi, and *not* in a true larynx situated at the summit of the windpipe. Lastly, the trachea of birds is always of considerable proportionate length, and it is often twisted or dilated at intervals, this structure, doubtless, having something to do with the production of vocal sounds.*

* The student desirous of fuller information as to the anatomy of Birds

Before passing on to the consideration of the divisions of Birds, a few words may be said as to the *migration* of birds. In temperate and cold climates comparatively few birds remain constantly in the same region in which they were hatched. Those which do so remain, are called "permanent birds" (*aves manentes*). Other birds, such as the Woodpeckers, wander about from place to place, without having any fixed direction. These are called "wandering birds" (*aves erraticæ*), and their irregular movements are chiefly conditioned by the scarcity or abundance of food in any particular locality. Other birds, however, at certain seasons of the year undertake long journeys, usually uniting for this purpose into large flocks. These birds—such as the swallows, for instance—are properly called "migratory birds" (*aves migratoriæ*). The movements of these birds are conditioned by the necessity of having a certain mean temperature, and consequently they leave the cold regions at the approach of winter, and return again for the warmer season.

CHAPTER LXVII.

DIVISIONS OF BIRDS.

1. GENERAL DIVISIONS OF AVES. 2. NATATORES. 3. GRALLATORES.

OWING to the extreme compactness and homogeneity of the entire class *Aves*, conditioned mainly by their adaptation to an aerial mode of life, the subject of their classification has been one of the greatest difficulties of the systematic Zoologist.

By Professor Huxley the Birds are divided into the following three orders :—

1. SAURURÆ.—In this order the caudal vertebræ are numerous, and there is no ploughshare-bone. The tail is longer than the body, and the metacarpal bones are not anchylosed together. This order includes only the single extinct bird the *Archæopteryx macrura*, in which the long lizard-like tail is only the most striking of several abnormalities.

2. RATITÆ. — This order comprises the Running birds,

should consult the masterly article by Owen on "Aves" in the 'Cyclopædia of Anatomy and Physiology,' or the second volume of the 'Vertebrata' of the same author, from which the preceding summary has been chiefly derived.

which cannot fly, such as the Ostriches, Emeus, and Cassowaries. It is characterised by the fact that the sternum has no median ridge or keel for the attachment of the great pectoral muscles. The sternum is therefore raft-like (from the Lat. *rates*, a raft), hence the name of the order.

3. CARINATÆ.—This comprises all the living Flying birds, and is characterised by the fact that the sternum is furnished with a prominent median ridge or keel (*carina*); hence the name of the order.

This is probably the nearest approach to a strictly natural classification of Birds which has yet been proposed; but the order *Carinatæ* is so disproportionately large as compared with the other two, that it would lead to considerable inconvenience if it were to be adopted here.

For the purposes of the present work it will be better to adhere, with some modifications, to the classification of the Birds originally proposed by Kirby, and since sanctioned by the adoption of other distinguished naturalists. In this more generally current, but certainly artificial, arrangement, the Birds are divided into the following seven orders, founded chiefly on the habits and mode of life, and on the resulting anatomical or structural peculiarities. To these an eighth order must be added for the reception of the Mesozoic bird, the *Archæopteryx*, the discovery of which dates from a recent period. Before entering upon a consideration of the individual orders, it will be as well to present to the student, synoptically and in an easily-remembered form, the leading differences between these eight orders.

1. *Natatores or Swimmers.*—These are characterised by the fact that the toes are united by a membrane or web; the legs are short, and are placed behind the point of equilibrium of the body. The body is closely covered with feathers, and with a thick coating of down next the skin. (*Ex.* Ducks, Geese, Pelicans, Gulls.)

2. *Grallatores or Waders.*—The Wading birds are characterised by the possession of long legs, which are naked or are not covered with feathers from the distal end of the tibia downwards. The toes are long, straight, and not united to one another by a membrane or web. (*Ex.* Curlews, Herons, Storks.)

3. *Cursores or Runners.*—The Cursorial birds have very short wings which are not used in flight, and the sternum is without a ridge or keel. The legs are exceedingly robust, and there are only two, or at most three developed toes, the hind-toe or hallux being always absent or quite rudimentary. The

order agrees with the *Ratitæ* of Huxley. (*Ex.* Ostrich, Emeu, Apteryx.)

4. *Rasores or Scratchers.*—The Rasorial birds have usually strong feet with powerful blunt claws adapted for scratching, but sometimes for perching. All the four toes are present. The upper mandible is vaulted, and the nostrils are pierced in a membranous space at its base, and are covered by a cartilaginous scale. (*Ex.* Fowls, Game-birds, Pigeons.)

5. *Scansores or Climbers.*—The Climbing birds are characterised by the structure of the foot, in which two toes are turned backwards and two forwards, so as to give the bird unusual facilities in climbing trees. (*Ex.* Parrots, Toucans, Woodpeckers.)

6. *Insessores or Perchers.*—The Insessorial or Passerine birds are characterised by having slender and short legs, with three toes before and one behind, the two external toes generally united by a very short membrane, and the whole foot being adapted for perching. This is by far the largest order of birds, and includes all our ordinary songsters, such as the Thrushes, Linnets, Larks, &c., together with the Swallows, Humming-birds, and many others.

7. *Raptores or Birds of Prey.*—The Birds of Rapine are characterised by their strong, curved, sharp-edged and sharp-pointed beak, adapted for tearing animal food; and by their robust legs armed with four toes, three in front and one behind, all furnished with long, strong, crooked claws or talons. (*Ex.* Eagles, Hawks, Owls.)

8. *Saururæ.*—The metacarpal bones are not anchylosed together, and the tail is longer than the body, and consists of numerous free vertebræ, without a terminal ploughshare-bone. The only member of this order is the extinct *Archæopteryx.*

ORDER I. NATATORES.—The order of the *Natatores*, or Swimmers, comprises a number of birds which are as much or even more at home in the water than upon the land. In accordance with their aquatic habit of life, the *Natatores* have a boat-shaped body, usually with a long neck. The legs are short, and placed behind the centre of gravity of the body, this position enabling them to act admirably as paddles, at the same time that it renders the gait upon dry land more or less awkward and shuffling. In all cases the toes are "webbed" or united by membrane to a greater or less extent (fig. 185, A). In many instances the membrane or web is stretched completely from toe to toe, but in others the web is divided or split up between the toes, so that the toes are fringed with membranous borders, but the feet are only imperfectly webbed. As their aquatic mode of life exposes them to great reductions of temperature, the body of the Natatorial birds is closely covered with feathers and with a thick coating of down next the skin. They are, further, prevented from becoming wet in the water by the great development of the coccygeal oil-gland, by means of which the lustrous plumage is kept constantly lubricated and waterproof. They are usually polygamous, each male consorting with several females; and the young are hatched in a condition not requiring any special assistance from the parents, being able to swim and procure food for themselves from the moment they are liberated from the egg.

Fig. 185.—Natatores. A, Foot of Cormorant (*Phalacrocorax*); B, Beak of the Bean-goose (*Anser segetum*).

The *Natatores* are divided into the following four families:—

Fam. 1. *Brevipennatæ.*—In this family of the swimming birds the wings are always short, and are sometimes useless as organs of flight, the tail is very short, and the legs are placed very far back, so as to render terrestrial progression very difficult or awkward. The family includes the Penguins, Auks, Guillemots, Divers, and Grebes. In the Penguins (*Spheniscidæ*) the wings are completely rudimentary, without quills, and covered with a scaly skin. They are useless, as far as flight is concerned, but they are employed by the bird as fins, enabling it to swim under water with great facility, and they are also used on the land as fore-legs. The feet are webbed, and the hinder toe is rudimentary or wanting. The Penguins live gregariously in the seas of the southern hemisphere, on the coasts of South Africa and South America, especially at Terra del Fuego, and in the solitary islands of the South Pacific. When on land the Penguins stand bolt upright, and as they usually stand on the shore in long lines, they are said to present a most singular appearance. The best-known species are the

Jackass Penguin (*Spheniscus demersus*) of the Falkland Islands, and the King Penguin (*Aptenodytes Patagonica*) of the Straits of Magalhaens. In the Auks (*Alcidæ*) the wings are better developed than in the Penguins, and they contain true quill-feathers; but they are still short as compared with the size of

Fig. 186.—Jackass Penguin (*Spheniscus demersus*).

the body, and are of more use as fins than for flight. The Great Auk or Gare-fowl (*Alca impennis*) is remarkable for being one of the birds which appear to have become entirely extinct within the human period, having been, in fact, destroyed by man himself. It used to abound in the arctic regions, and occasionally visited our own shores in the winter. The Little Auk (*Mergulus alba*) occurs still in abundance in the seas of the arctic regions. Other well-known members of this group are the Razor-bill, the Puffins (*Fratercula arctica*), and the Guillemots (*Uria*). The Guillemots have a short tail, narrow and pointed wings, short feet, and no hallux. Like the other members of the family, they inhabit northern and polar regions.

In the Divers (*Colymbidæ*), comprising the true Divers and the Grebes, the power of flight is pretty well developed, but the bird still is much more active in the water, swimming or diving, than on land. The Grebes are not uncommon in Britain, and are largely killed for making muffs, collars, and other articles of winter dress. They have the membrane between the toes deeply incised. They haunt the sea as well as lakes and rivers, and swim and dive admirably. In the Divers proper the front toes are completely united by a membrane. The Northern Diver or Loon (*Colymbus glacialis*) is a familiar example, and is found on the coasts of high northern latitudes.

Fam. 2. *Longipennatæ.*—This family of *Natatores* is characterised by the well-developed wings, the pointed, sometimes knife-like, sometimes hooked bill, and by never having the hallux united with the anterior toes by a membrane. The following are the more important groups coming under this head:—

a. Laridæ, or Gulls and Terns, having powerful wings, a free hinder toe, and the three anterior toes united by a membrane. The gulls form an exceedingly large and widely distributed group of birds; and the Terns or Sea-swallows are equally beautiful, if not quite so common. The Terns are distinguished by their long and pointed wings, forked tail, and comparatively short legs. They fly with great rapidity over the surface of the sea, from which they pick up their food.

b. Procellaridæ, or Petrels, closely resembling the true Gulls, but having a rudimentary hinder toe, and having the upper mandible strongly hooked. The smaller species of Petrel are well known to all sailors under the name of Storm-birds and Mother Carey's Chickens. They are nocturnal or crepuscular in their habits, breed in holes in the rocks, lay but one egg, and are almost all of small size and more or less sombre plumage. The largest member of the group is the gigantic Albatross (*Diomedea exulans*), not uncommonly found far from land in both the northern and southern oceans. The Albatross sometimes measures as much as fifteen feet from the tip of one wing to that of the other, and the flight is powerful in proportion.

Fam. 3. *Totipalmatæ*, characterised by having the hinder toe or hallux more or less directed inwards, and united to the innermost of the anterior toes by a membrane (fig. 185, A). In this family are the Pelicans, Cormorants, Gannets, Frigate-birds, Darters, and others. They all fly well, and have short legs, and amongst them are almost the only Natatorial Birds which ever perch upon trees.

The Pelicans (*Pelicanidæ*) are large birds, which subsist on fish, and are found in Europe, Asia, Africa, and the New World. They sometimes measure as much as from ten to fifteen feet between the tips of the wings, and most of the bones are pneumatic, so that the skeleton is extremely light. The lower mandible is composed of two flexible branches which serve for the support of a large "gular" pouch, formed by the loose unfeathered skin of the neck. The bill is long and straight, and the upper mandible is strongly hooked at the tip. The fish captured by the bird are temporarily deposited in this pouch, and the parent birds feed their young out of it.

In the Cormorants (*Phalacrocorax*) there is no pouch beneath the lower mandible, but the skin of the throat is very lax and distensible; the nail of the middle toe is serrated. They are widely distributed over the world, one species being very abundant in many parts of Europe. The Gannets (*Sula*) have a compressed bill, the margins of which are finely crenate or toothed. They occur abundantly on many parts of the coasts of northern Europe, one of the most noted of their stations being the Bass Rock at the mouth of the Firth of Forth. Another species (*Sula variegata*) is of greater importance to man, as being one of the birds from the accumulated droppings of which guano is derived. The Frigate-birds (*Tachypetes*) are chiefly remarkable for their extraordinary powers of flight, conditioned by their enormously long and powerful wings and long forked tail. They occur on the coasts of tropical America, and are often found at immense distances from any land. The Tropic-birds (*Phaeton*) inhabit intertropical regions, and are found far out at sea. They have short feeble feet, and long pointed wings.

The Darters or Snake-birds (*Plotus*) are somewhat aberrant members of this group, characterised by their elongated necks and long pointed bills. They occur in America, Africa, and Australia, and catch fish by suddenly darting upon them from above.

Fam. 4. *Lamellirostres.*—The last family of the *Natatores* is that of the *Lamellirostres*, including the Ducks, Geese, Swans, and Flamingoes; and characterised by the form of the beak (figs. 179, 185), which is flattened in form and covered with a soft skin. The edges of the bill are further furnished with a series of transverse plates or lamellæ, which form a kind of fringe or "strainer," by means of which these birds sift the mud in which they habitually seek their food. The bill is richly supplied with filaments of the fifth nerve, and doubtless serves as an efficient organ of touch. The feet are furnished with four toes, of which three are turned forwards, and are webbed, whilst the fourth is turned backwards, and is free. The trachea in the males is often enlarged or twisted in its lower part, and co-operates in the production of the peculiar clanging note of most of these birds. The body is heavy, and the wings only moderately developed.

The groups of the Ducks (*Anatidæ*), Geese (*Anserinæ*), and Swans (*Cygnidæ*), are too familiar to require much special notice. The *Anatidæ*, or true Ducks, have the hallux furnished with a very narrow membranous lobe, and the laminæ of the upper mandible generally projecting. As examples may be taken the Mallards and Teals (*Boschas*), the Widgeons (*Mareca*), the Shoveller (*Anas*), and the Pin-tail Ducks (*Dafila*). The Sea-Ducks (*Fuligulinæ*) frequent the sea chiefly, and have the hallux furnished with a wide membranous lobe. Good examples are the Eider-duck (*Somateria*), the Surf-duck (*Oidemia*), the Canvass-back Duck, and Pochard (*Fuligula*), and the Golden-eye (*Clangula*).

The *Anserinæ* are distinguished from the Ducks chiefly by their stronger and longer legs, and comparatively shorter wings. Good examples are the Grey Lag (*Anser ferus*), the Canada Goose (*A. Canadensis*), the Bean-goose (*A. Segetum*), and the Snow-goose (*A. hyperboreus*).

In the Swans the neck is extremely long, and the legs are short. In the Hooper Swan (*Cygnus ferus*) the sternal keel is double, and forms a cavity for the reception of a convoluted portion of the trachea. This is not the case, however, with the Mute Swan (*C. olor*), the Black Swan (*C. atratus*), or the Trumpeter Swan (*C. buccinator*), all well-known members of the group.

The Flamingoes, however, forming the group of the *Phœnicopteridæ*, require some notice; if only for the fact that the legs are so long and slender that they have often been placed in the order *Grallatores* on this account. The three anterior toes, however, are webbed or completely united by membrane, and the bill is lamellate, so that there can be little hesitation in leaving the Flamingo in its present position amongst the *Natatores*. The bill is singularly bent, both mandibles being suddenly curved downwards from the middle. The common Flamingo (*Phœnicopterus ruber*) occurs abundantly in various parts of southern Europe. It stands between three and four feet in height, the general plumage being rose-coloured, the wing-coverts red, and the quill-feathers of the wings black.

The tongue is fleshy, and one of the extravagances of the Romans during the later period of the Empire was to have dishes composed solely of Flamingoes' tongues. Other species occur in South America and Africa.

ORDER II. GRALLATORES.—The birds comprising the order of the *Grallatores*, or Waders, for the most part frequent the banks of rivers and lakes, the shores of estuaries, marshes, lagoons, and shallow pools, though some of them keep almost exclusively to dry land, preferring, however, moist and damp situations. In accordance with their semi-aquatic amphibious habits, the Waders are distinguished by the great length of their legs; the increase in length being mainly due to the great elongation of the tarso-metatarsus. The legs are also unfeathered from the lower end of the tibia downwards. The toes are elongated and straight (fig. 187, A), and are never completely palmate, though sometimes semi-palmate. There are three anterior toes, and usually a short hallux, but the latter may be wanting. The wings are long, and the power of flight usually considerable; but the tail is short, and the long legs are stretched out behind in flight to compensate for the brevity of the tail. The body is generally slender, and the neck and beak usually of considerable length (fig. 187, B). They are sometimes polygamous, sometimes monogamous, and the young of the former are able to run about as soon as they are hatched.

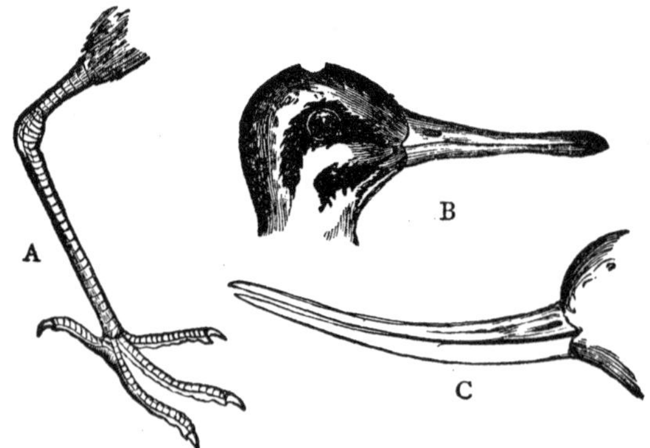

Fig. 187.—Grallatores. A, Leg and foot of the Curlew; B, Head of Snipe; C, Beak of the Avocet.

The most typical Waders—those, namely, which are semi-aquatic in their habits—spend most of their time wading about in shallow waters or marshes, feeding upon small fishes, worms, shell-fish, or insects. Others, such as the Storks, live mostly upon the land, and are more or less exclusively vegetable-feeders.

The *Grallatores* are divided into the four families of the *Macrodactyli*, the *Cultirostres*, the *Longirostres*, and the *Pressirostres*.

Fam. 1. *Macrodactyli.*—In this family the feet are furnished with four elongated, sometimes lobate, toes, and the wings are of moderate or less than average size. In many of their characters a considerable number of the birds of this family approach the Rasorial birds, and differ from the true Waders. The beak is mostly short, rarely longer than the head, and is compressed from side to side, or wedge-shaped. The legs are strong and not particularly lengthy; but the toes are often of great length, and are furnished with long claws. The neck is not very long, and the tail is very short. Some of them are strictly aquatic in their habits, and, like the Coots, approach in many respects to the *Natatores;* others, again, are exclusively terrestrial. The most familiar members of this family are the Rails (*Rallus*), Water-hens (*Gallinulæ*), the Coots (*Fulica*), and the Jacana (*Parra jacana*). The Water-hens and Coots are aquatic or semi-aquatic, swimming and diving with great ease. In the Coots the toes are semi-palmate, being bordered by membranous lobes, like the toes of the Grebes, but the toes are not fringed in the Gallinules. Amongst the Coots should probably be placed the *Notornis* (Owen), long supposed to be extinct, but recently proved to be still living in the Middle Island of New Zealand. The *Notornis* is much larger than the ordinary Coots, and is remarkable in the fact that, like many extinct and some living New Zealand birds, the wings are so rudimentary as to be useless for flight. The true Rails, comprising the common Land-rail (*Rallus aquaticus*), and the Corn-crake (*Crex pratensis*) of Britain, and the Marsh Hen (*Rallus elegans*), and Virginian Rail (*R. Virginianus*) of North America, live almost exclusively on land, though the former usually frequents damp or marshy places. In the Jacanas, lastly, the feet are furnished with excessively long and slender toes, which enable the bird to run about upon the leaves of aquatic plants; whilst the carpus is armed with formidable spurs. They are natives of South America, Africa, and India. Closely allied to the Jacanas are the Screamers (*Palamedea*) of South America, of which the Horned Screamer (*P. cornuta*) is the best known. It has a long frontal horn, and has spurs implanted on the edge of the wing.

Fam. 2. *Cultirostres.*—In this family of the *Grallatores* are some of the most typical and familiar forms contained in the entire order. The bill in this family is long—usually longer than the head—and is compressed from side to side; the legs are long and slender, having a considerable portion of the tibiæ unfeathered; and the feet have four toes, which are usually connected to a greater or less extent at their bases by membrane. In this family are the Cranes, Herons, Stork, Ibis, Spoonbill, and others of less importance.

Fig. 188.—Crested Heron (*Ardea cinerea*). Europe.

The Cranes (*Gruidæ*) are large and elegant birds, and are chiefly remarkable for their long migrations, which were noticed by many classical authors. In these journeys the Cranes usually fly in large flocks, led by a single leader, so that the whole assemblage assumes a wedge-like form; or they fly in long lines. The common Crane (*Grus cinerea*) breeds in the north of Europe and Siberia, and migrates southwards at the approach of winter. The Numidian Crane or Demoiselle inhabits Asia and Africa, the Stanley Cranes (*Anthropoides*) are natives of the East Indies, and the Crowned Cranes (*Balearica*) are African. The Herons (*Ardeidæ*) are familiarly known to every one in the person of the common Grey or Crested Heron

(*Ardea cinerea*, fig. 188). It was one of the birds most generally pursued in the now almost extinct sport of falconry. Various species of Heron are found over the whole world, both in temperate and hot climates. Here, also, belong the various species of Night Heron (*Nycticorax*), the Bitterns (*Botaurus*), and the Boat-bills (*Cancroma*).

The Ibises (*Tantalinæ*) form a group of beautiful birds, species of which occur in all the warm countries of the world. They are distinguished by their metallic colours, long, cylindrical, curved bill, and more or less naked head. One, the *Ibis religiosa*, was regarded by the ancient Egyptians as a deity, and was treated with divine honours, being often embalmed along with their mummies, or figured on their monuments.

The Storks (*Ciconinæ*) are large birds, of which one, the common Stork (*Ciconia alba*), is rarely found in Britain, but occurs commonly on the Continent, where it is often semi-domesticated. The Storks live in marshes, and feed on frogs, fishes, &c. Nearly related to the true Storks are the gigantic Marabout (*Ciconia Marabou*), and Adjutant (*C. Argala*) of Africa and India, which possess a sausage-shaped appendage in front of the neck.

The Spoonbills (*Plataleadæ*) are also large birds, very like the Storks, but the bill is flattened out so as to form a broad spoon-like plate. The common White Spoonbill (*Platalea leucorodia*) is found commonly on the Continent, but is of very rare occurrence in Britain.

Fam. 3. *Longirostres.*—The third family of Waders is that of the *Longirostres*, characterised by the possession of long, slender, soft bills, grooved for the perforations of the nostrils (fig. 187, B). The legs are sometimes rather short, sometimes of great length; the toes are of moderate length, and the hallux is usually short, and is sometimes absent. The bill in these birds serves as an organ of touch, being used as a kind of probe to feel for food in mud or marshy soil. To fulfil this purpose, the tip of the bill is furnished with numerous filaments of the fifth nerve. They feed mostly upon insects and worms, and are not strictly aquatic in their habits, mostly frequenting marshy districts, moors, fens, the banks of rivers or lakes, or the shores of the sea.

In this family of the Long-billed Waders are the various species of Snipe and Woodcock (*Scolopacidæ*), the Sandpipers (*Tringa*), the Curlews (*Numenius*), the Turnstones (*Strepsilas*), the Ruffs (*Machetes*), the Redshanks (*Totanus*), the Godwits (*Limosa*), and others which need no special notice.

Fam. 4. *Pressirostres.* — The members of this family are

characterised by the moderate length of the bill, which is seldom longer than the head, and has a compressed tip. The legs are long, but the toes are short, and are almost always partially connected together at their bases by membrane. The hallux is short, and is often wanting. The wings are long, and they can both fly powerfully and run with great swiftness. In this section are two very distinct sub-families, the *Charadriidæ* or Plovers, and the *Otidæ* or Bustards. In the former of these the legs are long and slender, the toes are united at their bases by a small membrane, and the hind-toe is very small and raised above the ground, or is entirely wanting. In this group are the true Plovers and Lapwings (*Charadrius* and *Vanellus*), the Pratincoles (*Glareola*), the Long-shanks (*Himantopus*), the Oyster-catcher (*Hæmatopus*), and the Thick-knee (*Œdicnemus*). In the *Otidæ*, or Bustards, the legs are long and the toes are short and furnished with stout claws. The hinder toe or hallux is entirely wanting; and these birds are chiefly interesting from the affinities which they exhibit to the *Rasores* on the one hand, and to the *Cursores* (Ostrich, &c.) on the other. The wings, however, are of ample size, and the tail is long, the reverse being the case in the *Cursores*. The Bustards are entirely confined to the Old World, and two species were formerly not uncommon in Britain. They are found in plains and downs, and rarely fly, but run with great swiftness, using the wings to accelerate their course. They are polygamous, and the males are generally brighter and more variegated in plumage than the females.

CHAPTER LXVIII.

CURSORES AND RASORES.

ORDER III. CURSORES.—The third order of Birds is that of the *Cursores*, or Runners, comprising the Ostriches, Rheas, Cassowaries, Emeus, and the singular *Apteryx* of New Zealand. In many respects the *Cursores* are to be looked upon as an artificial assemblage; but in the meanwhile it will be most convenient to consider them as forming a distinct division. The *Cursores* are characterised by the rudimentary condition of the wings, which are so short as to be useless for flight, and by the compensating length and strength of the legs. In accordance with this condition of the limbs, many of the bones retain their marrow, and the sternum (fig. 189, B) is destitute of the prominent ridge or keel, to which the great pectoral muscles are attached (hence the name of *Ratidæ*, applied by Huxley to the order). In the Ostrich, the pubic bones of the pelvis unite to form a symphysis pubis, as they do in no other bird; and in all, the pelvic arch possesses unusual strength and stability. The legs are extremely robust and powerful, and the hind-toe is entirely wanting, except in the *Apteryx*, in which it is rudimentary. The anterior toes are two or three in number, and are provided with strong blunt claws or nails. The plumage presents the remarkable peculiarity that the barbs of the feathers, instead of being connected to one another by hooked barbules, as is usually the case, are remote and disconnected from one another, presenting some resemblance to hairs

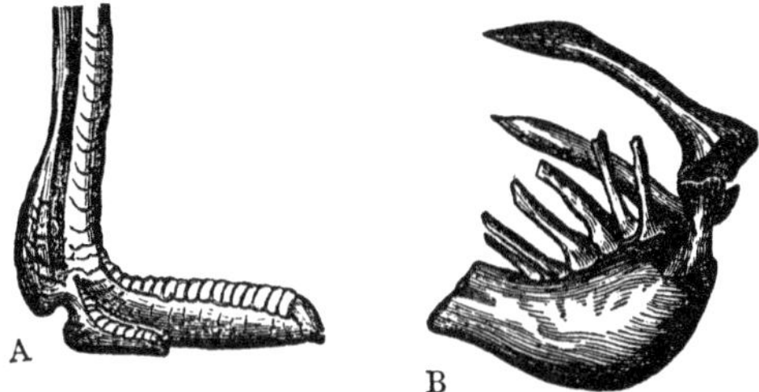

Fig. 189.—Cursores. A, Foot of the Ostrich (*Struthio camelus*); B, Sternum of the Emeu (*Dromaius Novæ-Hollandiæ*).

The order *Cursores* may be divided into the two families of the *Struthionidæ* and the *Apterygidæ*—the former characterised by the absence of the hallux, and comprising the Ostrich, Rhea, Emeu, and Cassowary, with several extinct forms; the latter comprising only the *Apteryx* of New Zealand, and characterised by the possession of a rudimentary hallux.

The African Ostrich (*Struthio camelus*) occurs in the desert plains of Africa and Arabia, and is the largest of all living birds, attaining a height of from six to eight feet. The head and neck are nearly naked, and the quill-feathers of the wings and tail have their barbs wholly disconnected, constituting the ostrich-plumes of commerce. The legs are extremely strong, and are terminated by two toes only (fig. 189, A), these consisting respectively of four and five phalanges, showing that it is the hallux and the innermost toe which are wanting. The internal one of the two toes is much the largest, and is clawed, the outer toe is small and clawless. The Ostriches run with extraordinary speed, and can outstrip the fastest horse. They are polygamous, each male consorting with several females, and they generally keep together in larger or smaller flocks. The eggs are of great size, averaging three pounds each in weight, and the hens lay their eggs in the same nest, this being nothing more than a hole scratched in the sand. The eggs appear to be hatched mainly by the exertions of both parents, relieving each other in the task of incubation, but also partly by the heat of the sun.*

The American Ostriches or *Rheas* are much smaller than the African Ostrich, and have the head feathered, whilst the feet are furnished with three toes each. The wings are rudimentary, and the phalanges are plumed and terminated by a spur. They inhabit the great plains of South America, and are polygamous.

The Emeu (*Dromaius Novæ-Hollandiæ*) is exclusively found in the Australian continent, and nearly equals the African Ostrich in size, attaining a height of from five to seven feet. The feet are furnished with three toes each, and the head is feathered. The throat, however, is naked, and the general plumage resembles long hairs, the feathers hanging down on both sides of the body from a central line or parting which runs down the middle of the back. The Emeus are monogamous, and the eggs are dark green in colour. The male Emeu is smaller than the female, and undertakes all the duties of incubation.

The last of the *Struthionidæ* is the Cassowary (*Casuarius galeatus*), which inhabits the Moluccan Islands and New Guinea, and was first brought alive to Europe by the Dutch. It stands about five feet in height, and possesses a singular horny crest upon its head. The head and neck are naked, with pendent wattles, and the feet have three toes each. The general plumage is black, and the feathers more or less closely resemble hairs. The wings are rudimentary, each with five naked pointed quills. The male is much the smallest, and sits upon the eggs. Besides the Galeated Cassowary, other species have been described from New Britain and North Australia.

The second family of the Cursorial birds is that of the *Apterygidæ*, comprising only the singular *Apteryx* of New Zealand. The beak in the *Apteryx* is long, slender, and slightly curved, the tip being obtuse, and the nostrils placed at the extremity of the upper mandible. The legs are comparatively short, and there is a rudimentary hind-toe or hallux, forming a kind of spur, furnished with a claw. The wings are

* Mr Sclater, however, states that the duty of incubation is entirely taken by the males.

entirely rudimentary, and are quite concealed by the feathers, each terminating in a sharp claw. The feathers are long and narrow, and the tail is short and inconspicuous. The *Apteryx* is wholly confined to New Zealand, and is nocturnal in its habits, living upon insects and worms. Three species of *Apteryx* have been described, of which *A. australis* (fig. 190) is the best known.

Fig. 190.—*Apteryx australis.* (Gould.)

Besides the above-mentioned living forms, the order *Cursores* comprises several gigantic extinct forms, which will be treated of when describing the geological distribution of Birds as a class.

ORDER IV. RASORES.—The fourth order of Birds is that of the *Rasores*, or Scratchers, often spoken of collectively as the "Gallinaceous" birds, from the old name of "Gallinæ," given to the order by Linnæus. The *Rasores* are characterised by the convex, vaulted upper mandible, having the nostrils pierced in a membranous space at its base. The nostrils are covered by a cartilaginous scale. The legs are strong and robust, mostly covered with feathers as far as the joint between the tibia and tarso-metatarsus. There are four toes, three in front and one behind, the latter being short, and placed at a higher evel than the other toes. All the toes terminate in strong blunt claws suitable for scratching (fig. 191, A). The food of the Scratchers or Gallinaceous birds consists chiefly of hard grains and seeds, and in accordance with this they have a capacious crop and an extremely strong and muscular gizzard. They mostly nidificate, or build their nests, upon the ground, and the more typical members of the order are polygamous. The males take no part in either nidification or incubation, and the young are generally "precocious," being able to run about and provide themselves with food from the moment they quit the egg. The young of the Pigeons and Doves, however, are brought forth in a comparatively helpless condition. The wings in the majority of the *Rasores* are more or less weak, and the flight is feeble and accompanied with a whirring sound. Many of the Pigeons, however, are capable of very powerful and sustained flight.

The order *Rasores* is divided into two sub-orders, called respectively the *Gallinacei* and the *Columbacei*, or sometimes, from the characters of the sounds which they utter, the *Clamatores* and the *Gemitores*.

Sub-order 1. *Gallinacei* or *Clamatores.*—This sub-order comprises the typical members of the order *Rasores*, such as the common Fowls, Turkeys, Partridge, Grouse, Pea-fowl, and a number of allied forms. Its characters are therefore those of the order itself, but it is especially distinguished from the *Columbacei* by being less fully adapted for flight. The body is

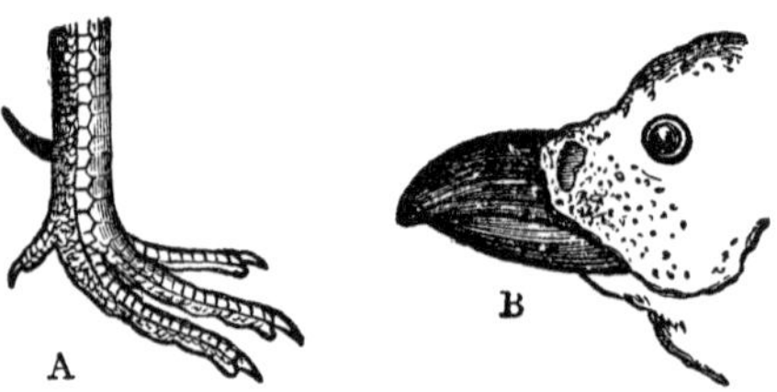

Fig. 191.—Rasores. A, Foot of Fowl (*Gallus Bankiva*); B, Head of Guinea-fowl.

much heavier comparatively speaking, the legs and feet are stronger, and the wings shorter and less powerful. On the whole, therefore, these birds are worse fliers than the *Columbacei*, and are better adapted for living upon the ground. The hallux is elevated above the anterior toes, and merely touches the ground in walking. The back of the tarsus, too, is usually furnished in the males with a spur (*calcar*), which is used as an offensive weapon, and has sometimes been looked upon as a rudimentary toe.* Lastly, the *Gallinacei* are mostly polygamous, and the males are usually much more brilliantly coloured than the females, this being an adaptive modification of the plumage to meet this peculiarity in their mode of life.†

The following are the most important families of the *Gallinacei* :—

The *Tetraonidæ*, or Grouse family, comprises the various species of Grouse (*Tetrao*), the Ruffed Grouse (*Bonasa*), the Cock of the Plains (*Centrocercus*), and the Ptarmigans (*Lagopus*).

The *Perdicidæ*, or Partridge family, comprises the Partridges (*Perdix*), the Francolins (*Francolinus*), the Quails (*Coturnix*), the Maryland Quail (*Ortyx*), the Tufted Quails (*Lophortyx*), &c.

The *Phasianidæ* or Pheasant family, comprises the Turkeys and Guinea-fowl (*Meleagrinæ*), the common Pheasant (*Phasianus Colchicus*), the Golden and Silver Pheasants, the common Fowl (*Gallus domesticus*), and the Pea-fowl (*Pavoninæ*). None of these birds—all of which can be domesticated, and most of which are of great value to man—are natives of this country, though they will all breed readily, and thrive even in confinement. The domestic Turkey (*Meleagris gallopavo*) is originally a native of North America, where it still occurs in a wild condition, having been brought to Europe about the beginning of the sixteenth century. The Guinea-fowl (*Numida Meleagris*) is originally an African bird. The common Pheasant (*Phasianus Colchicus*), though now regarded as an indigenous bird, truly belongs to Asia, and it is asserted that it was really brought to Europe from Colchis by the Greeks; hence its specific name. The common Fowl is certainly not a native of Europe, and it is almost as certainly a native of Asia or of some of the Asiatic islands; but its exact original habitat is uncertain, as is the species from which the domestic breeds are descended (commonly said to be the *Gallus Bankiva* of Java). The introduction of the Fowl into Europe is lost in the mists of antiquity, and it is wholly unknown whence the original stock may have been brought. The Pea-fowl (*Pavo*) are really natives of Thibet and Hindostan, and were originally brought to Greece by Alexander the Great. They were formerly much esteemed as food, but are now regarded merely from an ornamental point of view.

* In some cases (as in the Java Peacock) the female possesses spurs as well as the male; and sometimes (as in *Polyplectron*) there are two or more spurs on each leg of the male.

† The Guinea-fowl, Red Grouse, Ptarmigan, and Partridge are monogamous, in a state of nature at any rate.

The *Megapodidæ*, or Mound-birds, belong to India and Australia, and have very large feet and long claws. They build immense mounds, often six or eight feet high, and twenty or thirty feet in diameter. They lay their eggs in the centre of these mounds at a depth of two or three feet, and leave them to be hatched by the heat produced by the fermentation of the vegetable matter of the mass.

The *Cracidæ*, or Curassows, are large heavy birds, belonging to Central and South America, and to a great extent arboreal in their habits. The best-known species is the Crested Curassow (*Crax alector*) of Mexico and Brazil.

Fig. 192.—Columbidæ. Rock-pigeon (*Columba livia*).

The second sub-order of the *Rasores* is that of the *Columbacei* or *Gemitores*, comprising the Doves and Pigeons, and often raised to the rank of a distinct order under the name of *Columbæ*. The *Columbacei* are separated from the more typical members of the *Rasores* by being furnished with strong wings, so as to endow them with considerable powers of flight. In place, therefore, of being chiefly ground-birds, they are to a great extent arboreal in their habits, and in accordance with this the feet are slender, and are well adapted for perching. There are four toes, three in front and one behind, and the former are never united towards their bases by a membrane, though the base of the outer toe is sometimes united to that of the middle toe. The hallux is articulated on the same plane as the other toes, and touches the ground in walking. Lastly, they are all monogamous, and pair for life; in consequence of which fact, and of their being readily susceptible of domestication, they present an enormous number of varieties, often so different from one another that they would certainly be described as distinct species if found in a wild state. It seems certain, however, that all the common domestic breeds of Pigeons, however unlike one another, are really descended from the Rock-pigeon (*Columba livia*), which occurs wild in many parts of Europe, and has retained its distinguishing peculiarities unaltered for many centuries up to the present day. Finally, the young of the *Columbacei* are born in a naked and helpless state, whilst those of the *Gallinacei* are "precocious," and can take care of themselves from the moment of their liberation from the egg.

Of the various living birds included in this section, the true Pigeons (*Columbidæ*) are too well known to require any description; but the Ground-pigeons (*Gouridæ*) depart to some extent from this type, being ground-loving birds, more closely allied to the ordinary *Gallinacei*. The only other member of the sub-order which requires special notice, is the remarkable extinct bird, the Dodo (*Didus ineptus*), which seems certainly to belong here, though its size was gigantic, and some of its characters very anomalous. The Dodo may, properly speaking, be said to be extinct, since it no longer occurs in a living state; but it is not extinct in the sense that geologists speak of "fossils" as extinct; since it has been extirpated by man himself within quite a recent period—in fact not more than three centuries ago. The Dodo was an inhabitant of the Island of the Mauritius up to the commencement of the seventeenth century, and was a large bird, considerably over the size of a swan. All that remains nowadays to prove the existence of the Dodo are two or three old, but apparently faithful, oil-paintings, two heads, a foot, and some feathers, to which a few bones have recently been added. The Dodo owed its extermination to the fact that it was unable to fly. The body must have been extremely weighty, and the wings were rudimentary and completely useless as organs of flight. The legs were short and stout, the feet had four toes each, and the tail was extremely short, carrying, as well as the wings, a tuft of soft plumes. The beak (unlike that of any of the *Columbacei* except the little *Didunculus strigirostris*) was strongly arched towards the end, and the upper mandible had a strongly-hooked apex, not at all unlike that of a bird of prey. The nearest living ally of the Dodo appears to be the little *Didunculus* just alluded to, which inhabits the Navigator Islands, and is little bigger than a partridge.

It is worthy of notice that in the little island of Rodriguez, lying to the east of Mauritius, there existed one large wingless bird, the Solitaire or *Pezophaps*, which has likewise become extinct during the human period. In the Mauritius, also, occur the remains of the *Aphanapteryx*, a wingless bird believed to have strong Grallatorial affinities, and to have been exterminated by man. Other cases in which wingless birds have been, or are being, exterminated by man, lead us to the belief that the absence of wings is not compatible with the coexistence of birds and human beings. In other words, the sole protection possessed by birds against the destructive propensities of man is to be found in their power of flight.

CHAPTER LXIX.

SCANSORES AND INSESSORES.

ORDER V. SCANSORES.—The order of the Scansorial or Climbing birds is easily and very shortly defined, having no other distinctive and exclusive peculiarity except the fact that the feet are provided with four toes, of which two are turned backwards and two forwards. Of the two toes which are directed backwards, one, of course, is the hallux or proper hind-toe, and the other is the outermost of the normal three anterior toes. This arrangement of the toes (fig. 193, A) enables the *Scansores* to climb with unusual facility. Their powers of flight, on the other hand, are generally only moderate and below the average. Their food consists of insects or fruit. Their nests are usually made in the hollows of old trees, but some of them have the remarkable peculiarity that they build no nests of their own, but deposit their eggs in the nests of other birds. They are all monogamous.

The most important families of the *Scansores* are the Cuckoos (*Cuculidæ*), the Woodpeckers and Wry-necks (*Picidæ*), the Parrots (*Psittacidæ*), the Toucans (*Rhamphastidæ*), and the Trogons (*Trogonidæ*).

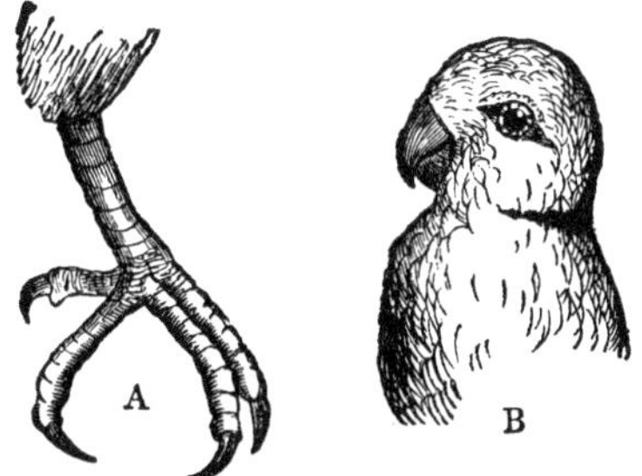

Fig. 193.—Scansores. A, Foot of Woodpecker (*Picus*); B, Head of Love-bird (*Agapornis*).

The *Cuculidæ*, or Cuckoos, are chiefly remarkable for the extraordinary fact that many of them, instead of nidificating and incubating for themselves, lay their eggs in the nests of other birds. The only bird not belonging to this family which has the same "parasitic" habit, is the Cow Bunting (*Molothrus pecoris*) of the United States. As a rule, only one egg is deposited in each nest, and the young Cuckoo which is hatched from it, is brought up by the foster-parent, generally at the expense of the legitimate offspring. The large Channel-bill (*Scythrops Novæ-Hollandiæ*) is said to possess the same curious habit, but many species of this group build nests for themselves in the ordinary manner.

The second family of the *Scansores* is that of the *Picidæ*, and comprises the Woodpeckers and Wry-necks. These birds feed chiefly upon insects, and the tongue is extensible, barbed at the point, and covered with a viscid secretion, so as to enable them to catch their prey by suddenly darting it out. The bill is strong and wedge-shaped, and the claws crooked. The tail-feathers terminate in points, and are unusually hard and stiff, assisting the bird in running up the trunks of trees.

The next family is that of the Parrots (*Psittacidæ*), the largest group of the *Scansores*, comprising over three hundred species. The bill in the Parrots is large and strong, and the upper mandible is considerably longer than the lower and is hooked at its extremity (fig. 193, B). The bill is used as a kind of third foot in climbing. At the base of the upper mandible is a "cere," in which the nostrils are pierced. The tongue is soft and fleshy. The feet are especially adapted for climbing, some, however, of the Parrots moving about actively on the ground. The colours of the plumage are generally extremely bright and gaudy; and they live for the most part upon

Fig. 194.—Psittacidæ. Purple-capped Lory (*Lorius domicella*).

fruits. The *Psittacidæ* are distributed throughout the tropics, and in the southern hemisphere as far south as the 52d parallel. They are monogamous, and make their nests in holes in the rocks. Their natural voice is harsh and grating. The true Parrots (*Psittacus*) are mostly inhabitants of tropical America, and their prevailing colour is green. The Cockatoos (*Plyctolophus*), the Love-birds (*Agapornis*), and the Lorikeets (*Trichoglossus*) belong to the Australian province. The Lories (*Lorius*) inhabit the East (fig. 194). The true Macaws (*Arainæ*) are exclusively American; and the true Parrakeets (*Pezoporinæ*) are exclusively confined to the eastern hemisphere, being especially characteristic of Australia.

In the next family of the *Scansores* are the Toucans (*Rhamphastidæ*), characterised by having a bill which is always very large, longer than the head, and sometimes of comparatively gigantic size. The mandibles are, however, to a very great extent hollowed out into air-cells, so that the weight of the bill is much less than would be anticipated from its size. The tongue is very long, notched at its side, or feathered with delicate lateral processes. The Toucans live chiefly upon fruits, and are all confined to the hotter regions of South America, frequenting the forests in considerable flocks.

The Trogons have short and weak feet, a short triangular bill, the gape bordered with strong bristles, and short wings. The plumage is soft and loose, and generally of the most gorgeous description. They inhabit the most retired recesses of the forests of the intertropical regions of both hemispheres, and show many decided points of affinity to the Goat-suckers.

ORDER VI. INSESSORES.—The sixth order of Birds is that of the *Insessores*, or Perchers—often spoken of as the *Passeres*, or "Passerine" Birds. They are defined by Owen as follows:—

"Legs slender, short, with three toes before and one behind, the two external toes united by a very short membrane" (fig. 195, A, B).

"The *Perchers* form the largest and by far the most numerous order of birds, but are the least easily recognisable by distinctive characters common to the whole group. Their feet, being more especially adapted to the delicate labours of nidification, have neither the webbed structure of those of the *Swimmers*, nor the robust strength and destructive talons which characterise the feet of the *Birds of Rapine*, nor yet the extended toes which enable the *Wader* to walk safely over marshy soils and tread lightly on the floating leaves of aquatic plants; but the toes are slender, flexible, and moderately elongated, with long, pointed, and slightly-curved claws.

"The Perchers in general have the females smaller and less brilliantly coloured than the males; they always live in pairs, build in trees, and display the greatest art in the construction of their nests. The young are excluded in a blind and naked state, and wholly dependent for subsistence during a certain period on parental care. The brain arrives in this order at its greatest proportionate size; the organ of voice here attains its greatest complexity, and all the characteristics of the bird, as power of flight, melody of voice, and beauty of plumage, are enjoyed in the highest perfection by one or other of the groups of this extensive and varied order."

The structure of the feet, then, gives the definition of the order, but the minor subdivisions are founded on the nature of the beak; this organ varying in form according to the nature of the food, "which may be small or young birds, carrion, insects, fruit, seeds, vegetable juices, or of a mixed kind" (Owen).

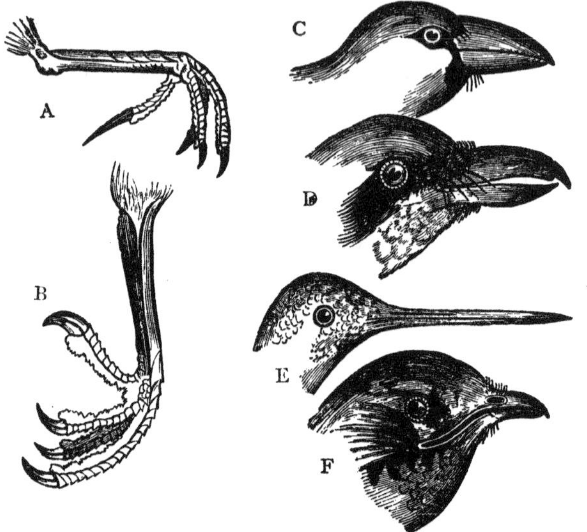

Fig. 195.—Insessores. A, Foot of Yellow Wagtail; B, Foot of Water-ouzel; C, Conirostral beak (Hawfinch); D, Dentirostral beak (Shrike); E, Tenuirostral beak (Humming-bird); F, Fissirostral beak (Swift).

In accordance with the form of the beak, the *Insessores* have been divided into four great sections or sub-orders, known as the *Conirostres*, *Dentirostres*, *Tenuirostres*, and *Fissirostres*.

Sub-order I. *Conirostres.*—In this section of the *Insessores* the beak is strong and on the whole conical, broad at the base and tapering with considerable rapidity to the apex (fig. 195, C). The upper mandible is not markedly toothed at its lower margin. Good examples of the Conirostral type of beak are to be found in the common Sparrow, Hawfinch, or Bullfinch. The greater number of the *Conirostres* are omnivorous; the remainder are granivorous, or feed on seeds and grains. The sub-order includes the families of the Horn-bills (*Bucerida*), the Starlings (*Sturnidæ*), the Crows (*Corvidæ*), the Cross-bills (*Loxiadæ*), and the Finches and Larks (*Fringillidæ*).

In the Horn-bills the conirostral shape of the beak is masked, partly by its being of very great size, and partly by the fact that above the upper mandible is placed a hollow appendage like a kind of helmet. Both the beak and the appendage above it are rendered light by the presence of numerous air-cells. The Horn-bills are exclusively confined to the warm countries of the eastern hemisphere, and are the largest of all the Insessorial birds, sometimes attaining the size of a goose. They live on fruits, and make their nests in the holes of trees. The best-known species is the Rhinoceros Bird (*Buceros Rhinoceros*) of India and the Indian Archipelago.

The family of the *Corvidæ*, or Crows, is an extremely extensive one, and includes a large number of very dissimilar-looking birds, all characterised by their long, strong, and compressed beaks, the tip of the upper mandible being slightly hooked and more or less notched. In this family are the Jays (*Garrulinæ*); the true Crows or *Corvinæ* (comprising the Rooks, Carrion-crows, Ravens, Jackdaws, Magpie, Chough, &c.), and the Birds of Paradise (*Paradiseidæ*). These last differ considerably from the ordinary *Corvidæ*, but can hardly be separated as a distinct family. They are amongst the most beautiful of all birds, and are entirely confined to New Guinea and the neighbouring islands. They feed upon insects and fruit, and are largely destroyed for the sake of their feathers. The natives who capture them usually cut off their legs; hence the notion formerly prevailed that the Birds of Paradise were destitute of these limbs. It is only the males which possess the brilliant plumage, the females being soberly dressed; and in accordance with this fact, it is stated that the Birds of Paradise are polygamous, being in this respect an exception to almost the entire order of the *Insessores*.*

The family of the Starlings (*Sturnidæ*) is not separated from that of the Crows by any important characters. Besides our common Starlings, it includes a number of other more or less singular birds, of which the Bower-birds of Australia are perhaps the most peculiar. These curious birds have the habit of building very elaborate bowers, often very beautifully constructed and of considerable size, in which they amuse themselves and apparently make love to one another. These bowers are wholly independent of their nests, which they construct elsewhere.

The last family of the *Conirostres* is that of the *Fringillidæ*, comprising the Finches, Linnets, and Larks. In these birds the bill is stout and conical, with a sharp apex, but not having the upper mandible toothed. The toes are adapted for perching, and are provided with long and curved claws, that of

* The Humming-birds are thought to be polygamous, and this is certainly the case with the Whydah Finch (*Vidua*).

the hinder toe being usually longer than the rest. They are almost all monogamous, and they build more or less elaborate nests. In this family are the true Finches (*Fringilla*), the Sparrows (*Pyrgita*), the Linnets and Goldfinches (*Carduelis*), the Whydah Finches (*Vidua*), the Grosbeaks (*Coccothraustes*), the Bullfinches (*Pyrrhula*), and many others, but their numbers are so great that any further notice of them is impossible here. It may be mentioned, however, that the Finches of the Old World are represented in the tropical parts of America by the Tanagers (*Tanagrinæ*), remarkable for their brilliant colours.

Fig. 196.—Head of the common Bullfinch (*Pyrrhula vulgaris*), showing the Conirostral beak.

The only remaining members of the *Conirostres* which require notice are the Cross-bills (*Loxiadæ*), which are sometimes placed with the Finches, and sometimes considered as a separate family. In these birds the structure of the beak is so peculiar that its Conirostral character is completely masked, and it has been looked upon as a deformity. Both mandibles, namely, cross one another towards the tip, giving the entire bill a most remarkable appearance. In point of fact, however, instead of being a deformity, the bill of the Cross-bills is a beautiful natural adaptation, enabling the bird with the greatest facility to tear in pieces the hard fir-cones, on the seeds of which it feeds.

Sub-order 2. *Dentirostres.*—The birds in this section are characterised by the fact that the upper mandible is provided with a distinct notch in its lower margin near the tip (fig. 195, D). They all feed chiefly upon insects. This sub-order includes the Shrikes (*Laniidæ*), the Fly-catchers (*Muscicapidæ*), the Thrushes (*Merulidæ*), the Tits (*Parinæ*), and the Warblers (*Sylviadæ*).

The *Muscicapidæ*, including the numerous species of Fly-catcher, are the most insectivorous of the *Dentirostres*. The gape is wide and bordered with bristles, and the legs are short and weak. They are mostly sedentary, catching their prey from a fixed point.

The Shrikes are highly predacious birds, which in many respects make a close approach to the true Birds of Prey. They feed, however, mostly upon worms and insects, and only occasionally destroy small birds or mice.

The great family of the Thrushes (*Merulidæ*) comprises not only the true Thrushes, Field-fares, and Blackbirds, but a number of exotic forms, of which the most familiar are the Orioles, so well known for their brilliant plumage and their beautifully-constructed nests.

In the *Sylviadæ*, amongst other forms, are the Wag-tails (*Motacillinæ*) and the Pipits (*Anthus*), the Titmice, Robins, Hedge-sparrow, Stone-chat, Redstarts, and other well-known British birds. The Titmice (*Parinæ*) are often placed in the sub-order of the *Conirostres*. The Nightingale also belongs to this family.

Sub-order 3. *Tenuirostres.*—The members of this sub-order are characterised by the possession of a long and slender beak, gradually tapering to a point (fig. 195, E). The toes are very long and slender, the hind-toe or hallux especially so. Most of the Tenuirostral birds live upon insects, and some of these present a near resemblance in many of their characters to the *Dentirostres*, but it is asserted that some live partially or wholly on the juices of flowers.

The chief families of the *Tenuirostres* are the Creepers (*Certhidæ*), the Honey-eaters (*Meliphagidæ*), the Humming-birds (*Trochilidæ*), the Sun-birds (*Promeropidæ*), and the Hoopoes (*Upupidæ*), of which only the Creepers and Humming-birds need any further notice.

The family *Certhidæ* includes several familiar British birds, such as the little brown Creeper (*Certhia familiaris*), the Nuthatch (*Sitta Europæa*), and the Wrens (*Troglodytes*). With these are a number of exotic forms, of which the singular Lyre-birds of Australia are the most remarkable.

The family of the *Trochilidæ*, or Humming-birds, includes the most fragile and brightly coloured of all the birds, some not weighing more than twenty grains when alive, and many exhibiting the most brilliant play of metallic colours. The Humming-birds are pre-eminently South American, but extend northwards as far even as the southern portions of Canada. The bill (fig. 195, E) is always very long and slender, as are the toes also. The tongue is bifid and tubular, and appears to be used either to catch insects within the corollas of flowers, or to suck up the juices of the flowers themselves. The plumage of the males is always brilliant, with metallic reflections, that of the female generally sombre. The legs are short and weak, but the wings are proportionately very long, and the flight is exceedingly rapid.

The Sun-birds represent in the Old World the Humming-birds of the western hemisphere, and the Australian Honey-eaters show also many points of resemblance to the *Trochilidæ.*

Sub-order 4. *Fissirostres.*—In this sub-order of the *Insessores* the beak is short but remarkably wide in its gape (fig. 195, F), and the opening of the bill is fenced in by a number of bristles (*vibrissæ*). This arrangement is in accordance with the habits of the *Fissirostres*, the typical members of which live upon insects and take their prey upon the wing. The most typical Fissirostral birds, in fact, such as the Swallows and Goat-suckers, fly about with their mouths widely opened; and the insects which they catch in this way are prevented from escaping partly by the bristles which border the gape, and partly by a viscid saliva which covers the tongue and inside of the mouth.

The typical *Fissirostres*, characterised by this structure of the beak, comprise three families — the Swallows and Martins (*Hirundinidæ*), the Swifts (*Cypselidæ*), and the Goat-suckers (*Caprimulgidæ*). These three families differ in many important respects from one another, but it would be inconvenient to separate them here. The Swifts, especially, are remarkable for the peculiarity that whilst the hallux is present, it is turned forwards along with the three anterior toes. The Goat-suckers, again, hunt their prey by night, and they are provided with the large eyes and thick soft plumage of all nocturnal birds. Besides the above, there remain the two families of the King-fishers and Bee-eaters, which are generally placed amongst the *Fissirostres*, though in very many respects the arrangement appears to be an unnatural one. These families are characterised by their stronger and longer bills, and by having the external toe nearly as long as the middle one, to which it is united nearly as far as the penultimate joint. In consequence of this peculiar conformation of the toes, these families were united by Cuvier into a single group under the name of *Syndactyli.*

The *Caprimulgidæ* are intermediate between the Owls and the Passerine Birds. Their plumage is lax and soft, and they have a hawking flight. The eyes and ears are large, the feet short and weak, and the gape of enormous size and bordered by vibrissæ. Amongst the more remarkable members of the family may be mentioned the Whip-poor-will (*Antrostomus vociferus*) of North America, the More-pork (*Podargus Cuvieri*) of Australia, and the extraordinary Guacharo Bird (*Steatornis Caripensis*) of the valley of Caripe in the West Indies.

The Bee-eaters (*Meropidæ*) live upon insects, chiefly upon various species of bees and wasps; but the King-fishers live upon small fish, which they capture by dashing into the water.

The common King-fisher (*Alcedo ispida*) is a somewhat rare native of Britain, and is perhaps the most beautiful of British birds. Some exotic King-fishers are of large size, and one of the most remarkable of them is the Laughing Jackass (*Dacelo gigas*) of Australia, so called from its extraordinary song, resembling a prolonged hysterical laugh. A very beautiful species is the Belted King-fisher (*Ceryle alcyon*) of North America.

The Bee-eaters are found chiefly in the warmer regions of the Old World, and their place is taken in America by the Motmots (*Momotus*).

CHAPTER LXX.

RAPTORES AND SAURURÆ.

ORDER VII. RAPTORES.—All the members of this order are characterised by the shape of the bill, which is "strong, curved, sharp-edged, and sharp-pointed, often armed with a lateral tooth" (Owen). The upper mandible is the longest (fig. 197, B), and is strongly hooked at the tip. The body is very muscular; the legs are robust, short, with three toes in front and one behind, all armed with long, curved, crooked claws or talons (fig. 197, A); the wings are commonly pointed,

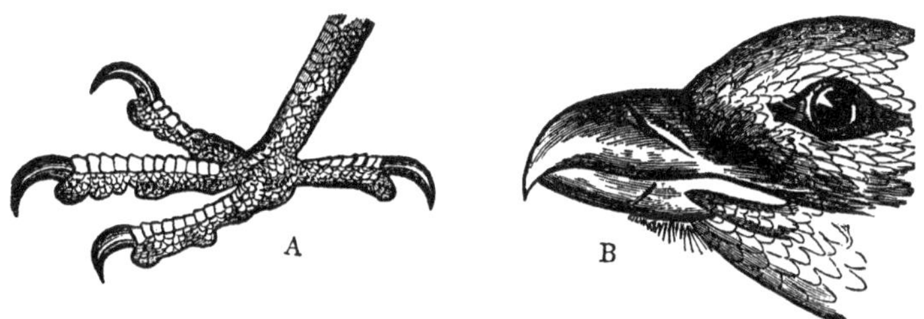

Fig. 197.—A, Foot of the Peregrine Falcon; B, Head of Buzzard.

and of considerable size, and the flight is usually rapid and powerful. The Birds of Rapine are monogamous, and the female is larger than the male. They build their nests generally in lofty and inaccessible situations, and rarely lay more than four eggs, from which the young are liberated in a naked and helpless condition.

The order *Raptores* is divided into two great sections—the *Nocturnal* Birds of Prey, which hunt by night, and have the eyes directed forwards; and the *Diurnal* Raptores, which catch their prey by day, and have the eyes directed laterally.

The section of the *Nocturnal Raptores* includes the single family of the *Strigidæ*, or Owls. In these birds the eyes are large, and are directed forwards. The plumage is exceedingly loose and soft, so that their flight (even when they are of large size) is almost noiseless; and it is generally spotted or barred with different shades of grey, brown, or yellow. The beak is short, strongly hooked, furnished with bristles at its base, and having the nostrils pierced in a membranous "cere" at the base of the upper mandible. The cranial bones are highly pneumatic, and the head is therefore of large size. The feathers of the face usually form an incomplete or complete "disc" or circle round each eye (fig. 198, B), and a circle of plumes is likewise placed round

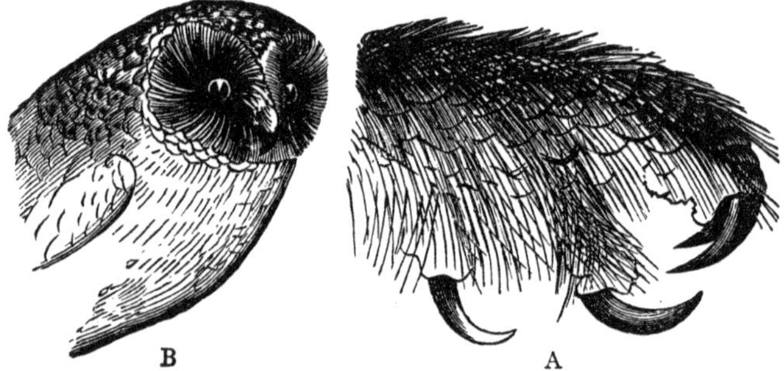

Fig. 198.—A, Foot of tawny Owl (*Ulula stridula*); B, Head of white Owl (*Strix flammea*).

each external meatus auditorius. Besides this auricular circle of feathers, the external meatus of the ear is likewise protected by a fold of skin. The legs are short and strong, and are furnished with four toes, all armed with strong crooked talons. The outer toe can be turned backwards, so that the foot has some resemblance to that of the *Scansores*. The tarso-metatarsus is densely feathered (fig. 198, A), and the plumes sometimes extend to the extremities of the toes. The œsophagus is not dilated into a crop; and the indigestible portions of the food are rejected by regurgitation from the stomach in the form of small pellets. The Owls hunt their prey in the twilight or on moonlight nights, and they live mostly upon field-mice and small birds, though they will also eat insects or frogs.

The section of the *Diurnal Raptores* includes the two groups of the *Accipitrinæ* (Falcons, Hawks, and Eagles), and the *Vulturidæ*, or Vultures. The eyes in this section are much smaller than in the preceding, and are placed laterally; and the plumage is not soft. As regards their power of flight, they show a decided advance upon the Nocturnal Birds of Prey. The wings are long and pointed; the sternal keel and pectoral muscles are greatly developed; and many of the members of this section exhibit a more rapid power of locomotion than is seen in any other division of the animal kingdom. The bill is long and strong, with a large "cere" at the base of the upper mandible, in which the nostrils are pierced. The tarso-metatarsus and toes are usually covered by scales, and are rarely feathered. Lastly, the œsophagus is dilated into a capacious crop, the gizzard is thin, the intestinal cæca are rudimentary, and the intestinal canal is generally short and wide.

In the *Accipitrinæ* or *Falconidæ* (fig. 197, B) the head and neck are always clothed with feathers, and the eyes are more or less sunk in the head, and provided with a superciliary ridge or eyebrow. It is to a great extent to the presence of this ridge that many of these birds owe their fearless and bold expression. In this family are the Falcons, Hawks, Buzzards, Kites, Harriers, and Eagles, most of which are so well known that any description is unnecessary.

Fig 199.—Head of Vulture (*Neophron percnopterus*)

In the *Vulturidæ* (fig. 199) the eyes are destitute of an eyebrow, and the head and neck are frequently naked, or covered only by a short down. In this family are the Bearded Vultures, the true Vultures, and the Condor.

The Bearded Vulture, or Lammergeyer (*Gypaëtos barbatus*), is the largest of European birds, measuring from nine to ten feet from the tip of one wing to that of the other. This powerful and rapacious bird inhabits the mountain-ranges of the south of Europe and the west of Asia, and feeds chiefly on goats, lambs, and deer, which it kills by precipitating down steep declivities. It is distinguished from the true Vultures by the fact that the head and neck are feathered.

The true Vultures have the head, and generally the neck also, naked, or covered with down. They are filthy and disgusting birds, which live almost entirely upon carrion, a peculiarity which renders them of great service in hot climates.

The Secretary Vultures (*Serpentarius*) are distinguished by their very long and slender legs, unfeathered tarso-metatarsus, and long wings armed with blunt spines. They are found in Africa and the Philippine Islands, and live upon Serpents and other Reptiles.

The last member of this section is the gigantic Condor (*Sarcorhampus gryphus*). This enormous bird has a stretch of wing of over fourteen feet, and is usually seen soaring in majestic circles at a great elevation in the air, rising, it is said, to a height of over twenty thousand feet. It inhabits the lofty mountain-ranges of the Andes, and lays its eggs at a height of from ten to fifteen thousand feet. It differs from the Vultures of the Old World chiefly in possessing a large fleshy protuberance or caruncle above the base of the beak.

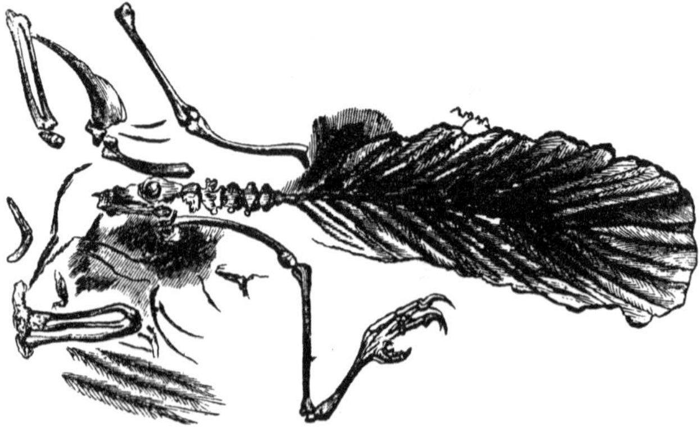

Fig 200.—*Archæopteryx macrura*, showing tail and tail-feathers, with detached bones.

ORDER VIII. SAURURÆ. — This order includes only the extinct bird, the *Archæopteryx macrura*, a single specimen of which—and that but a fragmentary one—has been discovered in the Lithographic Slates of Solenhofen (Upper Oolites). This extraordinary bird appears to have been about as big as a Rook; but it differs from all known birds in having two free claws belonging to the wing, and in having a long lizard-like tail, longer than the body, and composed of separate vertebræ. The tail was destitute of any ploughshare-bone, and each vertebra carried a single pair of quills. The metacarpal bones, also, were not anchylosed together as they are in all other known birds, living or extinct.

CHAPTER LXXI.

DISTRIBUTION OF AVES IN TIME.

As regards the geological distribution of Birds, there are many reasons why we should be cautious in reasoning upon merely negative evidence, and more than ordinarily careful not to infer the non-existence of birds during any particular geological epoch, simply because we can find no positive evidence for their presence. As Sir Charles Lyell has well remarked, "the powers of flight possessed by most birds would insure them against perishing by numerous casualties to which quadrupeds are exposed during floods;" and "if they chance to be drowned, or to die when swimming on water, it will scarcely ever happen that they will be submerged so as to become preserved in sedimentary deposits," since, from the lightness of the bones, the carcass would remain long afloat, and would be liable to be devoured by predacious animals. As, with a few utterly trivial exceptions, all the deposits in which fossils are found have been laid down in water, and more especially as they are for the most part marine, these considerations put forward by Sir Charles Lyell afford obvious ground against the anticipation that the remains of birds should be either of frequent occurrence or of a perfect character in any of the fossiliferous rocks. In accordance with these considerations, as a matter of fact, most of the known remains of birds are either fragmentary or belong to forms which were organised to live a terrestrial life, and were not organised for flight.

The earliest remains which have been generally referred to

birds are in the form of footprints impressed upon certain sandstones in the valley of the Connecticut River in the United States. These sandstones are almost certainly Triassic; and if the ornithic character of these footprints be admitted, then Birds date their existence from the commencement of the Mesozoic period, and, for anything we know to the contrary, may have existed during the Palæozoic epoch.

The evidence as to the ornithic character of the footprints in the American Trias is as follows:—

Firstly, The tracks are, beyond all question, those of a *biped*—that is to say, of an animal which walked upon two legs. No living animals walk habitually upon two legs except Man and Birds, and therefore there is a *primâ facie* presumption that the authors of these prints were birds.

Secondly, The impressions are mostly tridactylous—that is to say, formed by an animal with three toes on each foot, as is the case in many Waders and most Cursorial birds.

Thirdly, The impressions of the toes show the same numerical progression in the number of phalanges as exists in living birds—that is to say, the innermost of the anterior toes has three phalanges, the middle one has four, and the outermost toe has five phalanges.

Taking this evidence collectively, it would have seemed, till lately, tolerably certain that these impressions were formed by Birds. We must not, however, lose sight of the possibility that these impressions may have been formed by Reptiles more bird-like in their characters than any of the living forms with which we are acquainted. The recent researches of Huxley, Cope, and others, go to show that the *Dinosaurian* Reptiles possessed the power of walking temporarily or permanently on the hind-legs, and many curious affinities to the true Birds have been pointed out. It is therefore by no means impossible that these footprints of the Connecticut valley are truly Reptilian.*

The size and other characters of the above-mentioned impressions vary much, and they have certainly been produced by several different animals. In the largest hitherto discovered, each footprint is twenty-two inches long, and twelve inches wide, showing that the feet were four times as large as those of the African Ostrich. The animal, therefore, which

* The occurrence of many *four-toed* impressions on these same sandstones, and the further discovery of the bones of Dinosaurian Reptiles in the same beds, have rendered the Reptilian nature of these footprints almost certain; but some may possibly have been formed by Birds.

produced these impressions—whether Avian or Reptilian—must have been of gigantic size.

The first unmistakable remains of a bird have been found in the Solenhofen Slates of Bavaria, of the age of the Upper Oolites. A single unique specimen, consisting of bones and feathers, but unfortunately without the skull, is all that has hitherto been discovered; and it has been named the *Archæopteryx macrura.* The characters of this singular and aberrant bird, which alone constitutes the order *Saururæ,* have been already given, and need not be repeated here.

Other doubtful remains of birds have been alleged to occur in the Mesozoic series, but many of these certainly belong in reality to Pterodactyles. In the Cretaceous rocks, however, of the United States, occur the bones of several Wading Birds (*Laornis*, *Telmatornis*, and *Palæotringa*).

In the Tertiary rocks, however, there are, comparatively speaking, many remains of birds. In the Eocene rocks of France has been found a large bird, as big as an Ostrich, the so-called *Gastornis Parisiensis;* and in England, in the same formation, we have a small Vulture (*Lithornis vulturinus*), and a King-fisher (*Halcyornis toliapicus*). In the Eocene of Glaris, in Switzerland, occurs also the oldest known Insessorial or Passerine bird, the *Protornis Glarisiensis,* which was about as big as a lark.

Numerous remains of birds have likewise been found in the Miocene and Pliocene deposits. Amongst these we have Parrots, Trogons, Secretary Birds, Petrels, Cranes, Guillemots, &c. With the exception, however, of the Mesozoic *Archæopteryx,* by far the most remarkable remains of birds have been found in the Post-tertiary or Pleistocene deposits. All the remains now alluded to are those of gigantic wingless birds; and it is worthy of notice that they are exclusively found in regions now tenanted by smaller wingless birds, whilst there is reason to believe that some of them have been in existence during the human period. Most of the remains in question have been found in New Zealand, where there have been obtained the bones of several species of large wingless birds, referred by Owen to the genera *Dinornis*, *Palapteryx,* and *Aptornis.* The *Dinornis giganteus* must have been one of the most gigantic of the whole class of birds, the tibia measuring upwards of a yard in length, and the skeleton indicating a bird which stood at least ten feet in height. In another species, the *Dinornis elephantopus,* the "framework of the skeleton is the most massive of any in the whole class of birds," and "the toe-bones almost rival those of the Elephant"

(Owen). The feet were furnished with three anterior toes, and are of interest as presenting us with an undoubted bird big enough to produce the largest of the footprints of the Triassic Sandstones of Connecticut. There is reason to believe from the traditions of the Maories that the *Dinornis* was living at no very remote period, and that it has been exterminated by man.

In Madagascar, bones have been discovered of a bird as large as, or larger than, the *Dinornis giganteus*, which has been described under the name of the *Æpiornis maximus*. With the bones have been found eggs measuring from thirteen to fourteen inches in diameter, and computed to be as big as three ostrich-eggs, or one hundred and forty-eight hens' eggs. Unlike New Zealand, where there is the *Apteryx*, Madagascar itself has no living wingless birds; but in the neighbouring island of Mauritius, the Dodo has been exterminated less than three hundred years ago; and the little island of Rodriguez, in the same geographical province, has in a similar period lost the wingless Solitaire (*Pezophaps*).

DIVISION III. MAMMALIA.

CHAPTER LXXII.

General Characters of the Mammalia.

The last and highest class of the *Vertebrata*, that of the *Mammalia*, may be shortly defined as including *Vertebrate animals in which some part or other of the integument is always providea with hairs at some time of life; and the young are nourished, for a longer or shorter time, by means of a special fluid—the milk—secreted by special glands—the mammary glands.* These two characters are of themselves sufficient broadly to separate the Mammals from all other classes of the Vertebrate sub-kingdom. In addition, however, to these two leading peculiarities, the Mammals exhibit the following other characters of scarcely less importance:—

1. The skull articulates with the vertebral column by means of a double articulation, the occipital bone carrying two condyles, in place of the single condyle of the Reptiles and Birds.

2. The lower jaw or mandible consists of two halves or rami, united anteriorly by a symphysis, but not necessarily anchylosed; but these are each composed of a single piece, instead of being complex and consisting of several pieces, as in the Reptiles and Birds. Further, the lower jaw always articulates directly with the squamosal element of the skull; and is never united to an os quadratum, as in the *Sauropsida.*

3. The two hemispheres of the cerebral mass, or brain proper, are united together by a more or less extensively developed "corpus callosum" or commissure.

4. The heart consists—as in Birds—of four cavities or chambers, two auricles and two ventricles. The right and left sides of the heart are completely separated from one another, and there is no communication between the pulmonary and systemic circulations. The red blood-corpuscles are non-nucleated, and, with the exception of the *Camelidæ*, they are circular biconcave discs. There is only one aorta—the left—which turns over the left bronchus, and not over the right, as it does in Birds.

5. The cavities of the thorax and abdomen are completely separated from one another by a muscular partition—the diaphragm or midriff.

6. The respiratory organs are in the form of two lungs placed in the thorax, but none of the bronchi end in air-receptacles, distributed through the body, as in Birds.

7. The embryo mammal is invariably enveloped in an amnion, and an allantois is never wanting. The allantois, however, either disappears at an early period of life, or it develops the structure known as the "placenta." The placenta is a vascular organ which serves as a means of communication between the parent and the fœtus, but it will be noticed more particularly hereafter.

8. In no Mammal do the visceral arches and clefts of the embryo ever carry branchiæ, as they do in the Fishes and Amphibians.

These are the essential characters which distinguish the *Mammalia* as a class, but it will be necessary to consider these, and some other points, in a more detailed manner.

In the first place, with regard to the osteology of the Mammals, the following points should be noticed:—

With the exception of the Whales and Dolphins (*Cetacea*), and the Dugongs and Manatees (*Sirenia*), the vertebral column is divisible into the same regions as in man—namely, into a cervical, dorsal, lumbar, sacral, and caudal or coccygeal region (see fig. 126). In the *Cetacea* and *Sirenia* the dorsal region of the spine is followed by a number of vertebræ which compose the hinder extremity of the body, but which cannot be separated into lumbar, sacral, and caudal vertebræ.

In spite of the great difference which is observable in the length of the neck in different Mammals, the number of vertebræ in the cervical region is extraordinarily constant, being almost invariably seven, as in man. In this respect there is no difference between the Whale and the Giraffe. The only exceptions to this law are the *Manatus australis*, one of the Sea-cows, which has usually six cervical vertebræ, and the three-toed Sloth (*Bradypus tridactylus*), which is commonly regarded as possessing nine, though competent anatomists would refer the posterior two of these to the dorsal region.

The dorsal vertebræ are mostly thirteen in number, but they vary from ten to twenty-four. In Man there are twelve, in one of the Armadillos only ten, and in the three-fingered Sloth the maximum is attained. The lumbar vertebræ are usually six or seven in number, rarely fewer than four. In Man they are five in number, and they are reduced to two in the two-toed Sloth, one of the Ant-eaters, and the Duck-mole.

The first vertebra, or atlas, always bears two articular cavities for the reception of the two condyles of the occipital bone; and the second vertebra, or axis, usually has an "odontoid" process on which the head rotates. In the true Whales, however, in which the cervical vertebræ are anchylosed together to a greater or less extent, and the neck is immovable, the odontoid process is also wanting.

In almost all Mammals the spinous processes of the dorsal vertebræ are very largely developed for the attachment of the structure which is known as the *ligamentum nuchæ*. This is a great band of elastic fibrous tissue, which is attached in front to the occipital bone and spinous processes of the cervical vertebræ, and which relieves the muscles of the task of supporting the head, in those Mammals which progress with the body in a horizontal position. The development of the *ligamentum nuchæ* is consequently, as a rule, proportionate to the size of the head and the length of the neck. In Whales no such apparatus is necessary, owing to the fixation of the cervical vertebræ by anchylosis; and in Man, who walks erect, the *ligamentum nuchæ* can hardly be said to exist as a distinct structure, being merely represented by a band of fascia.

The number of lumbar and sacral vertebræ, as we have seen, varies in different Mammals; but ordinarily some of the vertebræ are anchylosed into a single bone, and have the iliac bones abutting against them, thus constituting the "sacrum" of human anatomists. In the *Cetacea* and *Sirenia*, in which the hind-limbs are wanting, and the pelvis rudimentary, there is no "sacrum."

The thoracic cavity or chest in Mammals is always enclosed by a series of ribs, the number of which varies with that of the dorsal vertebræ. In most cases each rib articulates by its head with the bodies of *two* vertebræ, and by its tubercle with the transverse process of one of these vertebræ (the lower one). In the *Monotremata* (*e.g.*, the Duck-mole), the ribs articulate with the body of the vertebra only; and in the Whales, the hindermost of the ribs, or all of them, articulate with the transverse processes only, and not with the centra at all.

There are usually no bony pieces uniting the ribs with the sternum or breast-bone in front, as in Birds; but the so-called "sternal ribs" of *Aves* are represented by the "costal cartilages" of the Mammals. In some cases, however, the cartilages of the ribs do become ossified and constitute sternal ribs. Sometimes, as in the Armadillos, there is a joint between the vertebral rib and costal cartilage. More rarely, as in the *Monotremes*, an intermediate piece is found between the vertebral and costal portions of the rib. Only the anterior ribs reach the sternum, and these are called the "true" ribs; the posterior ribs, which fall short of the breast-bone, being known as the "false" ribs.

The sternum or breast-bone is formed of several pieces placed one behind the other, but usually anchylosed together to form a single bone. It is placed upon the ventral surface of the body, and is united with the vertebral column by the ribs and their cartilages. It is generally a long and narrow bone, but in the *Cetácea* it is broad. It is only in some burrowing animals (such as the Moles) and in the true flying Mammals (the Bats), that the sternum is provided with any ridge or keel for the attachment of the pectoral muscles, as it is in Birds. The sternum is primitively composed of three pieces, an anterior piece or *præsternum*, a middle piece or *mesosternum*, and a posterior piece or *xiphisternum*. The præsternum is the "manubrium sterni" of human anatomy, and is the portion of the sternum which lies in front of the attachment of the second pair of ribs. All the other ribs are connected with the mesosternum. The *xiphisternum* is the "xiphoid cartilage" of human anatomy, and it commonly remains throughout life more or less unossified. In the Monotremes there is a T-shaped bone above or in front of the præsternum, but this is probably to be regarded as belonging to the shoulder-girdle, and as representing the "episternum" or "interclavicle" of the Reptiles.

The normal number of limbs in the *Mammalia* is four, two anterior and two posterior; and hence they are often spoken of as "quadrupeds," though all the limbs are not universally present, and other animals have four limbs as well. The anterior limbs are not known to be wanting in any Mammal, but the posterior limbs are absent in the *Cetacea* and *Sirenia*.

As regards the structure of the anterior limb, the chief points to be noticed concern the means by which it is connected with the trunk. The scapula or shoulder-blade is never absent, and it is in the form of a broad flat bone, applied to the outer aspect of the ribs, and much more developed than in the Birds. The coracoid bone, which forms such a marked feature in the scapular arch of *Aves*, is fused with the scapula, and only articulates with the sternum in the Duck-mole and Echidna (*Monotremata*). In all other Mammals the coracoid forms merely a process of the scapula, and does not reach the top of the breast-bone. The collar-bones or clavicles never unite in any Mammal to form a "furculum," as in Birds; but in the Monotremes they unite with an "interclavicle" placed in front of the sternum. The clavicles, in point of fact, are not present in a well-developed form in any Mammals except in those which use the anterior limbs in flight, in digging, or in prehension. The *Cetacea*, the Hoofed Quadrupeds (*Ungulata*), and some of the *Edentata*, have no clavicles. Most of the *Carnivora* and some Rodents possess a clavicle, but this is imperfect, and does not articulate with the top of the sternum. The Insectivorous Mammals, many of the Rodents, the Bats, and all the *Quadrumana*, have (with man) a perfect clavicle articulating with the anterior end of the sternum.

The humerus, or long bone of the upper arm (*brachium*), is never wanting, but is extremely short in the Whales, in which the anterior limbs are converted into swimming-paddles. In many Mammals, as in the Monkeys, and *Felidæ* (constituting the most typical group of the *Carnivora*), the median nerve and brachial or ulnar artery are protected on their way down the arm by a canal placed a little above the elbow, and formed by a process—the "supra-condyloid" process—which is sometimes present in man as an abnormality.

In the fore-arm of all Mammals the ulna and radius are recognisable, but they are not necessarily distinct; and the radius, as being the bone which mainly supports the hand, is the only one which is always well developed, the ulna being often rudimentary. In the *Cetacea* the ulna and radius are anchylosed together; and in most of the Hoofed Quadrupeds they are anchylosed towards their distal extremities. In the flying Mammals or Bats alone is the ulna ever altogether absent. The fore-arm attains its greatest perfection in man, in whom the radius can rotate upon the ulna, so as to allow the back of the hand to be placed upwards or downwards, these movements being known respectively as "pronation" and "supination." In the Monkeys only is there any approach to this power of rotation.

The fore-arm is succeeded by the small bones which compose the wrist or "carpus." These are eight in number in man, but vary in different Mammals from five to eleven.

The metacarpus in man and in most Mammals consists of five cylindrical bones, articulating proximally with the carpus, and distally with the phalanges of the fingers. The most remarkable modification of this normal state of things occurs in the Ruminants and in the Horse. In the Ruminants, in which the foot is cleft, and consists of two toes only, there are two metacarpal bones in the embryo; but these are anchylosed together in the adult; and form a single mass which is known as the "canon-bone" (fig. 201, *ca*). In the Horse, in which the foot consists of no more than a single digit, there is only a single metacarpal bone, on each side of which are two little bony spines — the so-called "splint-bones" — which are attached superiorly to the carpus. These are to be regarded as rudimentary metacarpals; but by Cuvier they were looked upon as imperfect fingers. In most of the other Ungulates there are at least three metacarpals, and in the Elephants there are five.

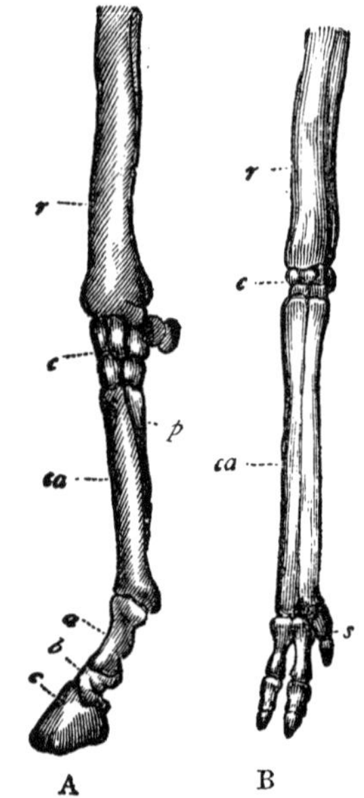

Fig. 201.—A, Fore-leg of the Horse: *r* Radius; *c* Carpus; *ca* Canon-bone; *p* Splint-bone; *a* First phalanx or "great pastern;" *b* Second phalanx or "small pastern;" *c* Ungual phalanx or "coffin-bone." B, Fore-limb of a Deer; *r* Radius; *c* Carpus; *ca* Canon-bone; *s* Supplementary toe.

The normal number of digits is five, but they vary from one to five. The middle finger is the longest and most persistent of the digits of the fore-limb; and in the Horse it is the only one which is left (fig. 201, A). The thumb is very frequently absent. In the Ruminants there are only two fingers which are functionally useful, these carrying the hoofs. In most Ruminants, however, there are two rudimentary and functionally useless digits in addition.

Normally each digit has three phalanges, except the thumb, which has only two. In the Whales and Dolphins (*Cetacea*), in which the anterior limbs form swimming-paddles, very like those of the *Ichthyosaurus* and *Plesiosaurus*, the phalanges are considerably increased in number as they are in those Reptiles. In all the Mammalia, too, except the *Cetacea*, it is the rule that the terminal phalanx in each digit should carry a nail, claw, or hoof.

The power of opposing the thumb to the other digits of the hand is found only in Man, and in a considerable number of the *Quadrumana*, but never so perfectly developed as in Man. In Man only does this power attain its full perfection, and it constitutes one of the most striking of the merely *anatomical* peculiarities by which Man is separated from the Monkeys.

As, however, this feature is purely adaptive, and is really to be regarded as of extremely small physiological value, we ought to learn from this that the difference between Man and the *Quadrumana* is to be sought in the mental powers of each, and not in any merely structural character.

Whilst the anterior limbs are never absent in any Mammal, the posterior limbs are occasionally wholly wanting, as in the *Cetacea* and *Sirenia*. Generally speaking, however, the posterior limbs are present, and the pelvic arch has much the same structure as in Man. The two halves of the pelvis—the ossa innominata—consist each of three pieces in the embryo—viz., the ilium, ischium, and pubes, which meet to form the cup-shaped cavity known as the "acetabulum," with which the head of the thigh-bone articulates. In the adult Mammal these three bones are anchylosed together, and the two ossa innominata unite in front by means of a symphysis pubis, constituted either by a cartilaginous union (synchondrosis), or by merely ligamentous attachment. In some Mammals, however, such as the Mole, and many of the Bats, the pubic bones remain disunited during life. As a rule, also, the ossa innominata are firmly united with the vertebral column. In the Cetaceans, in which the hind-limbs are wanting, and there is no sacrum, the innominate bones are rudimentary, and are not attached in any way to the spine.

The only other bones which are ever connected with the pelvis are two small bones which are directed upwards from the brim of the pelvic cavity in Marsupials and Monotremes. These are the so-called "Marsupial bones," regarded generally as not forming parts of the skeleton properly so called, but as being ossifications of the internal tendons of the "external oblique" muscles of the abdomen (fig. 204).

In those Mammals which possess hind-limbs, the normal composition of the member is of the following parts:—1. A thigh-bone or femur; 2. Two bones forming the shank, and known as the tibia and fibula; 3. A number of small bones constituting the ankle or tarsus; 4. The "root" of the foot, made up of the "metatarsus;" 5. The phalanges of the toes (see fig. 128).

The thigh-bone or femur articulates with the pelvis, usually at a very open angle. In Man it is distinguished by being the longest bone of the body, and by having the axis of its shaft nearly parallel to that of the vertebral column. In most Mammals the femur is relatively shorter, and the axis of its shaft deviates considerably from that of the spine, being sometimes at right angles, or even at an acute angle.

Of the bones of the leg proper the tibia corresponds to the radius in the fore-limb, as shown by its carrying the tarsus; and the fibula is the representative of the ulna. The articulation between the tibia and fibula on the one hand, and the femur on the other, constitutes the "knee-joint," which is usually defended in front by the "knee-pan" or patella, a large sesamoid bone developed in the tendons of the great extensor muscles of the thigh. The patella is of small size in the *Carnivora*, but does not appear to be wanting in any except the Marsupials. In many cases the tibia and fibula are anchylosed towards their distal extremities. In the Horse the fibula has much the same character as in Birds, being a long splint-like bone which only extends about half-way down the tibia. In the Ruminants the reverse of this obtains, the upper half of the fibula being absent, and only the lower half present.

The tibia articulates with the tarsus, consisting in Man of seven bones, but varying in different Mammals from four to nine.

The foot consists normally of five toes connected with the tarsus by means of five metatarsal bones, which closely resemble the metacarpals. In the Ruminants there are only two metatarsals, and these are anchylosed in the adult, and carry two toes. In the Horse there is only one metatarsal supporting a single toe. As a rule, the number of digits in the hind-limb or foot is the same as that in the fore-limb or hand; but this is not always the case. In the Lions, Tigers, Cats, and Dogs, the posterior limb carries only four toes, the innermost toe or hallux being wanting. In the *Quadrumana*, again, all the five toes are generally present, but the four outer toes are much longer than in Man, and the hallux is shorter than the other toes, and often opposable to them, so that the foot forms a kind of posterior hand. The hallux is also not uncommonly opposable in other cases.

The cranial bones are invariably connected with one another by sutures, and in no other examples than the Monotremes are these sutures obliterated in the adult. The differences of opinion which are entertained as to the fundamental structure of the skull are so enormous, that it will be best not to attempt here any detailed description of the skull of the Mammalia, more especially as there is as yet no universal agreement even as to the nomenclature to be employed. It is sufficient to remember that the skull is composed of a series of bony segments, which are usually regarded as modified vertebræ. The occipital bone carries two condyles for articulation with the first cervical vertebra. The lower jaw is composed of two halves or rami, which are distinct from another in the embryo, and may or may not be anchylosed together in the adult. However this may be, in no Mammal is the ramus of the lower jaw composed of several pieces, as it is in Birds and Reptiles, nor does it articulate with the skull by the intervention of an os quadratum. On the other hand, each ramus of the lower jaw in the Mammals is composed of only a single piece, and articulates with the squamosal element of the skull, or, in other words, with the squamous portion of the temporal bone.

Teeth are present in the great majority of Mammals; but they are only present in the embryo of the whalebone Whales, and are entirely absent in the genera *Echidna*, *Manis*, and *Myrmecophaga*. In the Duck-mole (*Ornithorhynchus*) the teeth are horny, and the same was the case in the extinct *Rhytina* amongst the *Sirenia*. In all other Mammals the teeth have their ordinary structure of dentine, enamel, and crusta petrosa, these elements being variously disposed in different cases. In no Mammals are the teeth ever anchylosed with the jaw; and in all, the teeth are implanted into distinct sockets or alveoli, which, however, are very imperfect in some of the Cetacea.

Many Mammals have only a single set of teeth throughout life, and these are termed by Owen "monophyodont." In most cases, however, the first set of teeth—called the "milk" or "deciduous" teeth—is replaced in the course of growth by a second set of "permanent" teeth. The deciduous and permanent sets of teeth do not necessarily correspond to one another; but no Mammal has ever *more* than these two sets. The Mammals with two sets of teeth are called by Owen "diphyodont."

In Man and in many other Mammals the teeth are divisible into four distinct groups, which differ from one another in position, appearance, and function; and which are known respectively as the *incisors*, *canines*, *præmolars*, and *molars* (fig. 202). "Those teeth which are implanted in the præmaxillary bones, and in the corresponding part of the lower jaw, are called 'incisors,' whatever be their shape or size. The tooth in the maxillary bone which is situated at or near to the suture with the præmaxillary, is the 'canine,' as is also that tooth in the lower jaw which, in opposing it, passes in front of its crown when the mouth is closed. The other teeth of the first set are the 'deciduous molars;' the teeth which displace and succeed them vertically are the 'præmolars;' the more posterior teeth, which are not displaced by vertical successors, are the 'molars' properly so called."—(Owen.) The deciduous dentition, therefore, of a diphyodont Mammal consists of only three kinds of teeth—incisors, canines, and molars. The incisor and canine teeth of the deciduous set are replaced by the teeth which bear the same names in the permanent set. The deciduous "molars," however, are replaced by the permanent "præmolars," and the "molars" of the permanent set of teeth are not represented in the deciduous series, only existing once, and not being replaced by successors.

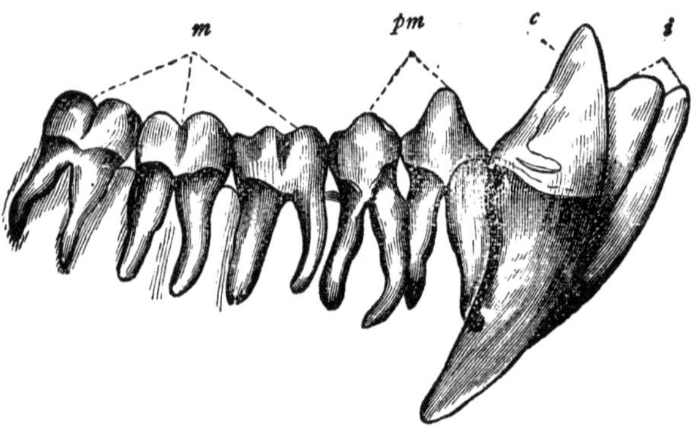

Fig. 202.—Teeth of the right side of the lower jaw of the Chimpanzee (after Owen). *i* Incisors; *c* Canine teeth: *pm* Præmolars; *m* Molars.

All these four kinds of teeth are not necessarily present in all Mammals, and, as will be afterwards seen, the characters of the teeth are amongst the most important of the distinctions by which the Mammalian orders are separated from one another. The variations which exist in the number of teeth in different Mammals are usually expressed by a "dental formula," which presents the "dentition" of both jaws in a condensed and easily-recognised form.

According to Owen, the typical permanent dentition of a diphyodont Mammal would be expressed by the following formula:—

$$i\,\frac{3-3}{3-3};\quad c\,\frac{1-1}{1-1};\quad pm\,\frac{4-4}{4-4};\quad m\,\frac{3-3}{3-3} = 44.$$

The four kinds of teeth are indicated in such a formula by the letters—incisors *i*, canines *c*, præmolars *pm*, molars *m*. The numbers in the upper line indicate the teeth in the upper jaw, those in the lower line stand for those in the lower jaw; and the number of teeth on each side of the jaw is indicated by the short dashes between the figures.

As regards the digestive system of the *Mammalia*, salivary glands are present in all except the true *Cetacea*. The alimentary canal has in most cases essentially the same structure as in man; and the same accessory glands are present,—namely, the liver and pancreas. Some very remarkable modifications occur in the structure of the stomach and in the termination of the intestine; but these will be noticed in speaking of the orders in which they occur. The cavity of the abdomen is always separated from that of the thorax by a complete muscular partition—the diaphragm—as is the case in no other Vertebrate animals. The abdomen contains the greater portion of the alimentary canal, the liver, spleen, pancreas, kidneys, and other organs. The thorax mainly holds the heart and lungs.

The heart is contained in a serous bag, the pericardium, and consists (as in Birds) of two auricles and two ventricles. The effete and deoxygenated blood is returned from the tissues by the veins, and is conducted by the two venæ cavæ to the right side of the heart into the right auricle. From the right auricle it passes into the right ventricle, whence it is propelled through the pulmonary artery to the lungs. Having been submitted to the action of the air, the blood, now arterialised, is carried by the pulmonary veins to the left auricle, and thence into the left ventricle. From the left ventricle the aerated blood is driven through the aorta and systemic vessels to all parts of the body. In Mammals, therefore, as in Birds, the pulmonary and systemic circulations are altogether distinct and separate from one another. The two sides of the heart—except in the fœtus and as an abnormality in adults—have no communication with one another except by means of the capillaries.

The red blood-corpuscles are never nucleated, and in all except the *Camelidæ* (in which they are oval) they are circular and discoid.

The lungs of Mammals differ from those of Birds in being freely suspended in the thoracic cavity, the greater part of which they fill, and in being enclosed freely in a serous sac (*pleura*) which envelops each lung. The lungs are minutely cellular throughout, and the bronchi never open on the surface of the lung into a series of air-receptacles communicating with one another, and placed in different parts of the body, as is the case in Birds.

There is no "inferior larynx" in any Mammal, and the upper aperture of the true larynx is always protected by an epiglottis.

The kidneys in Mammals are situated in the lumbar region, and exhibit a division of their substance into cortical and medullary portions.

There are two ovaries in all Mammals, and the oviducts are known as the "Fallopian tubes." Each oviduct dilates on its way to the surface into a uterine cavity, which opens into the vagina. In the Monotremes and Marsupials this primitive condition is retained throughout life, the uterus remaining double, and opening by two apertures into the cloaca or vagina. In most cases this condition is so far modified in the adult, that the two uteri have coalesced inferiorly, so as to have only a single opening into the vagina, whilst they separate into two horns or "cornua" superiorly. Only in the Monkeys and in Man have the two uteri completely coalesced to form a completely single cavity, into the "fundus" of which the Fallopian tubes open. In male Mammals there are always two testes present. In many Mammals the testes are permanently retained in the abdominal cavity and there is no scrotum. This is the case in the Monotremes, the Elephants, all the *Cetacea*, and many of the *Edentata*. Mostly, however, the testes at an early period of life are transferred from the abdomen to a pouch of integument called the "scrotum." Usually the scrotum is placed beneath the pubic arch and behind the penis, but this position is reversed in the Marsupials.

Mammary glands are present in all Mammals, and they are regarded by Huxley as an extreme modification of the cutaneous sebaceous glands. In the male Mammals the mammary glands are present, but, under all ordinary circumstances, they remain functionally useless and undeveloped. Considerable differences obtain as to the number and position of the mammary glands in different cases; but they are always placed on the inferior surface of the body, and their ducts in the great majority of cases open collectively upon a common elevation—the "teat" or "nipple." In the *Monotremata*, however, there are no nipples, the ducts of the mammary glands opening either into a pouch of the integument (*Echidna*) or upon a flat surface (*Ornithorhynchus*).

The young Mammal is nourished for a longer or shorter time by the milk secreted by the mammary glands of the mother. In ordinary cases the milk is obtained by voluntary suction on the part of the young animal; but in the Marsupials the young are at first unable to suck for themselves, and the milk is forced out of the gland by the contractions of a special muscle.

The nervous system of Mammals is chiefly remarkable for the great proportionate development of the cerebral mass as compared with the size of the spinal cord. In the higher Mammals, again, the hemispheres of the cerebrum are much more largely developed proportionately than the remaining parts of the brain. The brain of the Mammals is chiefly distinguished from that of the lower *Vertebrata* by the fact that the two hemispheres of the cerebellum are united by a transverse commissure—the *pons Varolii*—and the hemispheres of the brain are connected by a great commissure—the *corpus callosum*—which is, however, of small size in the lower *Mammalia*.

The senses, as a rule, attain great perfection in the Mammals; and the only sense which appears to be ever entirely wanting is that of vision. In one of the most familiar instances of this last-mentioned fact — namely, in the Mole — it has recently been shown that it is only in the adult that vision is lost, but that the organs of sight are well developed in the young. The sclerotic coat of the eye is never supported by a ring of bony plates as in Birds and many Reptiles. As a rule, in addition to the upper and lower eyelids there is a third perpendicular lid—the *membrana nictitans*—but this is wanting or quite rudimentary in Man and in the Monkeys.

An external ear or *concha* for collecting the vibrations of sound is usually present, but is wanting in the *Cetacea*, many of the Seals, and in some other cases.

The integument is furnished over a greater or less portion of its surface with the epidermic appendages known as "hairs." These are developed, much as feathers are, upon little eminences or papillæ of the derma, but they do not split up in the process of development as feathers do. In the *Manis* or Scaly Ant-eater the epidermic appendages are in the form of horny scales, and not uncommonly they are developed into long spines, as in the *Echidna*, Porcupine, and Hedgehog. In the Armadillos, again, the integument has the power of developing plates of bone over a greater or less extent of its surface. The only apparent exception to the universal presence of hairs in some part or other of the skin of all Mammals is constituted by the *Cetacea*, some of which are without hairs in the adult state. Some, however, of these (such as the Whales) possess a few bristles in the neighbourhood of the mouth even when fully grown. And the Dolphins, which are totally hairless when adult, exhibit tufts of hair on the muzzle in the fœtal state.

CHAPTER LXXIII.

CLASSIFICATION OF THE MAMMALIA.

NUMEROUS classifications of the *Mammalia* have been proposed, and it is a matter of regret that no one has been universally accepted by Zoologists. Here, it will be sufficient to describe briefly the three leading systems upon which the *Mammalia* have been divided into sub-classes; whilst the first will be adopted as sufficient for all practical purposes.

I. By many writers the class *Mammalia* is divided into two great primary divisions, the *Placentalia* or Placental Mammals, and the *Implacentalia* or Non-placental Mammals, according as the structure known as the "placenta" is present or absent. The placenta, as before said, is a vascular organ developed in the greater number of Mammals, by means of which the blood of the fœtus is brought into relation with the blood of the mother. The sub-class *Placentalia*, in which such a vascular connection between the mother and fœtus exists, comprises by far the largest number of the Mammals. The sub-class *Implacentalia*, in which no such vascular connection exists, comprises only the two orders of the *Monotremata* and the *Marsupialia.*

II. By Professor Owen the *Mammalia* are divided into four sub-classes, characterised by the structure of the brain, as follows:—

a. Lyencephala, characterised by the fact that the cerebral hemispheres are usually without folds, and leave the cerebellum, the olfactory lobes, and part of the optic lobes uncovered. The hemispheres are not connected together by a corpus callosum. (*Monotremata* and *Marsupialia.*)

b. Lissencephala, characterised by the fact that the cerebral hemispheres are smooth or are provided with few folds, and leave the cerebellum and part of the olfactory lobes exposed. A corpus callosum is present. (*Cheiroptera, Insectivora, Rodentia, Edentata.*)

c. Gyrencephala, characterised by the fact that the hemispheres of the cerebrum cover the greater part of the cerebellum and the olfactory lobes. A corpus callosum is present, and the surface of the cerebral hemispheres is generally thrown into numerous convolutions. (*Cetacea, Carnivora, Sirenia, Proboscidea, Ungulata, Quadrumana.*)

d. Archencephala, characterised by the fact that the cerebral hemispheres now completely overlap the cerebellum and olfactory lobes; the number of convolutions attains its maximum; and there is a corpus callosum. (Man.)

This is the primary classification of the *Mammalia* put forth by Owen, and there can be no question but that in many respects it expresses substantial and important differences. It will not be adopted here, partly because it is somewhat difficult to follow or to apply in practice, and partly because some of the characters upon which it is founded are denied by other eminent naturalists. Thus, in the definition of the sub-class *Lyencephala* it is stated as one of the essential characters that there is no corpus callosum or commissure between the hemispheres of the cerebrum. On the other hand, it is asserted by Flower and Huxley that a corpus callosum does exist in these animals, though it never attains to any high degree of development.

III. It was proposed by De Blainville, and the arrangement has been accepted by Huxley and Rolleston, to divide the *Mammalia* into the following three sub-classes, founded upon the nature of the reproductive organs:—

a. Ornithodelphia, characterised by the fact that the uterine enlargements of the oviducts do not coalesce even in their inferior portion to form a common uterine cavity, but open separately as in the Birds and Reptiles. Furthermore, the two uteri open, not into a distinct vagina, but into a cloacal cavity, into which the rectum and ureters also discharge themselves; so that the condition of parts is very much the same as it is in Birds.

This division includes only the Duck-mole (*Ornithorhynchus*) and the Porcupine Ant-eater (*Echidna*), forming collectively the single order of the *Monotremata*.

b. Didelphia, characteristed by the fact that the uterine dilatations of the oviducts continue distinct throughout life, opening into two distinct vaginæ, which in turn open into a urogenital canal, which is distinct from the rectum, though embraced by the same sphincter muscle.

This sub-class contains the *Marsupialia*, such as the Kangaroos, Opossums, Wombats, &c., most of which are almost entirely confined to Australia. They have many other characters in common, which will be spoken of hereafter.

III. *Monodelphia*, characterised by the fact that the uterine enlargements of the oviducts coalesce to a greater or less extent to form a single uterine cavity, which, however, generally shows its true composition by being divided superiorly into two cornua. The uterus opens again into a single vagina, which is always distinct from the rectum. This sub-class corresponds with the division of the "Placental" Mammals, and includes all the *Mammalia* except the Monotremes and Marsupials.

Before going on to consider the different orders of the *Mammalia* in detail, it may be as well very briefly to run over the leading characters by which the various orders are distinguished:

Order I. Monotremata, characterised by the fact that the ureters and ducts of the reproductive organs open into a common urogenital canal, which in turn opens, along with the rectum, into a "cloaca." The testes are abdominal, and are not lodged in a scrotum. The mammary glands have no nipples. The young is devoid of a placenta, but the female possesses no marsupial pouch, though the pelvis is furnished with "marsupial bones." In this order are only the Duck-mole and the Echidna.

Order II. Marsupialia, characterised by the fact that the uterine dilatations of the oviducts open with the ureters into a urogenital canal, which is distinct from the rectum, though embraced by the same sphincter muscle. The testes are not abdominal, but are lodged in a scrotum which is suspended by a narrow neck in front of the penis. The females are mostly furnished with a marsupial pouch in which the young are carried for some period after birth. The young are not provided with a placenta, and are born in a very imperfect state of development. Marsupial bones are present. In this order are the Kangaroos, Opossums, Wombats, &c.

Order III. Edentata or *Bruta*, characterised by the universal absence of the median incisors, and the general absence of all the incisors. The canines are usually wanting as well, and sometimes there are no molars either. There is only one set of teeth, and the teeth have neither complete roots nor are furnished with a covering of enamel. The toes are always furnished with claws. Placenta sometimes deciduate, sometimes non-deciduate. As examples of this order may be taken the Sloths, Armadillos, and the great Ant-eater.

Order IV. Sirenia, comprising the Dugongs and Manatees, characterised by being adapted to an aquatic life. Body fish-like, with a strong horizontal tail-fin. There is no sacrum, and the hind-limbs are invariably wanting, whilst the fore-limbs are converted into swimming-paddles There are, in the living forms at any rate, two sets of teeth, and the molars have flattened crowns adapted for a vegetable diet. There are two nostrils, and these are placed at the upper part of the snout. There are two mammæ, and these are placed on the chest, and not on the abdomen.

Order V. Cetacea, comprising the true Whales and Dolphins, characterised by being aquatic Mammals, with a horizontal tail-fin, no sacrum nor hind-limbs, and fore-limbs in the form of swimming-paddles. The nostrils are single or double, and are placed on the top of the head. The mammary glands are two in number, and are placed in the region of the groin. There is never more than one set of teeth, and in many cases the adult is destitute of teeth altogether. Placenta non-deciduate.

Order VI. Ungulata or *Hoofed Quadrupeds*, comprising the whole of the Ruminants, the Horses, and most of the old group of the Pachydermatous Mammals. This order is split up into many important sections, and, as a whole, it is simply characterised by the fact that there are never more than four full-sized toes to each limb, and that the extremities of the toes are furnished with expanded nails, constituting hoofs. There are no clavicles. Placenta non-deciduate.

Order VII. Hyracoidea, comprising only the single genus *Hyrax*, characterised by having no canines, but by having long curved incisors, which grow from permanent pulps, as in the Rodents. There are no clavicles. The front-feet have four toes, and the hind-feet three. The placenta is deciduate and zonary.

Order VIII. Proboscidea, comprising no other living Mammal except the Elephant, characterised by having no canines, but only molars and incisors, of which the latter grow from permanent pulps, and constitute defensive tusks. There are no clavicles. The feet are five-toed. The nose is prolonged into a proboscis. The mammæ are two in number. The placenta is deciduate and zonary.

Order IX. Carnivora, comprising all the well-known beasts of prey, such as Lions, Tigers, Dogs, Cats, &c., together with the aquatic Seals and Walruses. They are all characterised by always possessing the three different kinds of teeth—incisors, canines, and molars—the canines being usually of great length, and a greater or less number of the molars having sharp cutting edges. The clavicles are always rudimentary, the teats are abdominal, and the placenta is deciduate and zonary.

Order X. Rodentia, comprising the Beavers, Rats, Mice, Hares, Rabbits, Squirrels, and others, characterised by the absence of canines and the possession of no more than two incisors in the lower jaw, and usually no more than two in the upper jaw. The incisors are greatly developed, growing from permanent pulps, and continuing to grow during the life of the animal. Placenta deciduate and discoidal.

Order XI. Cheiroptera, comprising only the various Bats, and characterised by the fact that the four outer or ulnar fingers are greatly developed and elongated, and are united together by a leathery flying-membrane or "patagium," which is continued from the hand and arm to the side of the body and hind-limb. By means of this patagium the Bats possess the power of flight. Clavicles are always present. The teeth vary a good deal, but there are always canines. The placenta is deciduate and discoidal.

Order XII. Insectivora, comprising the Moles, Shrew-mice, and Hedgehogs, characterised by having the crowns of the molar teeth furnished with sharp and pointed cusps. Well-developed clavicles are present in almost all cases. The placenta is deciduate and discoidal.

Order XIII. Quadrumana, comprising the Lemurs, Apes, and Monkeys. Dentition usually the same as in man, or with an additional præmolar on each side of each jaw, or varying a good deal in the lower forms. The series of teeth is uneven and interrupted. The innermost digit of the fore-limb (*pollex*) is mostly opposable to the other fingers when present, but it may be wanting. The hallux is also opposable to the other toes of the hind-limb, so that the hind-feet constitute prehensile hands. Clavicles are always present. The placenta is deciduate and discoidal.

Order XIV. Bimana.—This order includes Man alone. The dental formula is—

$$i\,\frac{2-2}{2-2};\quad c\,\frac{1-1}{1-1};\quad pm\,\frac{2-2}{2-2};\quad m\,\frac{3-3}{3-3} = 32.$$

The teeth are nearly even, and are not interrupted by any interval (*diastema*). The pollex or thumb of the fore-limb is opposable to the other digits, but this is not the case with the hallux or great toe. The attitude of the body in progression is habitually erect. The placenta is deciduate and discoidal.

NON-PLACENTAL MAMMALS.

CHAPTER LXXIV.

MONOTREMATA AND MARSUPIALIA.

ORDER I. MONOTREMATA.—The first and lowest order of the *Mammalia* is that of the *Monotremata*, constituting by itself the division *Ornithodelphia*, and containing only two genera, both belonging to Australia—namely, the Duck-mole (*Ornithorhynchus*) and the Porcupine Ant-eater (*Echidna*).

The order is distinguished by the following characters:—The intestine opens into a "cloaca," which receives also the products of the urinary and generative organs, which discharge themselves into a urogenital canal, the condition of parts being very much the same as in Birds. The jaws are either wholly destitute of teeth (*Echidna*), or are furnished with horny plates which act as teeth. The pectoral arch has some highly bird-like characters, the most important of these being the extension of the coracoid bones to the anterior end of the sternum. The females possess no marsupial pouch, but the pelvis is furnished with the so-called "marsupial bones," believed to be ossifications of the internal tendons of the external oblique muscles of the abdomen. The testes of the male are abdominal throughout life, and there is therefore no scrotum, whilst the vasa deferentia open into the cloaca. The corpus callosum is very small, and has been asserted to be altogether wanting. There are no external ears. The mammary glands have no nipples, and their ducts open either into a kind of integumentary pouch (*Echidna*) or simply on a flat surface (*Ornithorhynchus*). The young are said to be destitute of a placenta, or, in other words, no vascular connection is established between the fœtus and the mother. The feet have five toes each, armed with claws, and the males carry perforated spurs on the back of the tarsus (attached to a supplementary tarsal bone).

The order *Monotremata* includes only the two genera *Orni-*

thorhynchus and *Echidna*—the one represented by a single species (*O. paradoxus*), and the other by two species (*E. hystrix* and *E. setosa*). All are exclusively confined to Australia and Tasmania.

The *Ornithorhynchus* or Duck-mole is one of the most extraordinary of Mammals. The body (fig. 203) resembles that of a mole or small otter, and is covered with a close, short, brown fur. The tail is broad and flattened. The jaws are produced to form a beak just like that of a duck in appearance; hence the name of "Duck-billed animal," often applied to it. The margins of the jaw are sheathed with horn, and furnished with transverse horny plates; but there are no teeth. The

Fig. 203.—*Ornithorhynchus paradoxus.*

sternum is of five pieces, not counting in the episternum, and there are sternal ribs. The nostrils are placed at the apex of the upper mandible. The legs are short, and the feet have five toes each, furnished with strong claws, which enable the animal to burrow with facility. The toes are also united by a membrane or web, so that the animal swims with great ease. The *Ornithorhynchus* is exclusively found in Australia and Tasmania, and inhabits streams and ponds. Its food consists chiefly, if not exclusively, of insects, and the animal makes very extensive burrows on the banks of the rivers which it frequents. The young are born quite blind, and nearly naked, and the method in which they obtain milk from the mother is somewhat obscure, as there are no nipples, nor is there any marsupial pouch. It is certain, however, that the beak of the young animal is extremely different from what it is in the adult condition. The young animal is totally hairless, the mandibles are soft and flexible, the tongue is not placed far back in the mouth (as it is in the adult), and the eye is at first covered by the skin.

The genus *Echidna* is represented by two species, *E. hystrix* and *E. setosa*, both belonging to the Australian province. The *Echidna hystrix* is the best-known species, and in some external respects is not unlike a large hedgehog, having the back covered with strong spines, interspersed with a general coating of bristly hairs. The snout has not the form of a duck's bill, as in the *Ornithorhynchus*, but the two mandibles are greatly elongated, and are enclosed in a continuous skin till close upon their extremities, where there is a small aperture for the protrusion of a very long and flexible tongue. The jaws are wholly devoid of teeth or anything in the place of teeth; and the nostrils are placed at the extremity of the cylindrical snout. The feet have five toes each, furnished with strong curved digging-claws, but the toes are not webbed. The *Echidna* measures from fifteen to eighteen inches in length, and is a nocturnal animal. It lives in burrows, and feeds upon insects, which it catches by protruding its long and sticky tongue.

ORDER II. MARSUPIALIA. — The order *Marsupialia* constitutes by itself the sub-class *Didelphia*, and forms with the *Monotremata* the division of the Non-placental Mammals. With the single exception of the genus *Didelphys*, which is American, all the *Marsupialia* belong to the Melanesian province; that is to say, they all belong to Australia, Van Diemen's Land, New Guinea, and some of the neighbouring islands.*

The following are the characters which distinguish the order:—

The skull is composed of distinct cranial bones united by sutures, and they all possess true teeth; whilst the angle of the lower jaw is almost always inflected. The pectoral arch has the same form as in the higher Mammals, and the coracoid no longer reaches the anterior end of the sternum. All possess the so-called "marsupial bones," attached to the brim of the pelvis. The corpus callosum is very small, and has been asserted to be absent. The young Marsupials are born in a very imperfect condition, of very small size, and at a stage when their development has proceeded to a very limited degree only. (In the Kangaroo the period of gestation is only about thirty-nine days, and in the *Didelphidæ* it is said to be only fifteen or seventeen days). It is believed that there is no placenta or vascular communication between the mother

* One Kangaroo (*Macropus Bruynii*) is found in the Indian Archipelago, along with five Phalangers, which differ from the Australian forms in having the tail partially or entirely naked or scaly. There are also Tree-Kangaroos, and the curious *Cuscus*, distinguished by a prehensile tail, large eyes, and slow progression.

and fœtus, parturition taking place before any necessity arises for such an arrangement. As the young are born in such an imperfect state of development, special arrangements are required to secure their existence. When born, they are therefore, in the great majority of cases, transferred by the mother to a peculiar pouch formed by a folding of the integument of the abdomen. This pouch is known as the "marsupium," and gives the name to the order. Within the marsupium are contained the nipples, which are of great length. Being for some time after their birth extremely feeble, and unable to perform the act of suction, the young within the pouch are nourished involuntarily, the mammary glands being provided with special muscles which force the milk into the mouths of the young. At a later stage the young can suckle by their own exertions, and they leave the pouch and return to it at will. In a few forms there is no complete marsupium as above described; but the structure of the nipples is the same, and the young are carried about by the mother, adhering to the lengthy teats.

The so-called "marsupial bones" (fig. 204) doubtless serve to support the marsupial pouch and its contained young, but this cannot be their sole function, since they occur in the Monotremes, in which there is no pouch. It is believed by Owen that the function of the marsupial bones is to assist in the action of the mammæ and testes, serving respectively as a fulcrum for the muscle spread over the mammary gland and for the cremaster.

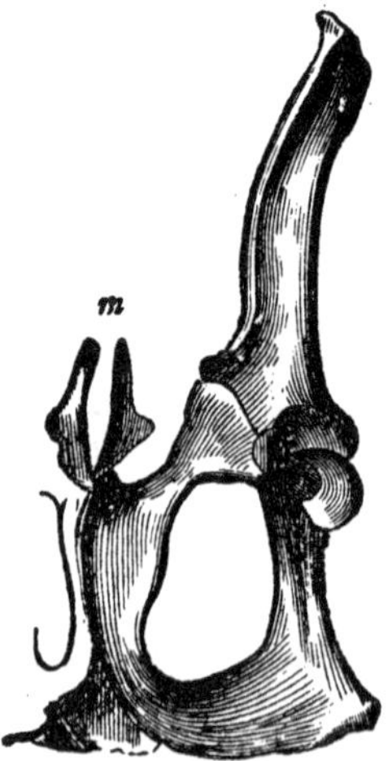

Fig. 204.—One side of the pelvis of a Kangaroo, showing the "marsupial bones" (*m*)—after Owen.

The oviducts open into vaginal tubes which open into a urogenital canal; but this does not open into a "cloaca," though embraced by a sphincter muscle common to it and to the rectum. In other words, the vagina is separated wholly or in great part into two distinct tubes. The testes are not abdominal throughout life as in the Monotremes, but are lodged in a scrotum. This, however, is placed in front of the penis, and not beneath the pubic arch as in most Mammals. From this unusual position of the scrotum, it is regarded by Owen as being the same structure as the marsupial pouch of the female, turned inside out. Though they form an extremely natural order, sharply separated from all the rest of the Mammals, the Marsupials form a large and varied group. In fact this order, from being the almost exclusive possessor of a continent as large as Australia, has to discharge in the economy of nature functions which are elsewhere discharged by several orders.

The *Marsupialia* are divided by Owen into the following sections:—

a. Rhizophaga.—In this section is the well-known Australian animal, the Wombat (*Phascolomys fossor*), often called by the colonists the "badger." The Wombat is a stout, heavy animal, which attains a length of from two to three feet. The legs are very short and stout, and the animal burrows with ease by means of strong curved digging-claws, with which the fore-feet are furnished. The tail in the Wombat is quite rudimentary, and the whole body is clothed with a brown woolly hair. In its dentition the Wombat presents a curious resemblance to the herbivorous Rodents. There are two incisors in each jaw, and these are long and rootless, growing from permanent pulps. There are no canines, so that the incisors and præmolars are separated by a considerable space. The dental formula is—

$$i\,\frac{1-1}{1-1};\ c\,\frac{0-0}{0-0};\ pm\,\frac{1-1}{1-1};\ m\,\frac{4-4}{4-4} = 24.$$

The præmolars and molars agree with the incisors in growing from permanent pulps, in which respect the Wombat differs from all the other Marsupials, and agrees with the herbivorous Rodents, with those *Edentata* which have teeth, and with the extinct *Toxodon.*—(Owen.)

The Wombat is a nocturnal animal, and feeds chiefly upon roots and grass.

b. Poephaga.—In this section are the Kangaroos (*Macropodidæ*) and the Kangaroo-rats or Potoroos (*Hypsiprymnus*), all strictly phytophagous. The Kangaroos are distinguished by the disproportionate length of the hind-limbs and disproportionate development of the posterior portion of the body as compared with the fore-limbs and fore part of the body. The hind-legs are exceedingly long and strong, and the feet are much elongated—the whole sole being applied to the ground. The hind-feet have four toes each, of which the central one is by far the largest, and the two inner toes are very small, and are united by a common integument. The tail is also extremely long and strong, and by the assistance of this organ and the powerful hind-limbs the Kangaroos are enabled to effect extraordinarily long and continuous leaps. In fact, leaping is the ordinary mode of progression in the typical Kangaroos; and when walking upon all fours their locomotion is slow and ungraceful. The anterior extremity of the body is

Fig. 205.—Koala or Kangaroo-bear (*Phascolarctos cinereus*)—after Gould.

very diminutive as compared with the posterior, and the fore-limbs are quite small, but have five well-developed toes armed with strong nails. The head is small, with large ears, and the dental formula is—

$$i\,\frac{3-3}{1-1};\ c\,\frac{0-0}{0-0};\ pm\,\frac{1-1}{1-1};\ m\,\frac{4-4}{4-4} = 28.$$

There are therefore six upper incisors, two lower incisors, and no functional canines (though rudimentary upper canines are present in the young of some of the Kangaroos, at any rate). The stomach is complex, and sacculated. The Kangaroos are all herbivorous, and mostly live, either scattered or gregariously, on the great grassy plains of Australia. The "Tree-kangaroos," however (constituting the genus *Dendrolagus*), live mostly in trees; and, in adaptation to this mode of life, the fore-legs are nearly as long and strong as the hind-legs, the tail is not used as a support, and the claws are long, curved, and pointed. They are natives of New Guinea. The "Rock

Kangaroos" form the genus *Petrogale*, and inhabit the mountainous regions of North-western Australia.

The Kangaroo-rats (*Hypsiprymnus*) differ from the true Kangaroos chiefly in their smaller size, and in the presence of well-developed upper canines (fig. 206, B), and in having scaly tails. They are diminutive nocturnal animals, and they live mostly upon roots.

c. Carpophaga.—Intermediate between the Kangaroos and the typical members of the present section (the Phalangers) is the *Phascolarctos*—the "native sloth" or "bear" of the Australian colonists and the "koala" of the natives (fig. 205). This curious animal is about two feet in length, having a stout body, covered with a dense bluish-grey fur. The tail is wanting; and the feet are furnished with strong curved claws, which enable the animal to pass the greater part of its existence in trees. In

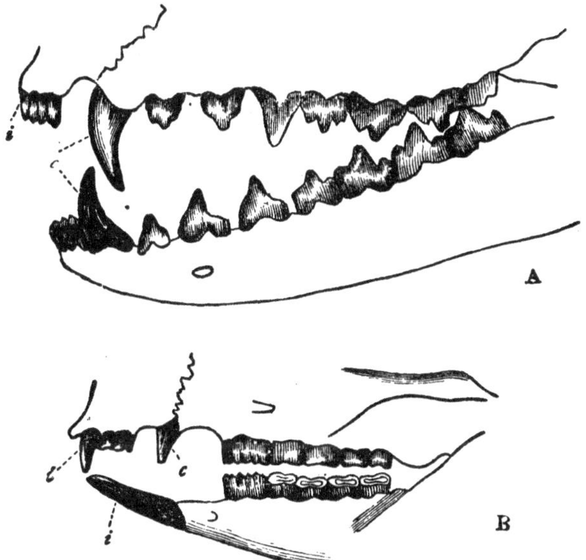

Fig. 206.—A, Dentition of a Carnivorous Marsupial (*Thylacinus*), showing **the long** and pointed canines and the trenchant molars and præmolars. B, Dentition of a herbivorous Marsupial (*Hypsiprymnus*), showing the flat-crowned molars. (After Owen.) *c* Canine teeth; *i i* Incisors.

this it is greatly assisted by the fact that all the feet are prehensile, the hallux being opposable, and the digits of the fore-limb divided into two sets, the thumb and index-finger being opposable to the other fingers. The koala is a slow animal which feeds on the foliage of the trees in which it spends its existence.

The typical group of the Carpophagous Marsupials is that of the *Phalangistidæ* or Phalangers, so called because the second and third digits of the hind-feet are joined together almost to their extremities. The family includes a number of small Marsupials, fitted for an arboreal existence, to which end the hallux is opposable and nail-less, whilst the four remaining toes of the hind-feet have long curved claws. The tail, too, is generally very long, and its tip is usually prehensile. The Phalangers are all small nocturnal animals which live upon fruits and other vegetable food. The best known of them is the Australian Opossum (*Phalangista vulpina*), which must not be confounded with the true or American Opossums, which belong to another section of the *Marsupialia*. The Phalangers, namely, are distinguished from the Opossums properly so called, amongst other characters, by their dentition, the canine teeth being always very small and functionally useless in the lower jaw, and sometimes in the upper jaw as well. The *Phalangista vulpina* is nocturnal and arboreal in its habits, and its flesh is esteemed a great delicacy by the native Australians, with whom opossum-hunting is a favourite pursuit.

The flying Phalangers or *Petauri* are closely allied to the true Phalangers, but differ in not having a prehensile tail, and in having a fold of skin extending on each side between the sides of the body and the fore and hind limbs. By the help of these lateral membranes the *Petauri* can take extensive leaps from tree to tree; but though called "flying" Phalangers, they have no power of flight properly so called. They are beautiful little animals, nocturnal in their habits, and having the body clothed with a soft and delicate fur.

d. Entomophaga. — In this section the jaws are always furnished with canine teeth, but these are not of very large size, and the animals composing the section are therefore not highly predacious, but "prey, for the most part, on the smaller and weaker classes of invertebrate animals." In this section are the Bandicoots (*Peramelidæ*), the American Opossums (*Didelphidæ*), and the Banded Ant-eater (*Myrmecobius*).

The Bandicoots* (*Peramelidæ*) are small Australian animals, which appear to fill the place of the Hedgehogs, Shrew-mice, and other small *Insectivora* of the Old World. The hind-

* The name "Bandicoot" properly belongs to the Great Rat (*Mus giganteus*) of India.

limbs in the Bandicoots are considerably longer than the fore-limbs, and their progression is therefore by a series of bounds. The fore-limbs have really five toes each, but only the central three of these are well developed, the outermost and innermost digits being rudimentary. The three functional toes are armed with long strong claws, with which the Bandicoots burrow with great ease. The marsupial pouch—and this is a singular point—opens, in some species at any rate, backwards instead of forwards. In the nearly-allied genus *Chœropus*, also from Australia, it appears that the two outer toes of the fore-feet are entirely absent.

The second family of this section—namely, the true Opossums or *Didelphidæ*—is remarkable in being the only group of the whole order which occurs out of the Australian province. The *Didelphidæ*, namely, are exclusively found in North and South America, where they are known as "Opossums." A considerable number of species is known, but they are mostly of small size, the largest measuring not more than from two to three feet, inclusive of the tail. The Virginian Opossum (*Didelphys Virginiana*) is the only member of the family which is found in North America, and it was the earliest Marsupial known to science. Most of the Opossums are carnivorous, feeding upon small quadrupeds and birds, but they also eat insects, and sometimes even fruit. One species (*Didelphys cancrivora*) lives chiefly upon Crabs; and the Yapock (*Cheironectes*) has webbed feet, and appears to lead a semi-aquatic life. All the *Didelphidæ* have the hallux nail-less and opposable to the other toes, so as to convert the hind-feet into prehensile hands, and all have a more or less perfectly prehensile tail, these being adaptations to an arboreal life. The marsupial pouch is sometimes not present in a complete form, but is merely represented by cutaneous folds of the abdomen concealing the nipples. In the *Didelphys dorsigera*, in which this peculiarity obtains, the young soon leave the nipples, and are then carried about on the back of the mother, to whom they cling by twining their prehensile tails round hers. The dentition of the Opossums is remarkable for the great number of the incisor teeth, the dental formula being—

$$i\,\frac{5-5}{4-4};\quad c\,\frac{1-1}{1-1};\quad pm\,\frac{3-3}{3-3};\quad m\,\frac{4-4}{4-4} = 50.$$

The Banded Ant-eater (*Myrmecobius fasciatus*) is a small but extremely elegant little animal, which inhabits Western and Southern Australia, and lives upon insects (fig. 207). The tail is bushy, and differs from that of the *Didelphidæ* in not being prehensile. The fore-feet have five toes armed with claws; the hind-feet have only four toes. The *Myrmecobius* is remarkable for the extraordinary number of molar teeth, in which it exceeds any existing Marsupial, and is only surpassed by some of the Armadillos. The dental formula is—

$$i\ \frac{4-4}{3-3};\quad c\ \frac{1-1}{1-1};\quad pm\ \frac{3-3}{3-3};\quad m\ \frac{6-6}{6-6} = 54.$$

e. Sarcophaga.—This is the last section of the existing Marsupials, and includes a number of predacious and rapacious forms, which fill the place held elsewhere by the true *Carnivora*. They are distinguished by the fact that the intestine is destitute of a cæcum, and by their strictly carnivorous dentition, the

Fig. 207.—*Myrmecobius fasciatus.*

canines being strong, long, and pointed, whilst the molars and præmolars have cutting edges furnished with three cusps (fig. 206, A). The best-known species of this section are the *Thylacinus cynocephalus* and the *Dasyurus ursinus*. The former of these is the largest of the rapacious Marsupials, being about as big as a shepherd's dog. It is a native of Van Diemen's Land, and is known to the colonists as the "hyæna." Its head is very large, and the back exhibits several transverse black bands. It lives in caverns and amongst the rocks in the wildest parts of the colony, and its numbers have been very much reduced by the constant war waged upon it by the settlers. The *Dasyurus ursinus* is also a native of Van Diemen's Land, where it is known as the "native devil." Though smaller than the Thylacine, the *Dasyurus* is extremely ferocious, and is capable of committing great havoc amongst animals even as large as sheep.

PLACENTAL MAMMALS.

CHAPTER LXXV.

EDENTATA.

ORDER III. EDENTATA OR BRUTA.—The lowest order of the placental or monodelphous Mammals is that of the *Edentata*, often known by the name of *Bruta*. The name *Edentata* is certainly not an altogether appropriate one, since it is only in two genera in the order that there are absolutely no teeth. The remaining members of the order have teeth, but these are always destitute of true enamel, are never displaced by a second set, and have no complete roots. Further, in none of the *Edentata* are there any median incisors, and in only one species (one of the Armadillos) are there any incisor teeth at all. Canine teeth, too, are almost invariably wanting. Clavicles are usually present, but are absent in the Scaly Ant-eater or *Manis*. All the toes are furnished with long and powerful claws. The mammary glands are usually pectoral, but are sometimes abdominal in position. The testes are abdominal in position. The skin is often covered with bony plates or horny scales.

The order *Edentata* is conveniently divided into two great sections, in accordance with the nature of the food, the one section being phytophagous, the other insectivorous. In the former section is the single group of the Sloths (*Bradypodidæ*). In the latter are the two groups of the Armadillos (*Dasypodidæ*), and the various species of Ant-eaters (the latter constituting Owen's group of the *Edentula*).

The order *Edentata* is but sparingly represented in modern times, and its geographical distribution is peculiar. The true Ant-eaters, the Armadillos and the Sloths, are entirely confined to South America, in which country a group of gigantic extinct Edentates existed in Post-tertiary times. The Scaly Ant-eater or *Manis* is common to Asia and Africa, and the genus *Orycteropus* is peculiar to South Africa.

The family *Bradypodidæ* comprises some exceedingly curious

animals which are exclusively confined to South America, inhabiting the vast primeval forests of that continent. The Sloths have a remarkably short and rounded face, and the body is covered with hair. The incisor teeth are altogether wanting, but there are always simple molars, and in the Two-toed Sloth or Unau the first tooth in each jaw on each side is so much larger than the others, and so much more pointed, that it has been regarded as a canine. The stomach is complex, somewhat resembling that of the Ruminants. The cervical vertebræ are generally regarded as being more than the normal seven in number in the Three-toed Sloth, and the long bones have no medullary cavities. The most striking peculiarities, however, about the Sloths are connected with their mode of life. The Sloths, in fact, are constructed to pass their life suspended from the under surface of the branches of the trees amongst which they live; and for this end their organisation is singularly adapted. The fore-limbs are much longer than the hind-limbs, and the bones of the fore-arm are unusually movable. All the feet, but especially the fore-feet, are furnished with enormously long curved claws (fig. 208), by the aid of which the animal is enabled to move about freely suspended back downwards from the branches. Not only is this the ordinary mode of progression amongst the Sloths, but even in sleep the animal appears to retain this apparently unnatural position.

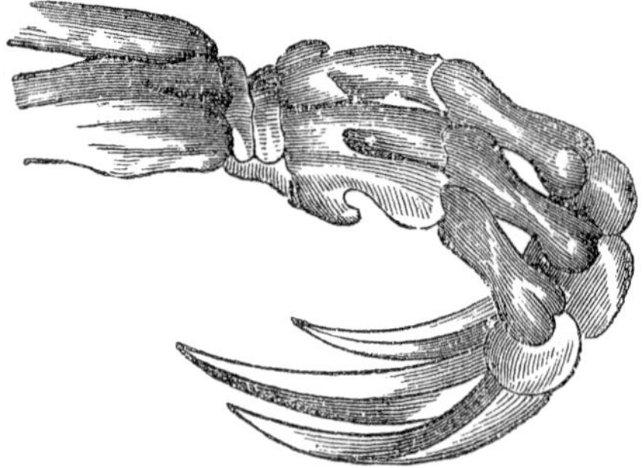

Fig. 208.—Hand of Three-toed Sloth (*Bradypus tridactylus*)—after Owen.

Owing to the disproportionate size of the fore-limbs as compared with the hind-limbs, and owing to the fact that the hind-feet are so curved as to render it impossible to apply the sole to the ground, the Sloth is an extremely awkward animal upon the ground, and it has therefore recourse to terrestrial progres sion only when absolutely compelled to do so Whilst the name of "Sloth" may thus appear to be a merited one from the point of view of a terrestrial Mammal, it is wholly undeserved when the animal is looked upon as especially adapted for an arboreal existence. In the Ai or Three-toed Sloth (*Bradypus tridactylus*) there are three toes to each foot, and these are short, completely rigid, and so enveloped in the integument as to leave nothing visible except the enormously long and crooked claws. The hand and foot are jointed to the arm and leg obliquely, so that the palm and sole cannot be applied to the ground, but are turned inwards. The ungual phalanges are also so articulated that the claws are bent inwards towards the palm or sole. There are sixteen pairs of ribs. The molars are rootless, growing from permanent pulps, and consisting of a simple cylinder of dentine enveloped in enamel. In the Unau (*Cholœpus*) the feet are two-toed, and there are twenty-three pairs of ribs, the greatest number known in the Mammals.

The second family of the *Edentata* is that of the *Dasypodidæ* or Armadillos. These are found exclusively in South America, as are the Sloths, but they are very different in their habits. The Armadillos are burrowing animals, furnished with strong digging-claws and well-developed collar-bones. The jaws are provided with numerous simple molars, which attain the enormous number of nearly one hundred in the great Armadillo (*Dasypus gigas*). The upper surface of the body is covered with a coat of mail, formed of hard bony plates or shields, united at their edges. A portion of this armour covers the head and shoulders, and another portion protects the hindquarters; whilst between these is a variable number of movable bands which run transversely across the body, and give the necessary flexibility to this singular dermoskeleton. In some species this flexibility is so great that the animal can roll itself up like a hedgehog. The tail is likewise mostly covered with bony scutes.

The Armadillos are confined entirely to South America, ranging from Mexico to Patagonia. In this country, also, have been found the remains of a gigantic armour-plated animal allied to the Armadillos, which will be subsequently described under the name of the *Glyptodon*. Amongst the best-known species of Armadillo are the Peba (*Dasypus Peba*), the Poyou (*D. sexcinctus*), the Tatouay (*D. Tatouay*), and the Great Armadillo (*D. gigas*). A somewhat aberrant form is the *Chlamyphorus* (fig. 209) of South America, the total length of which is only about six inches.

The remaining members of the *Edentata* are the various Ant-eaters, but these are so different from one another in their characters that they form three distinct families, also distinguished by their geographical distribution.

Fig. 209.—*Chlamyphorus truncatus.*

a. Myrmecophagidæ.—This family is exclusively confined to South America, as are the two preceding, and it contains only the Hairy or true Ant-eaters. These curious animals feed chiefly upon Ants and Termites, which they catch with their long sticky tongues. The jaws are wholly destitute of teeth; the body is covered with hair; there is a long tail; and the feet are armed with long and strong curved digging-claws. The toes are united by skin up to the bases of the claws, as in the Sloths; the ungual phalanges are articulated in the same way, and the palm and sole are similarly turned inwards.

The best-known species of this family is the Great Ant-eater (*Myrmecophaga jubata*). This singular animal attains a length of over four feet, and has an extremely long and bushy tail. The jaws are produced to form a long and slender snout, which is entirely enclosed in the skin, till just at its extremity, where there is an aperture for the protrusion of the thread-like tongue. The anterior feet have four, and the posterior feet five toes, all armed with strong curved claws, which, when not used in digging, are bent inwards, so that the animal walks on the sides of the feet. The animal is perfectly harmless and gentle when unmolested, and leads a solitary life. It lives mainly upon Termites, into the nests of which it forces its way by means of the powerful claws. When the Termites rush out to see what is the matter, the Ant-eater thrusts out its glutinous tongue, an action which can be repeated with marvellous rapidity. Two other species have been described, both much smaller than the preceding, and both arboreal in their habits, and furnished with prehensile tails.

b. Manidæ.—This family includes only the Scaly Ant-eaters or Pangolins, all exclusively confined to the Old World, and found in both Africa and Asia. The whole of the body, limbs, and tail in the *Manidæ* is covered with an armour of horny imbricated plates, overlapping like the tiles of a house, and apparently consisting of agglutinated hairs. The legs are short, and furnished with four or five toes each, ending in long and strong digging-claws; but there are no clavicles. The tongue resembles that of the Hairy Ant-eaters in being long and contractile, and capable of being exserted for a considerable distance beyond the mouth. It is covered with a glutinous saliva, and is the agent by which the animal catches ants and other insects. The jaws are wholly destitute of teeth. When threatened by danger, the Pangolins roll themselves up into a ball, like the hedgehogs. The tail is comparatively long, and is covered with scales. Though very strong for their size, none of the species attain a length of more than three or four feet, inclusive of the tail. The best-known species are the *Manis pentadactyla* of India, and the *Manis tetradactyla* of Africa.

c. Orycteropidæ.—The last family of the living Edentata is that of the *Orycteropidæ*, comprising only the single genus *Orycteropus.* This genus comprises only a single species, the *O. Capensis*, which is peculiar to South Africa, and is known by the Dutch colonists as the "Aardvark" or Ground-hog. The animal is nocturnal in its habits, and lives upon insects. The body is elongated, and the tail is long, the species attaining a total length of four feet or more. The legs are short, and the feet plantigrade, the anterior pair having four unguiculate toes, the posterior five. The claws are strong and curved, and enable the animal to construct extensive burrows. The skin is very thick, and is thinly covered with bristly hairs; and the tail is hairy. The head is elongated, and the mouth small—devoid of incisor and canine teeth, but furnished with a number of cylindrical molars $(\frac{7-7}{6-6})$. The crowns of the molars are flat, and they are composed of dentine traversed by numerous dichotomising pulp-cavities. The tongue is long, flat, and slender, and is covered by a sticky saliva, by the aid of which the animal catches insects. The head is long and attenuated, the snout truncated and callous, and the ears large, erect, and pointed.

CHAPTER LXXVI.

SIRENIA AND CETACEA.

ORDER IV. SIRENIA.—This order comprises no other living animals except the Dugongs and Manatees, which are often placed with the true *Cetaceans* (Whales and Dolphins) in a common order. There is no doubt, in fact, but that the *Sirenia* are very closely allied to the *Cetacea;* and though they are to be regarded as separate orders, yet they may be advantageously considered as belonging to a single section, which has been called *Mutilata*, from the constant absence of the hind-limbs.

The *Sirenia* agree with the Whales and Dolphins in their complete adaptation to an aquatic mode of life (fig. 210); especially in the presence of a powerful caudal fin, which differs from that of Fishes in being placed horizontally, and in being a mere expansion of the integuments, not supported by bony rays. The hind-limbs are wholly wanting, and there is no sacrum. The anterior limbs are converted into swimming-paddles or "flippers." The snout is fleshy and well developed, and the nostrils are placed on its upper surface, and not on the top of the head, as in the Whales. Fleshy lips are present, and the upper one usually carries a moustache. The skin is covered with fleshy bristles. The head is not disproportionately large, as in the true Whales, and is not so gradually prolonged into the body as it is in the latter. There may be only six cervical vertebræ. The teats are two in number and are "thoracic," *i.e.*, are placed on the chest. There are no clavicles, and the digits have no more than three phalanges each. The testes are retained throughout life within the abdomen, but vesiculæ seminales are present. The animal is diphyodont, the permanent teeth consisting of molars with flattened crowns adapted for bruising vegetable food, and incisors which are present in the young animal, at any rate. In the extinct *Rhytina* it does not appear that there were any incisor teeth.

The only existing *Sirenia* are the Manatees (*Manatus*) and the Dugongs (*Halicore*), often spoken of collectively as "sea-cows," and forming the family of the *Manatidæ*.

The Manatees are characterised by the possession of numerous molar teeth $(\frac{8-8}{8-8})$ and of two small upper incisors, which are wanting in the adult. The tail-fin is oblong or oval in shape, and the anterior limbs are furnished with nails to the

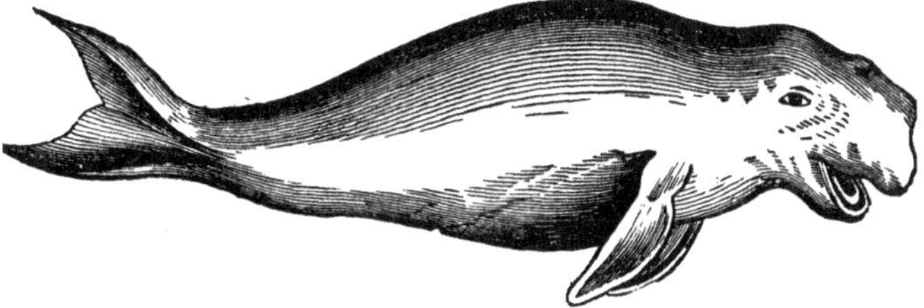

Fig. 210.—Sirenia. Dugong (*Halicore*).

four outer digits. They occur on the east coast of North America, especially in the Gulf of Mexico, and another species is found on the west coast of Africa. They are generally found in considerable numbers about the mouths of rivers and estuaries, and they appear to live entirely upon sea-weeds, aquatic plants, or the littoral vegetation. They are large, awkward animals, attaining a length of from eight to ten feet as a rule, but sometimes growing to a length of nearly twenty feet.

The Dugongs (*Halicore*, fig. 210) have $\frac{5-5}{5-5}$ molar teeth in the young condition, but only $\frac{2-2}{2-2}$ when old. Inferior incisors are present in the young animal, but are wanting in the adult. The upper jaw carries two permanent incisors, which are entirely concealed in the jaw in the females, but which increase in size in the males with the age of the animal, till they become pointed tusks. The anterior extremities are nail-less, and the tail-fin is crescentic in shape. In their general appearance and in their habits the Dugongs differ little from the Manatees, and they are often killed and eaten. They attain a length of from eight to ten, twelve, or more feet, and are found chiefly on the coasts of the Indian Ocean. The bones are remarkable for their extreme density, their texture being nearly as close as ivory.

The Manatees and Dugongs, as before said, are the only living *Sirenia*; but besides these there is a very singular form, the *Rhytina Stelleri*, which is now extinct, having been exterminated by man within a comparatively recent period. This remarkable animal was discovered about the middle of the eighteenth century in a little island (Behring's Island) off the coast of Kamtschatka. Upon this island the celebrated voyager Behring was wrecked, and he found the place inhabited by these enormous animals, which were subsequently described by M. Steller, who formed one of his party. The discovery, however, was fatal to the *Rhytina*, for the last appears to have been seen in the year 1768. The *Rhytina* was an animal of great size, measuring twenty-five feet in length, and twenty feet at its greatest circumference. There can hardly be said to have been any true teeth, but the jaws contained $\frac{1-1}{1-1}$ large lamelliform fibrous structures, which officiated as teeth, and may be looked upon as molars. The epidermis was extremely thick and fibrous, and hairs appear to have been wanting. There was a crescentic tail-fin, and the anterior limbs alone were present.

ORDER V. CETACEA.—In this order are the Whales, Dolphins, and Porpoises, all agreeing with the preceding in their complete adaptation to an aquatic life (figs. 213, 214). The body is completely fish-like in form; the anterior limbs are converted into swimming-paddles or "flippers;" the proximal bones of the fore-limbs are much reduced in length, and the succeeding bones are shortened and flattened, and are enveloped in a tendinous skin, thus reducing the limbs to oar-like fins; there are no external ears; the posterior limbs are completely absent; and there is a powerful, horizontally-flattened, caudal fin, sometimes accompanied by a dorsal fin as well. In all these characters the *Cetacea* agree with the *Sirenia*, except in the one last mentioned. On the other hand, the nostrils, which may be single or double, are always placed at the top of the head, constituting the so-called "blow-holes" or "spiracles;" and they are never situated at the end of a snout. The body is very sparingly furnished with hairs, or the adult may be completely hairless. The testes are retained throughout life within the abdomen, and there are no vesiculæ seminales. The teats are two in number and are placed upon the groin. The head is generally of disproportionately large size, and is never separated from the body by any distinct constriction or neck. The lumbar region of the spine is long, and, as in the *Sirenia*, there is no sacrum, and the pelvis is only present in a rudimentary form. There are no clavicles, and some of the digits may possess more than three phalanges each. Lastly, the adult is either destitute of teeth or is monophyodont—that is to say, possesses but a single set of teeth, which are never replaced by others. When teeth are present, they are usually conical and numerous, and they are always of one kind only.

The *Cetacea* may be divided into the three families of the *Balænidæ* or Whalebone Whales, the *Delphinidæ* or Dolphins and Porpoises, and the *Catodontidæ* or Sperm Whales. Of these, the *Balænidæ* are often spoken of as the "toothless" Whales, whilst the other two families are called the "toothed" Whales (*Odontoceti*).

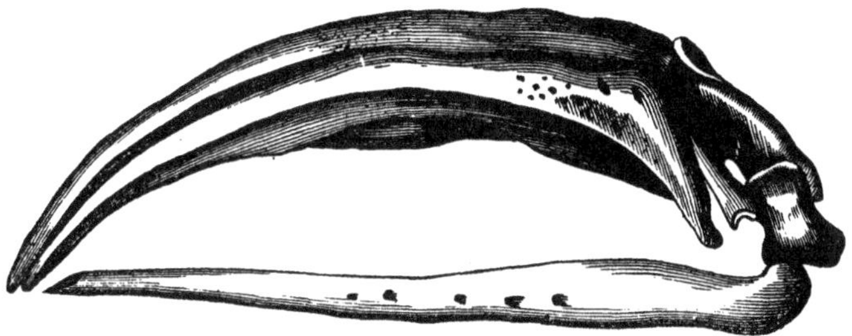

Fig. 211.—Skull of the Right Whale (*Balæna mysticetus*)—after Owen.

Fam. 1. *Balænidæ.*—The *Balænidæ* or Toothless Whales are characterised by the total absence of teeth in the adult (fig. 211). Teeth, however, are present in the fœtal Whale, but they never cut the gum. The place of teeth is supplied by a number of plates of whalebone or "baleen" attached to the palate; hence the name of "whalebone Whales" often given to this family. They are the largest of living animals, and may be divided into the two sections of the *Smooth* Whales, in which the skin is smooth and there is no dorsal fin (as in the Greenland Whale), and the *Furrowed* Whales, in which the skin is furrowed and a dorsal fin is present (as in the so-called Finner Whales and Hump-backed Whales).

The Greenland or "Right" Whale (*Balæna mysticetus*) will illustrate almost all the leading points of interest in the family. The Greenland Whale is the animal which is sought after in the whale-fishery of Europe, and hence the name of "Right" Whale often applied to it. It is an inhabitant of the Arctic seas, and reaches a length of from forty to sixty feet. Of this enormous length, nearly one-third is made up of the head, so that the eye looks as if it were placed nearly in the middle of the body. The skin is completely smooth, and is destitute of hairs in the adult. The fore-limbs are converted into "flippers" or swimming-paddles, but the main organ of progression is the tail, which often measures from twenty to twenty-five feet in breadth. The mouth is of enormous size, the upper jaw somewhat smaller than the lower, and both completely destitute of teeth. Along the middle of the palate runs a strong keel bordered by two lateral depressions, one on each side. Arranged transversely in these lateral depressions are an enormous number of horny plates, constituting what is known as the "baleen" plates, from which the whalebone of commerce is derived. The arrangement of the plates of baleen is somewhat as follows (fig. 212) :—Each plate is somewhat triangular in shape, the shortest side or base being deeply sunk in the palate. The outer edge of the plate is nearly straight, and is quite unbroken. The inner edge is slightly concave, and is furnished with a close fringe formed of detached fibres of whalebone. For simplicity's sake each baleen-plate has been regarded here as a single plate, but in reality each plate is composed of several pieces, of which the outermost is by far the largest, whilst the others gradually decrease in size towards the middle line of the palate. The large marginal plates are from eight to ten or fourteen feet in length, and there may be about two hundred on each side of the mouth.

The object of the whole series of baleen-plates with which the palate is furnished, is as follows :—The Whale is a strictly carnivorous or zoophagous animal, but owing to the absence of teeth, and the comparatively small calibre of the œsophagus, it lives upon very diminutive animals. The Whale, in fact, lives mostly upon the shoals of small Pteropodous Molluscs, *Ctenophora* and *Medusæ*, which swarm in the Arctic seas. To obtain these, the Whale swims with the mouth opened, and thus fills the mouth with an enormous mass of water. The baleen-plates have the obvious function of a "screening-apparatus." The water is strained through the numerous plates of baleen, and all the minute animals which it contains are arrested and collected together by the inner fibrous edges of the baleen-plates. When, by a repetition of this process, the Whale has accumulated a sufficient quantity of food within the central cavity of the mouth, it is enabled to swallow it, without taking the water at the same time.

We have now to speak of a phenomenon which has given rise to a considerable amount of controversy—namely, what is known as the "blowing" or "spouting" of the whale. In all the Cetaceans the nose opens by a single or double aperture (the latter in the *Balænidæ*) upon the top of the head, and these external apertures or nostrils are known as the "blow-holes" or "spiracles." The act known to the whalers as "blowing" consists in the expulsion from the blow-holes of a jet of what is apparently water, or at any rate looks like it. The act is performed by the whale upon rising to the surface, and it is usually by this that the whereabouts of the animal is discovered. The old view as to what takes place in the act of blowing is, that the whale is really occupied in getting rid of the surplus water which it has taken in at the mouth and strained through the baleen-plates. The modern and doubtless correct view, however, is, that the water which has been strained through the baleen really makes its escape at the side

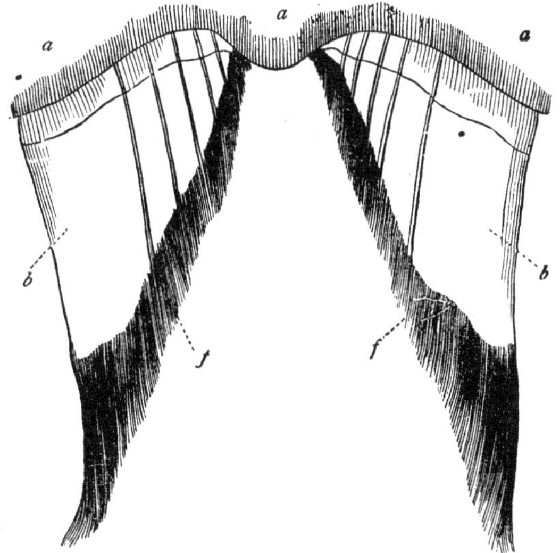

Fig. 212.—Diagram of the Baleen-plates of a Whale. *a a* Section of the palatal surface of the upper jaw, showing the strong median ridge or keel; *b b* Baleen-plates, sunk at their bases in the palate; *f f* Fibrous margin of Baleen-plates.

of the mouth, and does not enter the pharynx to be expelled through the nose. Upon this view the apparent column of water emitted from the blow-holes in the act of blowing consists really of the expired air from the lungs, the contained watery vapour of which is suddenly condensed on its entrance into the cold atmosphere. With the expired air there may be such water as may have gained access to the nose through the blow-hole, for the expulsion of which proper provision exists in the form of muscular diverticula of the nasal cavity. It is also possible that the column of air in being forcibly expelled from the blow-hole may take up with it some of the superincumbent water.

The skin in the Right Whale is perfectly smooth and naked, but it is underlaid by a thick layer of subcutaneous fat, which varies from eight to fifteen inches in thickness, and is known as the "blubber." The blubber serves partly to give buoyancy to the body, but more especially to protect the animal against the extreme cold of the medium in which it lives. It is the blubber which is chiefly the object of the whale-fishery, as it yields the whale-oil of commerce.

The whale which is captured in the Antarctic regions is not the same species as the Greenland Whale, and is termed the *Balæna australis*. It is much about the size of the Right Whale, averaging about fifty feet, but the head is proportionately smaller. This whale is an inhabitant of the greater part of the Pacific out of the regions of the tropics ; but it is chiefly captured when approaching land, which the females do for the purpose of bringing forth their young.

The only remaining members of the *Balænidæ* which require notice are the Rorquals and Hump-backed Whales, constituting the group of the " Furrowed " Whales. These are collectively distinguished by having the skin furrowed or plaited to a greater or less extent, whilst the baleen-plates are short, and there is a dorsal fin. The specific determination of these animals is a matter of great difficulty, but there would appear to be probably three well-marked genera :—1. The genus *Megaptera*, including the so-called Hump-backed Whales, in which the flippers are of great length, from one-third to one-fifth of the entire length of the body. 2. The genus *Balænoptera*, comprising the so-called Rorquals or Piked Whales, in which the flippers are of moderate size. 3. The Finner Whales proper (*Physalus*).

In all these genera there is a dorsal adipose fin, so that they are all "Finner" Whales. The *Balænopteræ* reach a gigantic size, being sometimes as much as eighty or one hundred feet in length. They are very active animals, however, and their whalebone is comparatively valueless, so that the whalers rarely meddle with them, though they are not uncommon, and are often driven ashore on our own coasts.

Fam. 2. *Catodontidæ.*—The family of the *Catodontidæ* or *Physeteridæ* comprises the Sperm Whales or Cachalots, with which we commence the series of the Toothed Whales (*Odontoceti*). They are characterised by the fact that the palate is destitute of baleen plates, and the lower jaw possesses a series (about fifty-four) of pointed conical teeth, separated by intervals, and sunk in a common alveolar groove, which is only imperfectly divided by septa. The upper jaw is also in reality furnished with teeth, but, with a single partial exception, these do not cut the gum.

Fig. 213.—Spermaceti Whale (*Physeter macrocephalus*).

The best-known species of this family is the great Cachalot or Spermaceti Whale (*Physeter macrocephalus*, fig. 213). This animal is of enormous size, averaging from fifty to seventy feet in length, but the females are a good deal smaller than the males. The head is disproportionately large, as in the *Balænidæ*, forming nearly one-third of the entire length of the body. The snout forms a broad truncated muzzle, and the nostrils are placed near the front margin of this. The Sperm Whales live together in troops or "schools," and they are found in various seas, especially in the North Pacific. They are largely sought after, chiefly for the substance known as "spermaceti;" but besides this they yield oil and the singular body called "ambergris." The spermaceti is a fatty substance, which has the power of concreting when exposed to the air, being in life a clear white oily liquid. It is not only diffused through the entire blubber, but is also contained in special cavities of the head. The sperm-oil yielded by the blubber is exceedingly pure, and is free from the unpleasant odour of ordinary whale-oil. The ambergris is a peculiar substance which is found in masses in the intestine, and is probably of the nature of a biliary calculus, since it is said to be composed of a substance very nearly allied to cholesterine. It is used both as a perfume itself, and to mix with other perfumes.

Fam. 3. *Delphinidæ.*—This family includes the Dolphins, Por poises, and Narwhal, and is characterised by usually possessing teeth in both jaws: the teeth being numerous, and conical in shape. The nostrils, as in the last family, are united, but they are placed further back, upon the top of the head. The single blow-hole or nostril is transverse and mostly crescentic or lunate in shape. The head is by no means so disproportionately large as in the former families, usually forming about one-seventh of the entire length of the body.

Fig. 214.—The common Dolphin (*Delphinus delphis*).

The most noticeable members of this family are the true Dolphins, the Porpoises, and the Narwhal.

The Dolphins have an elongated snout, separated from the head by a transverse depression. The common Dolphin (*Delphinus delphis*, fig. 214) is the best-known species. It averages from six to eight feet in length, and has the habit of swimming in flocks, often accompanying ships for many miles. The female, like most of the *Cetacea*, is uniparous. The Dolphin occurs commonly in all European seas, and is especially abundant in the Mediterranean.

The common Porpoise (*Phocæna communis*) is the commonest and smallest of all the *Cetacea*, rarely exceeding four feet in length. The head is blunt, and is not produced into a projecting muzzle. The Porpoise frequents the North Sea, and is commonly seen off our coasts. Another British species is the Grampus (*Phocæna orca*), but this is much larger, attaining a length of from eighteen to twenty feet. Nearly allied to the Grampus is the so-called "Caing" Whale, or, as it is sometimes termed, the "Bottle-nosed" Whale (*Globicephalus* or *Phocæna globiceps*). This species occurs not uncommonly round the Orkney and Shetland Islands, and attains a length of as much as twenty-four feet. It is gregarious in its habits, and is often killed for the sake of its oil.

Closely allied to the true Dolphins are two curious Cetaceans, belonging to different genera, but both inhabiting fresh waters. One of these is the Gangetic Dolphin (*Platanista Gangetica*), which inhabits the Ganges, especially near its mouth. This singular animal is characterised by the great length of its slender muzzle, and by the small size of the eyes. It attains the length of seven feet, and the blow-hole is a longitudinal fissure, and therefore quite unlike that of the typical *Del-*

phinidæ. The other fresh-water form is the *Inia Boliviensis*, which inhabits the rivers of Bolivia, and is found at a distance of more than two thousand miles from the sea. In its essential characters it differs little from its marine brethren, and it attains a length of from seven (female) to fourteen feet (male).

The last of the *Delphinidæ* is the extraordinary Narwhal or Sea-unicorn (*Monodon monoceros*). The Narwhal is an inhabitant of the Arctic seas, and attains a length of as much as fifteen feet, counting in the body alone. The dentition, however, is what constitutes the great peculiarity of the Narwhal. The lower jaw is altogether destitute of teeth, and the upper jaw in the females also exhibits no teeth externally, as a general rule at any rate, though there are two rudimentary incisors which do not cut the gum. In the males, the lower jaw is likewise edentulous, but the upper jaw is furnished with two molar teeth concealed in the gum, and with two incisors. Of these two upper incisors, that of the right side is generally rudimentary, and is concealed from view. The left upper incisor, on the other hand, is developed from a permanent pulp, and grows to an enormous size, continuing to increase in length throughout the life of the animal. It forms a tusk of from eight to ten feet in length, and it has its entire surface spirally twisted. As an abnormality, both the upper incisors may be developed in this way so as to form projecting tusks; and it is stated that the tusk is occasionally present in the female. The function of this extraordinary tooth is doubtless offensive.

CHAPTER LXXVII.

UNGULATA.

ORDER VI. UNGULATA.—The order of the *Ungulata*, or Hoofed Quadrupeds, is one of the largest and most important of all the divisions of the *Mammalia*. It comprises three entire old orders—namely, the *Pachydermata*, *Solidungula*, and *Ruminantia*.

The first of these old divisions—that of the *Pachydermata*—included the Elephants, Rhinoceros, Hippopotamus, Tapirs, and the Pigs, all characterised, as the name implies, by their thick integuments. The name is still used to express this fact, though the order is now abandoned, and is merged with that of the *Ungulata;* the Elephants alone being removed to a separate order under the name of *Proboscidea*.

The second old order—that of the *Solidungula* or Solipedes —included the Horse, Zebra, and Ass, all characterised by the fact that the foot terminates in a single toe, encased in an expanded hoof. The name *Solidungula* is still retained for these animals, as a section of the *Ungulata*.

The third old order—that of the *Ruminantia*—includes all those animals, such as Oxen, Sheep, Goats, Camels, Giraffes, Deer, and others, which chew the cud or "ruminate," and have two functional toes to each foot, encased in hoofs. The name *Ruminantia* is still retained for these animals, as constituting a most natural group of the *Ungulata*.

All these various animals, then, are now grouped together into the single order of the *Ungulata*, or Hoofed Quadrupeds, and the following are the characters of the order:—

All the four limbs are present, and that portion of the toe which touches the ground is always encased in a greatly-ex-

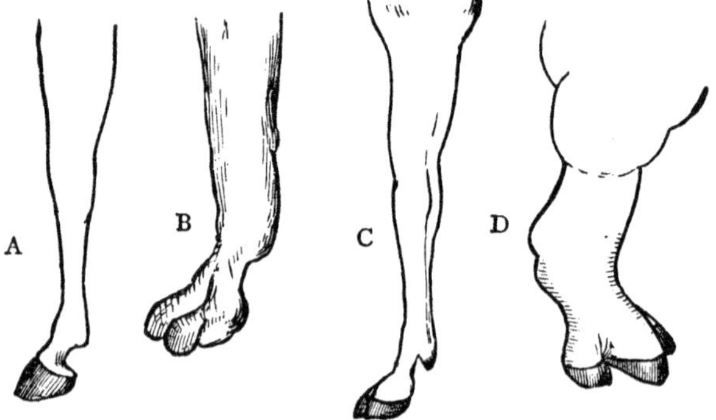

Fig. 215.—Ungulata. A, Perissodactyle foot of Zebra (*Solidungula*); B, Artiodactyle foot of Llama; C, Artiodactyle foot of Antelope; D, Perissodactyle foot of Rhinoceros.

panded nail, constituting a "hoof." There are never more than *four* full-sized toes to each limb. Owing to the encasement of the toes in hoofs, the limbs are useless for prehension, and only subserve locomotion; hence *clavicles* are always wanting in the entire order. There are always two sets of enamelled teeth, so that the animal is diphyodont. The molar teeth are massive and have broad crowns, adapted for grinding vegetable substances.

The order *Ungulata* is divided into two primary sections:— the *Perrissodactyla*, in which the toes or hoofs are odd in number (one or three), and the *Artiodactyla*, in which the toes are even in number (two or four).

SECTION A. PERISSODACTYLA.—The section of the *Perissodactyle* Ungulates includes the Rhinoceros, the Tapirs, the Horse and its allies, and some extinct forms, all agreeing in the following characters:—

The hind-feet are odd-toed in all (fig. 215, A, D), and the fore-feet in all except the Tapirs. The dorso-lumbar vertebræ are never less than twenty-two in number. The femur has a third trochanter. The horns, if present, are not paired. Usually there is only one horn, but if there are two, these are placed in the middle line of the head, one behind the other (fig. 216). In neither case are the horns ever supported by bony horn-cores. The stomach is simple, and is not divided into several compartments; and there is a large and capacious cæcum.

The three existing genera of Perissodactyle Ungulates—namely, the Horse, Tapir, and Rhinoceros—are widely removed from one another in many important characters; but the intervals between them are filled up by an extensive series of fossil forms, commencing in the Lower Tertiary Strata.

Fam 1. *Rhinocerid æ.*—This family comprises only a single genus, the genus *Rhinoceros*, unless, indeed, the little *Hyrax* is to be retained in this order. The Rhinoceroses are extremely large and bulky brutes, having a very thick skin, which is usually thrown into deep folds. The muzzle is rounded and blunt, and there are $\frac{7-7}{7-7}$ molars, with tuberculate crowns. There are no canines, but there are usually incisor teeth in both jaws. The skull is pyramidal, and the nasal bones are enormously developed. The feet are furnished with three toes each, encased in hoofs. The nasal bones support one or two horns, which are not paired. The horn is composed of longitudinal fibres, which are agglutinated together, and are of the nature of epidermic growths, somewhat analogous to hairs. When two horns are present, the hinder one is carried by the frontal bones, and is placed in the middle line of the head behind the anterior horn. The posterior horn is usually much shorter than the anterior one; and if not, it differs in shape. The Rhinoceroses live in marshy places, and subsist chiefly on the foliage of trees. They are exclusively confined at the present day to the warmer parts of the Old World; but an extinct species (*Rhinoceros tichorhinus*) formerly inhabited England, and ranged over the greater part of Europe. Of the one-horned species, the best known is the Indian Rhinoceros (*R.*

Indicus), which was probably the "Unicorn" of the ancients. Another species with one horn (*R. Sondaicus*) inhabits Java. Of the two-horned species, one (*R. Sumatrensis*) is found in Sumatra, and is remarkable for the comparative absence of cutaneous folds. The best known, however, is the African Rhinoceros (*R. bicornis*), which occurs abundantly in Cape Colony and in the southern parts of the African continent (fig. 216.)

Fig. 216.—Head of two-horned Rhinoceros (*R. bicornis*).

Fam. 2. *Tapiridæ.*—The Tapirs are characterised by the possession of a short movable proboscis or trunk. The skull is pyramidal, like that of the pigs, and the nasal bones project over the nasal cavity. The skin is hairy and very thick. The tail is extremely short. The fore-feet have *four* toes each, but these are unsymmetrical (the little toe being smaller than the rest and not touching the ground), and the hind-feet have only three toes, all encased in hoofs. The jaws are furnished with incisor teeth, $(\frac{3-3}{3-3})$, small canines, and $\frac{7-7}{6-6}$ molars.

Three species of Tapir are known, of which the most familiar is the American Tapir (*T. Americanus*), which inhabits the vast forests of South America. It is a large animal, something like a pig in shape, but brownish black in colour, and having a mane. It is nocturnal in its habits, and is strictly phytophagous. The proboscis is employed in conveying the food to the mouth, and the nostrils are placed at its extremity. It attains altogether a total length of from five to six feet. Another species, with longer hair (*T. villosus*), inhabits the Andes, and a still larger species (*T. Malayanus* is found in

Sumatra and Malacca. In this last, there is no mane, and the general colour is black; but the back, rump, and sides of the belly are white.

Nearly allied to the Tapirs is the fossil genus *Palæotherium*, found in the Eocene Rocks of France and other countries. Many species of the genus are known, all seeming to have possessed a short proboscis like that of the Tapirs. All the feet, however, were tridactylous.

Fam. 3. *Solidungula* or *Equidæ.*—This family comprises the Horses, Asses, and Zebras, characterised by the fact that the feet have only a single perfect toe each, enclosed in a single broad hoof, without supplementary hoofs (fig. 216, A). There is a discontinuous series of teeth in each jaw; and in the males, canines are present, but these are wanting in the females. The dental formula is—

$$i\,\frac{3-3}{3-3};\ c\,\frac{1-1}{1-1};\ pm\,\frac{3-3}{3-3};\ m\,\frac{3-3}{3-3}=40.$$

The skin is covered with hair, and the neck is furnished with a mane.

The family *Equidæ* is divided by Dr Gray into two sections or genera: *Equus*, comprising the Horse; and *Asinus*, comprising the Asses and Zebras.

The genus *Equus* is distinguished by the fact that the animal is not banded, and has no dorsal line; both the fore and hind legs have warts, and the tail is hairy throughout. The genus appears to contain no more than one well-marked species, as far as living forms are concerned—namely, the *Equus caballus.* From this single species appear to have descended all the innumerable varieties of horses which are employed by man. The native country of the horse appears to have been Central Asia, but all the known wild individuals at the present day appear to be descendants of domestic breeds.

The genus *Asinus* is characterised by the fact that there is always a distinct dorsal line, and the body is more or less banded; the fore-legs alone have warts, and the tail has a tuft of long hairs at its extremity. The Ass is probably a native of Asia (where the wild Ass is at present a native), and there appears to be little doubt but that the common Ass is merely the domesticated form of the wild Ass (*Equus onager*). Another well-known species is the mule-like "Djiggetai" (*Asinus hemionus*) of Central Asia. The striped members of this section are known as Zebras and Quaggas, and are natives of the southern parts of Africa.

SECTION B. ARTIODACTYLA.—In this section of the Ungulates the number of the toes is even—either two or four—and the third toe in each foot forms a symmetrical pair with the fourth (Fig. 215, B, C). The dorso-lumbar vertebræ are nineteen in number, and there is no third trochanter on the femur. If true horns are present, these are always in pairs, and are supported by a bony horn-core. The antlers of the Deer are also paired, but they are not to be regarded as true horns. The stomach is always more or less complex, or is divided into separate compartments, and the cæcum is comparatively small and simple.

The section *Artiodactyla* comprises the Hippopotamus, the Pigs, and the whole group of the Ruminants, including Oxen, Sheep, Goats, Antelopes, Camels, Llamas, Giraffes, Deer, &c. Besides these there is an extensive series of fossil forms commencing in the Eocene or Lower Tertiary period, and in many respects filling up the gaps between the living forms.

OMNIVORA.

1. *Hippopotamidæ.*—This group contains only the single genus *Hippopotamus*, characterised by the massive heavy body, the short blunt muzzle, the large head, and the presence of teeth of three kinds in both jaws. The incisors are $\frac{2-2}{2-2}$, the canines extremely large, $\frac{1-1}{1-1}$, and the molars, $\frac{7-7}{7-7}$ or $\frac{6-6}{6-6}$, with crowns adapted for grinding vegetable substances. The upper canines are short, but the lower canines are in the form of enormous tusks, with a chisel-shaped edge. The feet are massive, and are terminated by four hoofed toes each. The eyes and ears are small, and the skin is extremely thick, and is furnished with few hairs. The tail is very short.

Several extinct species of *Hippopotamus* are known, but there is only one well-established living form, the *Hippopotamus amphibius* or River-horse, and this is confined to the African continent. It is an enormously bulky and unwieldy animal, reaching a length of eleven or twelve feet. It is nocturnal in its habits, living upon grass and small shrubs, and it swims and dives with great facility. It is found in tolerable abundance in the rivers of Abyssinia, and occurs plentifully in South Africa. Another supposed species (*H. Liberiensis*) occurs on the west coast of Africa, but there is some doubt as to the specific distinctness of this.

2. *Suida.*—The group of the *Suida*, comprising the Pigs, Hogs, and Peccaries, is very closely allied to the preceding;

but the feet have only two functional toes, the other two toes being much shorter, and hardly touching the ground. All the three kinds of teeth are present, but they vary a good deal. The canines always are very large, and in the males they usually constitute formidable tusks projecting from the sides of the mouth. The incisors are variable, but the lower ones are always inclined forwards. The molars vary from three to seven on each side of the mouth ($\frac{3-3}{3-3}$ or $\frac{7-7}{7-7}$). The stomach is mostly slightly divided, and is not nearly so complex as in the Ruminants. The snout is truncated and cylindrical, fitted for turning up the ground, and is capable of considerable movement. The skin is more or less abundantly covered with hair, and the tail is very short, or represented only by a tubercle.

Of the true Swine, the best known and most important is the Wild Boar (*Sus scrofa*), from which it is probable that all our domestic varieties of swine have sprung. The Wild Boar formerly inhabited this country, and is still abundant in many of the forests of Europe. It is often hunted, and the size and sharpness of its canines render it a tolerably formidable adversary, as is also its congener, the Indian Hog (*Sus Indicus*). Another curious form, closely related to the Wild Boar, is the Babyroussa (*Sus Babyrussa*), which inhabits the Malayan Peninsula, and some of the islands of the Indian Archipelago. It is remarkable for the great size and backward curvature of the upper canines. The upper canines pierce the upper lip in the males, and their alveoli are directed upwards. The legs are very long and slender; hence the name "Hog-deer" sometimes applied to it.

The African Wart-hogs, forming the genus *Phacochœrus*, are distinguished by having a fleshy wart under each eye. They inhabit Abyssinia, the Guinea coast, and other parts of Africa.

The American Peccaries (*Dicotyles*) represent the Swine of the Old World. They are singular for having only three toes on the hind-feet, the outer of the two supplemental hoofs being wanting. The canines are not exserted, there is no tail, and there is a glandular pouch on the loins secreting a fetid fluid. They are exclusively confined to America, and the commonest species is the Collared Peccary (*Dicotyles torquatus*). They are not at all unlike small pigs either in their appearance or in their habits, and they are gregarious, generally occurring in small flocks.

Forming a kind of transition between the Swine and the true Ruminants, is the extinct group of the *Anoplotheridæ*, from the

Lower Tertiary Rocks. The *Anoplotheria* were slender in form, with long tails, and feet terminated by two hoofed toes each, sometimes with small accessory hoofs. The dentition consisted of six incisors in each jaw, small canines not larger than the incisors, and seven molars on each side, there being no interval or diastema between the molars and the canines.

RUMINANTIA.

The last section of the *Artiodactyle* Ungulates is the great and natural group of the *Ruminantia*, or Ruminant animals. This section comprises the Oxen, Sheep, Antelopes, Giraffes, Deer, Camels, &c., and is distinguished by the following characters :—

The foot is what is called "cloven," consisting of a symmetrical pair of toes encased in hoofs, and looking as if produced by the splitting into two equal parts of a single hoof. In addition to these functional toes, there are sometimes two smaller supplementary hoofs, placed at the back of the foot. The metacarpal bones of the two functional toes of the fore-limb, and the metatarsal bones of the same toes of the hind-limb, coalesce to form a single bone, known as the "canon-bone." The stomach is complex, and is divided into several compartments, this being in accordance with their mode of eating. They all, namely, ruminate or "chew the cud"—that is to say, they first swallow their food in an unmasticated or partially-masticated condition, and then bring it up again, after a longer or shorter time, in order to chew it thoroughly.

This process of rumination is so characteristic of this group, that it will be necessary to describe the structure of the stomach, as showing the mechanism by which this singular process is effected. The stomach (fig. 217) is divided into four compartments, which are usually so distinct from one another that they have generally been spoken of as so many separate stomachs. The gullet opens at a point situated between the first and second of these cavities or "stomachs." Of these the largest lies on the left side, and is called the "rumen" or "paunch" (fig. 217, *r*). This is a cavity of very large capacity, having its interior furnished with numerous hard papillæ or warts. It is the chamber into which the food is first received when it is swallowed, and here it is moistened and allowed to soak for some time. The second stomach, placed to the right of the paunch, is much smaller, and is known as the "reticulum" or "honeycomb-bag" (*h*). Its inner surface is reticulated, or is divided by ridges into a number of hexagonal or many-sided cells, somewhat resembling the cells of a honeycomb. The reticulum is small and globular, and it receives the food after it has lain a sufficient time in the paunch. The function of the reticulum is to compress the partially-masticated food into little balls or pellets, which are then returned to the mouth by a reversed action of the muscles of the œsophagus. After having been thoroughly chewed and prepared for digestion, the food is swallowed for the second time. On this occasion, however, the triturated food passes on into the third cavity (*p*), which is variously known as the "psalterium," "omasum," or (*Scotticé*) the "manyplies." The vernacular and the first of these technical names both refer to the fact that the inner lining of this cavity is thrown into a number of longitudinal folds, which are so close as to resemble the leaves of a book. The psalterium opens by a wide aperture into the fourth and last cavity, the "abomasum" (*a*), both appearing to be divisions of the pyloric

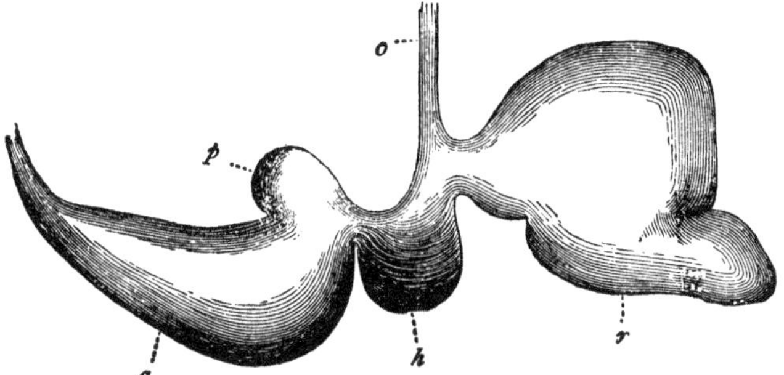

Fig. 217.—Stomach of a Sheep. *o* Gullet; *r Rumen* or Paunch; *h* Honeycomb-bag or *Reticulum*; *p* Manyplies or *Psalterium*; *a* Fourth Stomach or *Abomasum*.

portion of the stomach. The mucous membrane of the abomasum is thrown into a few longitudinal folds, and it secretes the true acid gastric juice. It terminates, of course, in the commencement of the small intestine—*i.e.*, the duodenum. The intestinal canal of Ruminants, as in most animals which live exclusively upon a vegetable diet, is of great relative length.

The dentition of the Ruminants presents peculiarities almost as great and as distinctive as those to be derived from the digestive system. In the typical Ruminants (*e.g.*, Oxen, Sheep, Antelopes) there are no incisor teeth in the upper jaw, their place being taken by a callous pad of hardened gum, against which the lower incisors impinge (fig. 218). There are also no upper canine teeth, and the only teeth in the upper jaw are six molars on each side. In the front of the lower jaw is a continuous and uninterrupted series of eight teeth, of which the central six are incisors, and the two outer ones are regarded by Owen as being canines. Upon this view, canine teeth are present in the lower jaw of the typical Ruminants, and they are only remarkable for being placed in the same series as the incisors, which they altogether resemble in shape, size, and direction. Behind this continuous series of eight

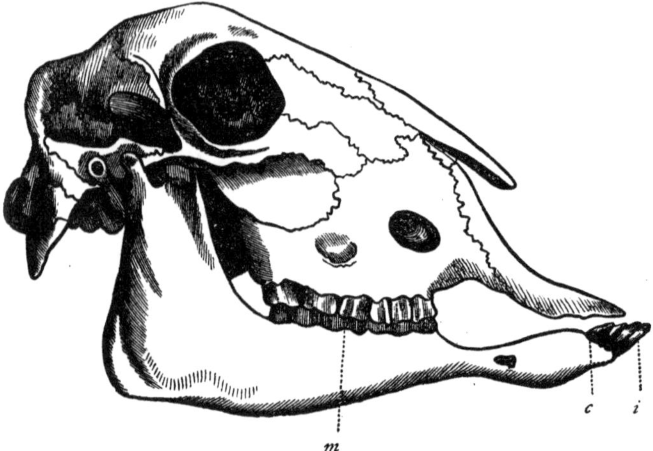

Fig. 218.—Skull of a hornless Sheep (after Owen). *i* Incisors; *c* Canines; *m* Molars and Præmolars.

teeth in the lower jaw, there is a vacant space, which is followed behind by six molars on each side. The præmolars and molars have their grinding-surfaces marked with two double crescents, the convexities of which are turned inwards in the upper, and outwards in the lower teeth.

The dental formula, then, for a typical Ruminant animal, is—

$$i\,\frac{0-0}{3-3};\ c\,\frac{0-0}{1-1};\ pm\,\frac{3-3}{3-3};\ m\,\frac{3-3}{3-3} = 32.$$

The departures from this typical formula occur in the *Camelidæ* and in some of the Deer. Most of the Deer conform in their dentition to the above formula, but a few forms (*e.g.*, the

Musk-deer) have canine teeth in the upper jaw. These upper canines, however, are mostly confined to the males; and if they occur in the females, they are of a small size. The dentition of the *Camelidæ* (Camels and Llamas) is still more aberrant, there being two canine-like upper incisors and upper canines as well. The lower canines also are more pointed and stand more erect than the lower incisors, so that they are easily recognisable. The group of the *Ruminantia* includes the families of the *Camelidæ* (Camels and Llamas), the *Moschidæ* (Musk-deer), the *Cervidæ* (Deer), the *Camelopardalidæ* (Giraffe), and the *Cavicornia* (Oxen, Sheep, Goats, Antelopes).

a. Camelidæ.—The Camels and Llamas constitute in many respects an aberrant group of the *Ruminantia*, especially in their dentition, the peculiarities of which have been spoken of above, and need not be repeated here. In their feet, too, the *Camelidæ* are peculiar. The feet are long and terminate in only two toes, which are covered by an imperfect nail-like hoof, covering no more than the upper surface of each toe. The two hinder supplementary toes, which are mostly present in the Ruminants, are here altogether wanting; and the soles of the feet are covered by a callous horny integument, by which the two toes of each foot are conjoined, and upon which the animal walks. The head in all the *Camelidæ* is destitute of horns, and the nostrils can be closed at the will of the animal.

The true Camels are peculiar to Asia and Africa, and two species are known, distinguished from one another by the possession of a double or single adipose hump on the back. The African or Arabian Camel (*Camelus Dromedarius*) is often called the Dromedary, and has only one hump on its back. The two toes are united together by the callous sole; and the chest, shoulders, and knees are furnished with callous pads, upon which they rest when they lie down. The hump is almost entirely composed of fat, and appears to act as a kind of reserve supply of food, as it is noticed to diminish much in size upon long journeys. The Camel can likewise support a very prolonged privation of water, as the paunch is furnished with large cells, which the animal fills when it has access to water, and then makes use of subsequently as occasion may require. The structure of the Camel adapts it admirably for locomotion in the sandy deserts of Arabia and Africa; and as it is very docile and good-tempered, it is almost exclusively employed as a beast of burden in the countries in which it occurs.

The Bactrian Camel (*C. Bactrianus*) is distinguished by the possession of two humps; but in other respects it does not differ from the Dromedary. The two species are said to breed together, and the hybrid offspring is stated to be occasionally fertile. The place of the Camels is taken in the New World by the Llama and Alpaca, with two other nearly-allied forms. These animals form the genus *Auchenia*, and are in many respects similar to the true Camels. They are distinguished, however, by having no hump upon the back, and by the fact that the two toes are not conjoined and supported by a callous pad, as in the Camels, but are separate, with separate pads, and with strong curved nails. The neck is long and the head comparatively small, whilst the upper lip is mobile and deeply cleft vertically. The Llamas are chiefly found in Peru and Chili, and considerable doubt exists as to the number of species. They live in flocks in mountainous regions, and are much smaller than the Camels in size. The true Llama is kept as a domesticated animal, and used as a beast of burden. The Alpaca is still smaller than the Llama, and is not very unlike a sheep, having a long woolly coat. It is partially domesticated, and the wool is largely imported into Europe.

b. Moschidæ.—The second group is that of the Musk-deer, characterised by the total absence of horns in both sexes, and by the presence of canines in both jaws, those in the upper jaw being in the form of tusks in the males, but being much smaller in the females.

The true Musk-deer (*Moschus moschiferus*) is an elegant little animal, which inhabits the elevated plains of central Asia. It is remarkable for the fact that the male has a glandular sac on the abdomen, by which the well-known perfume, musk, is secreted. The musk-gland is wanting in the Napu (*Moschus Javanicus*) of Java, and also in the little Kanchil (*Tragulus pygmæus*), which is the smallest of living Ruminants.

c. Cervidæ.—This family is of much greater importance than that of the *Moschidæ*, including as it does all the true Deer. They are distinguished from the other Ruminants chiefly by the nature of the horns. With the single exception of the Reindeer, these appendages are confined to the males amongst the *Cervidæ*, and do not occur in the females. They do not consist, as in the succeeding group, of a hollow sheath of horn surrounding a central bony core, nor are they permanently retained by the animal. On the other hand, the horns—or, as they are more properly called, the *antlers*—of the *Cervidæ* are deciduous, and are solid. They are bony throughout, and are usually more or less branched (fig. 219), and they are annually shed and annually reproduced at the breeding season. They increase in size and in the number of branches every time they are reproduced, until in the old males they may attain an enormous size. The first time they are produced, the horns are in the form of simple cylindrical shafts; the second year's horns have one or two "tynes," and so on. The antlers are carried upon the frontal bone, and are produced by a process not at all unlike that by which injuries of osseous structures are made good in man. At first the antlers are covered with a

Fig. 219.—Head of the Red-deer (*Cervus elaphus*).

sensitive hairy skin; but as development proceeds, the vessels of the skin are gradually obliterated, and the skin dies and peels off. In all the Deer there is a sebaceous gland, called the "lachrymal sinus," or "larmier," which is placed beneath each eye, and secretes a strongly-smelling waxy substance.

The *Cervidæ* are very generally distributed, but no member of the group has hitherto been discovered in either Australia or South Africa, their place in the latter continent seeming to be taken by the nearly-allied Antelopes (distinguished by their hollow horns).

Very many species of *Cervidæ* are known, and it is not possible to allude to more than two or three of the more familiar and important forms. Three species occur in Britain—namely, the Roebuck, Red-deer, and Fallow-deer, the last being a doubtful native. The Roebuck (*Capreolus capræa*) was once very generally distributed over Britain, but is almost confined to the wilder parts of Scotland at the present day. It is of small size, and the horns are without brow-tynes, and are of small size, with three terminal branches. The Red-deer, or Stag (*Cervus elephus*) is a much larger species, with well-developed spreading antlers. The Red-deer of Britain is represented in North America by a still larger species, known as the Wapiti (*Cervus Canadensis*).

The third British species is the Fallow-deer (*Dama platyceros*), characterised by the fact that the antlers are palmated—that is, dilated towards their extremities. It is a doubtful native, and is never found in a wild state at the present day. Allied to the Fallow-deer is a gigantic extinct species, the *Megaceros Hibernicus*, which inhabited Ireland, the Isle of Man, Scotland, and probably the greater part of Europe, up to a comparatively modern date, probably having survived into the human period. It is often, but incorrectly, spoken of as the Irish "Elk," but it is really a genuine Stag. The animal was of very great size, and was furnished with enormous spreading and palmate antlers, which measure from ten to twelve feet between the tips.

Of all the Deer, the largest living form is the true Elk (*Alces palmatus*), which is generally distributed over the northern parts of Europe, Asia, and America, being often spoken of as the Moose. The antlers in the Elk are of a very large size, and are very broad, terminating in a series of points along their outer edges.

The only completely domesticated member of the *Cervidæ* is the Reindeer (*Cervus tarandus*), which is remarkable for the fact that the female is furnished with antlers similar to, but smaller than, those of the males. At the present day the Reindeer is exclusively confined to the extreme north of Europe and Asia, abounding especially in Lapland. Remains, however, of the Reindeer are known to occur over the greater part of Europe, extending as far south, at any rate, as the Alps, and occurring also in Britain. From this fact, taken along with many others, the existence of an extremely cold climate over the greater part of Europe at a comparatively recent period may be safely inferred. The Reindeer lives chiefly upon moss and a peculiar kind of lichen (*Lichen rangiferina*), and they are extensively used by the Laplanders both as beasts of burden and as supplying food.

The so-called "Brockets," such as the Guazu-pita (*Subulo rufus*) of South America, have simple horns in the form of a stiletto. Lastly, the singular Muntjak of Java has the horns supported on long bony pedicles springing from the frontal bone; and the males have large upper canines.

d. Camelopardalidæ.—This family includes only a single living animal—the *Camelopardalis Giraffa*, or Giraffe—sometimes called the Camelopard, from the fact that the skin is spotted like that of the Leopard, whilst the neck is long, and gives it some distant resemblance to a Camel. There are no upper canines in the Giraffe, and both sexes possess two small frontal horns, which, however, are persistent, and remain permanently covered by a hairy skin, terminated by a tuft of long stiff bristles. The neck is of extraordinary length, but, nevertheless, consists of no more than the normal seven cervical vertebræ. The fore-legs appear to be much longer than the hind-legs, and all are terminated by two toes each, the supplementary toes being altogether wanting. The tongue is very long and movable, and is employed in stripping leaves off the trees. The Giraffe is the largest of all the Ruminants, measuring as much as from fifteen to eighteen feet in height. It is a harmless and inoffensive animal, but defends itself very effectually, if attacked, by kicking. It is found in Nubia, Abyssinia, and the Cape of Good Hope.

Remains of gigantic Ruminants allied to the Giraffe have been found in France and Greece (*Helladotherium*); but the *Sivatherium*, sometimes referred to this family, appears to have been more nearly allied to the true Antelopes.

e. Cavicornia.—The last family of the Ruminants is that of the *Cavicornia* or *Bovidæ*, comprising the Oxen, Sheep, Goats, and Antelopes. This family includes the most typical Ruminants, and those of most importance to man. The upper jaw in all the *Cavicornia* is wholly destitute of incisors and canines, the place of which is taken by the hardened gum, against which the lower incisors bite. There are six incisors and two canines in the lower jaw, placed in a continuous series, and the molars are separated by a wide gap from the canines. There are six molars on each side of each jaw. Both sexes have horns, or the males only may be horned, but in either case these appendages are very different to the "antlers" of the *Cervidæ*. The horns, namely, are persistent, instead of being deciduous, and each consists of a bony process of the frontal bone—or "horn-core"—covered by a sheath of horn. The feet are cleft, but are furnished with accessory hoofs placed on the back of the foot.

The *Cavicornia* comprise the three families of the *Antilopidæ*, *Ovidæ*, and *Bovidæ*. The Antelopes form an extremely large section, with very many species. They are characterised by their slender deer-like form, their long and slender legs, and their simple cylindrical annulated or twisted horns, which are sometimes confined to the males, but often occur in the females as well (fig. 220). The Antelopes must on no account be confounded with the true Deer, to which they present many points of similarity. The structure of the horns, however, is quite sufficient to distinguish them. The Antelopes are further distinguished by rarely having a beard or dew-lap, and by the general possession of "inguinal pores" and "lachrymal sinuses."

Fig. 220.—Head of the Koodoo (*Strepsiceros Koodoo*).

The inguinal pores are the apertures of two involutions of the integument of the groin, secreting a viscous substance, the use of which is unknown. The lachrymal sinuses, or "tearpits," have already been mentioned as occurring in the *Cervidæ*, and are not found in any of the *Cavicornia* except the Antelopes. Each consists of a sebaceous sac placed beneath the eye, and secreting a yellowish waxy substance. The function of these glands is uncertain, but it is probably sexual. The Antelopes are especially numerous, both in individuals and in species, in Africa, in which country they appear to take the place of the true Deer (only one species of Deer being indigenous to Africa). Amongst the better-known African species of Antelopes are the Springbok, Hartebeest, Gnu, Eland, and Gazelle. The only European Antelope is the Chamois (*Rupicapra tra-*

grus), which inhabits the Alps and other mountain-ranges of southern Europe. Amongst the more remarkable Antelopes may be mentioned the Prong-buck (*Antilope furcifer*) of N. America, in which there are no accessory hoofs, lachrymal sinuses, or inguinal pores; the females are hornless, and the horns of the male have a snag or branch in front. Another curious form is the Chickāra (*A. quadricornis*) of India, in which the females are hornless, but the males have four horns.

The Sheep and Goats (*Ovidæ*) have mostly horns in both sexes, and the horns are generally curved, compressed, and turned more or less backwards. The body is heavier, and the legs shorter and stouter, than in the true Antelopes. In the true Goats (*Capra*) both sexes have horns, and there are no lachrymal sinuses. The throat is furnished with long hair, forming a beard; and this appendage is usually present in both sexes, though sometimes in the males only. The goats live in herds, usually in mountainous and rugged districts. The domestic Goat (*Capra hircus*) is generally believed to be a descendant of a species which occurs in a wild state in Persia and in the Caucasus (the "Paseng," or *Capra ægagrus*). The true sheep (*Ovis*) are destitute of a beard, and the horns are generally twisted into a spiral. Horns may be present in both sexes, or in the males only.* Lachrymal sinuses are invariably absent. Numerous varieties of the domestic Sheep (*Ovis aries*) are known, but it is not certainly known from what wild species these were originally derived. The Merino Sheep (a Spanish breed) and the Thibet Sheep are particularly celebrated for their long and fine wool. With the exception of one species (the Big-horn, *Ovis montana*), all the Sheep appear to be originally natives of the Old World.

The true Oxen (*Bovidæ*) are distinguished by having simply rounded horns, which are not twisted in a spiral manner. There are no lachrymal sinuses. Most of the Oxen admit of being more or less completely domesticated, and some of them are amongst the most useful of animals, both as beasts of burden and as supplying food. The parent-stock of our numerous breeds of cattle is not known with absolute certainty; the nearest approach to British Wild Cattle being a celebrated breed which is still preserved in one or two places. These "Chillingham Cattle" are a fine wild breed, which at one time doubtless existed over a considerable part of Britain. They are pure white, with a black muzzle, the horns white, tipped

* In the Merino Sheep, and in some other breeds also, the males only are horned.

with black. Another large Ox, which formerly existed in Britain, and abounded over the whole of Europe, is the Aurochs or Lithuanian Bison (*Bos bison*). The Aurochs is of very large size, considerably exceeding the common Ox in bulk. It still occurs in the forests of the Caucasus in a wild state, but it no longer occurs wild in Europe, if we except a herd maintained by the Czar in one of the forests of Lithuania. Nearly allied to the Aurochs is the American Bison or Buffalo (*Bison Americanus*). This species formerly occurred in innumerable herds in the prairies of North America, but it has been gradually driven westwards, and has been much reduced in numbers. It has an enormous head, a shaggy mane, and a conical hump between the shoulders. Two other very well known forms are the Cape Buffalo (*Bubalus Caffer*) and the common Buffalo (*Bubalus bubalis*). The former of these occurs in southern and eastern Africa, and the latter is domesticated in India and in many parts of the south of Asia. The horns in both species are of large size, and their bases are confluent, so that the forehead is protected by a bony plate of considerable thickness.

Amongst the more remarkable Asiatic Oxen may be mentioned the Zebu (*Bos taurus*, var. *Indicus*), distinguished by the fatty hump over the withers at the back of the neck, and the Yak (*Bos grunniens*) of Thibet, remarkable for its long silky tail.

The last of the Oxen which deserves notice is the curious Musk-ox (*Ovibos moschatus*). This singular animal is at the présent day a native of Arctic America, and is remarkable for the great length of the hair. It is called the Musk-ox, because it gives out a musky odour. Like the Reindeer, the Musk-ox had formerly a much wider geographical range than it has at present; the conditions of climate which are necessary for its existence having at that time extended over a very much larger area than at present. The Musk-ox, in fact, in Post-tertiary times is known to have extended over the greater part of Europe, remains of it occurring abundantly in certain of the bone-caves of France.

CHAPTER LXXVIII.

HYRACOIDEA AND PROBOSCIDEA.

ORDER VII. HYRACOIDEA.—This is a very small order which has been constituted by Huxley for the reception of two or three little animals, which make up the single genus *Hyrax*. These have been usually placed in the immediate neighbourhood of the Rhinoceros, to which they have some decided affinities, and they are still retained by Owen in the section of the Perissodactyle Ungulates.

The order is distinguished by the following characters:—There are no canine teeth, and the incisors of the upper jaw are long and curved, and grow from permanent pulps, as they do in the Rodents (such as the Beaver, Rat, &c.) The molar teeth are singularly like those of the Rhinoceros. According to Huxley, the dental formula of the aged animal is—

$$i\,\frac{2-2}{2-2};\ c\,\frac{0-0}{0-0};\ pm\,\frac{4-4}{4-4};\ m\,\frac{3-3}{3-3} = 36.$$

The fore-feet are tetradactylous, the hind-feet tridactylous, and all the toes have rounded hoof-like nails, with the exception of the inner toes of the hind-feet, which have an obliquely-curved nail. There are no clavicles. The nose and ears are short, and the tail is represented by a mere tubercle. The placenta is deciduate and zonary, whereas in the Ungulates it is non-deciduate.

Several species of *Hyrax* are known, but they resemble one another in all essential particulars. They are all gregarious little animals, living in holes of the rocks, and capable of domestication. One species is said to be arboreal in its habits. The "coney" of Scripture is believed to be the *Hyrax Syriacus*, which occurs in the rocky parts of Syria and Palestine. Another species—the *Hyrax Capensis*, or "Klipdas" —occurs commonly in South Africa, and is known by the colonists as the "badger."

ORDER VIII. PROBOSCIDEA.—The eighth order of Mammals is that of the *Proboscidea*, comprising no other living animals except the Elephants, but including also the extinct *Mastodon* and *Deinotherium*.

The order is characterised by the total absence of canine teeth; the molar teeth are few in number, large, and transversely ridged or tuberculate; incisors are always present, and grow from persistent pulps, constituting long tusks (fig. 221). In living Elephants there are two of these tusk-like incisors in the upper jaw, and the lower jaw is without incisor teeth. In the *Deinotherium* this is reversed, there being two tusk-like lower incisors and no upper incisors. In the Mastodons, the incisors are usually developed in the upper jaw, and form tusks, as in the Elephants, but sometimes there are both upper and lower incisors, and both are tusk-like. The nose is prolonged into a cylindrical trunk, movable in every direction, highly sensitive, and terminating in a finger-like prehensile lobe (fig. 221). The nostrils are placed at the extremity of the proboscis. The feet are furnished with five toes each, but these are only partially indicated externally by the divisions of the hoof. The feet are furnished with a thick pad of integument, forming the palms of the hand and the soles of the feet. There are no clavicles. The testes are abdominal throughout life. There are two teats, and these are placed upon the chest. The placenta is deciduate and zonary.

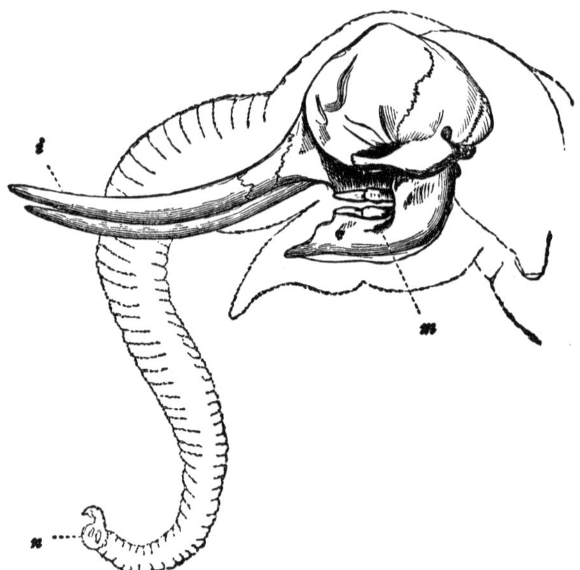

Fig. 221.—Skull of the Indian Elephant (*Elephas Indicus*). *i* Tusk-like upper incisors; *m* Lower jaw, with molars, but without incisors; *n* Nostrils, placed at the end of the proboscis. (After Owen.)

The recent Elephants are exclusively confined to the tropical regions of the Old World, in the forests of which they live in herds. Only two living species are known—the Asiatic Elephant (*Elephas Indicus*) and the African Elephant (*E. Africanus*). There can be no doubt, however, but that the Mammoth (*Elephas primigenius*) existed in Europe within the human period.

In both the living Elephants the "tusks" are formed by an

enormous development of the two upper incisors. The milk-tusks are shed early, and never attain any very great size. The permanent tusks grow throughout the life of the animal, and often reach six or seven feet in length, and from fifty to seventy pounds in weight. In the Indian Elephant, and its variety the Ceylon Elephant, the males alone have well-developed tusks, but both sexes have tusks in the African species, those of the males being the largest. The lower incisors are absent, and there are no other teeth in the jaws except the large molars, which are one or two in number on each side of each jaw. The molar teeth are of very large size, and are composed of a number of transverse plates of enamel united together by dentine. In the Indian Elephant the transverse ridges of enamel are narrow and undulating, whilst in the African Elephant they enclose lozenge-shaped intervals. The Indian Elephant is the only species which is now caught and domesticated, and as it will not breed in captivity, the demand for it is supplied entirely by the capture of adult wild individuals, which are taken chiefly by the assistance of those which have been already tamed. The Indian Elephant is distinguished by its concave forehead, its small ears, and the characters of the molars. Its skull is pyramidal, and it has five hoofs on the fore-feet, and only four on the hind-feet. Its colour is generally pale brown. (The so-called "White Elephants" are merely albinos.) The African Elephant, on the other hand, has a strongly convex forehead and great flapping ears. Its colour is darker, its skull is rounded, and it has four hoofs on the fore-feet, and only three on the hind-feet. The African Elephant is chiefly hunted for the sake of its ivory, and there is too much reason to believe that the pursuit will ultimately end in the destruction of these fine animals. A great deal, however, of the ivory of commerce comes from Siberia, and is really derived from the tusks of the now extinct Mammoth, which formerly inhabited the north of Asia in great numbers.

The Elephants are all phytophagous, living almost entirely on the foliage of shrubs and trees, which they strip off by means of the prehensile trunk. As the tusks prevent the animal from drinking in the ordinary manner, the water is sucked up by the trunk, which is then inserted into the mouth, into which it empties its contents.

Many species of fossil Elephants are known, but the most familiar of them is the Mammoth (*Elephas primigenius*). This enormous animal is now wholly extinct, but it formerly abounded in the northern parts of Asia and over the whole of Europe. It occurred also in Britain, and unquestionably existed in the earlier portion of the human period, its remains having been found in a great number of instances in connection with human implements. From its great abundance in Siberia, it might have been safely inferred that the Mammoth was able to endure a much colder climate than either of the living species. This inference, however, has been rendered a certainty by the discovery of the body of more than one Mammoth embedded in the frozen soil of Siberia. These specimens had been so perfectly preserved that even microscopical sections of some of the tissues could be made; and in one case even the eyes were preserved. From these specimens we know that the body of the Mammoth was covered with long woolly hair.

Closely allied to the true Elephants are the *Mastodons*, characterised by the fact that the crowns of the molar teeth have nipple-shaped tubercles placed in pairs. Generally speaking, the two upper incisors formed long curved tusks, as in the Elephants, but in some cases there were two lower incisors as well. The various species of *Mastodon* all belong to the later Tertiary and Post-tertiary periods.

The last of the *Proboscidea* is a remarkable extinct animal, the *Deinotherium*. This extraordinary animal has hitherto only been found in Miocene deposits, and little is known of it except its enormous skull. Molars and præmolars were present in each jaw, and the upper jaw was destitute of canines and incisors. In the lower jaw were two very large tusk-like incisors, which were not directed forwards as in the true Elephants, but were bent abruptly downwards (fig. 222). The animal must have attained an enormous size, and it is probable that the curved tusks were used either in digging up roots or in mooring the animal to the banks of rivers, for it was probably aquatic or semi-aquatic in its habits. It is placed by De Blainville in the *Sirenia*, being regarded as a Dugong with tusk-like lower incisors.

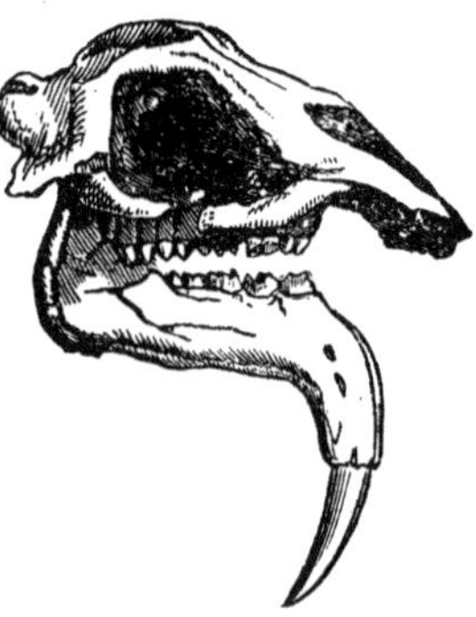

Fig. 222.—Skull of *Deinotherium giganteum.*

CHAPTER LXXIX.

CARNIVORA.

ORDER IX. CARNIVORA.—The ninth order of Mammals is that of the *Carnivora*, comprising the *Feræ*, or Beasts of Prey, along with the old order of the *Pinnipedia*, or Seals and Walruses, these latter being now universally regarded as merely a group of the *Carnivora* modified to lead an aquatic life.

The *Carnivora* are distinguished by always possessing two sets of teeth, which are simply covered by enamel, and are always of three kinds—incisors, canines, and molars—differing from one another in shape and size. The incisors are generally $\frac{3-3}{3-3}$ (except in some seals); the canines are always $\frac{1-1}{1-1}$, and are invariably much larger and longer than the incisors. The præmolars and molars are mostly furnished with cutting or trenchant edges; but they graduate from a cutting to a tuberculate form, as the diet is strictly carnivorous, or becomes more or less miscellaneous. In the typical Carnivores (such as the Lion and Tiger), the last tooth but one in the upper jaw and the last tooth in the lower jaw are known as the "carnassial" teeth, having a sharp cutting edge adapted for dividing flesh, and generally a more or less developed tuberculated heel or process. A varying number, however, of the molars and præmolars may be "tuberculate," their crowns being adapted for bruising rather than cutting. As a general rule, the shorter the jaw, and the fewer the præmolars and molars, the more carnivorous is the animal. The jaws are so articulated as to admit of vertical but not of horizontal movements; the zygomatic arches are greatly developed to give room for the powerful muscles of the jaws; and the orbits are not separated from the temporal fossæ. The intestine is comparatively short.

In all the *Carnivora* the clavicles are either altogether wanting, or are quite rudimentary. The toes are provided with sharp curved claws. The teats are abdominal; and the placenta is deciduate and zonular.

The order *Carnivora* is divided into three very natural sections:—

Section I. Pinnigrada or *Pinnipedia.* — This section comprises the Seals and Walruses, in which the fore and hind limbs are short, and are expanded into broad webbed swimming-paddles (fig. 223, B). The hind-feet are placed very far back, nearly in a line with the axis of the body, and they are more or less tied down to the tail by the integuments.

Section II. Plantigrada.—This section comprises the Bears and their allies, in which the whole, or nearly the whole, of the foot is applied to the ground, so that the animal walks upon the *soles* of the feet (fig. 223, A).

Section III. Digitigrada.—This section comprises the Lions, Tigers, Cats, Dogs, &c., in which the heel of the foot is raised entirely off the ground, and the animal walks upon the tips of the toes (fig. 223, C).

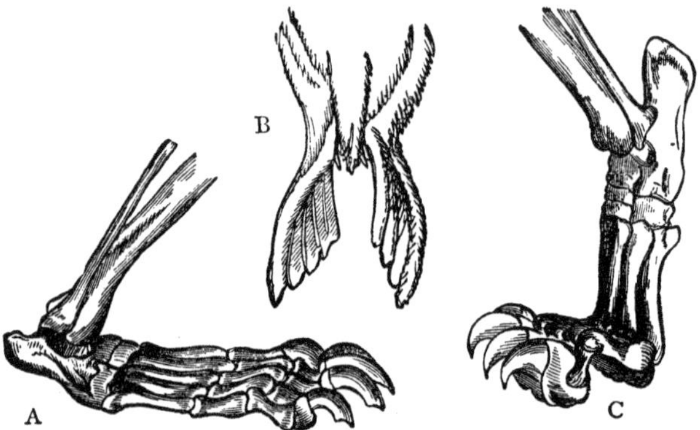

Fig. 223.—Feet of *Carnivora* (after Owen). A, *Plantigrada*, Foot of **Bear;** B, *Pinnigrada*, Hind-feet of Seal; C, *Digitigrada*, Foot of Lion.

SECTION I. PINNIGRADA OR PINNIPEDIA.—This section of the *Carnivora* comprises the amphibious Seals and Walruses, which differ from the typical Carnivores merely in points connected with their semi-aquatic mode of life. The body in these forms is elongated and somewhat fish-like in shape, covered with a short dense fur or harsh hairs, and terminated behind by a short conical tail. All the four limbs are present, but are very short, and the five toes of each foot are united together by the skin, so that the feet form powerful swimming-paddles. The hind-feet are of large size, and are placed far back, their axis nearly coinciding with that of the body (figs. 223, 224). From this circumstance, and from the fact that the integument often extends between the hind-legs and the sides of the short tail, the hinder end of the body forms an admirable swimming apparatus, similar in its action to the horizontal tail-fin of the *Cetacea* and *Sirenia*. The tips of the toes are furnished with strong claws, but their powers of terrestrial locomotion are very limited. On land, in fact, the Seals can only drag themselves along laboriously, chiefly by the contractions of the abdominal muscles. The ears are of small size, and are mostly only indicated by small apertures, which the animal has the power of closing when under water. The bones are light and spongy, and beneath the skin is a layer of fat or blubber. The dentition varies, but teeth of three kinds are always present, in the young animal at any rate. The canines are always long and pointed, and the molars are generally furnished with sharp cutting edges.

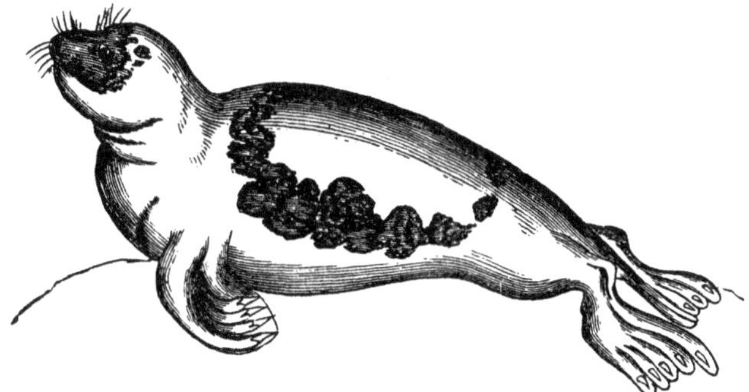

Fig. 224.—The Greenland Seal (*Phoca Grœnlandica*).

The section *Pinnigrada* includes the two families of the Seals (*Phocidæ*) and Walruses (*Trichecidæ*). The Seals are distinguished by having incisor teeth in both jaws, and by the fact that the canine teeth are not disproportionately developed. They form a very numerous family, of which species are found in almost every sea out of the limits of the tropics. They abound, however, especially in the seas of the Arctic and Antarctic regions. They live for the most part upon fish, and when awake, spend the greater part of their time in the water, only coming on land to bask and sleep in the sun and to suckle their young. They appear to be universally polygamous. The body is covered with a short fur, interspersed with long bristly hairs; and the lips are furnished with long whiskers, which act as organs of touch. The Seals are very largely captured foɪ the sake of tneir blubber.

The only common British Seal is the *Phoca vitulina*, which occurs not uncommonly on the northern shores of Scotland. It is yellowish-grey in colour, and measures from three to five feet in length. Other Seals attain a much greater length—the Great Seal measuring from eight to ten feet, and the Bottle-nosed Seal reaching a length of from twenty to twenty-five feet. The only Seals which possess external ears constitute the genus *Otaria*, and are almost exclusively confined to the seas of the southern hemisphere.

The second family of the Pinnigrade Carnivores is that of the *Trichecidæ*, comprising only the Walrus or Morse (*Trichecus rosmarus*). The chief peculiarity by which the Walrus is

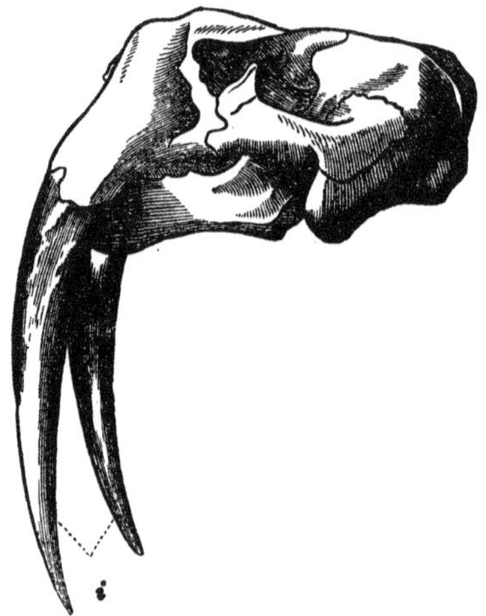

Fig. 225.—Skull of the Walrus (*Trichecus rosmarus* (after Owen).
i Tusk-like upper incisors.

distinguished from the true Seals is found in the dentition. According to Owen, there are six incisors in the upper jaw and four in the lower; but these are only present in the young animal, and soon disappear, with the exception of the outermost pair of upper incisors. The upper canines are enormously developed, growing from persistent pulps, and constituting two large pointed tusks, which attain a length of over fifteen inches (fig. 225). The direction of the tusks is downwards and slightly outwards, and they project considerably below the chin. The adult animal has usually three simple molars with flat crowns behind the tusks in the upper jaw; and four similar teeth on each side of the lower jaw; but the first of these has been regarded as a lower canine. The upper lip has many pellucid bristles as large as a straw in thickness. In the adult the incisors are obsolete, except the lateral pair in the upper jaw. Unlike the Seals, the Walrus is not polygamous.

Except as regards its dentition, the Walrus agrees in all essential respects with the Seals. It is a large and heavy animal, attaining a length of from ten to fifteen feet or upwards. The body is covered with short brownish or yellowish hair, and the face bears many long stiff bristles. There are no external ears. The chief use of the tusk-like canines appears to be that of assisting the unwieldy animal to get out of the water upon the ice; but they doubtless serve as weapons of offence and defence as well. The Walrus is hunted by whalers, both for its blubber, which yields an excellent oil, and for the ivory of the tusks. It is found, living in herds, in the Arctic seas, being especially abundant at Spitzbergen and Nova Zembla.

SECTION II. PLANTIGRADA.—The Carnivorous animals belonging to this section apply the whole or the greater part of the sole of the foot to the ground (fig. 223, A); and the portion of the sole so employed is destitute of hairs in most instances (the sole is hairy in the Polar Bear). From the structure of the foot, the *Plantigrada* have great power of rearing themselves up on the hind-feet. They approach the *Insectivora* in their comparatively slow movements and their nocturnal habits, and in possessing no cæcum. They mostly hybernate, and their feet are always pentadactylous.

The typical family of the Plantigrade *Carnivora* is that of the *Ursidæ* or Bears, in which the entire sole of the foot is applied to the ground in walking. The *Ursidæ* are much less purely carnivorous than the majority of the order, and in accordance with their omnivorous habits, the teeth do not exhibit the typical carnivorous characters. The incisors and canines have the ordinary carnivorous form, but the "carnassial" or sectorial molar has a tuberculate crown instead of a sharp cutting edge. The dental formula is—

$$i\,\frac{3-3}{3-3};\ c\,\frac{1-1}{1-1};\ pm\,\frac{4-4}{4-4};\ m\,\frac{2-2}{3-3}=42.$$

The claws are formed for digging, large, strong, and curved, but are not retractile. The tongue is smooth; the ears small, erect, and rounded; the tail short; the nose forms a movable truncated snout; and the pupil is circular.

As shown by their smooth tongues and tuberculate molars, the Bears are not peculiarly or strictly carnivorous. They eat flesh when they can obtain it, but a great part of their food is of a vegetable nature.

The Bears are very generally distributed over the globe, Australia alone having no representative of the family. The common Brown Bear (*Ursus Arctos*) was at one time an inhabitant of Britain, and also existed over the whole of Europe. At the present day the Brown Bear is only found in the great forests of the north of Europe and in Asia. It feeds on roots, fruits, honey, insects, and, when it can obtain them, upon other Mammals. It attains a great age, and hybernates during the winter months. Very nearly allied to the Brown Bear is the Black Bear of America (*Ursus Americanus*). Both are of some commercial value, being hunted for the sake of their skins, fat, and tongues. A much larger American species is the Grizzly Bear (*Ursus ferox*), found in many parts of the American continent. It is about twice as large as the ordinary Bear, but it is said to subsist to a great extent upon vegetable food, such as acorns. The most remarkable, however, of the bears is the great White Bear (*Thalassarctos maritimus*), which is exclusively a native of the Arctic regions. It is a very large and powerful animal, the fur of which is quite white. The paws are very long, and the soles of the feet are covered with coarse hair, giving the animal a firm foothold upon the ice. The Polar Bear differs from the other *Ursidæ* in being exclusively carnivorous, since vegetable food would be wholly unattainable. It is as much at home in the water as on land, and lives chiefly upon seals and fish, and upon the carcases of Cetaceans.

Other well-known Bears are the Syrian Bear (*Ursus Syriacus*) of Mount Lebanon, the Sloth Bear (*Prochilus labiatus*) of India, and the Malayan Bear (*Helarctos Malayanus*) of Borneo and Sumatra.

It is a singular fact that the bones of a bear—the *Ursus spelæus* or Cave Bear—have been found, in Britain and in many parts of Europe, along with the bones of other *Carnivora*, such as the Cave Lion and Cave Hyæna. The *Ursus spelæus* was a larger and more powerful animal than even the Polar Bear, and there can be no doubt that it existed in the earlier portion of the human period.

Nearly allied to the true Bears are several small animals, of which the Racoons (*Procyon*), the Coati (*Nasua*), the Wah

(*Ailurus*), and the Kinkajou (*Cercoleptes*) are the best known. The Racoons are natives of tropical and northern America, and have a decided external resemblance to the Bears. They have tolerably long tails, however, and sharp muzzles. The commonest species is the *Procyon lotor* of North America, which derives its specific name from its habit of washing its food before eating it. The place of the Racoon is taken in India by the Wah (*Ailurus fulgens*), which inhabits northern Hindostan. It is about the size of a large domestic cat, and is very prettily coloured, being chestnut brown above, and black inferiorly, with a white face and ears. The Kinkajous (*Cercoleptes*) are inhabitants of South America, and, as is the case with so many of the animals of this continent, they are adapted for an arboreal life, to which end their tails are prehensile. The Coatis (*Nasua*) are very closely allied to the Racoons, and are exclusively confined to the American continent. All the above-mentioned little animals (with the exception of the Wah) present a singularly close resemblance to the Lemurs of the Old World, and appear to be their representatives in the western hemisphere. In the genus *Paradoxurus* of the Indian Archipelago, the tail is capable of being rolled up, but its extremity is not prehensile. The "Benturongs" (*Arctictis*) have a long, hairy, and prehensile tail. They are nocturnal animals which are found in India, and are in some respects intermediate between the Racoons and the Civets.

The only remaining family of the *Plantigrada* is that of the *Melidæ* or Badgers, characterised by their elongated bodies and short legs, and by the fact that the carnassial tooth has a partly cutting edge, and is not wholly tuberculate as in the Bears.

The common Badger (*Meles taxus*), which may be regarded as the type of this group, occurs in Britain, and is one of the most inoffensive of animals. It is nocturnal in its habits, and is a very miscellaneous feeder, not refusing anything edible which may come in its way, though living mainly on roots and fruits. The Badger burrows with great ease, and can bite very severely. The European Badger is represented in the United States and Canada by the "Siffleur" (*Meles Labradoricus*), and in the hilly parts of India by the Indian Badger (*M. collaris*). The Glutton (*Gulo luscus*), often called the Wolverine, is of common occurrence in the northern parts of Europe, Asia, and America. It is from two to three feet in length, and though doubtless a tolerably voracious animal, it is certainly not so much so as to deserve the name of Glutton. The Grison (*Gulo vittatus*) is a closely-allied species, which is found in South America. The Ratels or Honey-badgers (*Mellivora*) are much like the common Badger in their habits and appearance, but they get their name from their fondness for honey. They are natives of southern and eastern Africa.

SECTION III. DIGITIGRADA.—In this section of the *Carnivora* the heel is raised above the ground, with the whole or the greater part of the metacarpus, so that the animals walk more or less completely on the tips of the toes (fig. 223, C). No absolute line, however, of demarcation can be drawn between the Plantigrade and Digitigrade sections of the *Carnivora*, since many forms (*e. g.*, *Mustelidæ* and *Viverridæ*) exhibit transitional characters, and it has even been proposed to place these in a separate section, under the name of *Semi-plantigrada.*

The first family of the *Digitigrada* is that of the *Mustelidæ* or Weasels, including a number of small Carnivores, with short legs, elongated worm-like bodies, and a peculiar gliding mode of progression (hence the name of *Vermiformes*, sometimes applied to the group). Amongst the best known of the *Mustelidæ* are the common Weasel (*Mustela vulgaris*), the Pole-cat (*Mustela putorius*), and the Ferret (*Mustela furo*), the last being supposed to be only an albino variety of one of the Pole-cats. It is really an African species, but has been long domesticated in Europe. Many of the *Mustelidæ* are of great commercial importance, furnishing beautiful and highly-valued furs. Amongst these are the Ermine (*Mustela erminea*), and the Sable (*Mustela zibellina*). It is asserted, however, that most of the Sable of commerce is derived from the Black Mink (*Putorius nigrescens*) and the Pine Marten (*Mustela Americana*) of the United States and Canada.

Almost all the Weasels have a very disagreeable odour, produced by the secretion of greatly-developed and modified sebaceous glands, placed in the neighbourhood of the anus, and known as the anal glands. In this respect, however, the nearly-allied genus *Mephitis*, comprising the American Skunk, is *facile princeps.* The Skunk is a pretty little animal, with a long bushy tail, and when unmolested it is perfectly harmless. If pursued or irritated, however, it has the power of ejecting the secretion of the anal glands to a greater or less distance with considerable force. The odour of this secretion is so powerful and persistent that no amount of washing will remove it from a garment, and its characters are said to be of the most intensely disagreeable description.

Also belonging to the family of the *Mustelidæ*, and very nearly allied to the Weasels, are the Otters (*Lutra*), distinguished by the possession of webbed feet adapted for swimming. The body is long, the legs short, and the tail long, stout, and horizontally flattened. The common Otter (*Lutra vulgaris*) is a native of Britain, frequenting the banks of streams and lakes. It lives upon fish, and is highly destructive to Salmon. A closely-allied form is the American Otter (*Lutra Canadensis*). In the Sea Otters (*Enhydra*) the tail is very short. They are found on both sides of the North Pacific, and yield a very valuable fur.

The second family of the Semi-plantigrade Carnivores is that of the *Viverridæ*, the Civets and Genettes. They are all of moderate size, with sharp muzzles and long tails, and more or less striped, or banded, or spotted. The carnassial molar is trenchant; the canines are long, sharp, and pointed; and the tongue is roughened by numerous prickly papillæ. The claws are semi-retractile, and the pupils can contract, on exposure to light, till they resemble a mere line. In most of their characters, therefore, the Civets are much more highly carnivorous than are any of the preceding families, and they approach in many respects very close to the typical group of the *Digitigrada* (viz., the *Felidæ*), having especially very close affinities with the Hyænas. All the species of the family are furnished with anal glands, which secrete the peculiar fatty substance known as "civet."

The true Civet-cat is the *Viverra civetta*, a native of Africa. It is a small nocturnal animal, which climbs trees with facility, and feeds chiefly upon small mammals, reptiles, and birds, but also upon roots and fruits. It furnishes the greater part of the "civet" of commerce, which was formerly in great repute both as a perfume and as a medicinal agent. It is a pomade-like substance with a strong musky odour, and is secreted by a deep double pouch beneath the anus. The Genette (*Viverra genetta*) is very closely related to the preceding, and is a native of Africa and southern Europe, being not uncommonly domesticated and kept like a cat. The anal pouch in the Genette is much reduced in size, and has hardly any perceptible secretion. Another nearly-allied species is the Ichneumon (*Herpestes*), which is kept as a domestic animal in Egypt, and lives upon Snakes, Lizards, the eggs of the Crocodile, and small Mammals.

Forming a transition between the *Viverridæ* and the *Felidæ* is the family of the *Hyænidæ*, distinguished by the fact that, alone of all the *Carnivora*, both pairs of feet have only four toes each. The hind-legs are shorter than the fore-legs, so that the trunk sinks towards the hind-quarters, and the tail is short. The tongue is rough and prickly. The head is extremely broad, the muzzle rounded, and the muscles of the jaw extremely powerful and well developed. The claws are non-retractile. All the molars are trenchant except the last upper molar, which is tuberculate. The upper carnassial has a small internal tubercle, and the lower carnassial is wholly trenchant. There is a deep glandular pouch beneath the anus.

There are two well-known species of Hyæna, and the whole group is exclusively confined to the Old World. The best-known species is the Striped Hyæna (*Hyæna striata*), which is found in North Africa, Asia Minor, Arabia, and Persia. It is an ill-conditioned ferocious beast, but will not attack man unless provoked. The Spotted Hyæna (*H. crocuta*) occurs solely in Africa, being especially abundant in Cape Colony. If the so-called Aardwolf (*Proteles*) is to be placed amongst the Hyænas, as is generally done, then the characters to be drawn from the feet are not invariable; since this singular animal has the fore-feet furnished with five toes, whilst the hind-feet are tetradactylous (as is the case in the Dogs). It is a nocturnal burrowing animal, and is found in South Africa. The singular "Hunting Dog" (*Lycaon pictus*), again, of South Africa, agrees, with the Hyænas in being tetradactylous, but has no mane, and approaches the Dogs in its dentition and osteology.

An extinct Hyæna, considerably larger than either of the living forms, formerly existed in Britain and in various parts of Europe. It is known as the Cave Hyæna (*H. spelæa*), its remains having been principally found in caves.

The next family is that of the *Canidæ*, comprising the Dogs, Wolves, Foxes, and Jackals. The members of this family are characterised by having pointed muzzles, smooth tongues, and non-retractile claws. The fore-feet have five toes each, the hind-feet have only four. The molar teeth are $\frac{6-6}{7-7}$, sometimes $\frac{7-7}{7-7}$, and of these, two or three on each side are tuberculate. The carnassial has a tolerably large heel or process.

The true Dogs (*i.e.*, the Dog and Wolf) have round pupils, and a tail which is of moderate length and rarely very hairy. The Foxes (*Vulpes*) have very long bushy tails, and the pupil contracts to a mere line.

The Dog (*Canis familiaris*) is only known to us at the present day as a domesticated animal. Such wild dogs as there are, are probably merely derived from the domestic dog; and the original stock, or stocks, from which our numerous varieties of dogs have sprung, is still uncertain. It is worth while remembering, however, all our varieties of dogs are capable of interbreeding; and there is a strong probability that the Wolf is the parent stock of at least *some* of our domestic breeds. The Dog, in fact, will interbreed with both the Wolf and the Jackal.

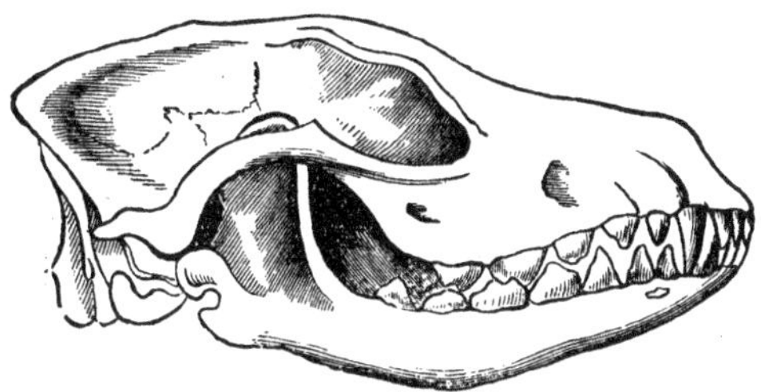

Fig. 226.—Skull of Jackal (*Canis aureus*).

The genus *Canis*, besides the Dog, contains the well-known Jackal (*Canis aureus*), and the Wolf (*Canis lupus*), and many writers place the Foxes in the same genus. The Foxes, however, are better considered as forming a separate genus (*Vulpes*), of which there are many species, all more or less like the common Fox (*Vulpes vulgaris*). One of the most remarkable species is the Arctic Fox, which abounds in the Arctic regions, and changes its colour with the season, being brown in summer and white in winter. The soles of its feet are hairy. Other well-known Foxes are the Red Fox (*V. fulvus*) of North America, the Deccan Fox (*V. Bengalensis*) of India, and the Caama (*V. Caama*) of Africa.

The Jackals have a round pupil and a dental formula like that of the Dogs. They inhabit Asia and Africa, are gregarious, hunt in packs, and burrow in the ground.

The last group of the *Digitigrada* is that of the *Felidæ* or Cat tribe, comprising the most typical members of the whole order of the *Carnivora*, such as the Lions, Tigers, Leopards, Cat, and Panthers. The members of this family all walk upon the tips of their toes, the soles of their feet being hairy, and the whole of the metacarpus and heel being raised above the ground (fig. 223, C). The jaws are short, and owing to this fact, and to the great size of the muscles concerned in mastication, the head assumes a short and rounded form, with an abbreviated and rounded muzzle. The molars and præmolars are fewer in number than in any other of the *Carnivora* (hence the shortness of the jaws), and they are all trenchant, except the last molar in the upper jaw, which is tuberculate. The upper carnassial has three lobes, and a blunt heel or internal process. The lower carnassial has two cutting lobes, and no internal process. According to Owen, the dental formula is—

$$i\,\frac{3-3}{3-3};\ c\,\frac{1-1}{1-1};\ pm\,\frac{3-3}{2-2};\ m\,\frac{1-1}{1-1} = 30.$$

The legs are nearly of equal size, and the hind-feet have only four toes each, whilst the fore-feet have five. All the toes are furnished with strong, curved, retractile claws, which, when not in use, are withdrawn within sheaths by the action of elastic ligaments, so as not to be unnecessarily blunted.

Fig. 227.—Skull of Lion (*Felis leo*).

The tongue is roughened and rendered prickly by the presence of horny papillæ, thus rendering it a most efficient rasp in licking the flesh from the bones of the prey. All the members of this group are exceedingly light upon their feet, and are excessively muscular, and they have all the habit of seizing their prey by suddenly springing upon it.

It is questionable if any good genera have hitherto been established in this family, and all the species may be considered as belonging to the single genus *Felis*.

The Lion (*Felis leo*) is too well known to require much special notice. Its colour is always uniform, generally a yellowish or reddish brown. The tail is terminated by a tuft of long hairs, and the male is usually furnished with a mane, which is very short, however, in an Indian form. The Lion is exclusively confined to the Old World, and is an inhabitant of Africa and all the southern parts of Asia. It is doubtful how far any valid species of Lions have as yet been established. The Lions are all nocturnal, and capture their prey by suddenly leaping upon it. They are by no means the generous and courageous animals they are generally considered to be; but, on the contrary, are cruel, cunning, and cowardly. They are enormously strong, and it is said that a full-grown Lion can run, and even leap, though carrying an ox in its jaws. Though now much restricted in its range, the Lion had formerly a much more extensive distribution, a form considerably larger than the modern species having formerly existed in Europe, and even in Britain (*Felis spelæa*).

In the tigers (*Felis tigris*), the tail is without a tuft of hairs at its extremity, and the skin is marked with stripes or spots. The Royal or Bengal Tiger is a native of southern Asia, but occurs also in Java and Sumatra. The skin is reddish yellow, marked with numerous transverse black stripes. It is a large and powerful animal, and, upon the whole, is probably a more dangerous opponent than even the Lion.

Of the large Spotted Cats, the largest is the Jaguar (*Felis onca*) which inhabits South America and the southern parts of North America. It is a very large and powerful animal, said to be able to carry a bullock without difficulty, and it can both swim and climb with great facility. Another American species is the Puma (*Felis concolor*), in which the colour is uniformly reddish brown. It is exclusively confined to America, and though of large size (nine feet in length, including the tail), it is a very cowardly species, and is seldom or never known to attack man.

The Leopard (*Felis leopardus*) is another well-known species, smaller than the Tiger, and marked with black spots in place of stripes. It is a native of all the warmer parts of the Old World.

The Panther is probably merely a variety of the Leopard, and the Ounce (*Felis uncia*) is generally believed to be only its immature form. Another allied form is the Cheetah or Hunting Leopard (*Felis jubata*), of southern Asia and Africa.

Of the smaller *Felidæ*, the best known are the Lynxes and the Cats, properly so called. Of these the Lynxes are distinguished by their short tails, and by the fact that the ears are furnished with a pencil of hairs. The best-known species are the European Lynx (*Felis lyncus*), the Caracal (*F. caracal*) of southern Asia and Africa, and the Canadian Lynx (*F. Canadensis*) of North America. In the true Cats (*Felis catus*), the tail is long, and the ears are not tufted. The Wild Cat formerly existed in Britain, but is now extinct, though it still occurs in

Europe, especially in the Hartz and Carpathian Mountains. It is a large and fierce animal, and appears to be quite a match for any man not possessing firearms. It seems tolerably certain that the Wild Cat is not the original stock of the Domestic Cat, the exact origin of which is uncertain. It has been supposed, however, that the Domestic Cat is descended from a small species (*Felis maniculata*) which occurs in Nubia.

CHAPTER LXXX.

RODENTIA.

ORDER X. RODENTIA.—The tenth order of *Mammalia* is that of the *Rodentia*, or Rodent Animals, often spoken of as *Glires*, comprising the Mice, Rats, Squirrels, Rabbits, Hares, Beavers, &c.

The *Rodentia* are characterised by the possession of two long curved incisor teeth in each jaw, separated by a wide interval from the molars. The lower jaw never has more than two of these incisors, and the upper jaw very rarely; but sometimes there are four upper incisors. There are no canine teeth, and the molars and præmolars are few in number (rarely more than four on each side of the jaw). The feet are usually furnished with five toes each, all of which are armed with claws; and the hallux, when present, does not differ in form from the other digits. The testes pass periodically from the abdomen into a temporary scrotum, and the placenta is discoidal and deciduate.

The most characteristic point about the Rodents is to be found in the structure of the incisors, which are adapted for continuous gnawing—hence the name of *Rodentia*. The incisor teeth are commonly two in each jaw, and they grow from persistent pulps, so that they continue to grow throughout the life of the animal. They are large, long, and curved (fig. 228, B), and are covered anteriorly by a plate of hard enamel. The back part of each incisor is composed only of the comparatively soft dentine, so that when the tooth is exposed to attrition, the soft dentine behind wears away more rapidly than the hard enamel in front. The result of this is that the crown of the tooth acquires by use a chisel-like shape, bevelled away behind, and the enamel forms a persistent cutting edge (fig. 228).

The gnawing action of the incisors is assisted by the articulation of the lower jaw, the condyle of which is placed longitudinally and not transversely, so that the jaw slides backwards and forwards. The molars, consequently, have flat crowns, the enamelled surfaces of which are always arranged in transverse ridges, in opposition to the antero-posterior movement of the jaw. The intestine is very long, and the cæcum voluminous

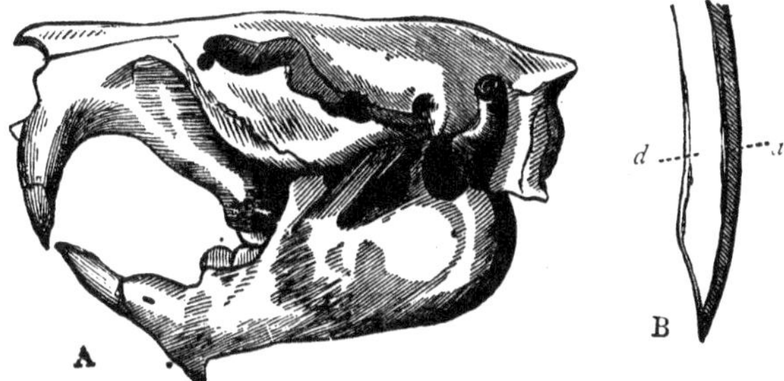

Fig. 228.—A, Skull of the Beaver (after Owen). B, Diagram of the incisor tooth of a Rodent, showing the chisel-shaped point; *a* Enamel; *d* Dentine.

(rarely wanting). The brain is nearly smooth, and without convolutions. The orbits are not separated from the temporal fossæ, and the eyes are directed laterally. The Rodents are almost all very small animals, and they are mostly very prolific. They subsist principally, if not entirely, upon vegetable matters, especially the harder parts of plants, such as the bark and roots. Many of them possess the power of building elaborate nests, and most of them hybernate. They are very generally distributed over the whole world, but no member of the order has hitherto been detected in rocks older than the Eocene Tertiary.

The Order *Rodentia* comprises a very large number of families, only the more important of which can be noticed here.

Fam. 1. *Leporidæ.*—In this family are the Hares (*Lepus timidus*) and Rabbits (*Lepus cuniculus*), distinguished amongst the Rodents by the possession of two small incisors in the upper jaw, placed behind the central chisel-shaped incisors, so that there are four upper incisors in all. The molars and præmolars are rootless, and the dental formula is—

$$i\frac{2-2}{1-1};\ c\frac{0-0}{0-0};\ pm\ \frac{3-3}{2-2};\ m\frac{3-3}{3-3}=28.$$

The clavicles are imperfect. The fore-legs are furnished with five toes, and are considerably shorter than the hind-legs, which have only four toes. The two orbits communicate by an aperture in the septum. Generally there is a short erect tail.

The common Hare (*Lepus timidus*) is dispersed over the whole of Europe, but is not met with in Sweden and Norway, its place there being taken by the Mountain-hare (white in winter), which occurs commonly in Scotland. As a rule, the Hares occur in temperate regions, but some are found in Africa, and one species (*Lepus glacialis*) is a native of the Arctic regions, whilst the common American Hare (*L. Americanus*) extends from Canada to Mexico. The Rabbit is also a native of temperate regions, but appears to thrive, to a more than average extent, in Australia.

In the Calling Hares or Pikas (*Lagomys*), the legs do not differ much in size, there is no visible tail, and the clavicles are nearly complete. They are found in Russia, Siberia, and North America.

Fam. 2. *Cavidæ.*—As examples of this family may be taken the Capybara (*Hydrochærus capybara*) and the Guinea-pig (*Cavia aperæa*). In this family the body is covered with hair, without spines, and the tail is rudimentary. The Capybara is the largest of living Rodents, attaining a length of three or four feet. It is a South American form, leading a semi-aquatic life, to which end the feet are incompletely webbed. It is a harmless stupid animal, and is not unlike a small pig in appearance. The *Cavia aperæa* is likewise a South American animal, and is believed to be the parent stock of the Guinea-pigs so often kept as domestic pets in Europe. To the same group as the Capybara belong the Agoutis (*Dasyprocta*) and the Pacas (*Cœlogenys*), all of which have eight rootless molars in each jaw, whilst the two former have four toes to the fore-feet, and three toes on the hind-feet. The various species of Agouti are found in South America and the West Indies, whilst the Pacas are exclusively South American. In this family, also, are usually placed the Chinchillas (*Chinchilla*) of Chili and Peru.

Fam. 3. *Hystricidæ.*—In this family are the well-known Porcupines, distinguished from the other Rodents by the fact that the body is covered with long spines or "quills," mixed with bristly hairs. They have four molars on each side of each jaw, and they possess imperfect clavicles.

The true Porcupines (*Hystrix*) have non-prehensile tails, which are mostly furnished with long hollow spines, but sometimes with scales and bristles. They are found in both the Old and New World, but the American species differ in several respects from those of the eastern hemisphere. They are mostly inhabitants of hot climates, with the exception of the common Porcupine (*H. cristata*), which occurs in southern Europe and in the north of Africa. In the genus *Atherura* of Asia and the Indian Archipelago, the tail is terminated by a bundle of flattened horny strips. In the genus *Erethizon*, represented by the Canada Porcupine (*E. dorsatum*) of North America, the quills are short, and are half hidden in the hair.

The nearly-allied genus *Cercolabes* is South American, and it is distinguished from the preceding by the possession of a long prehensile tail. In fact, *Cercolabes*, like so many of the inhabitants of this wonderful continent, is adapted for an arboreal life, instead of being confined to the ground.

Fam. 4. *Castoridæ.*—The best-known example of this family is the Beaver (*Castor fiber*). The distinctive peculiarities of the family are the possession of distinct clavicles, the possession of five toes to each foot, and the fact that the hinder feet are mostly webbed, adapting the animal to a semi-aquatic life.

The Beaver is a large Rodent, attaining a length of from two and a half to three feet. Naturally it is a social animal, living in societies, and this is still the case in America; but in northern Europe and Asia, where the animal has been much hunted, it leads a solitary life. When living in social communities the beavers build dams across the rivers, as well as habitations for themselves, by gnawing across the branches of trees or shrubs, and weaving them together, the whole being afterwards plastered with mud. In this last operation the tail, which is flattened and scaly, is employed very much as a mason uses his trowel. There is no doubt but that the Beaver shows extraordinary ingenuity in these and similar operations; but there can be equally little doubt as to the greatly-exaggerated stories which have been set afloat in this connection. The Beaver is hunted chiefly for the sake of the skin, but also for the substance known as *castoreum*. This is a fatty substance, secreted by peculiar glands, and employed as a therapeutic agent.

There are two other members of the *Castoridæ* which are likewise largely captured for the sake of their skins. One of these is the Musquash (*Fiber Zibethicus*), which inhabits North America, and the other is the Coypu (*Myopotamus coypus*), which inhabits burrows in the banks of rivers in Chili. In the Musquash the hind feet are not completely webbed, and the tail is moderate in size, and covered with short hairs and small rounded scales. In the Coypu the hind feet are webbed, but the tail is long, rounded, and furnished with scales and scattered hairs.

Fam. 5. *Muridæ.*—The fifth family of Rodents is that of

the *Muridæ*, comprising the Rats, Mice, and Lemmings. In this family the tail is long, always thinly haired, sometimes naked and scaly. The lower incisors are narrow and pointed, and there are complete clavicles. The hind-feet are furnished with five toes, the fore-feet with four, together with a rudimentary pollex.

The Rats (*Mus rattus* and *Mus decumanus*), the common Mouse (*Mus musculus*), the Field-mouse (*Mus sylvaticus*), and the Harvest-mouse (*Mus messorius*), are all well-known examples of this family, and are too familiar to require any description. The three first are also common in North America. Closely allied to the true Rats are the Hamsters (*Cricetus*, fig. 229), and the Voles (*Arvicola*); the latter represented by many species in both Europe and America.

A less familiar example of this family is the Lemming (*Myodes*

Fig. 229.—Common Hamster (*Cricetus vulgaris*).

lemmus). This curious little Rodent is found inhabiting the mountainous regions of Norway and Sweden. It is chiefly remarkable for migrating at certain periods, generally towards the approach of winter, in immense multitudes and in a straight line, apparently in obedience to some blind mechanical impulse. In these journeys the Lemmings march in parallel columns, and nothing will induce them to deviate from the straight line of march.

Fam. 6. *Dipodidæ.*—The sixth family of the Rodents, which is sufficiently important to need notice, is that of the *Dipodidæ* or Jerboas, mainly characterised by the disproportionate length of the hind-limbs as compared with the fore-limbs. The tail also is long and hairy, and there are complete clavicles. The Jerboas live in troops, and owing to the great length of the hind-legs, they can leap with great activity and to great distances. They are all of small size, and inhabit Russia, North Africa, and North America. The best-known members of this family are the common Jerboa (*Dipus Ægypticus*), which lives in societies and constructs burrows; the Jumping Hare (*Pedetes Capensis*) of South Africa, and the Jumping Mouse (*Meriones Hudsonicus*) of North America. Here also may be placed the Gerbilles (*Gerbillus*) of the Old World.

Fam. 7. *Myoxidæ.*—The members of this family are commonly known as Dormice, and they are often included in the following family of the Squirrels and Marmots. They only require to be mentioned, as they must not be confounded with the true Mice (*Muridæ*) on the one hand, or the Shrew-mice (*Soricidæ*) on the other; the latter, indeed, belonging to another order (*Insectivora*). The common Dormouse (*Myoxus avellanarius*) is a British species, and must be familiarly known to almost everybody. No species of this family have yet been described from the New World.

Fam. 8. *Sciuridæ.*—This is the last family of Rodents which calls for any special mention, and it comprises the true Squirrels, the Flying Squirrels, and the Marmots.

The true Squirrels (*Sciurus*) are familiarly known in the person of the common British species (*Sciurus vulgaris*), and the equally common Gray Squirrel (*S. cinereus*) of the United States. Numerous species more or less closely allied to these occur in other countries, and they are especially abundant in North America.

In the genera *Pteromys* and *Sciuropterus*, or Flying Squirrels, there is a peculiar modification by which the animal can take extended leaps from tree to tree. The skin, namely, extends in the form of a broad membrane between the hind and fore legs, and this acts as a kind of parachute, supporting the animal in the air. There is, however, no power whatever of true flight, and the structure is identically the same as what we have previously seen in the Flying Phalangers (*Petaurus*), which take the place of the Flying Squirrels on the Australian continent. The Flying Squirrels are found in southern Asia, Polynesia, the north-east of Europe, Siberia, and North America.

The Marmots (*Arctomys*), unlike the true Squirrels, are terrestrial in their habits, and live in burrows. Various intermediate forms, however, are known, by which a transition is effected between the typical Squirrels and the Marmots. Such, for example, are the Ground Squirrels (*Tamias*) of Europe, Asia, and North America. There are numerous species of this family inhabiting various parts of Europe and northern Asia, and generally distributed over the whole of North America.

Good examples are the Alpine Marmot (*A. Alpinus*) of Europe and the Prairie Dog (*Cynonys Ludovicianus*) of North America.

CHAPTER LXXXI.

CHEIROPTERA.

ORDER XI. CHEIROPTERA.*—This order is undoubtedly "the most distinctly circumscribed and natural group" in the whole class of the *Mammalia*. In many respects, however, it would be advantageous to regard the *Cheiroptera* as a sub-order of the next order (namely, the *Insectivora*) specially modified to lead an aerial life; just as the *Pinnigrada* are regarded as a mere section of the *Carnivora* specially modified to suit an aquatic life.

The *Cheiroptera* are essentially characterised by the fact that the anterior limbs are longer than the posterior, the digits of the fore-limb, with the exception of the pollex, being enormously elongated (fig. 230). These elongated fingers are united by an expanded membrane or "patagium," which is also extended between the fore and hind limbs and the sides of the body, and in many cases passes also between the hind-limbs and the tail. The patagium thus formed is naked, or nearly so, on both sides, and it serves for flight. Of the fingers of the hand, the pollex, and sometimes the next finger as well, is unguiculate, or furnished with a claw; but the other digits are destitute of nails. In the hind-limbs all the toes are unguiculate, and the hallux is not in any respect different from the other digits. Well-developed clavicles are always present, and the radius has no power of rotation upon the ulna. The mammary glands are two in number, and are placed upon the chest. There are teeth of three kinds, and the canines are always well developed. The molars are tuberculate or grooved in the frugivorous forms, and cuspidate in the insectivorous species. The ulna is sometimes quite rudimentary. The bones are not pneumatic. The testes are abdominal except during the breeding season. The stomach is complex and the intestine long in the fruit-eating Bats; but the reverse of this obtains amongst the Insectivorous forms.

* The *Cheiroptera* were placed by Linnæus in his order *Primates*, which contained also the Lemurs, the Apes, and Man.

The *Cheiroptera* are cosmopolitan in their distribution, and the oldest known species is from the Eocene rocks.

The Bats are all crepuscular and nocturnal in their habits, and are sometimes carnivorous, sometimes frugivorous. The eyes are small, but the ears are very large, and their sense of touch is most acute. During the day they retire to caves or crevices amongst the rocks, where they suspend themselves by means of the short thumbs, which are provided with curved claws. In their flight, though they can fly in the genuine and proper sense of the term, and can turn with great ease, they

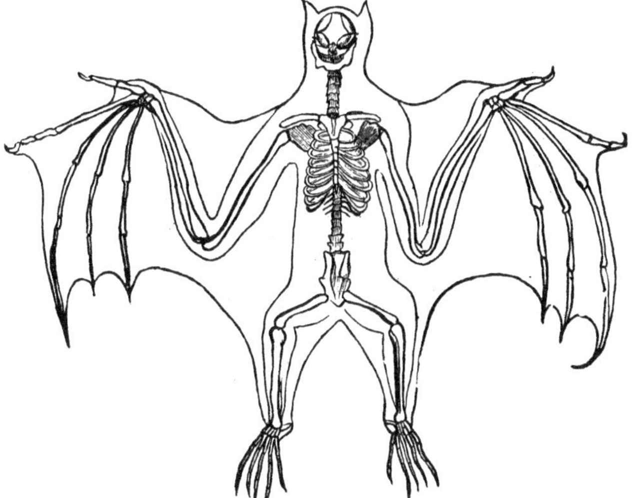

Fig. 230.—Skeleton of Fox-bat (*Pteropus*)—after Owen.

are by no means as rapid and as active as are the true birds. The tail is sometimes short, sometimes moderately long, and is usually included in a continuation of the leathery patagium, which stretches between the hind-legs, and is termed the "inter-femoral membrane." The body is covered with hair, but the patagium is usually hairless, or nearly so. Most of the Bats hybernate.

The *Cheiroptera* are conveniently divided into the two sections of the *Insectivora* and *Frugivora*, according as the diet consists of insects or of fruits.

SECTION A. INSECTIVORA.—In this section are the three families of the *Vespertilionidæ*, *Rhinolophidæ*, and *Phyllostomidæ*.

Fam. 1. *Vespertilionidæ*.—In this family are the ordinary Bats, distinguished by having a dentition very like that of the order of the Insectivorous Mammals, the molar teeth being furnished with small pointed eminences or cusps, adapted for crushing insects. The nose is not furnished with leaf-like appendages, and the tail is usually elongated, and enclosed in a large inter-femoral membrane. About fifteen species of this family have been described as British, but of these only two are at all common. Of these two, the Pipistrelle (*Vespertilio pipistrella*) is the commonest species, occurring over the whole of Britain. The long-eared Bat (*Plecotus auritus*) is also not uncommon, and is distinguished by its greatly elongated ears,

Fig. 231.—A, Head of Vampire-bat (*Alectops ater*). B, Head of Fox-bat (*Pteropus personatus*)—after Gray.

which are confluent above the forehead. The largest British species is the Noctule (*Vespertilio noctula*), which measures as much as fifteen inches in expanse of wing.

Fam. 2. *Rhinolophidæ*.—The second family of the Insectivorous Bats is that of the *Rhinolophidæ* or Horse-shoe Bats, which in most respects are very similar to the *Vespertilionidæ*, but are distinguished by the possession of a complex leaf-like apparatus appended to the nose. Of this family, two British species are known—the Greater and Lesser Horse-shoe Bats (*Rhinolophus ferrum-equinum* and *R. hipposideros*).

Fam. 3. *Phyllostomidæ*.—This is the only remaining family of the Insectivorous Bats, and comprises the well-known Vampire-bats (fig. 231, A), distinguished by having leaf-like nasal appendages, and by the fact that the ears are of small size; whereas in the preceding they are always very large (*Rhinolo-*

phus), and are often confluent above the forehead (*Megaderma*). They are all of large size, and are natives of South America. The Vampire-bat (*Phyllostoma spectrum*) has an expanse of wing of two feet and a half, and lives chiefly upon insects. It also has the habit of sucking the blood of sleeping animals, appearing sometimes to attack even man, though apparently never doing any substantial or lasting injury.

SECTION B. FRUGIVORA.—In the fruit-eating section of the *Cheiroptera* are only the *Pteropidæ* or the Fox-bats, so called from the resemblance of the head to that of a fox (fig. 231, B). The head in these bats is long and pointed. The ears are of moderate size, and the nose is destitute of any appendages. Cutting incisors and canines are present in both jaws, and the Fox-bats do not refuse to eat small birds or mammals. They live, however, mostly upon fruits, and the molars are therefore not cuspidate, but are furnished with blunt tubercular crowns. The tail is very short, or is entirely absent. The *Pteropidæ* are amongst the largest of the Bats, one species—the *Pteropus edulis*, or Kalong—attaining a length of from four to five feet from the tip of one wing to that of the other. The *Pteropidæ* are especially characteristic of the Pacific Archipelago—Java, Sumatra, Borneo, &c.—but they also occur in Asia, Australia, and Africa. They do not occur, however, in either North or South America.

CHAPTER LXXXII.

INSECTIVORA.

ORDER XII. INSECTIVORA.—The twelfth order of Mammals is that of the *Insectivora*, comprising a number of small Mammals which are very similar to the Rodents in many respects, but want the peculiar incisors of that order, and are likewise always furnished with clavicles.

In the *Insectivora*, all the three kinds of teeth are usually present, but the exact nature of the dentition varies considerably in different cases. The incisors and canines present little special, but the molars are always serrated with numerous small pointed eminences or cusps, adapted for crushing insects. With one exception, clavicles are always present in a complete form. All the feet are usually furnished with five toes; all the toes are furnished with claws; and the animal walks on the soles of the feet, or is plantigrade. The testes pass periodically from the abdomen into a temporary scrotum ; and the placenta is deciduate and discoidal. They are mostly nocturnal and subterranean, and generally hybernate. They are all of small size, and are found everywhere, except in the continents of South America and Australia, where their place is filled by Marsupials.

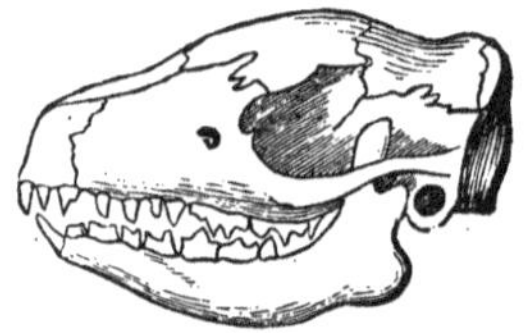

Fig. 232.—Insectivora. Skull of the common Hedgehog (*Erinaceus Europæus*).

The three leading families of the *Insectivora* are the *Talpidæ* or Moles, the *Soricidæ* or Shrew mice, and the *Erinaceidæ* or Hedgehogs.

Fam. 1. *Talpidæ.*—The body in this family is covered with hair ; the feet are formed for digging and burrowing, and the toes are furnished with strong curved claws. There are no external ears ; and the eyes in the adult are rudimentary, and more or less completely useless as organs of vision. There is a peculiar bone for the support of the muzzle. The clavicles are strong, the arm very short, the hand wide, and the palm always turned outwards and backwards. The fur is short and velvety, and the tail very short or wanting, in most cases.

Fig. 233.—European Mole (*Talpa Europæa*).

The common Mole (*Talpa Europæa*, fig. 233) is the only British species of the family, and a representative form (*Condylura*) occurs in North America. One of the most remarkable of the *Talpidæ* is the Golden Mole (*Chrysochloris aureus*) of Africa. In form and habits this species resembles the common Mole, but the hairs of the fur have the property of dispersing the rays of light, and thus of giving rise to beautiful metallic colours, such as are produced by the "setæ" of the Sea-mice (*Aphrodite*) amongst the Annelides. The star-nosed Moles (*Condylura*) are North American, and are distinguished by a fringe of elongated membranous caruncles surrounding the nostrils. The tail is moderately long.

Fam. 2. *Soricidæ.*—The *Soricidæ* or Shrew-mice are distinguished by having the body covered with hair, and the feet not adapted for digging; whilst there are external ears, and the eyes are well developed. Of all the *Insectivora*, no division is more abundant or more widely distributed than that of the Shrew-mice. In general form and appearance the Shrews very closely resemble the true Mice (*Muridæ*) and the Dormice (*Myoxidæ*), but they are in reality widely different, and must not be confounded with them. The common Shrew (*Sorex araneus*) and the Water-Shrew (*Sorex fodiens*) are both well-known species of this family. The smallest known Mammal is one of the Shrews (*Sorex Etruscus*), which is not more than two and a half inches in length, counting in the tail. Besides the true Shrews, the Shrew-moles (*Scalops*) and the Elephant Shrews (*Macroscelides*) are included in this family, the former being North American, whilst the latter are African.

Fam. 3. *Erinaceidæ.*—The last family of the *Insectivora* is that of the Hedgehogs, characterised by the fact that the upper part of the body is covered with prickly spines, the feet are not adapted for digging, and they have mostly the power of rolling themselves into a ball at the approach of danger. The common Hedgehog (*Erinaceus Europæus*) is in every way a typical example of this family, but is too well known to require any description. Other species of *Erinaceus* have been recorded from Africa and India.

The "Tenrecs" (*Centetes*) are natives of Madagascar, and may be regarded as Hedgehogs without the power of rolling themselves up into a ball. They have no tail, and have the skin beset with spines or spine-like bristles.

The "Banxrings" (*Tupaia*) are arboreal in their habits, and are confined to the Indian Archipelago. They must be regarded as the type of a distinct family of *Insectivora*. They have a long attenuated snout, with large eyes, a long body, and a close fur intermixed with soft hairs. The feet are plantigrade, five-toed, with naked soles and sickle-shaped claws. The tail is longer than the body, compressed and fringed at the sides.

GALEOPITHECIDÆ.

Before passing on to the *Quadrumana*, mention must be made here of a very singular animal which forms a kind of connecting link between the orders of the *Insectivora* and *Quadrumana*, having been sometimes placed in the one and sometimes in the other, or having been regarded as the type of a separate order. The order includes only the single genus *Galeopithecus*, comprising the so-called "Flying Lemurs." All the *Galeopitheci* inhabit the Indian Archipelago, but the best known is the *Galeopithecus volans* of Java, Sumatra, and Borneo. The most characteristic point in this singular animal is the presence of a flying membrane, presenting some superficial resemblance to the patagium of the Bats, but in reality very much the same as the integumentary expansions of the Flying Squirrels and Flying Phalangers. This membrane in the *Galeopithecus* extends as a broad expansion from the nape of the neck to the arms, from the arms to the hind-legs, and from the hind-legs to the tail, forming an inter-femoral membrane. The fingers are not elongated, and do not support a patagium, as in the Bats, so that the animals have no power of true flight, and can simply take extended leaps from tree to tree. The feet are furnished with five toes each, united by a membrane, but neither the hallux nor the pollex are opposable to the other digits. The dentition is complicated, and consists of incisors and molars, and, according to Owen, canines also, the dental formula being—

$$i\,\frac{2-2}{3-3};\ c\,\frac{1-1}{1-1};\ pm\,\frac{2-2}{2-2};\ m\,\frac{3-3}{3-3} = 34.$$

The six lower incisors are split into narrow strips, like the teeth of a comb. The *Galeopitheci* live chiefly upon small birds and insects, but also partially upon fruits. They are nocturnal animals, arboreal in their habits, and they sleep head-downwards, suspended by their prehensile tails.

CHAPTER LXXXIII.

QUADRUMANA.

ORDER XIII. QUADRUMANA.—The thirteenth order of Mammals is that of the *Quadrumana*, comprising the Apes, Monkeys, Baboons, Lemurs, &c., characterised by the following points:—

The hallux (innermost toe of the hind-limb) is separated from the other toes, and is opposable to them, so that the hind-feet become prehensile hands. The pollex (innermost toe of the fore-limbs) may be wanting, but when present, it also is usually opposable to the other digits, so that the animal becomes truly *quadrumanous*, or four-handed.

The incisor teeth generally are $\frac{2-2}{2-2}$, and the molars $\frac{3-3}{3-3}$, with broad and tuberculate crowns. Perfect clavicles are present. The teats are two in number, and are pectoral in position, and the placenta is discoidal and deciduate.

Fig. 234.—Green Monkey or Guenon (*Cercocebus sabæus*)—after Cuvier.

The Quadrumana are divided by Owen into three very natural groups, separated from one another by their anatomical characters and by their geographical distribution as follows :—

Section A. Strepsirhina.—The members of this section are characterised by the nostrils being curved or twisted, whilst the second digit of the hind-limb has a claw. This section includes the true Lemurs and a number of allied forms. It is chiefly referable to Madagascar as its geographical centre ; but it spreads westwards into Africa, and eastwards into the Indian Archipelago.

Section B. Platyrhina.—This section includes those Quadrumana in which the nostrils are placed far apart ; the thumbs of the fore-feet are either wanting, or, if present, are not opposable to the other digits; and the tail is generally prehensile. The Platyrhine Monkeys are exclusively confined to South America.

Section C. Catarhina.—In this section the nostrils are oblique, and placed close together. The thumb of the fore-limb (pollex), with one exception, is present, and is always opposable to the other digits. The Catarhine Monkeys are restricted entirely to the Old World, and, with the single exception of a Monkey which inhabits the rock of Gibraltar, they are exclusively confined to Africa and Asia. It is in the Catarhine section of the *Quadrumana* that we have the highest group of the Monkeys—that, namely, of the Anthropoid or Tail-less Apes.

Strepsirhina.

This section of the *Quadrumana*, as before said, is characterised by the possession of twisted or curved nostrils, placed at the end of the snout. The incisor teeth are generally much modified, and are in number $\frac{3-3}{3-3}$ as a rule; the lower incisors are produced and slanting; the præmolars are $\frac{3-3}{3-3}$ or $\frac{2-2}{2-2}$, and the molars are tuberculate. The second digit of the hind-limb has a claw, and both fore and hind feet have five toes each, all the thumbs being generally opposable. In the true Lemurs, all the digits, except the second toe of the hind-feet, are furnished with nails.

This section is often called that of the *Prosimiæ*, and it includes several families, of which the Aye-Ayes, Loris, and true Lemurs are the most important. In many works the *Galeopithecus* is also placed in this section.

The family of the Aye-Ayes (*Cheiromydæ*) includes only a single animal, the *Cheiromys Madagascariensis.* In appearance the Aye-Aye is not very unlike a large Squirrel, having a hairy body and a long bushy tail. There are no canines, and the molars are separated by a wide interval from the incisors. The incisors are ploughshare-shaped, and grow from permanent pulps, as in the Rodents. The fore-feet have five toes, armed with strong claws, but the pollex is scarcely opposable to the other digits. The middle-finger is about as long as the ring-finger, but only about half as thick, its last two joints being hairless. The hind-feet have also five toes, of which the hallux is opposable, and the second digit is furnished with a long claw. As far as is yet known, the *Cheiromys* is entirely confined to Madagascar.

In the *Nycticebidæ* are the Loris and the Slow Lemurs, in which there is no tail, or but a rudimentary one; the ears are short and rounded, and the eyes are large, and are placed close together. The species of this family are all of small size, and are exclusively confined to the eastern portion of the Old World, occurring in Java, Ceylon, the southern parts of Asia, and other localities in the same geographical area. They are nocturnal in their habits, living mostly on trees, and feeding upon insects; and from the slowness with which some of them progress, they are sometimes spoken of as "Slow Lemurs." The best-known species are the Slender Loris (*L. gracilis*) of Ceylon, and the *Nycticebus tardigradus* of the East Indies.

The largest and most important of the families of the *Strepsirhina* is that of the *Lemuridæ* or true Lemurs. In this family the muzzle is elongated, the feet are all furnished with opposable thumbs, and the nails on all the toes are flat, with the exception of the second toe of the hind-foot, in which there is a long and pointed claw. The body is covered with a soft fur, and the tail is usually of considerable length, and is covered with hair. They are easily domesticated; and though capable of biting pretty severely, their disposition is gentle and docile. They are mostly about the size of cats, and not unlike them in appearance, being often termed "Madagascar cats" by sailors. They are found almost exclusively in the great forests of Madagascar, moving about amongst the trees with great activity, by means of their prehensile tails. They appear to fill in Madagascar the place occupied by the higher *Quadrumana* upon the adjoining continent of Africa. The largest species is the Indri, which has very long hind-legs, and stands as much as three feet in height.

PLATYRHINA.

The section of the Platyrhine Monkeys is exclusively confined to South America, and one of its leading characters is to be found in the almost universal possession of a prehensile tail; this being an adaptive character by which they are suited to the arboreal life which so many of the South American Mammals are forced to lead. There are neither cheek-pouches nor natal callosities, and there is an additional præmolar, and sometimes a molar less than in man and the Old World Monkeys. The nostrils are simple, wide apart, and placed nearly at the extremity of the snout. The præmolars are $\frac{3-3}{3-3}$

in number, and have blunt tubercles. The thumbs of the fore-hands are either wanting altogether, or, if present, are not opposable, though versatile.

The Platyrhine Monkeys are divided into the two sections of the *Hapalidæ* and *Cebidæ*.

Fam. 1. *Hapalidæ.*—In this family the number of teeth is the same as in the Old World Monkeys and in Man, but there is an additional præmolar on each side of each jaw, and a molar less. According to Owen, the dental formula of the Marmoset is—

$$i\,\frac{2-2}{2-2};\ c\,\frac{1-1}{1-1};\ pm\,\frac{3-3}{3-3};\ m\,\frac{2-2}{2-2}=32.$$

The molars, however, are tuberculate, and though the number of teeth is the same as in the Catarhine Monkeys, in their other characters the Marmosets are genuine Platyrhines. The hind-feet have an opposable hallux with a flat nail, but all the other toes are unguiculate, and the pollex is hardly opposable. The tail is long, but is not prehensile.

The *Hapalidæ* are all small monkeys, mostly about as big as Squirrels, and they are exclusively South American, occurring especially in Brazil. The best-known species is the common Marmoset (*Hapale penicillata*), but several species are domesticated and kept as pets.

Fam 2. *Cebidæ.*—In this family are all the typical Platyrhine Monkeys, in which the dentition differs from that of the *Hapalidæ* in having an additional molar, so that the molars are the same as in the *Catarhina* and in Man, but the præmolars are more numerous. The dental formula is—

$$i\,\frac{2-2}{2-2};\ c\,\frac{1-1}{1-1};\ pm\,\frac{3-3}{3-3};\ m\,\frac{3-3}{3-3}=36.$$

There are neither cheek-pouches nor "callosities;" and the face is usually more or less naked, though sometimes whiskered. The tail is long, and is mostly prehensile; though in rare instances it is non-prehensile, and has its extremity clothed with hairs. The thumb of the fore-hand may be wanting, and, if present, is not opposable. All the fingers are furnished with flat nails. Their diet is miscellaneous, consisting partly of insects and partly of fruit.

The *Cebidæ* are exclusively confined to the warmer parts of South America, in the vast forests of which they are met with in large troops, climbing amongst the trees. The Spider Monkeys (*Ateles*), the Howling Monkeys (*Mycetes*), the Capuchin Monkey (*Cebus*), and the Squirrel Monkey (*Callithrix*), may serve as typical examples of this section of the *Quadru-*

mana. In *Ateles* the tail is long, slender, and powerfully prehensile; and the limbs are very long and slender. The pollex is absent, or is quite rudimentary. In *Mycetes* there is a bony drum which is formed by a convexity of the os hyoides and communicates with the larynx. The voice is thus rendered extraordinarily resonant. The pollex is not opposable, but is placed on a line with the other fingers.

CATARHINA.

The third and highest section of the *Quadrumana* is that of the *Catarhina* or Old World Monkeys. In this section the nostrils are oblique, and are placed close together, and the septum narium is narrow. The thumbs of all the feet are opposable, so that the animal is strictly quadrumanous. In *Colobus* alone the anterior thumbs (pollex) are wanting. The dental formula is the same as in man, viz.:—

$$i\,\frac{2-2}{2-2};\ c\,\frac{1-1}{1-1};\ pm\,\frac{2-2}{2-2};\ m\,\frac{3-3}{3-3} = 32.$$

The incisors, however, are projecting and prominent, and the canines—especially in the males—are large and pointed. Moreover, the teeth form an uneven series, interrupted by a diastema or interval. The tail is never prehensile, and is sometimes absent. Cheek-pouches* are often present, and the skin covering the *tubera ischii* is almost always callous and destitute of hair, constituting the so-called "natal callosities." With the single exception of a Monkey which inhabits the Rock of Gibraltar, all the *Catarhina* are natives of Africa and Asia.

There are three well-marked groups or tribes of the Catarhine Monkeys. In the first of these the tail is long, and there are both cheek-pouches and natal callosities. In this tribe is the genus *Semnopithecus*, all the species of which are natives of Asia and its islands. One of the best-known species is the Sacred Monkey of the Hindoos (*Semnopithecus entellus*). Closely allied to the *Semnopitheci* is the genus *Colobus*, in which alone, of all the Catarhine Monkeys, the pollex is either altogether absent or totally rudimentary. Closely allied to *Semnopithecus*, also, is the Proboscis Monkey or Kahau (*Presbytis nasalis*), distinguished by its elongated proboscidiform nose, short pollex, and long tail. Here also come the little Guenons (*Cercocebus* and *Cercopithecus*, fig. 234). Also referable to this

* The cheek-pouches are sacs or cavities in the cheeks, which open into the mouth, and serve to hold any superfluous food.

division is the genus *Macacus* or *Inuus* (comprising the Macaques), which includes most of the Monkeys which are ordinarily brought to this country. It is a Macaque which occurs at the Rock of Gibraltar, and is the only wild Monkey which is found in Europe at the present day. Most of the Macaques are Asiatic, and a good example is the Wanderoo (*M. Silenus*) of India.

The second tribe of the Catarhine Monkeys is that of the Baboons (*Cynocephalus* and *Papio*). In these forms the tail is mostly short, and is often quite rudimentary. The head is large, and the muzzle is greatly prolonged, having the nostrils at its extremity. The facial angle is about 30°, and the whole head has much the aspect of that of a large dog. The natal callosities are generally large and conspicuous, and usually of some bright colour. The Baboons are large strong animals, extremely unattractive in outward appearance, and of great ferocity. More than any other of the Monkeys, they employ the fore-limbs in terrestrial progression, running upon all-fours with the greatest ease. They are mainly inhabitants of Africa, and one of them, the Mandrill (*Cynocephalus Maimon*), attains very nearly the height of a man. The best-known species is the Chacma (*Cynocephalus porcarius*), the Derrias (*C. Hamadryas*), the Common Baboon (*C. papio*), and the Mandrill. The Derrias is found in Arabia and Abyssinia, and occurs both embalmed and sculptured upon ancient monuments in Egypt and Nubia. The Mandrill is rendered probably without exception the most disgustingly hideous of living beings by the possession of large blood-red natal callosities and of enormous cheek-protuberances striped with brilliant colours in alternate ribs.

The third family of the Catarhine Monkeys is that of the Anthropomorphous or Anthropoid Apes, so called from their making a nearer approach in anatomical structure to man than is the case with any other Mammal. The members of this family are Apes in which there is no tail, and cheek-pouches are absent, whilst in some cases there are also no natal callosities. The hind-limbs are short—shorter than the fore-limbs—and the animal can progress in an erect or semi-erect position. At the same time, the thumbs of the hind-feet (hallux) are opposable to the other digits, so that the hind-feet are prehensile hands. The spine shows a single curve, and articulates with the back part of the skull. The canine teeth of the males are long, strong, and pointed, but this is not the case with the females. The structure, therefore, of the canine teeth is to be regarded in the light of a sexual peculiarity, and not as having any connection with the nature of the food.

In this tribe are the Gibbons (*Hylobates*), the Orang-outang (*Simia satyrus*), the Chimpanzee, and the Gorilla.

The Gibbons form the genus *Hylobates*, and they belong to southern Asia and the Indian Archipelago. The anterior limbs are extremely long, and the hands nearly or quite reach the ground when the animal stands in an erect posture. There is no tail, but there are natal callosities. The body is covered with a thick fur. One of the best known of the Gibbons is the Siamang (*Hylobates syndactylus*), which has been sometimes regarded as making a nearer approach to man than any other of the Monkeys. It is a native of Sumatra. It is the largest of the Gibbons, and derives its specific name from the fact that the index and middle toes of the hind-foot are united to one another by skin as far as the nail joint. Another well-known species is the common Gibbon (*H. lar*).

In the Orang or "Mias" (*Simia satyrus*) there are neither cheek-pouches nor natal callosities, and the hips are covered with hair. As in the Gibbons, the arms are excessively long, reaching considerably below the knee when the animal stands in an erect posture. The hind-legs are very short, and there is no tail. When full grown the Orang stands about four feet high. It never progresses with the help of a stick, or walks erect at all, except along the branches of trees, supporting itself by a higher branch, or when attacked. When young, the head

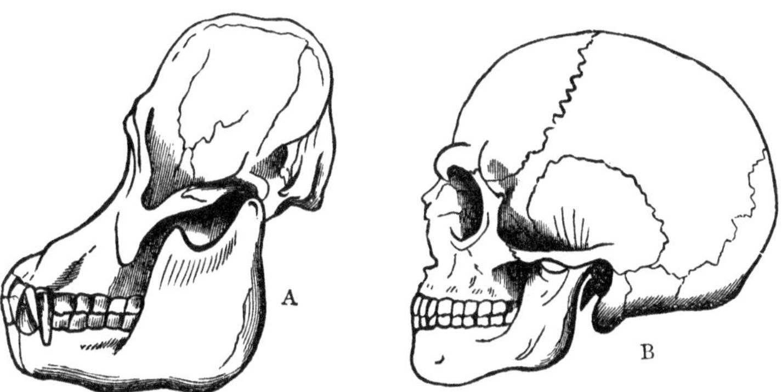

Fig. 235.—A, Skull of the Orang-outang. B, Skull of an adult European.

of the Orang is not very different from that of an average European child; but, as the animal grows, the facial bones become gradually produced, whilst the cranium remains in a tolerably stationary condition; great bony ridges are developed for the attachment of the muscles of the jaws and face; the incisors project; and ultimately the muzzle becomes as pronounced and well-marked a feature as in the typical *Carnivora* (fig. 235, A). The Orangs are inhabitants of Sumatra and Borneo.

The genus *Troglodytes* contains the Chimpanzee (*T. niger*) and the Gorilla (*T. Gorilla*). The Chimpanzee is a native of Western Africa, and has the arms much shorter, proportionately, than in the Gibbons and Orangs; still they are much longer than the hind-limbs, and they reach beneath the knee when the animal stands erect. The ears in the Chimpanzee are large, and the body is covered with dark-brown hair. The animal can stand erect, but the natural mode of progression is on all-fours. The hands are naked to the wrist, and the face is also naked, and is much wrinkled. The Chimpanzee lives in society in wooded districts, constructs huts, and can defend itself against even the Elephant.

The Gorilla is in most respects the same as the Chimpanzee, but is much larger, attaining a height of fully five feet. The hind-limbs are short, and the ears small. It is an enormously strong and ferocious animal, and is found in Lower Guinea and in the interior of equatorial Africa. It possesses a laryngeal sac, has a most appalling voice, and is polygamous. The Gorilla is now generally regarded as the most human of the Anthropoid Apes.

CHAPTER LXXXIV.

BIMANA.

ORDER XIV. BIMANA.—This, the last remaining order of the *Mammalia*, comprises Man (*Homo*) alone, and it will therefore require but little notice here, the peculiarities of Man's mental and physical structure properly belonging to other branches of science.

Zoologically, Man is distinguished from all other Mammals by his habitually erect posture and bipedal progression. The lower limbs are exclusively devoted to progression and to supporting the weight of the body. The anterior limbs are shorter than the posterior, and have nothing whatever to do with progression. The thumb is opposable, and the hands are prehensile, the fingers being provided with nails. The toes of the hind-limb are also furnished with nails, but the *hallux* is not opposable to the other digits, and the feet are therefore useless as organs of prehension. The foot is broad and plantigrade, and the whole sole is applied to the ground in walking.

The dentition consists of thirty-two teeth, and these form a nearly even and uninterrupted series, without any interval or diastema. The dental formula is—

$$i\,\frac{2-2}{2-2};\; c\,\frac{1-1}{1-1};\; pm\,\frac{2-2}{2-2};\; m\,\frac{3-3}{3-3}=32$$

The brain is more largely developed and more abundantly furnished with large and deep convolutions than is the case with any other Mammal. The mammæ are pectoral, and the placenta is discoidal and deciduate.

Man is the only terrestrial Mammal in which the body is not provided with a covering of hair.

The zoological or *anatomical* distinctions between Man and the other Mammals are thus seen to be of no very striking nature, and certainly *of themselves* would not entitle us to consider Man as forming more than a distinct *order*. When, however, we take into account the vast and illimitable *psychical* differences, both intellectual and moral — differences which *must* entail corresponding structural distinctions — between Man and the highest *Quadrumana*, it becomes a question whether the group *Bimana* should not have the value of a distinct sub-kingdom; whilst there can be little hesitation in giving Man, at any rate, a class to himself. At any rate, man's psychical peculiarities are as much an integral portion, or more, of his totality, as are his physical characters, and, as Dr Pritchard says,— "The sentiments, feelings, sympathies, internal consciousness, and mind, and the habitudes of mind and action thence resulting, are the real and essential characteristics of humanity."

CHAPTER LXXXV.

DISTRIBUTION OF MAMMALIA IN TIME.

As a matter of course, the remains of Mammals are scanty, and occupy but a small space in the geological record, since the greater number of the *Mammalia* are terrestrial, and the greater number of the stratified fossiliferous deposits are marine. The Mammals, too, are the most highly organised of the entire sub-kingdom of the *Vertebrata;* and therefore, in obedience to the well-known law of succession, they ought to make their appearance upon the globe at a later period than any of the lower classes of the *Vertebrata.* Such, in point of fact, is to a great extent the case; and if the geological record were perfect, the law would doubtless be carried out to its full extent.

It is in the upper portion of the Triassic Rocks—that is to say, not long after the commencement of the Mesozoic or Secondary epoch—that Mammals for the first time make their appearance; four species being now known in a zone of rocks which are placed at the summit of the Trias, just where this formation begins to pass into the Lias. The earliest of these—the oldest known of all the Mammals—appears at the upper part of the Upper Trias (Keuper) and also at its very summit (Penarth beds), and has been described under the name of *Microlestes antiquus.* The nearest ally of *Microlestes* amongst existing Mammals would seem to be the Marsupial and insectivorous *Myrmecobius*, or Banded Ant-eater of Australia. As only the teeth, however, of *Microlestes* have hitherto been discovered, it is impossible to decide positively whether this primeval Mammal was Marsupial or Placental.

The next traces of Mammals occur in the Stonesfield Slate (Lower Oolites), and here four species, all of small size, are known to occur. Most of these were Marsupial, but it is possible that one was placental. They form the genera *Amphilestes*, *Amphitherium*, *Phascolotherium*, and *Stereognathus.* After the Stonesfield Slate another interval succeeds, in which no Mammalian remains have hitherto been found; but in the freshwater formation of the Middle Purbeck—at the top, namely, of the Oolitic series—as many as fourteen small Mammals have been discovered. These constitute the genera *Plagiaulax*, *Triconodon*, and *Galestes.* Another gap then follows, no Mammal having hitherto been discovered in any portion of the Cretaceous series (with doubtful exceptions).

Leaving the Mesozoic and entering upon the Kainozoic period, remains of Mammals are never absent from any of the geological formations. From the base of the Eocene Rocks up to the present day remains of Mammals invariably occur, constantly increasing in number and importance, till we arrive at the fauna now in existence upon the globe.

The number of known fossil Mammals is so great, and they exhibit so many peculiarities and divergences from existing forms, that it will be impossible here to do more than simply point out the leading facts known as to the distribution of each order of Mammals in past time.

Order I. Monotremata.—The Monotremes are not known to be represented at all in past time; and this need not excite any surprise, seeing that the order is represented at the present day by no more than two genera, both confined to a single geographical region. Upon theoretical grounds, however, it may be expected that we shall ultimately discover that the antiquity of the order *Monotremata* is extremely high.

Order II. Marsupialia.—This is probably the oldest of the Mammalian orders; but owing to the detached and fragmentary condition of almost all Mammalian remains—consisting mostly of the ramus of the lower jaw, or of separate teeth—it is not possible to state this with absolute certainty. The *Microlestes* of the Trias, the oldest, or nearly the oldest, of the Mammals, was *probably* a Marsupial; but the evidence upon this point is not conclusive. In the Triassic rocks of America, also, perhaps at a lower horizon than that at which *Microlestes* occurs in Europe, has been found the jaw of a small Mammal, which is probably Marsupial, and has been named *Dromatherium* (fig. 236).

Fig. 236.—Lower jaw of *Dromatherium sylvestre* (after Emmons). From rocks, supposed to be of Triassic age, in North Carolina.

In the next mammaliferous horizon, however—namely, that of the Stonesfield Slate in the Lower Oolites—there is no doubt but that some of the Mammalian remains, if not all, belong to small Marsupials (fig. 237). From this horizon the two genera *Phascolotherium* and *Amphitherium*, are almost certainly referable to the *Marsupialia;* the latter seeming to be most nearly related to the living *Myrmecobius,* whilst the former finds its nearest living ally in the Opossums of America. The *Stereognathus* of the Stonesfield Slate is in a doubtful position. It may have been Marsupial; but, upon the whole, Professor Owen is inclined to believe that it was placental, hoofed, and herbivorous.

With the occurrence of small Marsupials in England within the Oolitic period, it is interesting to notice how the fauna of that time approached in other respects to that now inhabiting Australia. At the present day, Australia is almost wholly tenanted by Marsupials; upon its land-surface flourish *Araucariæ* and Cycadaceous plants, and in its seas swims the Port Jackson Shark (*Cestracion Philippi*); whilst the Molluscan genus *Trigonia* is nowadays exclusively confined to the Australian coasts. In England, at the time of the deposition of the

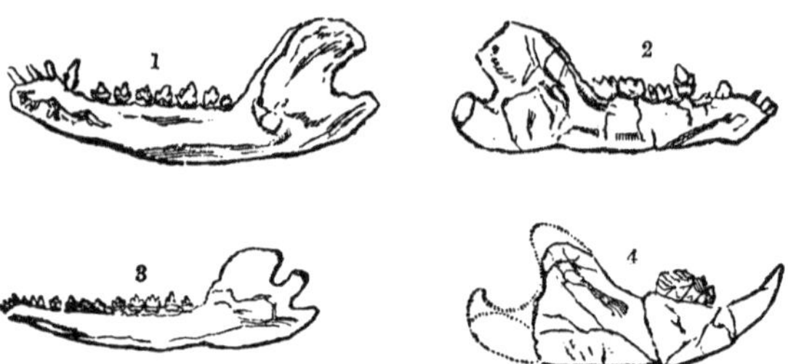

Fig. 237.—Oolitic Mammals, natural size. 1. Lower jaw and teeth of *Phascolotherium*; 2. of *Triconodon*; 3. of *Amphitherium*; 4. of *Plagiaulax*.

Stonesfield Slate, we must have had a fauna and flora very closely resembling what we now see in Australia. The small Marsupials, *Amphitherium* and *Phascolotherium*, prove that the Mammals were the same in order; cones of Araucarian pines, with tree-ferns and fronds of Cycads, occur throughout the Oolitic series; spine-bearing fishes, like the Port Jackson Shark, are abundantly represented by genera such as *Acrodus* and *Strophodus;* and, lastly, the genus *Trigonia*, now exclusively Australian, is represented in the Stonesfield Slate by species which differ little from those now existing.

In the middle Purbeck beds (Upper Oolite), where fourteen species of Mammals are known to exist, it is probable that all were Marsupial. All the Purbeck Mammalia were of small size, the largest being no bigger than a pole-cat or hedgehog. They form the genera *Plagiaulax*, *Triconodon*, and *Galestes*, of which *Plagiaulax* is believed to be most nearly allied to the living Kangaroo-rat (*Hypsiprymnus*) of Australia.

In the Tertiary series of rocks Marsupials are of rare occurrence; but an Opossum, closely allied to the existing American forms, has been discovered in the Eocene rocks of France (Gypseous series of Montmartre), and has been named the *Didelphys gypsorum*.

The next occurrence of Marsupials is in the later Tertiary (Pliocene) and in the Post-tertiary epoch; and here they are represented by some very remarkable forms. The remains in question have been found in the bone-caves of Australia—the country in which Marsupials now abound above every other part of the globe; and they show that Australia, at no distant geological period, possessed a Marsupial fauna, much resembling that which it has at present, but comparatively of a much more gigantic size. In the remains from the Australian bone-caves almost all the most characteristic living Marsupials of Australia and Van Diemen's Land are represented; but the extinct forms are usually of much greater size. We have Wombats, Phalangers, Flying Phalangers, and Kangaroos, with carnivorous Marsupials resembling the recent *Thylacinus* and *Dasyurus*. The two most remarkable of these extinct forms are *Diprotodon* and *Thylacoleo*. In most essential respects *Diprotodon* resembled the Kangaroos, the dentition, especially, showing many points of affinity. The hind-limbs, however, of *Diprotodon* were by no means so disproportionately long as in the Kangaroos. In size *Diprotodon* must have many times exceeded the largest of the living Kangaroos, since the skull measures three feet in length (fig. 238). *Thylacoleo* was a carnivorous and predacious Marsupial, equally gigantic when compared with living forms. *Thylacoleo*, in fact, must have been, on a moderate estimate, at least as large as a Lion; the largest living carnivorous Marsupial being no larger than a shepherd's dog. The flesh-tooth or carnassial molar of *Thylacoleo* measures two inches and a quarter across, or very nearly double the measurement of the same tooth in the largest existing Lion.

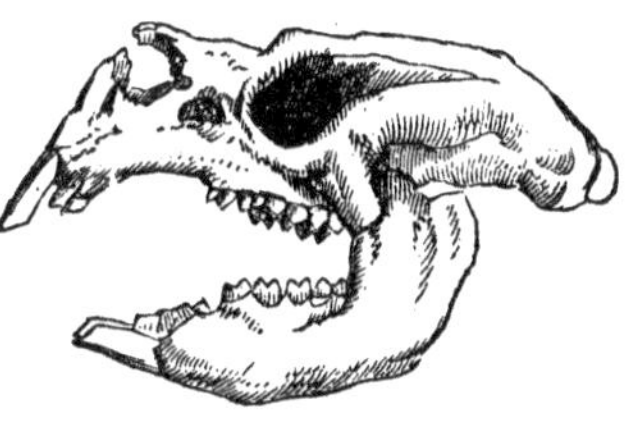
Fig. 238.—Skull of *Diprotodon Australis*.

Order III. Edentata.—The Edentates, like the Marsupials, are singularly circumscribed at the present day. No member of the order is at the present day indigenous in Europe. Tropical Asia and Africa have the Scaly Ant-eaters or Pangolins; and in Africa occurs the Edentate genus *Orycteropus*. South America, however, is the metropolis of the *Edentata*, the order being there represented by the Sloths, the Armadillos, and the true Ant-eaters. It is also in South America that by far the greater number of extinct Edentates have been found; and, as in the case of the Australian Marsupials, the fossil forms are gigantic in size compared with their living representatives.

The Sloths (*Bradypodidæ*) of the present day were represented in Post-tertiary times by a group of gigantic forms referable to the genera *Mylodon*, *Megalonyx*, and *Megatherium*. Of these, *Mylodon* attained a length of eleven feet, and *Mega-*

therium (fig. 239) was eighteen feet in length, with bones as massive, or more so, than the Elephant.

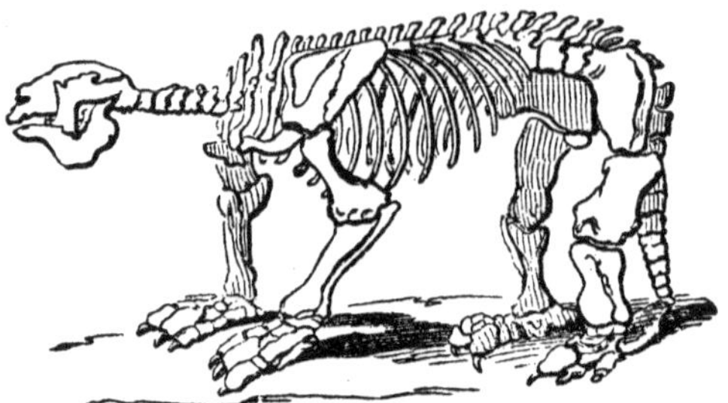

Fig. 239.—*Megatherium.* From the Upper Tertiaries of South America (Pleistocene).

In the same way the little banded Armadillos of South America were formerly represented by gigantic species, constituting the genus *Glyptodon.* The *Glyptodons* (fig. 240) differed from the living Armadillos in having no bands in their armour, so that they must have been unable to roll themselves up. It is rare at the present day to meet with any Armadillo over two or three feet in length; but the length of the *Glyptodon clavipes*, from the tip of the snout to the end of the tail, was more than nine feet.

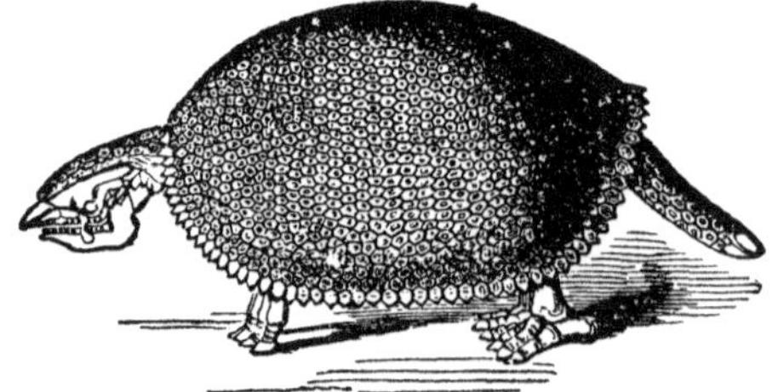

Fig. 240.—*Glyptodon clavipes.* Pleistocene deposits of South America.

All these gigantic South American Edentates occur in Post-tertiary deposits. Older, however, than any of these is the *Macrotherium.* This is a gigantic Edentate, intermediate in some respects between the *Pangolins* and *Orycteropus*, and found in certain lacustrine deposits of France, of Miocene age.

Order IV. Sirenia.—This order contains only the living Manatees and Dugongs, and is of little geological importance. The *Halitherium*, however, of the Eocene, Miocene, and Pliocene rocks, is a large form, intermediate between the African Manatee and the Dugong.

Order V. Cetacea.—The *Cetacea*, also, are of little geological importance. Remains of Dolphins (*Ziphius*) and of Whales (*Balænodon*) are found in Miocene deposits; and numerous ear-bones of Whales occur in the Red Crag of Suffolk (Pliocene). The most remarkable, however, of the extinct *Cetacea*, is the *Zeuglodon* of the American and Maltese Miocene deposits. This was an enormous toothed Whale, about seventy feet in length; but, unlike any of existing Cetaceans, it had the posterior teeth implanted by two distinct fangs or roots. By Owen, *Zeuglodon* is regarded as the type of a distinct family, intermediate between the *Cetacea* and the *Sirenia*. In *Saurocetes*, a large Cetacean from the Tertiary beds of Buenos Ayres, there were double-fanged teeth with conoid crowns. It was much smaller than *Zeuglodon*.

Order VI. Ungulata.—The hoofed Mammals are represented in past time by so many extinct forms that it will be wholly impossible here to do more than merely allude to some of the more important genera.

The earliest-known Ungulates occur in the Eocene rocks, where the order is represented by very numerous and interesting forms, the more important of which are *Pliolophus*, *Palæotherium*, and *Anoplotherium*.

Of the section of the Ungulates comprising the living Horse, Zebra, and Ass (*Solidungula*), the earliest fossil example is the *Anchitherium* of the Lower Miocene, and this was succeeded by the *Hipparion* of the Miocene rocks. This genus differed from the existing *Equidæ* in the presence of two small toes with hoofs, one on each side of the single functional toe, which alone remains in living horses. In the Pliocene period appear, for the first time, remains of horses which, like the present form, possessed only a single toe encased in a single hoof. It is interesting to observe that one of the Pliocene horses (*Equus curvidens*) occurs in South America; though this continent certainly possessed no native horse at the time of its discovery by the Spaniards. About twenty horses—one of them standing no more than two and a half feet in height—have been described from North America, in which continent no indigenous horse existed at the time of its discovery.

Of the *Rhinoceridæ*, a hornless species (*Acerotherium*) occurs in Miocene and Pliocene strata; but the best-known fossil species is the two-horned woolly Rhinoceros (*R. tichorhinus*). This curious species occurs in Post-pliocene deposits, and must have ranged over the greater part of Europe. It was adapted to a temperate climate, and, like the Mammoth, possessed a thick covering of mixed wool and hair. This has been demonstrated by the discovery of a frozen carcass in Siberia.

Of the *Hippopotamidæ*, the earliest-known species is the *Hippopotamus major* of the Pliocene period. This form agreed in all essential respects with the living *H. amphibius* of Africa, but it must have ranged over the whole of Southern Europe. *Hexaprotodon*, of the Tertiary deposits of the Sivalik Hills of India, is closely allied to *Hippopotamus*, but had six lower incisors.

Of the *Suida*, or Pig tribe, various extinct forms are known from the Eocene and Miocene rocks, where the family is represented by the genera *Chæropotamus*, *Anthracotherium*, *Hyopotamus*, and *Hippohyus*.

As regards the past existence of the Ruminants, the *Cervidæ*, or Stag tribe, is represented, for the first time in the Miocene period, by the genus *Dorcatherium*. The best-known species, however, of this family is the *Megaceros Hibernicus*, or so-called Irish Elk (fig. 241), which is not a true Elk, but is intermediate

Fig. 241.—The Irish Elk (*Megaceros Hibernicus*).

between the Fallow-deer and Reindeer. A fossil Camel (*C. Sivalensis*) has been discovered in the Tertiary deposits of the Sivalik Hills of India. Of the Giraffe family—represented at the present day by a single African species—a form has been discovered in the Pliocene rocks of Greece, and has been described under the name of *Helladotherium*. Somewhat similar forms have been found in the Pliocene deposits of the Sivalik Hills of India.

The earliest-known Antelopes are Miocene, but the largest and most extraordinary fossil examples of this family are two gigantic four-horned Antelopes, which occur in the Pliocene strata of the Sivalik Hills of India, and have been described under the names of *Sivatherium* and *Bramatherium*.

The *Bovidæ*, or Ox tribe, has hitherto only occurred in rocks not older than the Pliocene or Post-pliocene. At this latter period England alone possessed four oxen—viz., the Lithuanian Aurochs (*Bos bison* or *Bos priscus*), the Wild Bull or Urus (*Bos primigenius*), the *Bos antiquus*, and a small aboriginal species, the *Bos longifrons*, believed by Owen to be "the source of the domesticated cattle of the Celtic races before the Roman invasion."

Order VII. Hyracoidea.—This little order, represented at the present day by no more than the single genus *Hyrax*, is not known to have any fossil representatives.

Order VIII. Proboscidea.—This order, including no other living forms than the Elephants, came into existence in the Miocene period, where it is represented by all its three sections, *Deinotherium*, *Mastodon*, and *Elephas*.

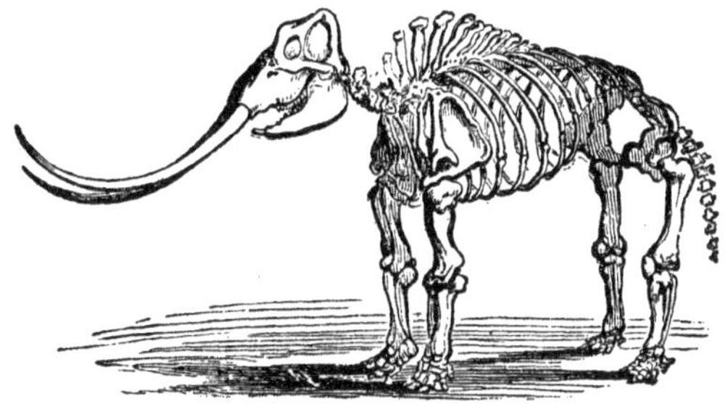

Fig. 242.—Skeleton of *Mastodon*.

The *Deinotherium* (fig. 222) was a gigantic Miocene Mammal, probably something like the living Elephants, but having no incisors in the upper jaw. In place of these, the *lower* jaw was furnished with two long tusk-like incisors, which were bent downwards.

In most essential respects the Mastodons (fig. 242) resemble the Elephants, but the molar teeth were furnished with nipple-shaped eminences. Usually there are two tusk-shaped upper incisors, but sometimes lower incisors are present as well

Four Mastodons occur in the Miocene of Europe, and three in that of India.

No Elephant has yet been discovered in the Miocene rocks of Europe, but six species are known from Miocene strata in India. In the Pliocene period Europe possessed its Elephants (viz., *E. priscus* and *E. meridionalis*); but the best known of the extinct Elephants, as well as the most modern, is the Mammoth (*E. primigenius*, fig. 243). The Mammoth enjoyed

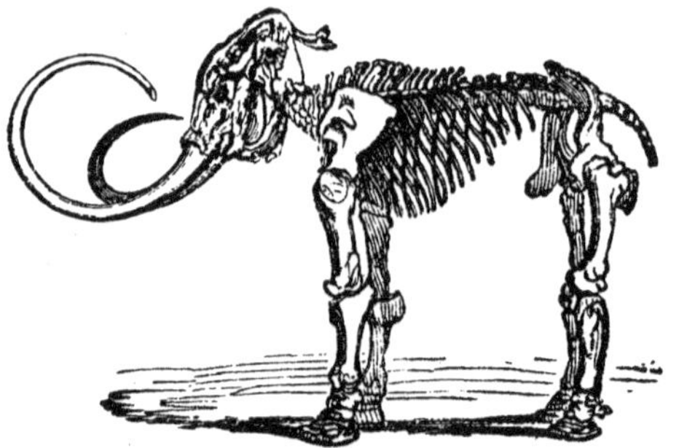

Fig. 243.—Skeleton of the Mammoth (*Elephas primigenius*).

a very extended geographical distribution, remains of it occurring in Britain, continental Europe, Siberia, and throughout a large portion of North America. There can also be no question but that the Mammoth existed in the earlier portion of the human period.

Order IX. Carnivora.—If the little *Microlestes* of the Upper Trias be Marsupial, as is most probably the case, then the order *Carnivora* is comparatively modern, the earliest undoubted remains having been found in the Eocene rocks. The tribe of the *Felidæ* is represented in the Miocene period by the large *Machairodus*, with sabre-shaped upper canines. Species of this genus must have been as large as a Lion. In the later Pliocene and Post-pliocene deposits occur the remains of a large Lion—the Cave-lion or *Felis spelæa*—along with which, in Britain and continental Europe, are the bones of a large Hyæna (*H. spelæa*) and a gigantic Bear (*Ursus spelæus*). Remains of Wolves, Foxes, Badgers, Otters, Pole-cats, Weasels, and other *Carnivora* are also found in various later Tertiary deposits, and in bone-caves.

Order X. Rodentia.—No Rodent animal is as yet known to have occurred earlier than the Eocene period. Here are found forms allied to the living Dormouse and Squirrel. In the

Miocene rocks occur numerous small Rodents. In the Pliocene and Post-pliocene deposits the order is also well represented, the most remarkable form being the Great Beaver (*Trogontherium*), which appears to have survived into the historical period.

Order XI. Cheiroptera.—The earliest-known indications of Bats are in the Eocene period, but the order is of no geological importance.

Order XII. Insectivora.—The Insectivorous Mammals, likewise, commenced their existence, so far as is known, in the Eocene period; and they, also, are of no importance from a geological point of view.

Order XIII. Quadrumana.—The earliest-known remains of *Quadrumana* occur in the Miocene period. Several genera are known, but the most important are *Pliopithecus* and *Dryopithecus*, both of which are European, and both of which belong to the section of the Catarhine Monkeys which are at present characteristic of the Old World. They appear to be most nearly allied to the recent Gibbons. It is interesting to notice that the American fossil Monkeys—from the later Tertiary deposits of South America—belong to the division of the *Quadrumana* now peculiar to that continent — to the section, namely, of the Platyrhine Monkeys.

Geographical Succession of Organic Forms.

A few words may be said here on a law which may be called the "law of the geographical succession of organic forms," and which is illustrated more completely by the *Mammalia* than by any other extinct animals. An examination, namely, of the facts of the geological distribution of Mammals leads to the striking generalisation that "the present distribution of organic *forms* dates back to a period anterior to the origin of existing *species*" (Lyell). In other words, though the extinct Mammals of the later geological deposits of any given country differ *specifically* from those now existing in the same country, they are nevertheless referable to the same *orders*, and are in every respect more closely allied to the present Mammalian fauna than to that of any other country. A few examples will render this perfectly clear.

Australia at the present day is an altogether peculiar zoological province, characterised by the abundance and variety of Marsupials which inhabit it. In the Post-tertiary deposits of Australia, however, we are presented with proofs that Marsupials were just as characteristic of Australia during late geological epochs as they are now. In the Post-pliocene period we know that Australia was occupied by Kangaroos, Kangaroo-rats, Wombats, Phalangers, and Carnivorous Marsupials, in every way *representing* the living Marsupials in zoological value, but *specifically* distinct, and generally of gigantic size.

In the same way, South America at the present day is especially characterised by a Mammalian fauna, containing many peculiar forms, the *Edentata* being especially conspicuous, and having a larger representation than in any other region. Similar but distinct forms, however, are found to have existed in South America anterior to the creation of any existing species. Thus, the modern Sloths of South America are represented by the colossal *Mylodon*, *Megalonyx*, *Scelidotherium*, and *Megatherium*. The little armour-plated Armadillos are represented by the equally colossal *Glyptodon*. The Llamas—representing in South America the Camels of the Old World—are represented by the curious extinct genus *Macrauchenia*. The Platyrhine Monkeys have their extinct representatives. Fossil Tapirs take the place of the two existing species; and the Peccaries are represented by at least five extinct species of *Dicotyles*.

Similarly, India is at present the only country in which four-horned Antelopes occur; and it is in the Sivalik Hills that there have been found the two gigantic four-horned Antelopes which constitute the genera *Sivatherium* and *Bramatherium*.

In Europe, again, the Mammalian fauna of the later Tertiary periods is much more closely allied to that now characterising the Old World, than to that of the New. We have the Lion, Bear, Wolf, Fox, and other well-known *Carnivora*. Elephants, Rhinoceroses, and Hippopotami, then as now, are characteristic Old World forms. The Ruminants are equally characteristic of the eastern hemisphere, though not exclusively confined to it, and they have numerous and varied representatives in later Tertiary deposits. The Giraffe is represented by the *Helladotherium*, and the Bactrian Camel by the *Merycotherium* of the Siberian Drift. The fossil *Quadrumana*, too, of Europe, all belong to the Catarhine section of the order.

It is unnecessary to pursue the subject further, but no law is more firmly established than this: "That with extinct as with existing Mammalia, particular forms were assigned to particular provinces; and that the same forms were restricted to the same provinces at a former geological period as they are at the present day" (Owen). It is to be borne in mind, however, that the law, as just stated, holds good for the later Tertiary period only, and does not apply, in any manner that admits of being traced, to the earlier geological epochs.

TABULAR VIEW OF THE CHIEF SUBDIVISIONS OF THE SUB-KINGDOM VERTEBRATA.

SUB-KINGDOM VI.—VERTEBRATA.

CLASS I. PISCES.
- Order 1. Pharyngobranchii.
- 2. Marsipobranchii.
- 3. Teleostei.
- 4. Ganoidei.
- 5. Elasmobranchii.
- 6. Dipnoi.

CLASS II. AMPHIBIA.
- Order 1. Labyrinthodonta.
- 2. Ophiomorpha.
- 3. Urodela.
- 4. Anoura.

CLASS III. REPTILIA.
- Order 1. Chelonia.
- 2. Ophidia.
- 3. Lacertilia.
- 4. Crocodilia.
- 5. Ichthyopterygia.
- 6. Sauropterygia.
- 7. Anomodontia.
- 8. Pterosauria.
- 9. Deinosauria.

CLASS IV. AVES.
- Order 1. Natatores.
- 2. Grallatores.
- 3. Cursores.
- 4. Rasores.
  - *Sub-order a.* Gallinacei.
  - *b.* Columbacei.
- 5. Scansores.
- 6. Insessores.
  - *Sub-order a.* Conirostres.
  - *b.* Dentirostres.
  - *c.* Tenuirostres.
  - *d.* Fissirostres.
- 7. Raptores.
- 8. Saururæ.

CLASS V. MAMMALIA.

Division A. Ornithodelphia.
Order 1. Monotremata.
Division B. Didelphia.
Order 2. Marsupialia.
Division C. Monodelphia.
Order 3. Edentata.
4. Sirenia.
5. Cetacea.
6. Ungulata.
Section Perissodactyla.
a. Multungula.
b. Solidungula.
Section Artiodactyla.
a. Omnivora.
b. Ruminantia.
7. Hyracoidea.
8. Proboscidea.
9. Carnivora.
a. Pinnigrada.
b. Plantigrada.
c. Digitigrada.
10. Rodentia.
11. Cheiroptera.
12. Insectivora.
13. Quadrumana.
a. Strepsirhina.
b. Platyrhina.
c. Catarhina.
14. Bimana.

GLOSSARY.

Abdomen (Lat. *abdo*, I conceal). The posterior cavity of the body, containing the intestines and others of the viscera. In many Invertebrates there is no separation of the body-cavity into thorax and abdomen, and it is only in the higher *Annulosa* that a distinct abdomen can be said to exist.

Aberrant (Lat. *aberro*, I wander away). Departing from the regular type.

Abnormal (Lat. *ab*, from; *norma*, a rule). Irregular; deviating from the ordinary standard.

Abomasum. The fourth cavity of the complex stomach of the Ruminants.

Abranchiate (Gr. *a*, without; *bragchia*, gills). Destitute of gills or branchiæ.

Acalephæ (Gr. *akalephe*, a nettle). Applied formerly to the Jelly-fishes or Sea-nettles, and other Radiate animals, in consequence of their power of stinging, derived from the presence of microscopic cells, called "thread-cells," in the integument.

Acanthocephala (Gr. *akantha*, a thorn; *kephale*, head). A class of parasitic worms in which the head is armed with spines.

Acanthometrina (Gr. *akantha*; and *metra*, the womb). A family of *Protozoa*, characterised by having radiating siliceous spines.

Acanthopterygii (Gr. *akantha*, spine; *pterux*, wing). A group of bony fishes with spinous rays in the front part of the dorsal fin.

Acarina (Gr. *akari*, a mite). A division of the *Arachnida*, of which the Cheese-mite is the type.

Acephalous (Gr. *a*, without; *kephale*, head). Not possessing a distinct head.

Acetabula (Lat. *acetabulum*, a cup). The suckers with which the cephalic processes of many *Cephalopoda* (Cuttle-fishes) are provided.

Acetabulum. The cup-shaped socket of the hip-joint in Vertebrates.

Acrita (Gr. *akritos*, confused). A term sometimes employed as synonymous with *Protozoa* or the lowest division of the animal kingdom.

Actinomeres (Gr. *aktin*, a ray; *meros*, a part). The lobes which are mapped out on the surface of the body of the *Ctenophora*, by the ctenophores, or comb-like rows of cilia.

Actinosoma (Gr. *aktin*; and *soma*, body). Employed to designate the entire body of any *Actinozoön*, whether this be simple (as in the Sea-anemones), or composed of several zoöids (as in most Corals).

Actinozoa (Gr. *aktin*; and *zoön*, an animal). That division of the *Cœlenterata* of which the Sea-anemones may be taken as the type.

Adelarthrosomata (Gr. *adelos*, hidden; *arthros*, joint; *soma*, body). An order of the *Arachnida*.

Agamic (Gr. *a*, without; *gamos*, marriage). Applied to all forms of reproduction in which the sexes are not directly concerned.

Allantoidea. The group of *Vertebrata* in which the fœtus is furnished with an allantois, comprising the Reptiles, Birds, and Mammals.

Allantois (Gr. *allas*, a sausage). One of the "membranes" of the fœtus in certain Vertebrates.

ALVEOLI (Lat. dim. of *alvus*, belly). Applied to the sockets of the teeth.

AMBULACRA (Lat. *ambulacrum*, a place for walking). The perforated spaces or "avenues" through which are protruded the tube-feet, by means of which locomotion is effected in the *Echinodermata*.

AMBULATORY (Lat. *ambulo*, I walk). Formed for walking. Applied to a single limb, or to an entire animal.

AMETABOLIC (Gr. *a*, without; *metabole*, change). Applied to those insects which do not possess wings when perfect, and which do not, therefore, pass through any marked metamorphosis.

AMNION (Gr. *amnos*, a lamb). One of the fœtal membranes of the higher Vertebratès.

AMNIOTA. The group of *Vertebrata* in which the fœtus is furnished with an amnion, comprising the Reptiles, Birds, and Mammals.

AMŒBA (Gr. *amoibos*, changing). A species of Rhizopod, so called from the numerous changes of form which it undergoes.

AMŒBIFORM. Resembling an *Amœba* in form.

AMORPHOZOA (Gr. *a*, without; *morphe*, shape; *zoön*, animal). A name sometimes used to designate the *Sponges*.

AMPHIBIA (Gr. *amphi*, both; *bios*, life). The Frogs, Newts, and the like, which have gills when young, but can always breathe air directly when adult.

AMPHICŒLOUS (Gr. *amphi*, at both ends; *koilos*, hollow). Applied to vertebræ which are concave at both ends.

AMPHIDISCS (Gr. *amphi*, at both ends; *diskos*, a quoit, or round plate). The spicula which surround the gemmules of *Spongilla*, and resemble two toothed wheels united by an axle.

AMPHIOXUS (Gr. *amphi*, at both ends; *oxus*, sharp). The Lancelet, a little fish, which alone constitutes the order *Pharyngobranchii*.

AMPHIPNEUSTA (Gr. *amphi*, both; *pneo*, I breathe). Applied to the "perennibranchiate" Amphibians which retain their gills through life.

AMPHIPODA (Gr. *amphi*; and *pous*, a foot). An order of *Crustacea*.

ANAL (Lat. *anus*, the vent). Connected with the anus, or situated near the anus.

ANALLANTOIDEA. The group of *Vertebrata* in which the embryo is not furnished with an allantois.

ANALOGOUS. Applied to parts which perform the same function.

ANAMNIOTA. The group of *Vertebrata* in which the embryo is destitute of an amnion.

ANARTHROPODA (Gr. *a*, without; *arthros*, a joint; *pous*, foot). That division of *Annulose* animals in which there are no articulated appendages.

ANCHYLOSIS or ANKYLOSIS (Gr. *ankulos*, crooked). The union of two bones by osseous matter, so that they become one bone, or are immovably joined together.

ANDROGYNOUS (Gr. *anēr*, a man; *gune*, a woman). Synonymous with hermaphrodite, and implying that the two sexes are united in the same individual.

ANDROPHORES (Gr. *anēr*, a man; and *phero*, I carry). Applied to medusiform gonophores of the *Hydrozoa*, which carry the spermatozoa, and differ in form from those in which the ova are developed.

ANNELIDA (a Gallicised form of *Annulata*). The Ringed Worms, which form one of the divisions of the *Anarthropoda*.

ANNULATED. Composed of a succession of rings.

ANNULOIDA (Lat. *annulus*, a ring; Gr. *eidos*, form). The sub-kingdom comprising the *Echinodermata* and the *Scolecida* (=*Echinozoa*).

ANNULOSA (Lat. *annulus*). The sub-kingdom comprising the *Anarthropoda* and the *Arthropoda* or *Articulata*, in all of which the body is more or less evidently composed of a succession of rings.

ANOMODONTIA (Gr. *anomos*, irregular; *odous*, tooth). An extinct order of Reptiles, often called *Dicynodontia*.

ANOMURA (Gr. *anomos*, irregular; *oura*, tail). A tribe of Decapod *Crustacea*, of which the Hermit-crab is the type.

ANOPLURA (Gr. *anoplos*, unarmed; *oura*, tail). An order of Apterous Insects.

ANOURA (Gr. *a*, without; oura, tail). The order of *Amphibia* comprising the Frogs and Toads, in which the adult is destitute of a tail. Often called *Batrachia*.

ANTENNÆ (Lat. *antenna*, a yard-arm). The jointed horns or feelers possessed by the majority of the *Articulata*.

ANTENNULES (dim. of *antennæ*). Applied to the smaller pair of antennæ in the *Crustacea*.

ANTIBRACHIUM (Gr. *anti*, in front of; *brachion*, the arm). The fore-arm of the higher Vertebrates, composed of the *radius* and *ulna*.

ANTLERS. Properly the branches of the horns of the Deer tribe (*Cervidæ*), but generally applied to the entire horns.

ANTLIA (Lat. *antlia*, a pump). The spiral trunk or proboscis with which Butterflies and other Lepidopterous Insects suck up the juices of flowers.

APHANIPTERA (Gr. *aphanos*, inconspicuous; *pteron*, a wing). An order of Insects comprising the Fleas.

APLACENTALIA. The section of the *Mammalia*, comprising the two divisions of the *Didelphia* and *Monodelphia*, in which the young is not furnished with a placenta.

APODA (Gr. *a*, without; *podes*, feet). Applied to those fishes which have no ventral fins. Also to the footless *Cæciliæ* amongst the *Amphibia*.

APODAL. Devoid of feet.

APODEMATA (Gr. *apodaio*, I portion off). Applied to certain chitinous septa which divide the tissues in *Crustacea*.

APTERA (Gr. *a*, without; *pteron*, a wing). A division of Insects, which is characterised by the absence of wings in the adult condition.

APTEROUS. Devoid of wings.

APTERYX (Gr. *a*, without; *pterux*, a wing). A wingless bird of New Zealand, belonging to the order *Cursores*.

ARACHNIDA (Gr. *arachne*, a spider). A class of the *Articulata*, comprising Spiders, Scorpions, and allied animals.

ARBORESCENT. Branched like a tree.

ARCHÆOPTERYX (Gr. *archaios*, ancient; *pterux*, wing). The singular fossil bird which alone constitutes the order of the *Saururæ*.

ARCHENCEPHALA (Gr. *archo*, I overrule; *egkephalos*, brain). The name applied by Owen to his fourth and highest group of *Mammalia*, comprising Man alone.

ARENACEOUS. Sandy, or composed of grains of sand.

ARTICULATA (Lat. *articulus*, a joint). A division of the animal kingdom, comprising Insects, Centipedes, Spiders, and Crustaceans, characterised by the possession of jointed bodies or jointed limbs. The term *Arthropoda* is now more usually employed.

ARTIODACTYLA (Gr. *artios*, even; *daktulos*, a finger or toe). A division of the hoofed quadrupeds (*Ungulata*) in which each foot has an even number of toes (two or four).

ASCIDIOIDA (Gr. *askos*, a bottle; *eidos*, a form). A synonym of *Tunicata*, a class of Molluscous animals, which have the shape, in many cases, of a two-necked bottle.

ASEXUAL. Applied to modes of reproduction in which the sexes are not concerned.

ASIPHONATE. Not possessing a respiratory tube or siphon. (Applied to a division of the *Lamellibranchiate* Molluscs.)

ASTEROID (Gr. *aster*, a star; and *eidos*, form). Star-shaped, or possessing radiating lobes or rays like a star-fish.

ASTEROIDEA. An order of *Echinodermata*, comprising the Star-fishes, characterised by their rayed form.

ASTOMATOUS (Gr. *a*, without; *stoma*, mouth). Not possessing a mouth.

ATLAS (Gr. the god who holds up the earth). The first vertebra of the neck, which articulates with and supports the skull.

ATRIUM (Lat. a hall). Applied to the great chamber or "cloaca," into which the intestine opens in the *Tunicata*.

AURELLA (Lat. *aurum*, gold). Applied to the chrysalides of some *Lepidoptera*, on account of their exhibiting a golden lustre.
AURICLE (Lat. dim. of *auris*, ear). Applied to one of the cavities of the heart, by which blood is driven into the ventricle.
AUTOPHAGI (Gr. *autos*, self; *phago*, I eat). Applied to birds whose young can run about and obtain food for themselves as soon as they escape from the egg.
AVES (Lat. *avis*, a bird). The class of the Birds.
AVICULARIUM (Lat. *avicula*, dim. of *avis*, a bird). A singular appendage, often shaped like the head of a bird, found in many of the *Polyzoa*.
AXIS (Gr. *axon*, a pivot). The second vertebra of the neck, upon which the skull and atlas usually rotate.
AZYGOS (Gr. *a*, without; *zugos*, yoke). Single; without a fellow.

BACTERIUM (Gr. *bakterion*, a staff). A kind of staff-shaped filament which appears in organic infusions after they have been exposed to the air.
BALANIDÆ (Gr. *balanos*, an acorn). A family of sessile *Cirripides*, commonly called "Acorn-shells."
BALEEN (Lat. *balæna*, a whale). The horny plates which occupy the palate of the true or "whale-bone" Whales.
BATIDES (Gr. *batos*, a bramble). The family of the *Elasmobranchii* comprising the Rays.
BATRACHIA (Gr. *batrachos*, a frog). Often loosely applied to any of the *Amphibia*, but sometimes restricted to the Amphibians as a class, or to the single order of the *Anoura*.
BIFID. Cleft into two parts; forked.
BILATERAL. Having two symmetrical sides.
BIMANA (Lat. *bis*, twice; *manus*, a hand). The order of *Mammalia* comprising Man alone.
BIPEDAL (Lat. *bis*, twice; *pes*, foot). Walking upon two legs.
BIRAMOUS (Lat. *bis*, twice; *ramus*, a branch). Applied to a limb which is divided into two branches (*e.g.*, the limbs of Cirripedes).
BIVALVE (Lat. *bis*, twice; *valvæ*, folding-doors). Composed of two plates or valves; applied to the shell of the *Lamellibranchiata* and *Brachiopoda*, and of the carapace of certain *Crustacea*.
BLASTOIDEA (Gr. *blastos*, a bud; and *eidos*, form). An extinct order of *Echinodermata*, often called *Pentremites*.
BRACHIOPODA (Gr. *brachion*, an arm; *pous*, the foot). A class of the *Molluscoida*, often called "Lamp-shells," characterised by possessing two fleshy arms continued from the sides of the mouth.
BRACHIUM (Gr. *brachion*, arm). Applied to the upper arm of Vertebrates
BRACHYURA (Gr. *brachus*, short; *oura*, tail). A tribe of the Decapod *Crustaceans* with short tails (*i.e.*, the Crabs).
BRACTS. (*See* Hydrophyllia.)
BRADYPODIDÆ (Gr. *bradus*, slow; *podes*, feet). The family of *Edentata* comprising the Sloths.
BRANCHIA (Gr. *bragchia*, the gill of a fish). A respiratory organ adapted to breathe air dissolved in water.
BRANCHIATE. Possessing gills or branchiæ.
BRANCHIFERA (Gr. *bragchia*, gill; and *phero*, I carry). A division of *Gasteropodous Molluscs*, in which the respiration is aquatic, and the respiratory organs are mostly in the form of distinct gills.
BRANCHIO-GASTEROPODA (= Branchifera).
BRANCHIOPODA (Gr. *bragchia*; and *pous*, foot). A legion of *Crustacea*, in which the gills are supported by the feet.
BRANCHIOSTEGAL (Gr. *bragchia*, gills; *stego*, I cover). Applied to a membrane and rays by which the gills are protected in many fishes.
BREVILINGUIA (Lat. *brevis*, short; *lingua*, tongue). A division of the *Lacertilia*.
BREVIPENNATÆ (Lat. *brevis*, short; *penna*, a wing). A group of the Natatorial Birds.

Bronchi (Gr. *brogchos*, the windpipe). The branches of the windpipe (*trachea*), by which the air is conveyed to the vesicles of the lung.

Bruta (Lat. *brutus*, heavy, stupid). Often used to designate the Mammalian order of the *Edentata*.

Bryozoa (Gr. *bruon*, moss; *zoön*, animal). A synonym of *Polyzoa*, a class of the *Molluscoida*.

Buccal (Lat. *bucca*, mouth or cheeks). Connected with the mouth.

Bursiform (Lat. *bursa*, a purse; *forma*, shape). Shaped like a purse; subspherical.

Byssiferous. Producing a byssus.

Byssus (Gr. *bussos*, flax). A term applied to the silky filaments by which the *Pinna*, the common Mussel, and certain other bivalve *Mollusca*, attach themselves to foreign objects.

Caducibranchiate (Lat. *caducus*, falling off; Gr. *bragchia*, gill). Applied to those Amphibians in which the gills fall off before maturity is reached.

Caducous. Applied to parts which fall off or are shed during the life of the animal.

Cæcal (Lat. *cæcus*, blind). Terminating blindly, or in a closed extremity.

Cæcum (Lat. *cæcus*). A tube which terminates blindly.

Cæspitose (Lat. *cæspes*, a turf). Tufted.

Cainozoic. (*See* Kainozoic.)

Calcar (Lat. a spur). Applied to the "spurs" of Rasorial Birds; and also to the rudiments of the hind-limbs in certain Snakes.

Calcareous (Lat. *calx*, lime). Composed of carbonate of lime.

Calice. The little cup in which the polype of a coralligenous Zoophyte (*Actinozoön*) is contained.

Calycophoridæ (Gr. *kalux*, a cup; and *phero*, I carry). An order of the Oceanic *Hydrozoa*, so called from their possessing bell-shaped swimming organs (*nectocalyces*).

Calyx (Lat. *calyx*, a cup). Applied to the cup-shaped body of *Vorticella* (*Protozoa*), or of a *Crinoid* (*Echinodermata*).

Campanularida (Lat. *campanula*, a bell). An order of Hydroid Zoophytes.

Canine (Lat. *canis*, a dog). The eye-tooth of Mammals, or the tooth which is placed at or close to the præmaxillary suture in the upper jaw, and the corresponding tooth in the lower jaw.

Capitulum (Lat. dim. of *caput*, head). Applied to the body of a Barnacle (*Lepadidæ*), from its being supported upon a stalk or peduncle.

Carapace. A protective shield. Applied to the upper shell of Crabs, Lobsters, and many other *Crustacea*; also to the case with which certain of the *Infusoria* are provided. Also the upper half of the immovable case in which the body of a Chelonian is protected.

Carinatæ (Lat. *carina*, a keel). Applied by Huxley to all those birds in which the sternum is furnished with a median ridge or keel.

Carnivora (Lat. *caro*, flesh; *voro*, I devour). An order of the *Mammalia*.

Carnivorous (Lat. *caro*, flesh; *voro*, I devour). Feeding upon flesh.

Carnose (Lat. *caro*). Fleshy.

Carpophaga (Gr. *karpos*, fruit; *phago*, I eat). A section of the *Marsupialia*.

Carpus (Gr. *karpos*, the wrist). The small bones which intervene between the fore-arm and the metacarpus.

Catarhina (Gr. *kata*, downwards; *rhines*, nostrils). A group of the *Quadrumana*.

Caudal (Lat. *cauda*, the tail). Belonging to the tail.

Cavicornia (Lat. *cavus*, hollow; *cornu*, a horn). The "hollow-horned" Ruminants, in which the horn consists of a central bony "horn-core" surrounded by a horny sheath.

Centrum (Gr. *kentron*, the point round which a circle is described by a pair of compasses). The central portion or "body" of a vertebra.

Cephalic (Gr. *kephale*, head). Belonging to the head.

Cephalo-branchiate (Gr. *kephale*; and *bragchia*, gill). Carrying gills upon the head. Applied to a section of the *Annelida*, which, like the *Serpulæ*, have tufts of external gills placed upon the head.

CEPHALOPHORA (Gr. *kephale;* and *phero,* I carry). Used synonymously with *Encephala*, to designate those *Mollusca* which possess a distinct head.

CEPHALOPODA (Gr. *kephale;* and *podes,* feet). A class of the *Mollusca*, comprising the Cuttle-fishes and their allies, in which there is a series of arms ranged round the head.

CEPHALOTHORAX (Gr. *kephale;* and *thorax,* chest). The anterior division of the body in many *Crustacea* and *Arachnida*, which is composed of the coalesced head and chest.

CERE. The naked space found at the base of the bill of some birds.

CERVICAL (Lat. *cervix,* neck). Connected with the region of the neck.

CESTOIDEA (Gr. *kestos,* a girdle). An old name for the *Tæniada*, a class of intestinal worms with flat bodies like tape (hence the name Tapeworms).

CESTRAPHORI (Gr. *kestra,* a weapon; *phero,* I carry). The group of *Elasmobranchii* represented at the present day by the Port Jackson Shark.

CETACEA (Gr. *kētos,* a whale). The order of Mammals comprising the Whales and Dolphins.

CHÆTOGNATHA (Gr. *chaite,* bristle; *gnathos,* jaw). An order of the *Anarthropoda*, comprising only the oceanic genus *Sagitta*.

CHEIROPTERA (Gr. *cheir,* hand; *pteron,* a wing). The order of Mammals comprising the Bats.

CHELÆ (Gr. *chele,* a claw). The prehensile claws with which some of the limbs are terminated in certain *Crustacea*, such as the Crab, Lobster, &c.

CHELATE. Possessing chelæ; applied to a limb.

CHELICERÆ (Gr. *chele,* a claw; and *keras,* a horn). The prehensile claws of the Scorpion, supposed to be homologous with antennæ.

CHELONIA (Gr. *chelone,* a tortoise). The order of Reptiles comprising the Tortoises and Turtles.

CHELONOBATRACHIA (Gr. *chelone,* a tortoise; *batrachos,* a frog). Sometimes applied to the Amphibian order of the *Anoura* (Frogs and Toads).

CHILOGNATHA (Gr. *cheilos,* a lip; and *gnathos,* a jaw). An order of the *Myriapoda*.

CHILOPODA (Gr. *cheilos;* and *podes,* feet). An order of the *Myriapoda*.

CHITINE (Gr. *chiton,* a coat). The peculiar chemical principle, nearly allied to horn, which forms the exoskeleton in many Invertebrate animals, especially in the *Arthropoda* (*Crustacea, Insecta,* &c.)

CHLOROPHYLL (Gr. *chloros,* green; and *phyllos,* a leaf). The green colouring matter of plants.

CHROMATOPHORES (Gr. *chroma,* complexion, or colour; and *phero,* I carry). Little sacs which contain pigment-granules, and are found in the integument of Cuttle-fishes.

CHRYSALIS (Gr. *chrusos,* gold). The motionless pupa of butterflies and moths, so called because sometimes exhibiting a golden lustre.

CHYLAQUEOUS FLUID. A fluid consisting partly of water derived from the exterior, and partly of the products of digestion (chyle), occupying the body-cavity or perivisceral space in many Invertebrates (*Annelides, Echinoderms,* &c.), and sometimes having a special canal-system for its conduction (chylaqueous canals).

CHYLE (Gr. *chulos,* juice). The milky fluid which is the result of the action of the various digestive fluids upon the food.

CHYLIFIC (Gr. *chulos,* juice [chyle]; and Lat. *facio,* I make). Producing chyle. Applied to one of the stomachs, when more than one is present. The word is of mongrel origin; and "chylopoietic" is more correct.

CHYME (Gr. *chumos,* juice). The acid pasty fluid produced by the action of the gastric juice upon the food.

CHYME-MASS. The central, semi-fluid sarcode in the interior of an *Infusorian*.

CILIA (Lat. *cilium,* an eyelash). Microscopic, hair-like filaments, which have the power of lashing backwards and forwards, thus creating currents in the surrounding or contiguous fluid, or subserving locomotion in the animal which possesses them.

CILIOGRADA (Lat. *cilium;* and *gradior*, I walk). Synonymous with *Ctenophora*, an order of *Actinozoa.*

CINCLIDES (Gr. *kiglis*, a lattice). Special apertures in the column-walls of some Sea-anemones (*Actiniæ*), which probably serve for the emission of the cord-like "craspeda."

CIRRI (Lat. *cirrus*, a curl). Tendril-like appendages, such as the feet of Barnacles and Acorn-shells (*Cirripedes*), the lateral processes on the arms of *Brachiopoda*, &c.

CIRRIFEROUS or CIRRIGEROUS. Carrying cirri.

CIRRIPEDIA, CIRRHIPEDIA, or CIRRHOPODA (Lat. *cirrus*, a curl; and *pes*, a foot). A sub-class of *Crustacea* with curled jointed feet.

CIRROSTOMI (Lat. *cirrus*, a tendril; Gr. *stoma*, mouth). Sometimes used to designate the *Pharyngobranchii.*

CLADOCERA (Gr. *klados*, a branch; *keras*, a horn). An order of *Crustacea* with branched antennæ.

CLAVATE (Lat. *clavus*, a club). Club-shaped.

CLAVICLE (Lat. *clavicula*, a little key). The "collar-bone," forming one of the elements of the pectoral arch of Vertebrates.

CLOACA (Lat. a sink). The cavity into which the intestinal canal and the ducts of the generative and urinary organs open in common, in some Invertebrates (*e.g.*, in Insects), and also in many Vertebrate animals.

CLYPEIFORM (Lat. *clypeus*, a shield; and *forma*, shape). Shield-shaped; applied, for example, to the carapace of the King-crab.

CNIDÆ (Gr. *knide*, a nettle). The urticating cells, or "thread-cells," whereby many *Cœlenterate* animals obtain their power of stinging.

COCCOLITHS (Gr. *kokkos*, a berry; *lithos*, stone). Minute oval or rounded bodies, which are found either free or attached to the surface of coccospheres.

COCCOSPHERES (Gr. *kokkos;* and *sphaira*, a sphere). Spherical masses of sarcode, enclosed in a delicate calcareous envelope, and bearing coccoliths upon their external surface. Both coccospheres and coccoliths are embedded in a diffused plasmodium of sarcode, the whole constituting a low *Rhizopodic* organism.

COCCYGEAL. Connected with the coccyx.

COCCYX (Gr. *kokkux*, a cuckoo). The terminal portion of the spinal column in man, so called from its resemblance to a cuckoo's beak.

COCOON (French *cocon*, the cocoon of the silk-worm; connected with Fr. *coque*, shell, which is derived from the Lat. *concha*). The outer covering of silky hairs with which the pupa or chrysalis of many insects is protected.

CODONOSTOMA (Gr. *kodon*, a bell; *stoma*, mouth). The aperture or mouth of the disc (nectocalyx) of a *Medusa*, or of the bell (gonocalyx) of a medusiform gonophore.

CŒLENTERATA (Gr. *koilos*, hollow; *enteron*, the bowel). The sub-kingdom which comprises the *Hydrozoa* and *Actinozoa*. Proposed by Frey and Leuckhart in place of the old term *Radiata*, which included other animals as well.

CŒNENCHYMA (Gr. *koinos*, common; *enchuma*, tissue). The common calcareous tissue which unites together the various corallites of a compound corallum.

CŒNŒCIUM (Gr. *koinos*, common; *oikos*, house). The entire dermal system of any *Polyzoön:* employed in place of the terms polyzoary or polypidom.

CŒNOSARC (Gr. *koinos*, common; *sarx*, flesh). The common organised medium by which the separate polypites of a compound *Hydrozoön* are connected together.

COLEOPTERA (Gr. *koleos*, a sheath; *pteron*, wing). The order of Insects (Beetles) in which the anterior pair of wings are hardened, and serve as protective cases for the posterior pair of membranous wings.

COLUBRINA (Lat. *coluber*, a snake). A division of the *Ophidia.*

COLUMBACEI (Lat. *columba*, a dove). The division of Rasorial Birds comprising the Doves and Pigeons.

COLUMELLA (Lat. dim. of *columna*, a column). In Conchology, the central axis round which the whorls of a spiral univalve are wound. Amongst the *Actinozoa*, it is the central axis or pillar which is found in the centre of the thecæ of many corals.

COLUMN. Applied to the cylindrical body of a Sea-anemone (*Actinia*); also to the jointed stem or peduncle of the stalked *Crinoids*.

COMMISSURAL (Lat. *committo*, I solder together). Connecting together; usually applied to the nerve-fibres which unite different ganglia.

CONCHA (Lat. a shell). The external ear by which sounds are collected and transmitted to the internal ear.

CONCHIFERA (Lat. *concha*, a shell; *fero*, I carry). Shell-fish. Applied in a restricted sense to the bivalve Molluscs, and used as a synonym for *Lamellibranchiata*.

CONDYLE (Gr. *kondulos*, a knuckle). The surface by which one bone articulates with another. Applied especially to the articular surface or surfaces by which the skull articulates with the vertebral column.

CONIROSTRES (Lat. *conus*, a cone; *rostrum*, a beak). The division of Perching Birds with conical beaks.

COPEPODA (Gr. *kope*, an oar; *podes*, feet). An order of *Crustacea*.

CORACOID (Gr. *korax*, a crow; *eidos*, form). One of the bones which enters into the composition of the pectoral arch in Birds, Reptiles, and Monotremes. In most Mammals it is a mere process of the scapula, having, in man, some resemblance in shape to the beak of a crow.

CORALLIGENOUS. Producing a corallum.

CORALLITE. The corallum secreted by an *Actinozoön* which consists of a single polype; or the portion of a composite corallum which belongs to, and is secreted by, an individual polype.

CORALLUM (from the Latin for red coral). The hard structures deposited in, or by, the tissues of an *Actinozoön*—commonly called a "coral."

CORIACEOUS (Lat *corium*, hide). Leathery.

CORPUS CALLOSUM (Lat. the "firm body"). The great band of nervous matter which unites the two hemispheres of the cerebrum in the Mammals.

CORPUSCULATED (Lat. *corpusculum*, a little body or particle). Applied to fluids which, like the blood, contain floating solid particles or "corpuscles."

CORTICAL LAYER. The layer of consistent sarcode, which in the *Infusoria* encloses the chyme mass, and is surrounded by the cuticle. Sometimes called the "parenchyma of the body."

COSTÆ (Lat. *costa*, a rib). Applied amongst the *Crinoidea* to designate the rows of plates which succeed the inferior or basal portion of the cup (pelvis). Amongst the *Corals* the "costæ" are vertical ridges which occur on the outer surface of the theca, and mark the position of the septa within.

COSTAL (Lat. *costa*, a rib). Connected with the ribs.

CRANIUM (Gr. *kranion*, the skull). The bony or cartilaginous case in which the brain is contained.

CRASPEDA (Gr. *kraspedon*, a margin or fringe). The long, convoluted cords, containing thread-cells, which are attached to the free margins of the mesenteries of a Sea-anemone.

CREPUSCULAR (Lat. *crepusculum*, dusk). Applied to animals which are active in the dusk on twilight.

CRINOIDEA (Gr. *krinos*, a lily; *eidos*, form). An order of *Echinodermata*, comprising forms which are usually stalked, and sometimes resemble lilies in shape.

CROCODILIA (Gr. *krokodeilos*, a crocodile). An order of Reptiles.

CROP. A partial dilatation of the gullet, technically called "ingluvies."

CRUSTACEA (Lat. *crusta*, a crust). A class of articulate animals, comprising Crabs, Lobsters, &c., characterised by the possession of a hard shell or crust, which they cast periodically.

CTENOCYST (Gr. *kteis*, a comb; *kustis*, a bag or cyst). The sense-organ (probably auditory) which occurs in the *Ctenophora*.

CTENOID (Gr. *kteis*, a comb; *eidos*, form). Applied to those scales of fishes, the hinder margins of which are fringed with spines or comb-like projections.

CTENOPHORA (Gr. *kteis*, a comb; and *phero*, I carry). An order of *Actinozoa*, comprising oceanic creatures, which swim by means of "ctenophores," or bands of cilia arranged in comb-like plates.
CURSORES (Lat. *curro*, I run). An order of *Aves*, comprising birds destitute of the power of flight, but formed for running vigorously (*e.g.*, the Ostrich and Emeu).
CUSPIDATE. Furnished with small pointed eminences or "cusps."
CUTICLE (Lat. *cuticula*, dim. of *cutis*, skin). The pellicle which forms the outer layer of the body amongst the *Infusoria*. The outer layer of the integument generally.
CUTIS (Lat. skin). The inferior vascular layer of the integument, often called the *cutis vera*, the *corium*, or the *derma*.
CYCLOID (Gr. *kuklos*, a circle; *eidos*, form). Applied to those scales of fishes which have a regularly circular or elliptical outline with an even margin.
CYCLOSTOMI. Sometimes used to designate the Hag-fishes and Lampreys, forming the order *Marsipobranchii*.
CYST (Gr. *kustis*, a bladder or bag). A sac or vesicle.
CYSTICA. The embryonic forms (scolices) of certain intestinal worms (Tapeworms), which were described as a distinct order, until their true nature was discovered.
CYSTOIDEA (Gr. *kustis*, a bladder; and *eidos*, form). An extinct order of *Echinodermata*.

DECAPODA (Gr. *deka*, ten; *podes*, feet). The division of *Crustacea* which have ten ambulatory feet; also the family of Cuttle-fishes, in which there are ten arms or cephalic processes.
DECIDUOUS (Lat. *decido*, I fall off). Applied to parts which fall off or are shed during the life of the animal.
DECOLLATED (Lat. *decollo*, I behead). Applied to univalve shells, the apex of which falls off in the course of growth.
DEINOSAURIA (Gr. *deinos*, terrible; *saura*, lizard). An extinct order of Reptiles.
DENDRIFORM, DENDRITIC, DENDROID (Gr. *dendron*, a tree). Branched like a tree, arborescent.
DENTIROSTRES (Lat. *dens*, a tooth; *rostrum*, a beak). The group of Perching Birds in which the upper mandible of the beak has its lower margin toothed.
DERMA. (*See* Cutis.)
DERMAL (Gr. *derma*, skin). Belonging to the integument.
DERMOSCLERITES (Gr. *derma*, skin; *skleros*, hard). Masses of spicules which occur in the tissues of some of the *Alcyonidæ* (*Actinozoa*).
DESMIDIÆ. Minute fresh-water plants, of a green colour, without a siliceous epidermis.
DEUTEROZOÖIDS (Gr. *deuteros*, second; *zoön*, animal; *eidos*, form). The zoöids which are produced by gemmation from zoöids.
DEXTRAL (Lat. *dextra*, the right hand). Right-handed; applied to the direction of the spiral in the greater number of univalve shells.
DIAPHRAGM (Gr. *diaphragma*, a partition). The "midriff," or the muscle which in *Mammalia* forms a partition between the cavities of the thorax and abdomen.
DIASTEMA (Gr. *dia*, apart; *histemi*, to place). A gap or interval, especially between teeth.
DIASTOLE (Gr. *diastello*, I separate or expand). The expansion of a contractile cavity such as the heart, which follows its contraction or "systole."
DIATOMACEÆ (Gr. *diatemno*, I sever). An order of minute plants, which are provided with siliceous envelopes.
DIBRANCHIATA (Gr. *dis*, twice; *bragchia*, gill). The order of *Cephalopoda* (comprising the Cuttle-fishes, &c.) in which only two gills are present.
DICYNODONTIA (Gr. *dis*, twice; *kuon*, dog; *odous*, tooth). An extinct order of Reptiles.
DIDELPHIA (Gr. *dis*, twice; *delphus*, womb). The subdivision of Mammals comprising the Marsupials.

DIGIT (Lat. *digitus*, a finger). A finger or toe.
DIGITIGRADA (Lat. *digitus; gradior*, I walk). A subdivision of the *Carnivora.*
DIGITIGRADE. Walking upon the tips of the toes, and not upon the soles of the feet.
DIMEROSOMATA (Gr. *dis; meros*, part; *soma*, body). An order of *Arachnida*, comprising the true Spiders, so called from the marked division of the body into two regions, the cephalothorax and abdomen. The name *Araneida* is often employed for the order.
DIMYARY (Gr. *dis*, twice; *muon*, muscle). Applied to those bivalve Molluscs (*Lamellibranchiata*) in which the shell is closed by two adductor muscles.
DIŒCIOUS (Gr. *dis*, twice; *oikos*, house). Having the sexes distinct; applied to species which consist of male and female individuals.
DIPHYODONT (Gr. *dis*, twice; *phuo*, I generate; *odous*, tooth). Applied to those Mammals which have two sets of teeth.
DIPHYOZOÖIDS. Detached reproductive portions of adult *Calycophoridæ*, an order of oceanic *Hydrozoa.*
DIPNOI (Gr. *dis*, twice; *pnoe*, breath). The order of fishes represented by the *Lepidosiren.*
DIPTERA (Gr. *dis*, twice; *pteron*, wing). An order of Insects characterised by the possession of two wings.
DISCOID (Gr. *diskos*, a quoit; *eidos*, form). Shaped like a round plate or quoit.
DISCOPHORA (Gr. *diskos*, a quoit; *phero*, I carry). This term is applied to the *Medusæ*, or Jelly-fishes, from their form; and is sometimes used to designate the order of the Leeches (*Hirudinea*), from the suctorial discs which these animals possess.
DISSEPIMENTS (Lat. *dissepio*, I partition off). Partitions. Used in a restricted sense to designate certain imperfect transverse partitions, which grow from the septa of many corals.
DISTAL. Applied to the quickly growing end of the hydrosoma of a *Hydrozoön*; the opposite, or "proximal," extremity growing less rapidly, and being the end by which the organism is fixed, when attached at all.
DIURNAL (Lat. *dies*, day). Applied to animals which are active during the day.
DIVERTICULUM (Lat. *diverticulum*, a by-road). A lateral tube with a blind extremity springing from the side of another tube.
DORSAL (Lat. *dorsum*, back). Connected with the back.
DORSIBRANCHIATE (Lat. *dorsum*, the back; Gr. *bragchia*, gill). Having external gills attached to the back; applied to certain *Annelides* and *Molluscs.* The term is of mongrel composition, and "notobranchiate" is more correctly employed.

ECDERON (Gr. *ek*, out; *deros*, skin). The outer plane of growth of the external integumentary layer (viz., the ectoderm, or epidermis).
ECDYSIS (Gr. *ekdusis*, a stripping off). A shedding or moulting of the skin.
ECHINOCOCCI (Gr. *echinos*, a hedgehog; *kokkos*, a berry). The larval forms (scolices) of the tapeworm of the dog (*Tænia echinococcus*), commonly known as "hydatids."
ECHINODERMATA (Gr. *echinos;* and *derma*, skin). A class of animals comprising the Sea-urchins, Star-fishes, and others, most of which have spiny skins.
ECHINOIDEA (Gr. *echinos*; and *eidos*, form). An order of *Echinodermata*, comprising the Sea-urchins.
ECHINULATE. Possessing spines.
ECTOCYST (Gr. *ektos*, outside; *kustis*, a bladder). The external investment of the cœnœcium of a *Polyzoön.*
ECTODERM (Gr. *ektos;* and *derma*, skin). The external integumentary layer of the *Cœlenterata.*
ECTOSARC (Gr. *ektos*; *sarx*, flesh). The outer transparent sarcode-layer of certain *Rhizopods*, such as the *Amœba.*
EDENTATA (Lat. *e*, without; *dens*, tooth). An order of *Mammalia* often called *Bruta.*

EDENTULOUS. Toothless; without any dental apparatus. Applied to the mouth of any animal, or to the hinge of the bivalve Molluscs.

EDRIOPHTHALMATA (Gr. *hedraios*, sitting; *ophthalmos*, eye). The division of *Crustacea* in which the eyes are sessile, and are not supported upon stalks.

ELASMOBRANCHII (Gr. *elasma*, a plate; *bragchia*, gill). An order of Fishes, including the Sharks and Rays.

ELYTRA (Gr. *elutron*, a sheath). The chitinous anterior pair of wings in Beetles, which form cases for the posterior membranous wings. Also applied to the scales or plates on the back of the Sea-mouse (*Aphrodite*).

EMBRYO (Gr. *en*, in; *bruo*, I swell). The earliest stage at which the young animal is recognisable in the impregnated ovum.

ENCEPHALON (Gr. *egcephalos*, brain). The portion of the cerebro-spinal nervous axis contained within the cranium.

ENCEPHALOUS (Gr. *en*, in; *kephale*, the head). Possessing a distinct head. Usually applied to all the *Mollusca* proper, except the *Lamellibranchiata*.

ENCYSTATION (Gr. *en*, in; *kustis*, a bag). The transformation undergone by certain of the *Protozoa*, when they become motionless, and surround themselves by a thick coating or cyst.

ENDERON (Gr. *en*, in; *deros*, skin). The inner plane of growth of the outer integumentary layer (viz., the ectoderm, or epidermis).

ENDOCYST (Gr. *endon*, within; *kustis*, a bag). The inner membrane or integumentary layer of a *Polyzoön*. In *Cristatella*, where there is no "ectocyst," the endocyst constitutes the entire integument.

ENDODERM (Gr. *endon*; and *derma*, skin). The inner integumentary layer of the *Cœlenterata*.

ENDOPODITE (Gr. *endon*; and *pous*, foot). The inner of the two secondary joints into which the typical limb of a *Crustacean* is divided.

ENDOSARC (Gr. *endon*; and *sarx*, flesh). The inner molecular layer of sarcode in the *Amœba*, and other allied *Rhizopods*.

ENDOSKELETON (Gr. *endon*; and *skeletos*, dry). The internal hard structures, such as bones, which serve for the attachment of muscles, or the protection of organs, and which are not a mere hardening of the integument.

ENSIFORM (Lat. *ensis*, a sword; *forma*, shape). Sword-shaped.

ENTOMOPHAGA (Gr. *entoma*, insects; *phago*, I eat). A section of the Marsupialia.

ENTOMOSTRACA (Gr. *entoma*, insects; *ostrakon*, a shell). Literally, shelled insects—applied to a division of *Crustacea*.

ENTOZOA (Gr. *entos*, within; *zoön*, animal). Animals which are parasitic in the interior of other animals.

EOCENE (Gr. *eos*, dawn; *kainos*, new or recent). The lowest division of the Tertiary rocks, in which species of existing shells are to a small extent represented.

EPIDERMIS (Gr. *epi*, upon; *derma*, the true skin). The outer non-vascular layer of the skin, often called the scarf-skin or *cuticle*.

EPIMERA (Gr. *epi*, upon; *mēron*, thigh). The lateral pieces of the dorsal arc of the somite of a *Crustacean*.

EPIPODIA (Gr. *epi*, upon; *pous*, the foot). Muscular lobes developed from the lateral and upper surfaces of the "foot" of some *Molluscs*.

EPIPODITE (Gr. *epi*, upon; *pous*, foot). A process developed upon the basal joint, or "protopodite," of some of the limbs of certain *Crustacea*.

EPISTERNA (Gr. *epi*, upon; *sternon*, the breast-bone). The lateral pieces of the inferior or ventral arc of the somite of a *Crustacean*.

EPISTOME (Gr. *epi*; and *stoma*, mouth). A valve-like organ which arches over the mouth in certain of the *Polyzoa*.

EPITHECA (Gr. *epi*; and *theke*, a sheath). A continuous layer surrounding the thecæ in some Corals, and being the external indications of tabulæ.

EPIZOA (Gr. *epi*, upon; *zoön*, animal). Animals which are parasitic upon other animals. In a restricted sense, a division of *Crustacea* which are parasitic upon fishes.

EQUILATERAL (Lat. *æquus*, equal; *latus*, side). Having its sides equal. Usually applied to the shells of the *Brachiopoda*. When applied to the spiral shells of the *Foraminifera*, it means that all the convolutions of the shell lie in the same plane.

EQUIVALVE (Lat. *æquus*, equal; *valvæ*, folding-doors). Applied to shells which are composed of two equal pieces or valves.

ERRANTIA (Lat. *erro*, I wander). An order of *Annelida*, often called *Nereidea*, distinguished by their great locomotive powers.

EURYPTERIDA (Gr. *eurus*, broad; *pteron*, wing). An extinct sub-order of *Crustacea*.

EXOPODITE (Gr. *exo*, outside; *pous*, foot). The outer of the two secondary joints into which the typical limb of a *Crustacean* is divided.

EXOSKELETON (Gr. *exo*, outside; *skeletos*, dry). The external skeleton, which is constituted by a hardening of the integument, and is often called a "dermoskeleton."

FASCICULATED (Lat. *fasciculus*, a bundle). Arranged in bundles.

FAUNA (Lat. *Fauni*, the rural deities of the Romans). The general assemblage of the animals of any region or district.

FEMUR. The thigh-bone, intervening between the pelvis and the bones of the leg proper (*tibia* and *fibula*).

FIBULA (Lat. a brooch). The outermost of the two bones of the leg in the higher *Vertebrata;* corresponding to the *ulna* of the fore-arm.

FILIFORM (Lat. *filum*, a thread; *forma*, shape). Thread-shaped.

FISSILINGUIA (Lat. *findo*, I cleave; *lingua*, tongue). A division of *Lacertilia*, with bifid tongues.

FISSION (Lat. *findo*, I cleave). Multiplication by means of a process of self-division.

FISSIPAROUS (Lat. *findo;* and *pario*, I produce). Giving origin to fresh structures by a process of fission.

FISSIROSTRES (Lat. *findo*, I cleave; *rostrum*, beak). A sub-order of the Perching Birds.

FLAGELLUM (Lat. for whip). The lash-like appendage exhibited by many *Infusoria*, which are therefore said to be "flagellate."

FLORA (Lat. *Flora*, the goddess of flowers). The general assemblage of the plants of any region or district.

FOOT-JAWS. The limbs of *Crustacea*, which are modified to subserve mastication.

FOOT-SECRETION. The term applied by Mr Dana to the sclerobasic corallum of certain *Actinozoa*.

FOOT-TUBERCLES. The unarticulated appendages of the *Annelida*, often called parapodia.

FORAMINIFERA (Lat. *foramen*, an aperture; *fero*, I carry). An order of *Protozoa*, usually characterised by the possession of a shell perforated by numerous pseudopodial apertures.

FRUGIVOROUS (Lat. *frux*, fruit; *voro*, I devour). Living upon fruits.

FURCULUM (Lat. dim. of *furca*, a fork). The "merry-thought" of birds, or the V-shaped bone formed by the united clavicles.

FUSIFORM (Lat. *fusus*, a spindle; and *forma*, shape). Spindle-shaped, or pointed at both ends.

GALLINACEI (Lat. *gallina*, a fowl). Sometimes applied to the whole order of the Rasorial Birds, but properly restricted to that section of the order of which the common Fowl is a typical example.

GANGLION (Gr. *gagglion*, a knot). A mass of nervous matter containing nerve-cells, giving origin to nerve-fibres.

GANOID (Gr. *ganos*, splendour, brightness). Applied to those scales or plates which are composed of an inferior layer of true bone covered by a superior layer of polished enamel.

GANOIDEI. An order of Fishes.

GASTEROPODA (Gr. *gaster*, stomach; *pous*, foot). The class of the *Mollusca* comprising the ordinary univalves, in which locomotion is usually effected by a muscular expansion of the under surface of the body (the "foot").

GEMMÆ (Lat. *gemma*, a bud). The buds produced by any animal, whether detached or not.
GEMMATION. The process of producing new structures by budding.
GEMMIPAROUS (Lat. *gemma*, a bud; *pario*, I produce). Giving origin to new structures by a process of budding.
GEMMULES (Lat. dim. of *gemma*). The ciliated embryos of many *Cœlenterata;* also the seed-like reproductive bodies or "spores" of *Spongilla*.
GEPHYREA (Gr. *gephura*, a bridge). A class of the *Anarthropoda*, comprising the Spoon-worms (*Sipunculus*) and their allies.
GIZZARD. A muscular division of the stomach in Birds, Insects, &c.
GLADIUS (Lat. a sword). Applied to the horny endoskeleton or "pen" of certain Cuttle-fishes.
GLENOID (Gr. *glene*, a cavity; *eidos*, form). A shallow cavity; applied especially to the shallow articular cavity in the shoulder-blade to which the head of the humerus is jointed.
GNATHITES (Gr. *gnathos*, a jaw). The masticatory organs of *Crustacea*.
GONOBLASTIDIA (Gr. *gonos*, offspring; *blastidion*, dim. of *blastos*, a bud). The processes which carry the reproductive receptacles, or "gonophores," in many of the *Hydrozoa*.
GONOCALYX (Gr. *gonos*; and *kalux*, cup). The swimming-bell in a medusiform gonophore, or the same structure in a gonophore which is not detached.
GONOPHORE (Gr. *gonos;* and *phero*, I carry). The generative buds, or receptacles of the reproductive elements, in the *Hydrozoa*, whether these become detached or not.
GONOSOME (Gr. *gonos;* and *soma*, body). Applied as a collective term to the reproductive zoöids of a *Hydrozoön*.
GONOTHECA (Gr. *gonos;* and *theke*, a case). The chitinous receptacle within which the gonophores of certain of the *Hydrozoa* are produced.
GRALLATORES (Lat. *grallæ*, stilts). The order of the long-legged Wading Birds.
GRANIVOROUS (Lat. *granum*, a grain or seed; *voro*, I devour). Living upon grains or other seeds.
GRAPTOLITIDÆ (Gr. *grapho*, I write; *lithos*, stone). An extinct sub-class of the *Hydrozoa*.
GREGARINIDA (Lat. *gregarius*, occurring in numbers together). A class of the *Protozoa*.
GUARD. The cylindrical fibrous sheath with which the internal chambered shell (phragmacone) of a *Belemnite* is protected.
GYMNOLÆMATA (Gr. *gumnos*, naked; *laimos*, the throat). An order of the *Polyzoa* in which the mouth is devoid of the valvular structure known as the "epistome."
GYMNOPHIONA (Gr. *gumnos*, naked; *ophis*, a snake). The order of the *Amphibia* comprising the snake-like *Cæciliæ*.
GYMNOPHTHALMATA (Gr. *gumnos;* and *ophthalmos*, the eye). Applied by Edward Forbes to those *Medusæ* in which the eye-specks at the margin of the disc are unprotected. The division is now abandoned.
GYMNOSOMATA (Gr. *gumnos;* and *soma*, the body). The order of *Pteropoda* in which the body is not protected by a shell.
GYNOPHORES (Gr. *gune*, woman; *phero*, I carry). The generative buds, or gonophores, of *Hydrozoa*, which contain ova alone, and differ in form from those which contain spermatozoa.
GYRENCEPHALA (Gr. *guroo*, I wind about; *egkephalos*, brain). Applied by Owen to a section of the Mammalia in which the cerebral hemispheres are abundantly convoluted.

HÆMAL (Gr. *haima*, blood). Connected with the blood-vessels, or with the circulatory system.
HÆMATOCRYA (Gr. *haima*, blood; *cruos*, cold). Applied by Owen to the "cold-blooded" Vertebrates—viz., the Fishes, Amphibia, and Reptiles.
HÆMATOTHERMA (Gr. *haima*, blood; *thermos*, warm). Applied by Owen to the "warm-blooded" Vertebrates—viz., Birds and Mammals.
HALLUX (Lat. *allex*, the thumb or great toe). The innermost of the fiv

digits which normally compose the *hind* foot of a Vertebrate animal. In man, the great toe.

HALTERES (Gr. *halteres*, weights used by athletes to steady themselves in leaping). The rudimentary filaments or "balancers" which represent the posterior pair of wings in the *Diptera*, an order of Insects.

HAUSTELLATE (Lat. *haurio*, I drink). Adapted for sucking or pumping up fluids; applied to the mouth of certain *Crustacea* and *Insecta*.

HECTOCOTYLUS (Gr. *hekaton*, a hundred; *kotulos*, a cup). The metamorphosed reproductive arm of certain of the male Cuttle-fishes. In the *Argonaut* the arm becomes detached, and was originally described as a parasitic worm.

HELMINTHOID (Gr. *helmins*, an intestinal worm). Worm-shaped, vermiform.

HEMELYTRA (Gr. *hemi*, half; *elutron*, a sheath). The wings of certain Insects, in which the apex of the wing is membranous, whilst the inner portion is chitinous, and resembles the elytron of a beetle.

HEMIMETABOLIC (Gr. *hemi*, half; *metabole*, change). Applied to those Insects which undergo an incomplete metamorphosis.

HEMIPTERA (Gr. *hemi*; and *pteron*, wing). An order of Insects in which the anterior wings are sometimes "hemelytra."

HERMAPHRODITE (Gr. *Hermes*, Mercury; *Aphrodite*, Venus). Possessing the characters of both sexes combined.

HETEROCERCAL (Gr. *heteros*, diverse; *kerkos*, tail). Applied to the tail of Fishes when it is unsymmetrical, or composed of two unequal lobes.

HETEROGANGLIATE (Gr. *heteros*, diverse; *gagglion*, a knot). Possessing a nervous system in which the ganglia are scattered and unsymmetrical (as in the *Mollusca*, for example).

HETEROMORPHIC (Gr. *heteros*; *morphe*, form). Differing in form or shape.

HETEROPHAGI (Gr. *heteros*, other; *phago*, I eat). Applied to Birds the young of which are born in a helpless condition, and require to be fed by the parents for a longer or shorter period.

HEXAPOD (Gr. *hexa*, six; *pous*, foot). Possessing six legs; applied to the *Insecta*.

HILUM (Lat. *hilum*, a little thing). A small aperture (as in the gemmules of sponges), or a small depression (as in *Noctiluca*).

HIRUDINEA (Lat. *hirudo*, a horse-leech). The order of *Annelida* comprising the Leeches.

HISTOLOGY (Gr. *histos*, a web; *logos*, a discourse). The study of the tissues; more especially of the minuter elements of the body.

HOLOCEPHALI (Gr. *holos*, whole; *kephale*, head). A sub-order of the *Elasmobranchii* comprising the *Chimæræ*.

HOLOMETABOLIC (Gr. *holos*, whole; *metabole*, change). Applied to Insects which undergo a complete metamorphosis.

HOLOSTOMATA (Gr. *holos*, whole; *stoma*, mouth). A division of *Gasteropodous Molluscs*, in which the aperture of the shell is rounded, or "entire."

HOLOTHUROIDEA (Gr. *holothourion*; and *eidos*, form). An order of *Echinodermata* comprising the Trepangs.

HOMOCERCAL (Gr. *homos*, same; *kerkos*, tail). Applied to the tail of Fishes when it is symmetrical, or composed of two equal lobes.

HOMOGANGLIATE (Gr. *homos*, like; *gagglion*, a knot). Having a nervous system in which the ganglia are symmetrically arranged (as in the *Annulosa*, for example).

HOMOLOGOUS (Gr. *homos*; and *logos*, a discourse). Applied to parts which are constructed upon the same fundamental plan.

HOMOMORPHOUS (Gr. *homos*; and *morphe*, form). Having a similar external appearance or form.

HUMERUS. The bone of the upper arm (*brachium*) in the Vertebrates.

HYALINE (Gr. *hualos*, crystal). Crystalline or glassy.

HYDATIDS (Gr. *hudatis*, a vesicle). The vesicle containing the larval forms (*Echinococci*) of the tapeworm of the dog.

HYDRAFORM. Resembling the common fresh-water polype (*Hydra*) in form.

HYDROCAULUS (Gr. *hudra*, a water-serpent; and *kaulos*, a stem). The main stem of the cœnosarc of a *Hydrozoön*.
HYDROCYSTS (Gr. *hudra;* and *kustis*, a cyst). Curious processes attached to the cœnosarc of the *Physophoridæ*, and termed "feelers" (*fühler* and *taster* of the Germans).
HYDRŒCIUM (Gr. *hudra;* and *oikos*, a house). The chamber into which the cœnosarc in many of the *Calycophoridæ* can be retracted.
HYDROIDA (Gr. *hudra;* and *eidos*, form). The sub-class of the *Hydrozoa*, which comprises the animals most nearly allied to the *Hydra*.
HYDROPHYLLIA (Gr. *hudra;* and *phyllon*, a leaf). Overlapping appendages or plates which protect the polypites in some of the oceanic *Hydrozoa* (*Calycophoridæ* and *Physophoridæ*). They are often termed "bracts," and are the "*deckstücke*" of the Germans.
HYDRORHIZA (Gr. *hudra;* and *rhiza*, root). The adherent base or proximal extremity of any *Hydrozoön*.
HYDROSOMA (Gr. *hudra;* and *soma*, body). The entire organism of any *Hydrozoön*.
HYDROTHECA (Gr. *hudra;* and *theke*, a case). The little chitinous cups in which the polypites of the *Sertularida* and *Campanularida* are protected.
HYDROZOA (Gr. *hudra;* and *zoön*, animal). The class of the *Cœlenterata*, which comprises animals constructed after the type of the *Hydra*.
HYMENOPTERA (Gr. *humen*, a membrane; *pteron*, a wing). An order of Insects (comprising Bees, Ants, &c.) characterised by the possession of four membranous wings.
HYOID (Gr. *U; eidos*, form). The bone which supports the tongue in Vertebrates, and derives its name from its resemblance in man to the Greek letter U.
HYPOSTOME (Gr. *hupo*, under; *stoma*, mouth). The upper lip, or "labrum," of certain *Crustacea* (*e.g.*, Trilobites).
HYRACOIDEA (Gr. *hurax*, a shrew; *eidos*, form). An order of the *Mammalia* constituted for the reception of the single genus *Hyrax*.

ICHTHYODORULITE (Gr. *ichthus*, fish; *dorus*, spear; *lithos*, stone). The fossil fin-spines of Fishes.
ICHTHYOMORPHA (Gr. *ichthus; morphe*, shape). An order of Amphibians, often called *Urodela*, comprising the fish-like Newts, &c.
ICHTHYOPHTHIRA (Gr. *ichthus; phtheir*, a louse). An order of *Crustacea* comprising animals which are parasitic upon Fishes.
ICHTHYOPSIDA (Gr. *ichthus; opsis*, appearance). The primary division of *Vertebrata*, comprising the Fishes and Amphibia. Often spoken of as the *Branchiate Vertebrata*.
ICHTHYOPTERYGIA (Gr. *ichthus; pterux*, wing). An extinct order of Reptiles.
ICHTHYOSAURIA (Gr. *ichthus; saura*, lizard). Synonymous with *Ichthyopterygia*.
ILIUM. The haunch-bone, one of the bones of the pelvic arch in the higher Vertebrates.
IMAGO (Lat. an image or apparition). The perfect insect, after it has undergone its metamorphoses.
IMBRICATED. Applied to scales or plates which overlap one another like tiles.
INCISOR (Lat. *incido*, I cut). The cutting teeth fixed in the intermaxillary bones of the *Mammalia*, and the corresponding teeth in the lower jaw.
INEQUILATERAL. Having the two sides unequal, as in the case of the shells of the ordinary bivalves (*Lamellibranchiata*). When applied to the shells of the *Foraminifera*, it implies that the convolutions of the shell do not lie in the same plane, but are obliquely wound round an axis.
INEQUIVALVE. Composed of two unequal pieces or valves.
INFUNDIBULUM (Lat. for funnel). The tube formed by the coalescence or apposition of the epipodia in the *Cephalopoda*—commonly termed the "funnel," or "siphon."
INFUSORIA (Lat. *infusum*, an infusion). A class of *Protozoa*, so called because they are often developed in organic infusions.
INGUINAL (Lat. *inguen*, groin). Connected with, or situated upon, the groin.

INOPERCULATA (Lat. *in*, without; *operculum*, a lid). The division of pulmonate *Gasteropoda* in which there is no shelly or horny plate (operculum) by which the shell is closed when the animal is withdrawn within it.

INSECTA (Lat. *inseco*, I cut into). The class of Articulate animals commonly known as Insects.

INSECTIVORA (Lat. *insectum*, an insect; *voro*, I devour). An order of Mammals.

INSECTIVOROUS. Living upon Insects.

INSESSORES (Lat. *insedeo*, I sit upon). The order of the Perching Birds, often called *Passeres*.

INTERAMBULACRA. The rows of plates in an *Echinoderm* which are not perforated for the emission of the "tube-feet."

INTERMAXILLÆ, or PRÆMAXILLÆ. The two bones which are situated between the two superior maxillæ in *Vertebrata*. In man, and some monkeys, the præmaxillæ anchylose with the maxillæ, so as to be irrecognisable in the adult.

INTUSSUSCEPTION (Lat. *intus*, within; *suscipio*, I take up). The act of taking foreign matter into a living being.

INVERTEBRATA (Lat. *in*, without; *vertebra*, a bone of the back). Animals without a spinal column or backbone.

ISCHIUM (Gr. *ischion*, the hip). One of the bones of the pelvic arch in Vertebrates.

ISOPODA (Gr. *isos*, equal; *podes*, feet). An order of *Crustacea* in which the feet are like one another and equal.

JUGULAR (Lat. *jugulum*, the throat). Connected with, or placed upon, the throat. Applied to the ventral fins of fishes when they are placed beneath or in advance of the pectorals.

KAINOZOIC (Gr. *kainos*, recent; *zoe*, life). The Tertiary period in Geology, comprising those formations in which the organic remains approximate more or less closely to the existing fauna and flora.

KERATODE (Gr. *keras*, horn; *eidos*, form). The horny substance of which the skeleton of many sponges is made up.

KERATOSA. The division of Sponges in which the skeleton is composed of keratode.

LABIUM (Lat. for lip). Restricted to the lower lip of Articulate animals.

LABRUM (Lat. for lip). Restricted to the upper lip of Articulate animals.

LABYRINTHODONTIA (Gr. *labyrinthos*, a labyrinth; *odous*, tooth). An extinct order of *Amphibia*, so called from the complex microscopic structure of the teeth.

LACERTILIA (Lat. *lacerta*, a lizard). An order of *Reptilia* comprising the Lizards and Slow-worms.

LÆMODIPODA (Gr. *laimos*, throat; *dis*, twice; *podes*, feet). An order of *Crustacea*, so called because they have two feet placed far forwards, as it were under the throat.

LAMELLIBRANCHIATA (Lat. *lamella*, a plate; Gr. *bragchia*, gill). The class of *Mollusca*, comprising the ordinary bivalves, characterised by the possession of lamellar gills.

LAMELLIROSTRES (Lat. *lamella*, a plate; *rostrum*, beak). The flat-billed Swimming Birds (*Natatores*), such as Ducks, Geese, Swans, &c.

LARVA (Lat. a mask). The insect in its first stage after its emergence from the egg, when it is usually very different from the adult.

LARYNX. The upper part of the windpipe, forming a cavity with appropriate muscles and cartilages, situated beneath the hyoid bone, and concerned in Mammals in the production of vocal sounds.

LENTICULAR (Lat. *lens*, a bean). Shaped like a biconvex lens.

LEPIDOPTERA (Gr. *lepis*, a scale; *pteron*, a wing). An order of Insects, comprising Butterflies and Moths, characterised by possessing four wings which are usually covered with minute scales.

LEPIDOTA (Gr. *lepis*, a scale). Formerly applied to the order *Dipnoi*, containing the Mud-fishes (*Lepidosiren*).

LEPTOCARDIA (Gr. *leptos*, slender, small; *cardia*, heart). The name given by Müller to the order of Fishes comprising the Lancelet, now called *Pharyngobranchii*.

LIGAMENTUM NUCHÆ (Lat. *nucha*, the nape of the neck). The band of elastic fibres by which the weight of the head in *Mammalia* is supported.

LINGUAL (Lat. *lingua*, the tongue). Connected with the tongue.

LINGULA (Lat. *ligula*, a little tongue). The upper flexible portion of the labium or lower lip in Insects.

LISSENCEPHALA (Gr. *lissos*, smooth; *egkephalos*, brain). A primary division of *Mammalia*, according to Owen, in which the cerebral hemispheres are smooth or have few convolutions.

LITHOCYSTS (Gr. *lithos*, a stone; *kustis*, a cyst). The sense-organs or "marginal bodies" of the *Lucernarida* or *Steganophthalmate Medusæ*.

LONGIPENNATÆ (Lat. *longus*, long; *penna*, wing). A group of the Natatorial Birds.

LONGIROSTRES (Lat. *longus*; *rostrum*, beak). A group of the Wading Birds.

LOPHOPHORE (Gr. *lophos*, a crest; and *phero*, I carry). The disc or stage upon which the tentacles of the *Polyzoa* are borne.

LOPHYROPODA (Gr. *lophouros*, having stiff hairs; and *podes*, feet). An order of *Crustacea*.

LORICA (Lat. a breast-plate). Applied to the protective case with which certain *Infusoria* are provided.

LORICATA (Lat. *lorica*, a cuirass). The division of Reptiles comprising the *Chelonia* and *Crocodilia*, in which bony plates are developed in the skin (*derma*).

LUCERNARIDA (Lat. *lucerna*, a lamp). An order of the *Hydrozoa*.

LUMBAR (Lat. *lumbus*, loin). Connected with the loins.

LUNATE (Lat. *luna*, moon). Crescentic in shape.

LYENCEPHALA (Gr. *luo*, I loose; *egkephalos*, brain). A primary division of Mammals, according to Owen.

MACRODACTYLI (Gr. *makros*, long; *daktulos*, a finger). A group of the Wading Birds.

MACRURA (Gr. *makros*, long; *oura*, tail). A tribe of Decapod *Crustaceans* with long tails (*e.g.*, the Lobster, Shrimp, &c.)

MADREPORIFORM. Perforated with small holes, like a coral; applied to the tubercle by which the ambulacral system of the *Echinoderms* mostly communicates with the exterior.

MALACOSTRACA (Gr. *malakos*, soft; *ostrakon*, shell). A division of *Crustacea*. Originally applied by Aristotle to the entire class *Crustacea*, because their shells were softer than those of the *Mollusca*.

MALLOPHAGA (Gr. *mallos*, a fleece; *phago*, I eat). An order of Insects which are mostly parasitic upon birds.

MAMMALIA (Lat. *mamma*, the breast). The class of Vertebrate animals which suckle their young.

MANDIBLE (Lat. *mandibulum*, a jaw). The upper pair of jaws in Insects; also applied to one of the pairs of jaws in *Crustacea* and Spiders, to the beak of Cephalopods, the lower jaw of Vertebrates, &c.

MANTLE. The external integument of most of the Mollusca, which is largely developed, and forms a cloak in which the viscera are protected. Technically called the "pallium."

MANUBRIUM (Lat. a handle). The polypite which is suspended from the roof of the swimming-bell of a *Medusa*, or from the gonocalyx of a medusiform gonophore amongst the *Hydrozoa*.

MANUS (Lat. the hand). The hand of the higher Vertebrates.

MARSIPOBRANCHII (Gr. *marsipos*, a pouch; *bragchia*, gill). The order of Fishes comprising the Hag-fishes and Lampreys, with pouch-like gills.

MARSUPIALIA (Lat. *marsupium*, a pouch). An order of Mammals in which the females mostly have an abdominal pouch in which the young are carried.

MASTAX (Gr. mouth). The muscular pharynx or "buccal funnel" into which the mouth opens in most of the *Rotifera*.

MASTICATORY (Lat. *mastico*, I chew). Applied to parts adapted for chewing.

MAXILLÆ (Lat. jaws). The inferior pair or pairs of jaws in the *Arthropoda* (Insects, Crustacea, &c.) The upper jaw-bones of Vertebrates.

MAXILLIPEDES (Lat. *maxillæ*, jaws; *pes*, the foot). The limbs in *Crustacea* and *Myriapoda* which are converted into masticatory organs, and are commonly called "foot-jaws."

MEDULLA (Lat. marrow). Applied to the marrow of bones, or to the spinal cord, with or without the adjective "*spinalis*."

MEDUSÆ. An order of *Hydrozoa*, commonly known as Jelly-fishes (*Discophora*, or *Acalephæ*), so called because of the resemblance of their tentacles to the snaky hair of the Medusa. Many *Medusæ* are now known to be merely the gonophores of *Hydrozoa*.

MEDUSIFORM. Resembling a *Medusa* in shape.

MEDUSOID. Like a *Medusa;* used substantively to designate the medusiform gonophores of the *Hydrozoa*.

MEMBRANA NICTITANS (Lat. *nicto*, I wink). The third eyelid of Birds, &c.

MENTUM (Lat. the chin). The basal portion of the labium or lower lip in Insects.

MEROSTOMATA (Gr. *mēron*, thigh; *stoma*, mouth). An order of *Crustacea* in which the appendages which are placed round the mouth, and which officiate as jaws, have their free extremities developed into walking or prehensile organs.

MESENTERIES (Gr. *mesos*, intermediate; *enteron*, intestine). In a restricted sense, the vertical plates which divide the somatic cavity of a Sea-anemone (*Actinia*) into chambers.

MESOPODIUM (Gr. *mesos*, middle; *pous*, foot). The middle portion of the "foot" of Molluscs.

MESOSTERNUM (Gr. *mesos*, intermediate; *sternon*, the breast-bone). The middle portion of the sternum, intervening between the attachment of the second pair of ribs and the xiphoid cartilage (*xiphisternum*).

MESOTHORAX (Gr. *mesos;* and *thorax*, the chest). The middle ring of the thorax in Insects.

MESOZOIC (Gr. *mesos;* and *zoe*, life). The Secondary period in Geology.

METACARPUS (Gr. *meta*, after; *karpos*, the wrist). The bones which form the "root of the hand," and intervene between the wrist and the fingers.

METAMORPHOSIS (Gr. *meta*, implying change; *morphe*, shape). The changes of form which certain animals undergo in passing from their younger to their fully-grown condition.

METAPODIUM (Gr. *meta*, after; *pous*, the foot). The posterior lobe of the foot in *Mollusca;* often called the "operculigerous lobe," because it develops the operculum when this structure is present.

METASTOMA (Gr. *meta*, after; *stoma*, mouth). The plate which closes the mouth posteriorly in the *Crustacea*.

METATARSUS (Gr. *meta*, after; *tarsos*, the instep). The bones which intervene between the bones of the ankle (*tarsus*) and the digits in the hind-foot of the higher Vertebrates.

METATHORAX (Gr. *meta*, after; *thorax*, the chest). The posterior ring of the thorax in Insects.

MIMETIC (Gr. *mimetikos*, imitative). Applied to organs or animals which resemble each other in external appearance, but not in essential structure.

MOLARS (Lat. *mola*, a mill). The "grinders" in man, or the teeth in diphyodont Mammals which are not preceded by milk-teeth.

MOLLUSCA (Lat. *mollis*, soft). The sub-kingdom which includes the Shell-fish proper, the *Polyzoa*, the *Tunicata*, and the Lamp-shells; so called from the generally soft nature of their bodies.

MOLLUSCOIDA (*Mollusca;* Gr. *eidos*, form). The lower division of the *Mollusca*, comprising the *Polyzoa*, *Tunicata*, and *Brachiopoda*.

MONADS (Gr. *monas*, unity). Microscopical organisms of an extremely simple character, developed in organic infusions.

MONOCULOUS. Possessed of only one eye.

MONODELPHIA (Gr. *monos*, single; *delphus*, womb). The division of *Mammalia* in which the uterus is single.

MONŒCIOUS (Gr. *monos*, single; *oikos*, house). Applied to individuals in which the sexes are united.

MONOMYARY (Gr. *monos*, single; *muon*, muscle). Applied to those bivalves (*Lamellibranchiata*) in which the shell is closed by a single adductor muscle).

MONOPHYODONT (Gr. *monos; phuo*, I generate; *odous*, tooth). Applied to those Mammals in which only a single set of teeth is ever developed.

MONOTHALAMOUS (Gr. *monos;* and *thalamos*, chamber). Possessing only a single chamber. Applied to the shells of *Foraminifera* and *Mollusca*.

MONOTREMATA (Gr. *monos; trema*, aperture). The order of Mammals comprising the Duck-mole and *Echidna*, in which the intestinal canal opens into a "cloaca" common to the ducts of the urinary and generative organs.

MULTILOCULAR (Lat. *multus*, many; *loculus*, a little purse). Divided into many chambers.

MULTIVALVE. Applied to shells which are composed of many pieces.

MULTUNGULA (Lat. *multus*, many; *ungula*, hoof). The division of Perissodactyle Ungulates, in which each foot has more than a single hoof.

MYELON (Gr. *muelos*, marrow). The spinal cord of Vertebrates.

MYRIAPODA (Gr. *murios*, ten thousand; *podes*, feet). A class of *Arthropoda* comprising the Centipedes and their allies, characterised by their numerous feet.

NACREOUS (Fr. *nacre*, mother-of-pearl, originally Oriental). Pearly; of the texture of mother-of-pearl.

NATATORES (Lat. *nare*, to swim). The order of the Swimming Birds.

NATATORY (Lat. *nare*, to swim). Formed for swimming.

NAUTILOID. Resembling the shell of the *Nautilus* in shape.

NECTOCALYX (Gr. *necho*, I swim; *kalux*, cup). The swimming-bell or "disc" of a *Medusa* or Jelly-fish.

NEMATELMIA (Gr. *nema*, thread; *helmins*, a worm). The division of *Scolecida* comprising the Round-worms, Thread-worms, &c.

NEMATOCYSTS (Gr. *nema*, thread; *kustis*, a bag). The thread-cells of the *Cœlenterata*. (*See* Cnidæ.)

NEMATOIDEA (Gr. *nema*, thread; *eidos*, form). An order of *Scolecida* comprising the Thread-worms, Vinegar-eels, &c.

NEMATOPHORES (Gr. *nema*, thread; *phero*, I carry). Cæcal processes found on the cœnosarc of certain of the *Sertularida*, containing numerous thread-cells at their extremities.

NEMERTIDA (Gr. *Nemertes*, proper name). A division of the *Turbellarian Worms*, commonly called "Ribbon-worms."

NERVURES (Lat. *nervus*, a sinew). The ribs which support the membranous wings of insects.

NEURAL (Gr. *neuron*, a nerve). Connected with the nervous system.

NEURAPOPHYSIS (Gr. *neuron*, a nerve; *apophusis*, a projecting part). The "spinous process" of a vertebra, or the process formed at the point of junction of the neural arches.

NEUROPODIUM (Gr. *neuron*, a nerve; *pous*, the foot). The ventral or inferior division of the "foot-tubercle" of an *Annelide;* often called the "ventral oar."

NEUROPTERA (Gr. *neuron;* and *pteron*, a wing). An order of Insects characterised by four membranous wings with numerous reticulated nervures (*e.g.*, Dragon-flies).

NEUTER (Lat. neither the one nor the other). Having no fully-developed sex.

NIDIFICATION (Lat. *nidus*, a nest; *facio*, I make). The building of a nest.

NOCTURNAL (Lat. *nox*, night). Applied to animals which are active by night.

NORMAL (Lat. *norma*, a rule). Conforming to the ordinary standard.

NOTOBRANCHIATA (Gr. *notos*, the back; and *bragchia*, gill). Carrying the gills upon the back; applied to a division of the *Annelida*.

NOTOCHORD (Gr. *notos*, back; *chorde*, string). A cellular rod which is developed in the embryo of Vertebrates immediately beneath the spinal cord, and which is usually replaced in the adult by the vertebral column. Often it is spoken of as the "chorda dorsalis."

NOTOPODIUM (Gr. *notos*, the back; and *pous*, the foot). The dorsal division of one of the foot-tubercles or parapodia of an *Annelide;* often called the "dorsal oar."

NUCLEATED. Possessing a nucleus or central particle.

NUCLEOLUS. 1. The minute solid particle in the interior of the nucleus of some cells. 2. The minute spherical particle attached to the exterior of the "nucleus," or ovary, of certain *Infusoria*, performing the functions of a testicle.

NUCLEUS (Lat. *nucleus*, a kernel). 1. The solid or vesicular body found in many cells. 2. The solid rod, or band-shaped body, found in the interior of many of the *Protozoa*, and having, in certain of them, the functions of an ovary. 3. The "madreporiform tubercle" of the *Echinodermata*. 4. The embryonic shell which is retained to form the apex of the adult shell in many of the *Mollusca*.

NUDIBRANCHIATA (Lat. *nudus*, naked; and Gr. *bragchia*, gill). An order of the *Gasteropoda* in which the gills are naked.

NYMPHS. The active pupæ of certain Insects.

OCCIPITAL. Connected with the *occiput*, or the back part of the head.

OCEANIC. Applied to animals which inhabit the open ocean (= pelagic).

OCELLI (Lat. diminutive of *oculus*, eye). The simple eyes of many Echinoderms, Spiders, Crustaceans, Molluscs, &c.

OCTOPODA (Gr. *octo*, eight; *pous*, foot). The tribe of Cuttle-fishes with eight arms attached to the head.

ODONTOCETI (Gr. *odous*, tooth; *ketos*, whale). The "toothed" Whales, in contradistinction to the "whalebone" Whales.

ODONTOID (Gr. *odous; eidos*, form). The "odontoid process" is the centrum or body of the first cervical vertebra (*atlas*). It is detached from the atlas, and is usually anchylosed with the second cervical vertebra (*axis*), and it forms the pivot upon which the head rotates.

ODONTOPHORE (Gr. *odous*, tooth; *phero*, I carry). The so-called "tongue" or masticatory apparatus of *Gasteropoda*, *Pteropoda*, and *Cephalopoda*.

ŒSOPHAGUS. The gullet or tube leading from the mouth to the stomach.

OLIGOCHÆTA (Gr. *oligos*, few; *chaite*, hair). An order of *Annelida*, comprising the Earth-worms, in which there are few bristles.

OMASUM (Lat. bullock's tripe). The third stomach of Ruminants, commonly called the *psalterium*, or many-plies.

OMNIVOROUS (Lat. *omnia*, everything; *voro*, I devour). Feeding indiscriminately upon all sorts of food.

OPERCULATA (Lat. *operculum*, a lid). A division of pulmonate *Gasteropoda*, in which the shell is closed by an operculum.

OPERCULUM. A horny or shelly plate developed in certain *Mollusca* upon the hinder part of the foot, and serving to close the aperture of the shell when the animal is retracted within it; also the lid of the shell of a *Balanus* or Acorn-shell; also the chain of flat bones which cover the gills in many fishes.

OPHIDIA (Gr. *ophis*, a serpent). The order of Reptiles comprising the Snakes.

OPHIDOBATRACHIA (Gr. *ophis; batrachos*, a frog). Sometimes applied to the order of Snake-like Amphibians comprising the *Cæciliæ*.

OPHIOMORPHA (Gr. *ophis; morphe*, shape). The order of *Amphibia* comprising the *Cæciliæ*.

OPHIUROIDEA (Gr. *ophis*, snake; *oura*, tail; *eidos*, form). An order of *Echinodermata* comprising the Brittle-stars and Sand-stars.

OPISTHOBRANCHIATA (Gr. *opisthen*, behind; *bragchia*, gill). A division of *Gasteropoda* in which the gills are placed on the posterior part of the body.

OPISTHOCŒLOUS (Gr. *opisthen*, behind; *koilos*, hollow). Applied to vertebræ, the bodies of which are hollow or concave behind.

ORAL (Lat. *os*, mouth). Connected with the mouth.

ORNITHODELPHIA (Gr. *ornis*, a bird; *delphus*, womb). The primary division of Mammals comprising the *Monotremata*.

ORTHOPTERA (Gr. *orthos*, straight; *pteron*, wing). An order of Insects.
OSCULA (Lat. diminutive of *os*, mouth). 1. The large apertures by which a sponge is perforated ("exhalant apertures"). 2. The suckers with which the *Tæniada* (Tape-worms and Cystic Worms) are provided.
OSSICULA (Lat. diminutive of *os*, bone). Literally small bones. Often used to designate any hard structures of small size, such as the calcareous plates in the integument of the Star-fishes.
OSTRACODA (Gr. *ostrakon*, a shell). An order of small Crustaceans which are enclosed in bivalve shells.
OTOLITHS (Gr. *ous*, ear; and *lithos*, stone). The calcareous bodies connected with the sense of hearing, even in its most rudimentary form.
OVARIAN VESICLES or CAPSULES. The generative buds of the *Sertularida*.
OVARY (OVARIUM). The organ by which ova are produced.
OVIPAROUS (Lat. *ovum*, an egg; and *pario*, I bring forth). Applied to animals which bring forth eggs, in contradistinction to those which bring forth their young alive.
OVIPOSITOR (Lat. *ovum*; and *pono*, I place). The organ possessed by some insects, by means of which the eggs are placed in a position suitable for their development.
OVISAC. The external bag or sac in which certain of the Invertebrates carry their eggs after they are extruded from the body.
OVOVIVIPAROUS (Lat. *ovum*, egg; *vivus*, alive; *pario*, I produce). Applied to animals which retain their eggs within the body until they are hatched.
OVUM (Lat. an egg). The germ produced within the ovary, and capable under certain conditions of being developed into a new individual.

PACHYDERMATA (Gr. *pachus*, thick; *derma*, skin). An old Mammalian order constituted by Cuvier for the reception of the Rhinoceros, Hippopotamus, Elephant, &c.
PALÆONTOLOGY (Gr. *palaios*, ancient; and *logos*, discourse). The science of fossil remains or of extinct organised beings.
PALÆOZOIC (Gr. *palaios*, ancient; and *zoe*, life). Applied to the oldest of the great geological epochs.
PALLIOBRANCHIATA (Lat. *pallium*; and Gr. *bragchia*, gill). An old name for the *Brachiopoda*, founded upon the belief that the system of tubes in the mantle constituted the gills.
PALLIUM (Lat. *pallium*, a cloak). The mantle of the *Mollusca*. *Pallial*: relating to the mantle. *Pallial line* or *impression*; the line left in the dead shell by the muscular margin of the mantle. *Pallial shell*; a shell which is secreted by, or contained within, the mantle, such as the "bone" of the Cuttle-fishes.
PALPI (Lat. *palpo*, I touch). Processes supposed to be organs of touch, developed from certain of the oral appendages in Insects, Spiders, and Crustacea, and from the sides of the mouth in the Acephalous Molluscs.
PAPILLA (Lat. for nipple). A minute soft prominence.
PARAPODIA (Gr. *para*, beside; *podes*, feet). The unarticulated lateral locomotive processes or "foot-tubercles" of many of the *Annelida*.
PARIETAL (Lat. *paries*, a wall). Connected with the walls of a cavity or of the body.
PARIETOSPLANCHNIC (Lat. *paries*; Gr. *splagchna*, viscera). Applied to one of the nervous ganglia of the *Mollusca*, which supplies the walls of the body and the viscera.
PARTHENOGENESIS (Gr. *parthenos*, a virgin; and *gignomai*, to be born). Strictly speaking, confined to the production of new individuals from virgin females by means of ova without the intervention of a male. Sometimes used also to designate asexual reproduction by gemmation or fission.
PATAGIUM (Lat. the border of a dress). Applied to the expansion of the integument by which Bats, Flying Squirrels, and other animals support themselves in the air.
PATELLA. The knee-cap or knee-pan. A sesamoid bone developed in the tendon of insertion of the great extensor muscles of the thigh.

PECTINATE (Lat. *pecten*, a comb). Comb-like; applied to the gills of certain *Gasteropods;* hence called *Pectinibranchiata.*

PECTORAL (Lat. *pectus*, chest). Connected with, or placed upon, the chest.

PERENNIBRANCHIATA (Lat. *perennis*, perpetual; Gr. *bragchia*, gill). Applied to those *Amphibia* in which the gills are permanently retained throughout life.

PEDAL (Lat. *pes*, the foot). Connected with the foot of *Mollusca.*

PEDICELLARIÆ (Lat. *pedicellus*, a louse). Certain singular appendages found in many *Echinoderms*, attached to the surface of the body, and resembling a little beak or forceps supported on a stalk.

PEDICLE (Lat. dim of *pes*, the foot). A little stem.

PEDIPALPI (Lat. *pes*, foot; and *palpo*, I feel). An order of *Arachnida* comprising the Scorpions, &c.

PEDUNCLE (Lat. *pedunculus*, a stem or stalk). In a restricted sense applied to the muscular process by which certain *Brachiopods* are attached, and to the stem which bears the body (capitulum) in Barnacles.

PEDUNCULATE. Possessing a peduncle.

PELAGIC (Gr. *pelagos*, sea). Inhabiting the open ocean.

PELVIS (Lat. for basin). Applied, from analogy, to the basal portion of the cup (*calyx*) of *Crinoids.* The bony arch with which the hind-limbs are connected in Vertebrates.

PERGAMENTACEOUS (Lat. *pergamena*, parchment). Of the texture of parchment.

PERICARDIUM (Gr. *peri*, around; *kardia*, heart). The serous membrane in which the heart is contained.

PERIDERM (Gr. *peri*, around; and *derma*, skin). The hard cuticular layer which is developed by the cœnosarc of certain of the *Hydrozoa.*

PERIGASTRIC (Gr. *peri*, around; and *gaster*, stomach). The perigastric space is the cavity which surrounds the stomach and other viscera, corresponding to the abdominal cavity of the higher animals.

PERIOSTRACUM (Gr. *peri;* and *ostrakon*, shell). The layer of epidermis which covers the shell in most of the *Mollusca.*

PERIPLAST (Gr. *peri;* and *plasso*, I mould). The intercellular substance or matrix in which the organised structures of a tissue are embedded.

PERISOME (Gr. *peri;* and *soma*, body). The coriaceous or calcareous integument of the *Echinodermata.*

PERISSODACTYLA (Gr. *perissos*, uneven; *daktulos*, finger). Applied to those Hoofed Quadrupeds (*Ungulata*) in which the feet have an uneven number of toes.

PERISTOME (Gr. *peri;* and *stoma*, mouth). The space which intervenes between the mouth and the margin of the calyx in *Vorticella;* also the space between the mouth and the tentacles in a sea-anemone (*Actinia*); also the lip or margin of the mouth of a univalve shell.

PERIVISCERAL (Gr. *peri;* and Lat. *viscera*, the internal organs). Applied to the space surrounding the viscera.

PETALOID. Shaped like the petal of a flower.

PHALANGES (Gr. *phalanx*, a row). The small bones composing the digits of the higher *Vertebrata.* Normally each digit has three phalanges.

PHARYNGOBRANCHII (Gr. *pharugx*, pharynx; *bragchia*, gill). The order of Fishes comprising only the Lancelet.

PHARYNX. The dilated commencement of the gullet.

PHRAGMACONE (Gr. *phragma*, a partition; and *konos*, a cone). The chambered portion of the internal shell of a *Belemnite.*

PHYLACTOLÆMATA (Gr. *phulasso*, I guard; and *laimos*, throat). The division of *Polyzoa* in which the mouth is provided with the arched valvular process known as the "epistome."

PHYLLOCYSTS (Gr. *phullon*, leaf; and *kustis*, a cyst). The cavities in the interior of the "hydrophyllia" of certain of the Oceanic *Hydrozoa.*

PHYLLOPODA (Gr. *phullon*, leaf; and *pous*, foot). An order of *Crustacea.*

PHYOGEMMARIA (Gr. *phuo*, I produce; and Lat. *gemma*, bud). The small gonoblastidia of *Velella*, one of the *Physophoridæ.*

PHYSOGRADA (Gr. *phusa*, bellows or air-bladder; and Lat. *gradior*, I walk).

Applied formerly to the *Physophoridæ*, an order of Oceanic *Hydrozoa*, in which a "float" is present.

PHYSOPHORIDÆ (Gr. *phusa*, air-bladder; and *phero*, I carry). An order of Oceanic *Hydrozoa*.

PHYTOID (Gr. *phuton*, a plant; and *eidos*, form). Plant-like.

PHYTOPHAGOUS (Gr. *phuton*, a plant; and *phago*, I eat). Plant-eating, or herbivorous.

PINNATE (Lat. *pinna*, a feather). Feather-shaped, or possessing lateral processes.

PINNIGRADA (Lat. *pinna*, a feather; *gradior*, I walk). The group of *Carnivora*, comprising the Seals and Walruses, adapted for an aquatic life. Often called *Pinnipedia*.

PINNULÆ (Lat. dim. of *pinna*). The lateral processes of the arms of *Crinoids*.

PISCES (Lat. *piscis*, a fish). The class of Vertebrates comprising the Fishes.

PLACENTA (Lat. a cake). The "after-birth," or the organ by which a vascular connection is established in the higher *Mammalia* between the mother and the fœtus.

PLACENTAL. Possessing a placenta, or connected with the placenta.

PLACOID (Gr. *plax*, a plate; *eidos*, form). Applied to the irregular bony plates, grains, or spines which are found in the skin of various fishes (*Elasmobranchii*).

PLAGIOSTOMI (Gr. *plagios*, transverse; *stoma*, mouth). The Sharks and Rays, in which the mouth is transverse, and is placed on the under surface of the head.

PLANARIDA (Gr. *planē*, wandering). A sub-order of the *Turbellaria*.

PLANTIGRADE (Lat. *planta*, the sole of the foot; *gradior*, I walk). Applying the sole of the foot to the ground in walking.

PLANULA (Lat. *planus*, flat). The oval ciliated embryo of certain of the *Hydrozoa*.

PLASTRON. The lower or ventral portion of the bony case of the Chelonians.

PLATYELMIA (Gr. *platus*, broad; and *helmins*, an intestinal worm). The division of *Scolecida* comprising the Tape-worms, &c.

PLATYRHINA (Gr. *platus*, broad; *rhines*, nostrils). A group of the *Quadrumana*.

PLEURA (Gr. the side). The serous membrane covering the lung in the air-breathing Vertebrates.

PLEURON (Gr. *pleuron*, a rib). The lateral extensions of the shell of *Crustacea*.

PLUTEUS (Lat. a pent-house). The larval form of the *Echinoidea*.

PNEUMATIC (Gr. *pneuma*, air). Filled with air.

PNEUMATOCYST (Gr. *pneuma*, air; and *kustis*, cyst). The air-sac or float of certain of the Oceanic *Hydrozoa* (*Physophoridæ*).

PNEUMATOPHORE (Gr. *pneuma*, air; and *phero*, I carry). The proximal dilatation of the cœnosarc in the *Physophoridæ* which surrounds the pneumatocyst.

PNEUMOSKELETON (Gr. *pneuma*; and *skeletos*, dry). The hard structures which are connected with the breathing organs (*e.g.*, the shell of Molluscs).

PODOPHTHALMATA (Gr. *pous*, foot; and *ophthalmos*, eye). The division of Crustacea in which the eyes are borne at the end of long foot-stalks.

PODOSOMATA (Gr. *pous*, foot; *soma*, body). An order of *Arachnida*.

POEPHAGA (Gr. *poē*, grass; *phago*, I eat). A group of the Marsupials.

POLLEX (Lat. the thumb). The innermost of the five normal digits of the anterior limb of the higher Vertebrates. In man, the thumb.

POLYCYSTINA (Gr. *polus*, many; and *kustis*, a cyst). An order of *Protozoa*, with foraminated siliceous shells.

POLYGASTRICA (Gr. *polus*; and *gaster*, stomach). The name applied by Ehrenberg to the *Infusoria*, under the belief that they possessed many stomachs.

POLYPARY. The hard chitinous covering secreted by many of the *Hydrozoa*.

POLYPE (Gr. *polus*, many; *pous*, foot). Restricted to the single individual of a simple *Actinozoön*, such as a Sea-anemone, or to the separate zoöids of a compound *Actinozoön*. Often applied indiscriminately to any of the *Cœlenterata*, or even to the *Polyzoa*.

POLYPIDE. The separate zoöid of a *Polyzoön*.
POLYPIDOM. The dermal system of a colony of a *Hydrozoön*, or *Polyzoön*.
POLYPITE. The separate zoöid of a *Hydrozoön*.
POLYSTOME (Gr. *polus*, many; and *stoma*, mouth). Having many mouths; applied to the *Acinetæ* amongst the *Protozoa*.
POLYTHALAMOUS (Gr. *polus*; and *thalamos*, chamber). Having many chambers; applied to the shells of *Foraminifera* and *Cephalopoda*.
POLYZOA (Gr. *polus*; and *zoön*, animal). A division of the *Molluscoida*, comprising compound animals, such as the Sea-mat. Sometimes called *Bryozoa*.
POLYZOARIUM. The dermal system of the colony of a *Polyzoön* (= Polypidom).
PORCELLANOUS. Of the texture of porcelain.
PORIFERA (Lat. *porus*, a pore; and *fero*, I carry). Sometimes used to designate the *Foraminifera*, or the *Sponges*.
POST-ANAL. Situated behind the anus.
POST-ŒSOPHAGEAL. Situated behind the gullet.
POST-ORAL. Situated behind the mouth.
PRÆ-MAXILLÆ. (*See* Intermaxillæ.)
PRÆMOLARS (Lat. *præ*, before; *molares*, the grinders). The molar teeth of Mammals which succeed the molars of the milk-set of teeth. In man, the bicuspid teeth.
PRÆ-ŒSOPHAGEAL. Situated in front of the gullet.
PRÆ-STERNUM. The anterior portion of the breast-bone, corresponding with the *manubrium sterni* of human anatomy, and extending as far as the point of articulation of the second rib.
PRESSIROSTRES (Lat. *pressus*, compressed; *rostrum*, beak). A group of the Grallatorial Birds.
PROBOSCIDEA (Lat. *proboscis*, the snout). The order of Mammals comprising the Elephants.
PROBOSCIS (Lat. or Gr. the snout). Applied to the spiral trunk of *Lepidopterous Insects*, to the projecting mouth of certain *Crinoids*, and to the central polypite in the *Medusæ*.
PROCŒLOUS (Gr. *pro*, front; *koilos*, hollow). Applied to vertebræ, the bodies of which are hollow or concave in front.
PROGLOTTIS (Gr. for the tip of the tongue). The generative segment or joint of a Tape-worm.
PRO-LEGS. The false abdominal feet of Caterpillars.
PRONATION (Lat. *pronus*, lying on the face, prone). The act of turning the palm of the hand downwards.
PROPODIUM (Gr. *pro*, before; *pous*, foot). The anterior part of the foot in Molluscs.
PROSCOLEX (Gr. *pro*, before; *scolex*, worm). The first embryonic stage of a Tape-worm.
PROSOBRANCHIATA (Gr. *proson*, in advance of; *bragchia*, a gill). A division of Gasteropodous Molluscs in which the gills are situated in advance of the heart.
PROSOMA (Gr. *pro*, before; *soma*, body). The anterior part of the body.
PROTHORAX (Gr. *pro*; and *thorax*, chest). The anterior ring of the thorax of insects.
PROTOPHYTA (Gr. *protos*: and *phuton*, plant). The lowest division of plants.
PROTOPLASM (Gr. *protos*; and *plasso*, I mould). The elementary basis of organised tissues. Sometimes used synonymously for the "sarcode" of the *Protozoa*.
PROTOPODITE (Gr. *protos*, first; and *pous*, foot). The basal segment of the typical limb of a *Crustacean*.
PROTOZOA (Gr. *protos*: and *zoön*, animal). The lowest division of the animal kingdom.
PROVENTRICULUS (Lat. *pro*, in front of; *ventriculus*, dim. of *venter*, belly). The cardiac portion of the stomach of Birds.
PROXIMAL (Lat. *proximus*, next). The slowly-growing, comparatively-fixed extremity of a limb or of an organism.

PSALTERIUM (Lat. a stringed instrument). The third stomach of Ruminants. (*See* Omasum.)

PSEUDEMBRYO (Gr. *pseudos,* falsity; *embruon,* embryo). The larval form of an *Echinoderm.*

PSEUDOBRANCHIA (Gr. *pseudos,* falsity; *bragchia,* gill). A supplementary gill found in certain fishes, which receives arterialised blood only, and does not, therefore, assist in respiration.

PSEUDOHÆMAL (Gr. *pseudos,* falsity; and *haima,* blood). Applied to the vascular system of *Annelida.*

PSEUDO-HEARTS. Certain contractile cavities connected with the atrial system of *Brachiopoda,* and long considered to be hearts.

PSEUDO-NAVICELLÆ (Gr. *pseudos,* false; and *Navicula,* a genus of Diatoms). The embryonic forms of the *Gregarinidæ,* so called from their resemblance in shape to the *Navicula.*

PSEUDOPODIA (Gr. *pseudos;* and *pous,* foot). The extensions of the body-substance which are put forth by the *Rhizopoda* at will, and which serve for locomotion and prehension.

PSEUDOVA (G. *pseudos;* Lat. *ovum,* egg). The egg-like bodies from which the young of the viviparous *Aphis* are produced.

PTEROPODA (Gr. *pteron,* wing; and *pous,* foot). A class of the *Mollusca* which swim by means of fins attached near the head.

PTEROSAURIA (Gr. *pteron,* wing; *saura,* lizard). An extinct order of Reptiles.

PUBIS (Lat. *pubes,* hair). The share-bone; one of the bones which enter into the composition of the pelvic arch of Vertebrates.

PULMOGASTEROPODA (=Pulmonifera).

PULMONARIA. A division of *Arachnida* which breathe by means of pulmonary sacs.

PULMONATE. Possessing lungs.

PULMONIFERA (Lat. *pulmo,* a lung; and *fero,* I carry). The division of *Mollusca* which breathe by means of a pulmonary chamber.

PUPA (Lat. a doll). The stage of an insect immediately preceding its appearance in a perfect condition. In the pupa-stage it is usually quiescent—when it is often called a "chrysalis;" but it is sometimes active—when it is often called a "nymph."

PYLORUS (Gr. *puloros,* a gatekeeper). The valvular aperture between the stomach and the intestine.

PYRIFORM (Lat. *pyrus,* a pear; and *forma,* formed). Pear-shaped.

QUADRUMANA (Lat. *quatuor,* four; *manus,* hand). The order of Mammals comprising the Apes, Monkeys, Baboons, Lemurs, &c.

RADIATA (Lat. *radius,* a ray). Formerly applied to a large number of animals which are now placed in separate sub-kingdoms (*e.g.,* the *Cœlenterata,* the *Echinodermata,* the *Infusoria,* &c.)

RADIOLARIA (Lat. *radius,* a ray). A division of *Protozoa.*

RADIUS (Lat. a spoke or ray). The innermost of the two bones of the forearm of the higher Vertebrates. It carries the thumb, when present, and corresponds with the tibia of the hind-limb.

RAMUS (Lat. a branch). Applied to each half or branch of the lower jaw or mandible of Vertebrates.

RAPTORES (Lat. *rapto,* I plunder). The order of the Birds of Prey.

RASORES (Lat. *rado,* I scratch). The order of the Scratching Birds (Fowls, Pigeons, &c.)

RATITÆ (Lat. *rates,* a raft). Applied by Huxley to the Cursorial Birds, which do not fly, and have therefore a raft-like sternum without any median keel.

RECTUM (Lat. *rectus,* straight). The terminal portion of the intestinal canal, opening at the surface of the body at the anus.

REPTILIA (Lat. *repto,* I crawl). The class of the *Vertebrata* comprising the Tortoises, Snakes, Lizards, Crocodiles, &c.

RETICULARIA (Lat. *reticulum,* a net). Employed by Dr Carpenter to designate those *Protozoa*, such as the *Foraminifera*, in which the pseudopodia run into one another and form a network.

RETICULUM (Lat. a net). The second division of the complex stomach of Ruminants, often called the "honeycomb bag."

REVERSED. Applied to spiral univalves, in which the direction of the spiral is the reverse of the normal—*i.e.*, *sinistral*.

RHIZOPHAGA (Gr. *rhiza*, root; *phago*, I eat). A group of the Marsupials.

RHIZOPODA (Gr. *rhiza*, a root; and *pous*, foot). The division of *Protozoa* comprising all those which are capable of emitting pseudopodia.

RHYNCHOLITES (Gr. *rhunchos*, beak; and *lithos*, stone). Beak-shaped fossils, consisting of the mandibles of *Cephalopoda*.

RODENTIA (Lat. *rodo*, I gnaw). An order of the Mammals; often called *Glires* (Lat. *glis*, a dormouse).

ROSTRUM (Lat. *rostrum*, beak). The "beak" or suctorial organ formed by the appendages of the mouth in certain insects.

ROTATORIA (= Rotifera).

ROTIFERA (Lat. *rota*, wheel; and *fero*, I carry). A class of the *Scolecida* (*Annuloida*) characterised by a ciliated "trochal disc."

RUGOSA (Lat. *rugosus*, wrinkled). An extinct order of Corals.

RUMEN (Lat. the throat). The first cavity of the complex stomach of Ruminants; often called the "paunch."

RUMINANTIA (Lat. *ruminor*, I chew the cud). The group of Hoofed Quadrupeds (*Ungulata*) which "ruminate" or chew the cud.

SACRUM. The vertebræ (usually anchylosed) which unite with the haunch-bones (*ilia*) to form the pelvis.

SAND-CANAL (= STONE-CANAL). The tube by which water is conveyed from the exterior to the ambulacral system of the *Echinodermata*.

SARCODE (Gr. *sarx*, flesh; *eidos*, form). The jelly-like substance of which the bodies of *Protozoa* are composed. It is an albuminous body containing oil-granules, and is sometimes called "animal protoplasm."

SARCOIDS (Gr. *sarx*; and *eidos*, form). The separate amœbiform particles which in the aggregate make up the "flesh" of a Sponge.

SAURIA (Gr. *saura*, a lizard). Any lizard-like Reptile is often spoken of as a "Saurian;" but the term is sometimes restricted to the Crocodiles alone, or to the Crocodiles and Lacertilians.

SAUROBATRACHIA (Gr. *saura*; *batrachos*, frog). Sometimes applied to the order of the tailed Amphibians (*Urodela*).

SAUROPSIDA (Gr. *saura*; and *opsis*, appearance). The name given by Huxley to the two classes of the Birds and Reptiles collectively.

SAUROPTERYGIA (Gr. *saura*; *pterux*, wing). An extinct order of Reptiles, called by Huxley *Plesiosauria*, from the typical genus *Plesiosaurus*.

SAURURÆ (Gr. *saura*; *oura*, tail). The extinct order of Birds comprising only the *Archæopteryx*.

SCANSORES (Lat. *scando*, I climb). The order of the Climbing Birds (Parrots, Woodpeckers, &c.)

SCAPHOGNATHITE (Gr. *skaphos*, boat; and *gnathos*, jaw). The boat-shaped appendage (epipodite) of the second pair of maxillæ in the Lobster; the function of which is to spoon out the water from the branchial chamber.

SCAPULA (Lat. for shoulder-blade). The shoulder-blade of the pectoral arch of Vertebrates; in a restricted sense, the row of plates in the cup of *Crinoids*, which give origin to the arms, and are usually called the "axillary radials."

SCLERENCHYMA (Gr. *skleros*, hard; and *enchuma*, tissue). The calcareous tissue of which a coral is composed.

SCLERITES (Gr. *skleros*). The calcareous spicules which are scattered in the soft tissues of certain *Actinozoa*.

SCLEROBASIC (Gr. *skleros*, hard; *basis*, pedestal). The coral which is produced by the outer surface of the integument in certain *Actinozoa* (*e.g.*, Red Coral), and forms a solid axis which is invested by the soft parts of the animal. It is called "foot-secretion" by Mr Dana.

SCLERODERMIC (Gr. *skleros;* and *derma,* skin). Applied to the corallum which is deposited within the tissues of certain *Actinozoa,* and is called "tissue-secretion" by Mr Dana.
SCLEROTIC (Gr. *skleros,* hard). The outer dense fibrous coat of the eye.
SCOLECIDA (Gr. *skolēx,* worm). A division of the *Annuloida.*
SCOLEX (Gr. *skolēx*). The embryonic stage of a Tape-worm, formerly known as a "Cystic Worm."
SCUTA (Lat. *scūtum,* a shield). Applied to any shield-like plates; especially to those which are developed in the integument of many Reptiles.
SELACHIA or SELACHII (Gr. *selachos,* a cartilaginous fish, probably a shark). The sub-order of *Elasmobranchii,* comprising the Sharks and Dog-fishes.
SEPIOSTAIRE. The internal shell of the Cuttle-fish, commonly known as the "cuttle-bone."
SEPTA. Partitions.
SERPENTIFORM. Resembling a serpent in shape.
SERTULARIDA (Lat. *sertum,* a wreath). An order of *Hydrozoa.*
SESSILE (Lat. *sedo,* I sit). Not supported upon a stalk or peduncle; attached by a base.
SETÆ (Lat. bristles). Bristles, or long stiff hairs.
SETIFEROUS. Supporting bristles.
SETIGEROUS (= Setiferous).
SETOSE. Bristly.
SILICEOUS (Lat. *silex,* flint). Composed of flint.
SINISTRAL (Lat. *sinistra,* the left hand). Left-handed; applied to the direction of the spiral in certain shells, which are said to be "reversed."
SINUS (Lat. *sinus,* a bay). A dilated vein or blood-receptacle.
SIPHON (Gr. *siphon,* a tube). Applied to the respiratory tubes in the *Mollusca;* also to other tubes of different functions.
SIPHONOPHORA (Gr. *siphon;* and *phero,* I carry). A division of the *Hydrozoa,* comprising the Oceanic forms (*Calycophoridæ* and *Physophoridæ*).
SIPHONOSTOMATA (Gr. *siphon;* and *stoma,* mouth). The division of *Gasteropodous Molluscs,* in which the aperture of the shell is not "entire," but possesses a notch or tube for the emission of the respiratory siphon.
SIPHUNCLE (Lat. *siphunculus,* a little tube). The tube which connects together the various chambers of the shell of certain *Cephalopoda* (*e.g.,* the Pearly Nautilus).
SIPHUNCULOIDEA (Lat. *siphunculus,* a little siphon). A class of *Anarthropoda* (*Annulosa*).
SIRENIA (Gr. *seiren,* a mermaid). The order of *Mammalia* comprising the Dugongs and Manatees.
SOLIDUNGULA (Lat. *solidus,* solid; *ungula,* a hoof). The group of Hoofed Quadrupeds comprising the Horse, Ass, and Zebra, in which each foot has only a single solid hoof. Often called *Solipedia.*
SOMATIC (Gr. *soma,* body). Connected with the body.
SOMATOCYST (Gr. *soma;* and *kustis,* a cyst). A peculiar cavity in the cœnosarc of the *Calycophoridæ* (*Hydrozoa*).
SOMITE (Gr. *soma*). A single segment in the body of an Articulate animal.
SPERMARIUM. The organ in which spermatozoa are produced.
SPERMATOPHORES (Gr. *sperma,* seed; *phero,* I carry). The cylindrical capsules of the *Cephalopoda,* which carry the spermatozoa; sometimes called the "moving filaments of Needham."
SPERMATOZOA (Gr. *sperma,* seed; and *zoōn,* animal). The microscopic filaments which form the essential generative element of the male.
SPICULA (Lat. *spiculum,* a point). Pointed needle-shaped bodies.
SPINNERETS. The organs by means of which Spiders and Caterpillars spin threads.
SPIRACLES (Lat. *spiro,* I breathe). The breathing-pores, or apertures of the breathing-tubes (tracheæ) of Insects. Also the single nostril of the Hagfishes, the "blow-hole" of Cetaceans, &c.
SPLANCHNOSKELETON (Gr. *splagchna,* viscera; *skeletos,* dry). The hard

structures occasionally developed in connection with the internal organs or viscera.

SPONGE-PARTICLES. (*See* Sarcoids.)

SPONGIDA (Gr. *spoggos*, a sponge). The division of *Protozoa* commonly known as sponges.

SPORES (Gr. *spora*, seed). Germs, usually of plants; in a restricted sense, the reproductive "gemmules" of certain Sponges.

SPOROSACS (Gr. *spora*, seed; and *sakkos*, a bag). The simple generative buds of certain *Hydrozoa*, in which the medusoid structure is not developed.

SQUAMATA (Lat. *squama*, a scale). The division of Reptiles comprising the *Ophidia* and *Lacertilia* in which the integument develops horny scales, but there are no dermal ossifications.

STATOBLASTS (Gr. *statos*, stationary; *blastos*, bud). Certain reproductive buds developed in the interior of *Polyzoa*, but not liberated until the death of the parent organism.

STEGANOPHTHALMATA (Gr. *steganos*, covered; and *ophthalmos*, the eye). Applied by Edward Forbes to certain *Medusæ*, in which the sense-organs ("marginal bodies") are protected by a sort of hood. The *Steganophthalmata* are now separated from the true *Medusidæ*, and placed in a separate division under the name *Lucernarida*.

STELLERIDA (Lat. *stella*, star). Sometimes employed to designate the order of the Star-fishes.

STELLIFORM. Star-shaped.

STEMMATA (Gr. *stemma*, garland). The simple eyes, or "ocelli," of certain animals, such as Insects, Spiders, and Crustacea.

STERNUM (Gr. *sternon*). The breast-bone.

STIGMATA. The breathing-pores in *Insects* and *Arachnida*.

STOLON (Gr. *stolos*, a sending forth). Offshoots.—The connecting processes of sarcode, in *Foraminifera*; the connecting tube in the social *Ascidians*; the processes sent out by the cœnosarc of certain *Actinozoa*.

STOMAPODA (Gr. *stoma*, mouth; *pous*, foot). An order of *Crustacea*.

STOMATODE (Gr. *stoma*). Possessing a mouth. The *Infusoria* are thus often called the Stomatode *Protozoa*.

STREPSIPTERA (Gr. *strepho*, I twist; and *pteron*, wing). An order of Insects in which the anterior wings are represented by twisted rudiments.

STREPSIRHINA (Gr. *strepho*, I twist; *rhines*, nostrils). A group of the Quadrumana, often spoken of as *Prosimiæ*.

STROBILA (Gr. *strobilos*, a top, or fir-cone). The adult Tapeworm with its generative segments or proglottides; also applied to one of the stages in the life history of the *Lucernarida*.

STYLIFORM (Lat. *stylus*, a pointed instrument; *forma*, form). Pointed in shape.

SUB-CALCAREOUS. Somewhat calcareous.

SUB-CENTRAL. Nearly central, but not quite.

SUB-PEDUNCULATE. Supported upon a very short stem.

SUB-SESSILE. Nearly sessile, or without a stalk.

SUPINATION (Lat. *supinus*, lying with the face upwards). The act of turning the hand with the palm upwards.

SUTURE (Lat. *suo*, I sew). The line of junction of two parts which are immovably connected together. Applied to the line where the whorls of a univalve shell join one another; also to the lines made upon the exterior of the shell of a chambered *Cephalopod* by the margins of the septa.

SWIMMERETS. The limbs of *Crustacea*, which are adapted for swimming.

SYMPHYSIS (Gr. *sumphusis*, a growing together). Union of two bones in which there is no motion, or but a very limited amount.

SYNAPTICULÆ (Gr. *sunapto*, I fasten together). Transverse props sometimes found in Corals, extending across the loculi like the bars of a grate.

SYSTOLE (Gr. *sustello*, I contract). Applied to the contraction of any contractile cavity, especially the heart.

TABULÆ (Lat. *tabula*, a tablet). Horizontal plates or floors found in some Corals, extending across the cavity of the "theca," from side to side.

TACTILE (Lat. *tango*, I touch). Connected with the sense of touch.

TÆNIADA (Gr. *tainia*, a ribbon). The division of *Scolecida* comprising the Tapeworms.

TÆNIOID (Gr. *tainia;* and *eidos*, form). Ribbon-shaped, like a Tape-worm.

TARSO-METATARSUS. The single bone in the leg of Birds produced by the union and anchylosis of the lower or distal portion of the tarsus with the whole of the metatarsus.

TARSUS (Gr. *tarsos*, the flat of the foot). The small bones which form the ankle (or "instep" of man), and which correspond with the wrist (*carpus*) of the anterior limb.

TECTIBRANCHIATA (Lat. *tectus*, covered; and Gr. *bragchia*, gills). A division of *Opisthobranchiate Gasteropoda* in which the gills are protected by the mantle.

TEGUMENTARY (Lat. *tegumentum*, a covering). Connected with the integument or skin.

TELEOSTEI (Gr. *teleios*, perfect; *osteon*, bone). The order of the "Bony" Fishes.

TELSON (Gr. *telson*, a limit). The last joint in the abdomen of *Crustacea;* variously regarded as a segment without appendages, or as an azygos appendage.

TENUIROSTRES (Lat. *tenuis*, slender; *rostrum*, beak). A group of the Perching Birds characterised by their slender beaks.

TERGUM (Lat. for back). The dorsal arc of the somite of an Arthropod.

TERRICOLA (Lat. *terra*, earth; and *colo*, I inhabit). Employed occasionally to designate the Earth-worms (*Lumbricidæ*).

TEST (Lat. *testa*, shell). The shell of *Mollusca*, which are for this reason sometimes called "*Testacea;*" also, the calcareous case of *Echinoderms;* also, the thick leathery outer tunic in the *Tunicata*.

TESTACEOUS. Provided with a shell or hard covering.

TESTIS (Lat. *testis*, the testicle). The organ in the male animal which produces the generative fluid or semen.

TETRABRANCHIATA (Gr. *tetra*, four; *bragchia*, gill). The order of *Cephalopoda*, characterised by the possession of four gills.

THALASSICOLLIDA (Gr. *thalassa*, sea; *kolla*, glue). A division of *Protozoa*.

THECA (Gr. *theke*, a sheath). A sheath or receptacle.

THECOSOMATA (Gr. *theke;* and *soma*, body). A division of *Pteropodous Molluscs*, in which the body is protected by an external shell.

THERIOMORPHA (Gr. *ther*, beast; *morphe*, shape). Applied by Owen to the order of the Tail-less Amphibians (*Anoura*).

THORAX (Gr. a breastplate). The chest.

THREAD-CELLS. (*See* Cnidæ.)

THYSANURA (Gr. *thusanoi*, fringes; and *oura*, tail). An order of Apterous Insects.

TIBIA (Lat. a flute). The shin-bone, being the innermost of the two bones of the leg, and corresponding with the *radius* in the anterior extremity.

TOTIPALMATÆ (Lat. *totus*, whole; *palma*, the palm of the hand). A group of Wading Birds in which the hallux is united to the other toes by membrane, so that the feet are completely webbed.

TRACHEA (Gr. *tracheia*, the rough wind-pipe). The tube which conveys air to the lungs in the air-breathing Vertebrates.

TRACHEÆ. The breathing-tubes of insects and other articulate animals.

TRACHEARIA. The division of *Arachnida* which breathe by means of tracheæ.

TREMATODA (Gr. *trema*, a pore). An order of *Scolecida*.

TRICHOCYSTS (Gr. *thrix*, hair; and *kustis*, a cyst). Peculiar cells found in certain *Infusoria*, and very nearly identical with the "thread-cells" of *Cœlenterata*.

TRILOBITA (Gr. *treis*, three; *lobos*, a lobe). An extinct order of *Crustaceans*.

TRITOZOÖID (Gr. *tritos*, third; *zöon*, animal; and *eidos*, form). The zoöid produced by a deuterozoöid; that is to say, a zoöid of the third generation.

TROCHAL (Gr. *trochos*, a wheel). Wheel-shaped; applied to the ciliated disc of the *Rotifera*.

TROCHANTER (Gr. *trecho*, I turn). A process of the upper part of the thigh-bone (*femur*) to which are attached the muscles which rotate the limb. There may be two, or even three, trochanters present.

TROCHOID (Gr. *trochos*, a wheel; and *eidos*, form). Conical with a flat base; applied to the shells of *Foraminifera* and *Univalve Molluscs*.

TROPHI (Gr. *trophos*, a nourisher). The parts of the mouth in insects which are concerned in the acquisition and preparation of food. Often called "instrumenta cibaria."

TROPHOSOME (Gr. *trepho*, I nourish; and *soma*, body). Applied collectively to the assemblage of the nutritive zoöids of any *Hydrozoön*.

TRUNCATED (Lat. *trunco*, I shorten). Abruptly cut off; applied to univalve shells, the apex of which breaks off, so that the shell becomes "decollated."

TUBICOLA (Lat. *tuba*, a tube; and *cola*, I inhabit). The order of *Annelida* which construct a tubular case in which they protect themselves.

TUBICOLOUS. Inhabiting a tube.

TUNICATA (Lat. *tunica*, a cloak). A class of *Molluscoida* which are enveloped in a tough leathery case or "test."

TURBELLARIA (Lat. *turbo*, I disturb). An order of *Scolecida*.

TURBINATED (Lat. *turbo*, a top). Top-shaped; conical, with a round base.

ULNA (Gr. *olene*, the elbow). The outermost of the two bones of the fore-arm, corresponding with the *fibula* of the hind-limb.

UMBELLATE (Lat. *umbella*, a parasol). Forming an umbel—*i.e.*, a number of nearly equal *radii* all proceeding from one point.

UMBILICUS (Lat. for navel). The aperture seen at the base of the axis of certain univalve shells, which are then said to be "perforated" or "umbilicated."

UMBO (Lat. the boss of a shield). The beak of a bivalve shell.

UMBRELLA. The contractile disc of one of the *Lucernarida*.

UNCINATE (Lat. *uncinus*, a hook). Provided with hooks or bent spines.

UNGUICULATE (Lat. *unguis*, nail). Furnished with claws.

UNGULATA (Lat. *ungula*, hoof). The order of *Mammals* comprising the Hoofed Quadrupeds.

UNGULATE. Furnished with expanded nails constituting hoofs.

UNILOCULAR (Lat. *unus*, one; and *loculus*, a little purse). Possessing a single cavity or chamber. Applied to the shells of *Foraminifera* and *Mollusca*.

UNIVALVE (Lat. *unus*, one; *valvæ*, folding-doors). A shell composed of a single piece or valve.

URODELA (Gr. *oura*, tail; *delos*, visible). The order of the tailed Amphibians (Newts, &c.)

URTICATING CELLS (Lat. *urtica*, a nettle). (*See* Cnidæ.)

VACUOLES (Lat. *vacuus*, empty). The little cavities formed in the interior of many of the *Protozoa* by the presence of little particles of food, usually surrounded by a little water. These are properly called "food-vacuoles," and were supposed to be stomachs by Ehrenberg. Also the clear spaces which are often seen in the tissues of many *Cœlenterata*.

VARICES (Lat. *varix*, a dilated vein). The ridges or spinose lines which mark the former position of the mouth in certain univalve shells.

VASCULAR (Lat. *vas*, a vessel). Connected with the circulatory system.

VELUM (Lat. a sail). The membrane which surrounds and partially closes the mouth of the "disc" of *Medusæ*, or medusiform gonophores.

VENTRAL (Lat. *venter*, the stomach). Relating to the inferior surface of the body.

VENTRICLE (Lat. dim of *venter*, stomach). Applied to one of the cavities of the heart, which receives blood from the auricle.

VERMES (Lat. *vermis*, a worm). Sometimes employed at the present day in the same, or very nearly the same, sense as *Annuloida*, or as *Annuloida* plus the *Anarthropoda*.
VERMIFORM (Lat. *vermis*, worm ; and *forma*, form). Worm-like.
VERTEBRA (Lat. *verto*, I turn). One of the bony segments of the vertebral column or back-bone.
VERTEBRATA. (Lat. *vertebra*, a bone of the back, from *vertere*, to turn). The division of the Animal Kingdom roughly characterised by the possession of a back-bone.
VESICLE (Lat. *vesica*, a bladder). A little sac or cyst.
VIBRACULA (Lat. *vibro*, I shake). Long filamentous appendages found in many *Polyzoa*.
VIBRIONES (Lat. *vibro*, I shake). The little moving filaments developed in organic infusions.
VIPERINA (Lat. *vipera*, a viper). A group of the Snakes.
VIVIPAROUS (Lat. *vivus*, alive; and *pario*, I bring forth). Bringing forth young alive.

WHORL. The spiral turn of a univalve shell.

XIPHISTERNUM (Gr. *xiphos*, sword; *sternon*, breast-bone). The inferior or posterior segment of the sternum, corresponding with the "xiphoid cartilage" of human anatomy.
XIPHOSURA (Gr. *xiphos*, a sword; and *oura*, tail). An order of *Crustacea*, comprising the *Limuli* or King-Crabs, characterised by their long sword-like tails.
XYLOPHAGOUS (Gr. *xulon*, wood; and *phago*, I eat). Eating wood; applied to certain *Mollusca*.

ZOÖID (Gr. *zoön*, animal; and *eidos*, like). The more or less completely independent organisms, produced by gemmation or fission, whether these remain attached to one another or are detached and set free.
ZOOPHYTE (Gr. *zoön*, animal; *phuton*, plant). Loosely applied to many plant-like animals, such as Sponges, Corals, Sea-anemones, Sea-mats, &c.
ZOOSPORES (Gr. *zoön*, animal; and *spora*, seed). The ciliated locomotive germs of some of the lowest forms of plants (*Protophyta*).

INDEX.

www.ingramcontent.com/pod-product-compliance
Lightning Source LLC
LaVergne TN
LVHW021048110826
845150LV00001B/11

* 9 7 8 1 4 2 5 5 6 8 5 7 3 *